Springer Undergraduate Texts
in Mathematics and Technology

For other titles published in this series, go to
http://www.springer.com/series/7438

Wilhelm Forst • Dieter Hoffmann

Optimization—
Theory and Practice

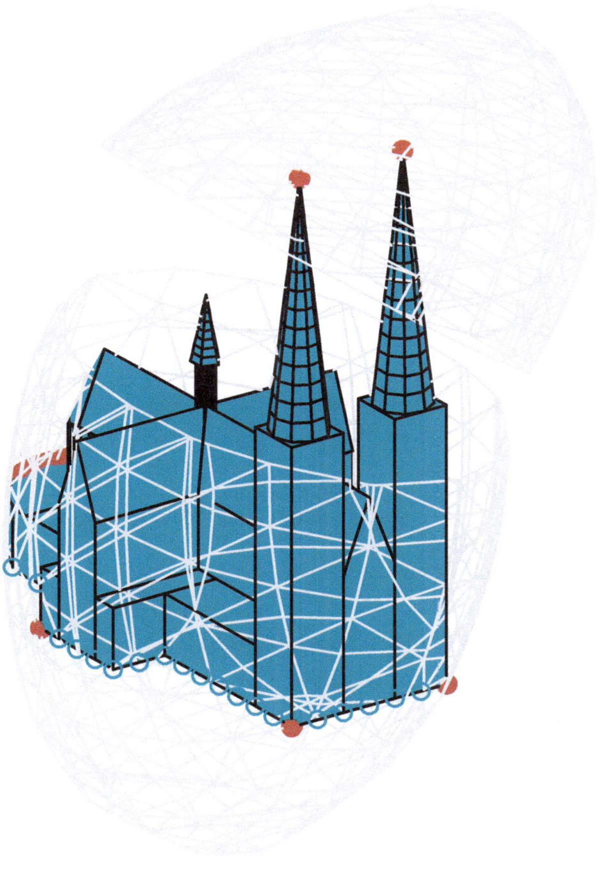

Springer

Wilhelm Forst
Universität Ulm
Fak. Mathematik und
Wirtschaftswissenschaften
Inst. Numerische Mathematik
Helmholtzstr. 18
89069 Ulm
Germany
wilhelm.forst@uni-ulm.de

Dieter Hoffmann
Universität Konstanz
FB Mathematik und Statistik
Fach 198
78457 Konstanz
Germany
dieter.hoffmann@uni-konstanz.de

Series Editors
Jonathan M. Borwein
Computer Assisted Research Mathematics
and its Applications, CARMA
School of Mathematical & Physical Sciences
University of Newcastle
Callaghan NSW 2308
Australia
jonathan.borwein@newcastle.edu.au

Helge Holden
Department of Mathematical Sciences
Norwegian University Science
and Technology
Alfred Getz vei 1
NO-7491 Trondheim
Norway
holden@math.ntnu.no

MATLAB® is a registered trademarks of The MathWorks, Inc. For MATLAB® product information, please contact: The MathWorks, Inc., 3 Apple Hill Drive, Natick, MA, 01760-2098 USA, E-mail: info@mathworks.com, Web: www.mathworks.com.

Maple® is a registered trademark of Maplesoft. For Maple™ product information, please contact: Maplesoft, 615 Kumpf Drive, Waterloo, ON, Canada, N2V 1K8, Email: info@maplesoft.com.

ISSN 1867-5506 e-ISSN 1867-5514
ISBN 978-1-4939-3933-6 ISBN 978-0-387-78977-4 (eBook)
DOI 10.1007/978-0-387-78977-4
Springer New York Dordrecht Heidelberg London

Mathematics Subject Classification (2010): Primary 90-01; Secondary 68W30, 90C30, 90C46, 90C51

Printed on acid-free paper

Springer is part of Springer Science+Business Media (www.springer.com)

To my children

Martin, Christina and *Christopher*

WILHELM FORST

To my grandsons

Léon, Étienne, Gabriel, Nicolas and *Luca*

who may some day want to read this book

DIETER HOFFMANN

Contents

Preface

This self-contained book on optimization is designed to serve a variety of purposes. It will prove useful both as a textbook for undergraduate and first-year graduate-level courses as well as a reference book for mathematicians, engineers and applied scientists interested in a careful exposition of this fascinating and useful branch of mathematics. Students, mathematicians and practitioners alike can profit from an approach which treats central topics of optimization in a concise, comprehensive and modern way. The book is fresh in conception and lucid in style and will appeal to anyone who has a genuine interest in optimization. The mutually beneficial interaction of theory and practice is presented in a stimulating manner. As JOHANN WOLFGANG VON GOETHE states: *"All theory, dear friend, is gray, but the golden tree of life springs ever green."*

Optimization is not only important in its own right but nowadays forms an integral part of a great number of applied sciences such as operations research, management science, economics and finance, and all branches of math-oriented engineering. Constrained optimization models are used in numerous areas of application and are probably the most widely used mathematical models in operations research and management science.

This book is the outgrowth of many years of teaching optimization in the mathematics departments of the Universities of Konstanz and Ulm (Germany) and W.F.'s teaching experiences with a first English version of parts of this book during his stay as a guest professor at the Universidad Nacional de Trujillo (Peru).

As the title suggests, one major aim of our book is to give a modern and well-balanced treatment of the subject by not only focusing on theory but also including algorithms and instructive examples from different areas of application. We put particular emphasis on presenting the material in a systematic and mathematically convincing way.

We introduce theory and methods at an elementary level but with an accurate and concise formulation of the theoretical tools needed. We phrase ideas and theorems clearly and succinctly, and complement them with detailed proofs and explanations of the ideas behind the proofs. We are convinced that many readers will find our book easy to read and value its accessibility.

Since it is intended as an introduction, the mathematical prerequisites have been kept to a minimum, with only some knowledge of multidimensional calculus, linear algebra and basic numerical methods needed to fully understand the concepts presented in the book. For example, the reader should know a little about the *convergence rate* of iteration methods and have a basic understanding of the *condition number* for matrices, which is a measure of the sensitivity of the error in the solution to a system of linear equations relative to changes in the inputs.

From the wide range of material that could be discussed in optimization lectures, we have selected aspects that every student interested in optimization should know about, as well as some more advanced issues. It is, however, not expected that the whole book will be 'covered' in one term. In practice, we have found that a good half is manageable.

This book provides a systematic, thorough and insightful discussion of *optimization of continuous nonlinear problems.* It contains in one volume of reasonable size a very clear presentation of key ideas and basic techniques. The way they are introduced, discussed and illustrated is designed to emphasize the connections between the various concepts.

In addition, we have included several features which we believe will greatly help the readers to get the most out of this book: First of all, every method considered is motivated and explained. Abstract statements are made into working knowledge with the help of detailed explanations on how to proceed in practice.

BENJAMIN FRANKLIN already knew: *"A good example is the best sermon!"* Therefore, the text offers a rich collection of detailed analytical and numerical examples which bring to life central concepts from diverse areas of science and applications. These selected examples have intentionally often been kept simple so that they can still be verified by hand easily. Counterexamples are presented whenever we want to show that certain assumptions cannot simply be dropped.

Additionally, the reader will find many elaborate two-colored graphics which help to facilitate understanding via geometric insight. Often the idea of a result has a simple geometric background — or, as the saying goes: *"A picture is worth a thousand words!"*

Another feature of the book is that it presents the concepts and examples in a way that invites the readers to think for themselves and to critically assess the observations and methods presented. In short, the book is written for the *active reader.*

Furthermore, we have also included *more than one hundred additional exercises*, which are partly supplemented by hints or *Matlab*® / *Maple*® code fragments. Here, the student has ample opportunity to practice concepts, statements and procedures, passing from routine problems to demanding extensions of the main text. In these exercises and particularly in appendix C, we will describe relevant features of the *Matlab*® *Optimization Toolbox* and demonstrate its use with selected examples.

Today each student uses a notebook computer the way we old guys used slide rules and log tables. Using a computer as a tool in teaching and learning allows one to concentrate on the essential ideas. Nowadays the various opportunities of current computer technology should be part of any lecture on optimization. Nevertheless, we would like to stress that neither knowledge of nor access to *Matlab*® or *Maple*® is really needed to fully understand this text. On the other hand, readers who follow the proposed way will benefit greatly from our approach.

Since many of the results presented here are nowadays classical — the novelty lies in the arrangement and streamlined presentation — we have not attempted to document the origin of every item. Some of the sources which served as valuable stimuli for our lectures years ago have since fallen into oblivion.

We are, however, aware of the fact that it took many brilliant men numerous years to develop what we teach in one term. In section 1.2, we have therefore compiled a *historical survey* to give our readers an idea of how optimization has developed through the centuries to an important branch of mathematics. Interior-point methods, for example, show that this process is still going on with ever new and exciting results!

We are certain that our selection of topics and their presentation will appeal to students, researchers and practitioners. The emphasis is on insight, not on giving the latest refinements of every method. We introduce the reader to optimization in a gradual and 'digestible' manner. Much of the material is 'classical' but streamlined proofs are given here in many instances. The systematic organization, structure and clarity of our book together with its insightful illustrations and examples make it an ideal introductory textbook. Written in a concise and straightforward style, this book opens the door to one of the most fascinating and useful branches of mathematics.

The book is divided into eight chapters. Here is a sketch of the *content:*

In *Chapter 1*, *Examples of Optimization Problems* are given. It is, however, not only intended as a general motivating introduction, but also as a demonstration of the workings and uses of mathematical tools like *Matlab*® and

Maple® and thus serves to encourage our readers not to neglect practice over theory. In particular the graphics features of these tools are important and very helpful in many instances. In addition, the first chapter includes — as already mentioned — a detailed *Historical Overview.*

Chapter 2 on *Optimality Conditions* studies the key properties of constrained problems and necessary and sufficient optimality conditions for them. As an introduction, we give a summary of the 'classic' results for unconstrained and equality constrained problems. In order to formulate optimality conditions for problems with *inequality constraints,* some simple aids from *Convex Analysis* are introduced. The LAGRANGE *multiplier rule* is generalized by the KARUSH–KUHN–TUCKER *conditions.* This core chapter presents the essential tools via a *geometrical point of view* using *cones.*

Chapter 3 introduces basics on *Unconstrained Optimization Problems.* The corresponding methods seek a local minimum (or maximum) in the absence of restrictions. Optimality criteria are studied and above all algorithmic methods for a wide variety of problems. Even though most optimization problems in 'real life' have restrictions to be satisfied, the study of unconstrained problems is useful for two reasons: First, they occur directly in some applications and are thus important in their own right. Second, unconstrained problems often originate as a result of transformations of constrained optimization problems. Some methods, for example, solve a general problem by converting it into a sequence of unconstrained problems.

Chapter 4 presents *Linearly Constrained Optimization Problems.* Here the problems have general objective functions but linear constraints. Section 4.1 gives a short introduction to *linear optimization* which covers the simplest — yet still highly important — kinds of constrained optimization problems, where the objective function *and* all constraints are linear. We consider two methods as examples of how these types of problems can be solved: the *revised simplex algorithm* and the *active set method. Quadratic problems,* which we treat in section 4.2, are linearly constrained with a quadratic objective function. Quadratic optimization is an important field in its own right, since it forms the basis of several algorithms for nonlinearly constrained problems. This section contains, for example, the BARANKIN–DORFMAN existence theorem as well as a lucid description of the GOLDFARB–IDNANI method. In contrast to the active set method this is a *dual method* which has the advantage that it does not need a primally feasible starting point. In section 4.3, we give an outline of selected *projection methods.* Besides classical gradient-based methods we present a sequential quadratic feasible point method.

In the next chapter, *Nonlinearly Constrained Optimization Problems* are treated. In section 5.1, a constrained optimization problem is replaced by an unconstrained one. There are two different approaches to this: In *exterior penalty methods* a term is added to the objective function which 'penalizes' a violation of constraints. In *interior penalty methods* a barrier term prevents points from leaving the interior of the feasible region. In section 5.2, *sequential quadratic programming methods* are introduced. The strategy here is to convert a usually nonlinear problem into a sequence of quadratic optimization problems which are easier to solve.

Chapter 6 presents *Interior-Point Methods for Linear Optimization.* The development of the last 30 years has been greatly influenced by the aftermath of a "scientific earthquake" triggered in 1979 by the findings of KHACHIYAN (1952–2005) and in 1984 by those of KARMARKAR. Efficient interior-point methods have in the meantime been applied to large classes of nonlinear optimization problems and are still topics of current research.

Chapter 7 treats basics of *Semidefinite Optimization.* This type of optimization differs from linear optimization in that it deals with problems over the cone of symmetric positive semidefinite matrices S_+^n instead of nonnegative vectors. It is a branch of *convex optimization* and covers many practically useful problems. The wide range of uses has quickly made semidefinite optimization very popular — besides, of course, the fact that such problems can be solved efficiently via polynomially convergent interior-point methods, which had originally only been developed for *linear* optimization.

The last chapter on *Global Optimization* deals with the computation and characterization of *global* optimizers of — in general — nonlinear functions. It is an important task since many real-world questions lead to global rather than local problems. Although this chapter is relatively short, we are convinced that it will suffice to give our readers a useful and informative introduction to this fascinating topic with all the necessary mathematical precision.

The book concludes with three *Appendices:*

- *A Second Look at the Constraint Qualifications*

  The GUIGNARD constraint qualification in section 2.2 seems to somewhat come out of the blue. The correlation between the regularity condition and the corresponding 'linearized' problem discussed here makes the matter more transparent.

- *The* FRITZ JOHN *Condition*

  This is — in a certain sense — a weaker predecessor of the KARUSH–KUHN–TUCKER conditions. No regularity condition is required, but in consequence an additional multiplier $\lambda_0 \geq 0$ must be attached to the objective function. In addition, an arbitrary number of constraints is possible. The striking picture on the title page also falls into this area, the *Minimum Volume Enclosing Ellipsoid of Cologne Cathedral (in an abstract 3D model).*

- *Optimization Software for Teaching and Learning*

  This part of the appendix gives a short overview of the software for the main areas of application in our book. We do not speak about professional software in modeling and solving large optimization problems. In our courses, we use *Matlab*® and *Maple*® and optimization tools like SeDuMi.

Each chapter starts with a summary, is divided into several sections and ends with numerous exercises which reinforce the material discussed in the chapter. Exercises are carefully chosen both in content and in difficulty to aid understanding and to promote mastery of the subject.

Acknowledgments

We would like to acknowledge many people who, knowingly or unknowingly, have helped us:

Vita Rutka and Michael Lehn each held the tutorial accompanying the lecture for some terms and provided a number of interesting problems and instructive exercises which we have gladly included in our book.

D.H. would like to thank Rainer Janssen for stimulating discussions and an oftentimes helpful look at the working conditions.

We are especially grateful to Andreas Borchert for his valuable advice concerning our electronic communication between Konstanz and Ulm.

Largely based on [Wri] and [Klerk], Julia Vogt and Claudia Lindauer wrote very good master's theses under our supervision which formed the basis of chapters 6 and 7.

The book has also benefited from Markus Sigg's careful readings. His comments and suggestions helped to revise a first preliminary version of the text. He also provided valuable assistance to D.H. with some computer problems.

It has been our very good fortune to have had Julia Neumann who translated essential parts of the text and thereby contributed to opening the book to a broader audience. She displayed great skill, courage and intuition in turning — sometimes clumsy — German into good English.

The editorial staff of Springer, especially Ann Kostant and Elizabeth Loew, deserve thanks for their belief in this special project and the careful attention to our manuscript during the publication process. We would also like to thank the copyeditor who did an excellent job revising our text and providing helpful comments.

Lastly, we would like to say, particularly to experts: We would appreciate personal feedback and welcome any comments, recommendations, spottings of errors and suggestions for improvement. Commendations will, of course, also be accepted!

For additions, updates and *Matlab*® and *Maple*® sources, please refer to the publisher's website for the book.

Ulm
Konstanz
October 2009

Wilhelm Forst
Dieter Hoffmann

Chapter 1

1

Introduction

The general theory has naturally developed out of the study of special problems. It is therefore useful to get a first impression by looking at the 'classic' problems. We will have a first look at some elementary examples to get an idea of the kind of problems which will be stated more precisely and treated in more depth later on. Consequently, we will often not go into too much detail in this introductory chapter.

Section 1.2 gives a historical survey of this relatively young discipline and points out that mathematics has at all times gained significant stimulations from the study of optimization problems.

1.1 Examples of Optimization Problems

Overdetermined Systems of Linear Equations

Let $m, n \in \mathbb{N}$ with $m > n$, $A \in \mathbb{R}^{m \times n}$ and $b \in \mathbb{R}^m$. We consider the *overdetermined linear system*

$$Ax = b \quad \left(x \in \mathbb{R}^n\right) . \tag{1}$$

W. Forst and D. Hoffmann, *Optimization—Theory and Practice*,
Springer Undergraduate Texts in Mathematics and Technology,
DOI 10.1007/978-0-387-78977-4_1,

Generally it has *no* solution.

However, if A is quadratic ($m = n$) and regular, it is well known that the equation $Ax = b$ has a unique solution

$$x = A^{-1}b\,.$$

The overdetermined case often appears in practical applications, for instance, if we have more observations or measurements than given independent parameters. In this case, it is useful to *minimize* the *'residual vector'*

$$r := r(x) := b - Ax$$

in some sense to be stated more precisely. In most cases we are satisfied with the *minimization of* $\|Ax - b\|$ with respect to a given norm $\|\;\|$ on $\mathbb{R}^m$, and often state the problem in the shortened form

$$\|Ax - b\| \longrightarrow \min_x$$

or in like manner.

So one considers, for instance, the following **alternative problem** for (1):

(i) $\|Ax - b\|_2 \longrightarrow \min\limits_x$

(ii) $\|Ax - b\|_\infty = \max\limits_{1\le\mu\le m} \left| \sum\limits_{\nu=1}^{n} a_{\mu\nu}x_\nu - b_\mu \right| \longrightarrow \min\limits_x$

or more general than (i)

(iii) $1 \le p < \infty: \quad \|Ax - b\|_p = \left(\sum\limits_{\mu=1}^{m} \left| \sum\limits_{\nu=1}^{n} a_{\mu\nu}x_\nu - b_\mu \right|^p \right)^{1/p} \longrightarrow \min\limits_x$

Ad (i): Linear Gauss Approximation Problem

We consider the functions f_μ defined by $f_\mu(x) := \sum\limits_{\nu=1}^{n} a_{\mu\nu}x_\nu - b_\mu$ for $x \in \mathbb{R}^n$ and $\mu = 1, \ldots, m$. Then our problem is equivalent to the linear least squares problem

$$F(x) := \frac{1}{2}\sum_{\mu=1}^{m} f_\mu(x)^2 = \frac{1}{2}\|Ax - b\|_2^2 \longrightarrow \min_x\,,$$

whose objective function is differentiable. Necessary and sufficient optimality conditions are given by the **Gauss** normal equations (cf. exercise 2)

$$A^TAx = A^Tb\,.$$

Ad (ii): Linear Chebyshev Approximation Problem

This problem (initially with a nondifferentiable objective function) is a special minimax problem and therefore equivalent to the 'linear program'

$$\eta \longrightarrow \min_{x\in\mathbb{R}^n,\eta\in\mathbb{R}}$$

with respect to the constraints

$$\left|\sum_{\nu=1}^{n} a_{\mu\nu}x_\nu - b_\mu\right| \le \eta \quad (\mu = 1,\ldots,m)$$

or alternatively

$$-\eta \le \sum_{\nu=1}^{n} a_{\mu\nu}x_\nu - b_\mu \le \eta \quad (\mu = 1,\ldots,m)\,.$$

Example 1

Consider the following overdetermined system of linear equations $A\,x = b$:

```
> restart: with(linalg): x := vector(2):
  A := matrix(5,2,[1,1,1,-1,1,2,2,1,3,1]);
  b := vector([3,1,7,8,6]);
```

$$A := \begin{bmatrix} 1 & 1 \\ 1 & -1 \\ 1 & 2 \\ 2 & 1 \\ 3 & 1 \end{bmatrix}, \quad b := [3, 1, 7, 8, 6]$$

The GAUSS approximation problem has the following normal equations:

```
> B := evalm(transpose(A)&*A): c := evalm(transpose(A)&*b):
  geneqns(B,x,c);
```

$$\{16\,x_1 + 7\,x_2 = 45,\ 7\,x_1 + 8\,x_2 = 30\}$$

These can be solved, for example, with the *Maple*® command linsolve :

```
> x := linsolve(B,c);
```

$$x := \left[\frac{150}{79}, \frac{165}{79}\right]$$

The command leastsqrs makes it even simpler:

```
> leastsqrs(A,b), evalf(%,4);
```

$$\left[\frac{150}{79}, \frac{165}{79}\right], \quad [1.899, 2.089]$$

We get the following approximation error:

```
> evalm(A&*x-b); norm(%,2); evalf(%,4);
```

$$\left[\frac{78}{79}, \frac{-94}{79}, \frac{-73}{79}, \frac{-167}{79}, \frac{141}{79}\right], \quad \frac{1}{79}\sqrt{68019}, \quad 3.302$$

It is also possible to solve this problem with the help of *Matlab*®. To do that, we utilize the function lsqlin from the *Optimization Toolbox* and demonstrate the use of some of the options which this command offers. In the interactive modus the following four command lines have to be entered successively:

```
A = [1, 1, 1, 2, 3; 1,-1, 2, 1, 1]';
b = [3, 1, 7, 8, 6]';
options = optimset('Display','iter','LargeScale','on');
[x,resnorm,residual,exitflag,output,lambda] = ...
   lsqlin(A,b,[],[],[],[],[],[],[],options)
```

The last command line does not fit into one line. We therefore continue it in the next line after having typed in "...". Distinctly simpler but also less informative is the *Matlab*® command $x = A \setminus b$. Since we have already discussed in detail how to solve this example with *Maple*®, we will abstain from giving the corresponding *Matlab*® output.

If we apply the maximum norm instead of the euclidean norm, the linear CHEBYSHEV approximation problem is equivalent to a *linear* optimization problem whose objective function and 'restrictions' can be calculated with *Maple*® in the following way:

```
> ObjFunc := eta;
  x := vector(2): convert(evalm(A&*x-b),list);
  Restr := convert(map(z->(-eta<=z,z<=eta),%),set);
```

$$ObjFunc := \eta$$

$$[x_1 + x_2 - 3,\ x_1 - x_2 - 1,\ x_1 + 2\,x_2 - 7,\ 2\,x_1 + x_2 - 8,\ 3\,x_1 + x_2 - 6]$$

$$Restr := \{-\eta \le x_1 + x_2 - 3 \le \eta,\ -\eta \le x_1 - x_2 - 1 \le \eta,\\ -\eta \le x_1 + 2\,x_2 - 7 \le \eta,\ -\eta \le 2\,x_1 + x_2 - 8 \le \eta,\ -\eta \le 3\,x_1 + x_2 - 6 \le \eta\}$$

We apply the simplex algorithm which is available under *Maple*®:

```
> with(simplex): minimize(ObjFunc,Restr); assign(%):
```

$$\left\{x_2 = \frac{21}{8}, \eta = \frac{15}{8}, x_1 = \frac{7}{4}\right\}$$

We verify that η is the approximation error:

```
> evalm(A&*x-b); 'eta' = norm(%); # maximum norm!
```

$$\left[\frac{11}{8}, \frac{-15}{8}, 0, \frac{-15}{8}, \frac{15}{8}\right], \quad \eta = \frac{15}{8}$$

The linear optimization problem we have just looked at can be solved with the *Matlab*® function linprog , for example, by writing the following sequence of commands in a *Matlab*® script called Cheb_Linprog.m :

```
f = [0; 0; 1]
A = [1, 1, 1, 2, 3,-1,-1,-1,-2,-3; 1,-1, 2, 1, 1,-1, 1,-2,-1,-1;
     -1,-1,-1,-1,-1,-1,-1,-1,-1,-1]'
b = [3  1  7  8  6 -3 -1 -7 -8 -6]'
disp('Medium Scale, Simplex:');
options = ...
   optimset('Display','iter','LargeScale','off','Simplex','on');
[x,fval,exitflag,output,lambda] = linprog(f,A,b,[],[],[],[],[],options)
```

This sequence of commands can then be executed in the Command Window with the command Cheb_Linprog .

As already mentioned, the linear CHEBYCHEV approximation problem is a special minimax problem. These kinds of problems can be solved with the *Matlab*® function fminimax in the following way:

```
x0 = [0; 0]; % Starting guess
disp('Minimize absolute values:');
options = optimset('GradObj','on','Display','iter', ...
                  'MinAbsMax',5); % Minimize absolute values
[x,fval,maxfval,exitflag,output] = ...
   fminimax(@ObjFunc,x0,[],[],[],[],[],[],[],options)
```

fminimax requires an m-file for the objective function and its gradient:

```
function [F,G] = ObjFunc(x)
   A = [1 1; 1 -1; 1 2; 2 1; 3 1];
   b = [3 1 7 8 6]';
   F = A*x-b;
   if nargout > 1
       G = A';
   end
```

◁

Nonlinear Least Squares Problems

More generally, we consider optimization problems of the form

$$F(x) := \frac{1}{2}\sum_{\mu=1}^{m} f_\mu(x)^2 \longrightarrow \min_x .$$

To simplify matters, we assume that the given functions $f_\mu\colon \mathbb{R}^n \longrightarrow \mathbb{R}$ $(\mu = 1, \ldots, m)$ are twice differentiable on $\mathbb{R}^n$. Problems of this kind appear, for example, when dealing with curve or data fitting. In addition, systems of nonlinear equations can also be reduced to this form and can then be solved by methods of nonlinear optimization.

Example 2

We wish to solve the following simple homogeneous system of nonlinear equations:

$$x_1^3 - x_2 - 1 = 0, \quad x_1^2 - x_2 = 0$$

```
> restart: Digits := 5: f_1 := x_1^3-x_2-1; f_2 := x_1^2-x_2;
```

$$f_1 := x_1^3 - x_2 - 1\,, \quad f_2 := x_1^2 - x_2$$

We calculate its solution numerically using the *Maple*® command fsolve :

```
> fsolve({f_1,f_2},{x_1,x_2});
```

$$\{x_1 = 1.4656,\ x_2 = 2.1479\}$$

The following figure shows that the nonlinear system has a unique solution:

```
> with(plots): implicitplot({f_1,f_2},x_1=-1..3,x_2=-2..4);
```

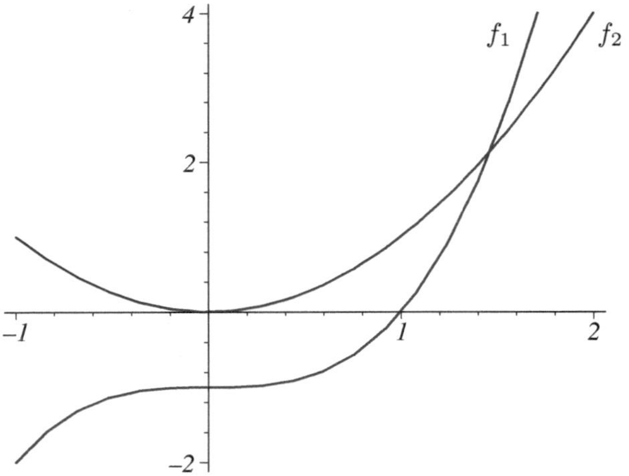

Next, we determine the local extreme points of the objective function corresponding to the underlying least squares problem:

```
> F   := 1/2*(f_1^2+f_2^2);
  D1F := diff(F,x_1); D2F := diff(F,x_2);
  solve({D1F,D2F},{x_1,x_2}):
  B := proc(z) local u;
    # This Boolean function selects the real stationary points
    subs(z,[x_1,x_2]); u := map(evalc@Im,%);
    is(u[1]=0 and u[2]=0)
  end proc:
  sol := allvalues([%%]): rsol := select(B,sol):
  'rsol' = evalf(rsol);
```

$$F := \frac{1}{2}\left(x_1^3 - x_2 - 1\right)^2 + \frac{1}{2}\left(x_1^2 - x_2\right)^2$$

$$D1F := 3\left(x_1^3 - x_2 - 1\right)x_1^2 + 2\left(x_1^2 - x_2\right)x_1\,, \quad D2F := -x_1^3 + 2\,x_2 + 1 - x_1^2$$

$$rsol = \{x_2 = -0.50000, x_1 = 0.\}, \{x_2 = -0.12963, x_1 = 0.66667\}, \{x_1 = 1.4655, x_2 = 2.1477\}$$

Obviously F has three (real) *stationary points*[1], of which $rsol[3]$ is the solution to the system of nonlinear equations:

```
> subs(rsol[3],F): 'F' = simplify(%);
```

$$F = 0$$

For the other two solutions we get:

```
> subs(rsol[1],F), subs(rsol[2],F);
```

$$\frac{1}{4}, \frac{961}{2916}$$

We evaluate the *Hessian* to decide of which kind the stationary points are:

```
> with(linalg): H := hessian(F,[x_1,x_2]):
  seq(subs(rsol[k],eval(H)),k=1..2),subs(evalf(rsol[3]),eval(H));
```

$$\begin{bmatrix} 1 & 0 \\ 0 & 2 \end{bmatrix}, \begin{bmatrix} \frac{65}{27} & \frac{-8}{3} \\ \frac{-8}{3} & 2 \end{bmatrix}, \begin{bmatrix} 50.101 & -9.3741 \\ -9.3741 & 2 \end{bmatrix}$$

```
> definite(%[1],positive_def),definite(%[2],positive_semidef),
  definite(%[2],negative_semidef),definite(%[3],positive_def);
```

[1] It is well known that these zeros of the first derivative are candidates for local extreme points.

true, *false*, *false*, *true*

$rsol[1]$ and $rsol[3]$ give local minima, and $rsol[2]$ is a saddle point because the Hessian is indefinite. We see that not every *local* minimum of F is a solution to the given nonlinear system. By means of the contour lines of F we finally visualize the geometric details:

```
> Points := pointplot([seq(subs(evalf(rsol[k]),[x_1,x_2]),k=1..3)],
    symbol=circle,symbolsize=17,color=black):
  Levels:= contourplot(F,x_1=-1.5..2,x_2=-1.5..5,axes=box,
    levels=[0.1*k $ k=1..10],grid=[100,100]):
  display(Points, Levels);
```

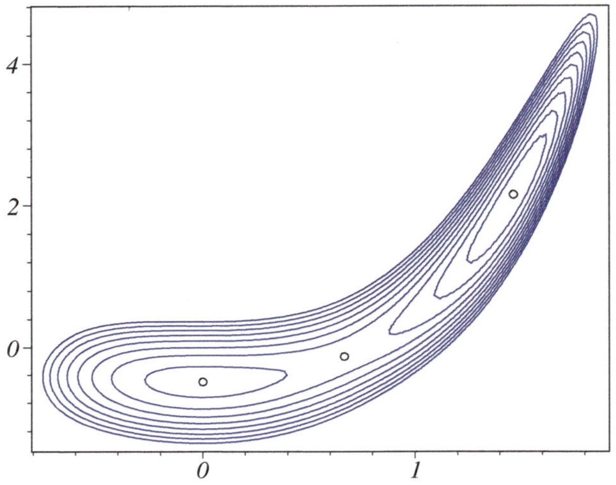

◁

Chebyshev Approximation

We explain the mathematical problem by means of the following simple example:

$$\max_{0 \le t \le 1} \left| e^t - a - b\,t \right| \longrightarrow \min_{a,\,b}$$

Consequently, we are looking for the linear polynomial function which is the best approximation of the exponential function in the interval $[0, 1]$ with respect to the maximum norm. This problem is equivalent to

$$\eta \longrightarrow \min_{a,\,b,\,\eta}$$

subject to the constraints

$$-\eta \le e^t - a - b\,t \le \eta \quad \text{for all} \quad t \in [0, 1]\,.$$

There occur infinitely many constraints. The objective function and the constraint functions depend only on a finite number of variables. This type of problem is called a *semi-infinite* (linear) optimization problem. It will not be discussed any further in later chapters.

Maple® provides the command numapprox[minimax] to solve it, which we will apply as a 'blackbox':

```
> restart: with(numapprox):
  p := minimax(exp,0..1,1,1,'eta'); 'eta' = eta;
```

$$p := x \to 0.8941114810 + 1.718281828\,x\,, \quad \eta = 0.105978313$$

```
> plot(exp-p,0..1,color=blue,thickness=2,tickmarks=[3,3],
    title="Error Function");
```

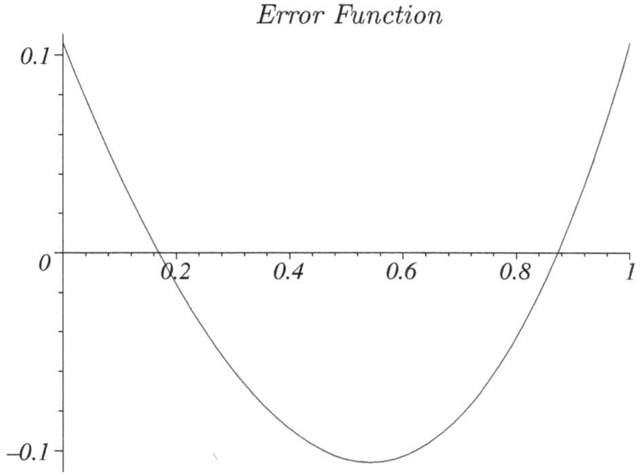

The error function alternately attains its maximal deviation η at 0, at a point $\tau \in (0, 1)$ and at 1. Therefore, we get the following equations for a, b, τ and η:

```
> eta := 'eta': f := exp: p := t -> a+b*t:
  eq1 := f(0)-p(0) = eta; eq2 := f(tau)-p(tau) = -eta;
  eq3 := f(1)-p(1) = eta; eq4 := D(f-p)(tau) = 0;
```

$$eq1 := 1 - a = \eta\,, \quad eq2 := e^{\tau} - a - b\,\tau = -\eta\,,$$
$$eq3 := e - a - b = \eta\,, \quad eq4 := e^{\tau} - b = 0$$

Maple® returns the following rounded values of the exact solution:

```
> solve({eq1,eq2,eq3,eq4},{a,b,tau,eta}): evalf(%);
```

$$\{\eta = .1059334158,\, b = 1.718281828,\, a = 0.8940665842,\, \tau = 0.5413248543\}$$

Comparing the results, we see that numapprox[minimax] computes the coefficient a and the maximal deviation η only with low accuracy.

Facility Location Problems

This kind of problem occurs in various applications. To illustrate this, we consider two planar location problems. Here, it is very important that distance measuring and the optimality criterion are suitable for the given problem.

Example 3 Optimal Location of an Electricity Transformer

To simplify matters, assume that the electricity supply (for instance, in a thinly populated region) is provided by connecting each customer *directly* with the transformer. We are looking for the location of the transformer which minimizes the total net length. To measure the distance, we use the euclidean norm, and so the objective function has the form

$$d(x,y) := \sum_{\mu=1}^{m} \sqrt{(x-x_\mu)^2 + (y-y_\mu)^2} \longrightarrow \min_{x,\,y},$$

where the coordinates $(x,\, y)$ describe the sought location of the transformer and $(x_\mu,\, y_\mu)$ the given locations of the customers.

```
> restart: Digits := 5: with(plots): x := vector(2):
```

Locations of the customers:

```
> Points := [[0,0],[5,-1],[4,6],[1,3]];
```

$$Points := [[0,\, 0],\, [5,\, -1],\, [4,\, 6],\, [1,\, 3]]$$

Objective function:

```
> d  := add(sqrt((x[1]-z[1])^2+(x[2]-z[2])^2),z=Points):
  P1 := op(map(z->z[1],Points)): # abscissas of the points
  P2 := op(map(z->z[2],Points)): # ordinates of the points
  r1 := x[1]=min(P1)..max(P1);
  r2 := x[2]=min(P2)..max(P2);     # defines graphic window
```

$$r1 := x_1 = 0..5\,, \quad r2 := x_2 = -1..6$$

```
> p1 := plot3d(d,r1,r2,style=patchcontour,axes=box):
  display(p1,orientation=[-50,25]); display(p1,orientation=[-90,0]);
```

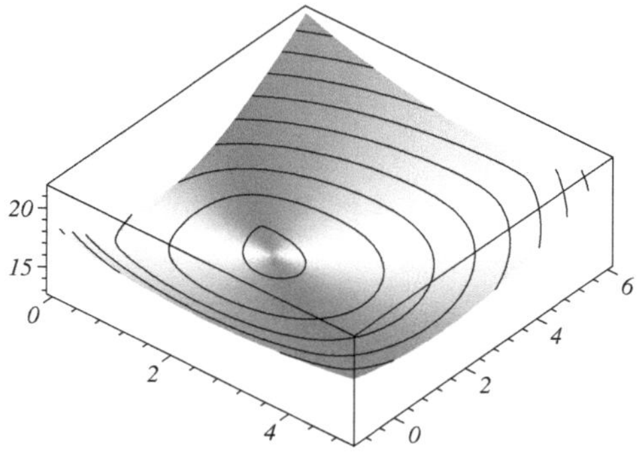

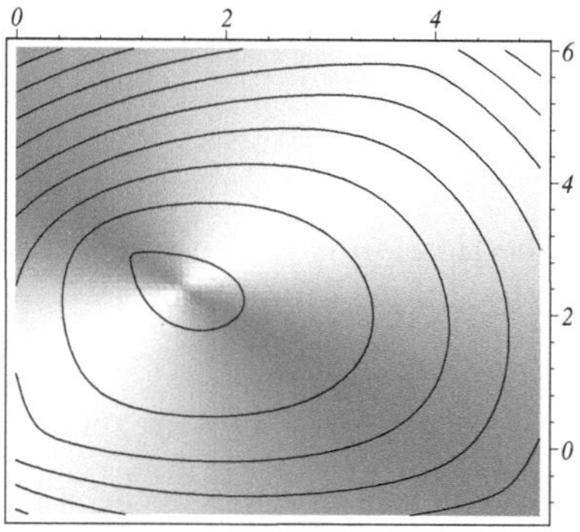


```
> p2 := pointplot(Points,symbol=circle,symbolsize=18,axes=frame):
  p3 := contourplot(d,r1,r2,levels=30,grid=[70,70],color=gray):
  display(p2,p3,color=black);
```

In contrast to the foregoing figure, we now get a real 2D graphic. Clicking on the surmised minimum, causes *Maple*® to return a rough approximation of the optimal location. We forego its reproduction in this book. The exact minimum can be calculated by analytical means:

```
> with(linalg): g := grad(d,x):
  #solve(convert(g,set),{x[1],x[2]}); # symbolic calculation fails!
  Solution := fsolve(convert(g,set),{x[1],x[2]});
```

$$Solution := \{x_1 = 1.6000,\ x_2 = 2.4000\}$$

```
> subs(Solution,eval(g)); # insert computed results
```

$$[0.,\ 0.00005]$$

As decimal numbers are well known to us, we detect the exact solution without the help of solve :

```
> Solution := {x[1]=8/5,x[2]=12/5}:
  subs(Solution,eval(g)): simplify(%); # symbolic calculation!
```

$$[0,\ 0]$$

By pure chance we have guessed the exact solution! ☺

```
> H := hessian(d,x):
  subs(Solution,eval(H)): simplify(%); evalf(%);
  definite(%,positive_def);
```

$$\begin{bmatrix} \frac{75}{676}\sqrt{13}+\frac{25}{51}\sqrt{2} & -\frac{25}{338}\sqrt{13}+\frac{25}{51}\sqrt{2} \\ -\frac{25}{338}\sqrt{13}+\frac{25}{51}\sqrt{2} & \frac{25}{507}\sqrt{13}+\frac{25}{51}\sqrt{2} \end{bmatrix}, \quad \begin{bmatrix} 1.0933 & 0.42656 \\ 0.42656 & 0.87103 \end{bmatrix}, \quad true$$

So the Hessian is positive definite in [8/5,12/5]. As d is strictly convex, there lies the unique minimum.

```
> U := [8/5,12/5]:
  subs({x[1]=U[1],x[2]=U[2]},d): simplify(%); evalf(%);
```

$$2\sqrt{13}+4\sqrt{2}, \quad 12.868$$

```
> p4 := plot([seq([z,U],z=Points)],color=blue,thickness=2):
  p5 := plot([U],style=point,symbol=circle,symbolsize=24,color=blue):
  display(p2,p3,p4,p5,axes=frame);
```

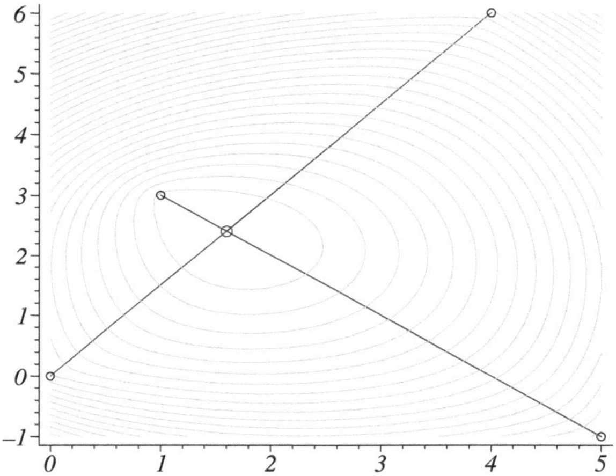

We see that the optimal location of the electricity transformer is the intersection point of the two diagonals. ◁

Example 4 Optimal Location of a Rescue Helicopter

Now, we are looking for the optimal location of a rescue helicopter. As in the foregoing case, the distance is measured in a straight line. The helicopter should reach its destination in minimal time. Therefore, we now have to minimize the maximum distance and get the objective function

$$d_{\max}(x, y) := \max_{1\le\mu\le m} \sqrt{(x-x_\mu)^2+(y-y_\mu)^2} \longrightarrow \min_{x,y}.$$

Coordinates of the destinations:

```
> restart: with(plots): x := vector(2):
  Points := [[0,0],[5,-1],[4,6],[1,3]];
  r1 := x[1]=-1.5..7.5: r2 := x[2] = -2..7: # drawing window
```

$$Points := [[0, 0], [5, -1], [4, 6], [1, 3]]$$

Objective function:

```
> d_max := max(seq(sqrt((x[1]-z[1])^2+(x[2]-z[2])^2),z=Points)):
  p1 := plot3d(d_max,r1,r2,style=patchcontour,axes=box,grid=[70,70]):
  display(p1,orientation=[-40,20]); display(p1,orientation=[-90,0]);
```

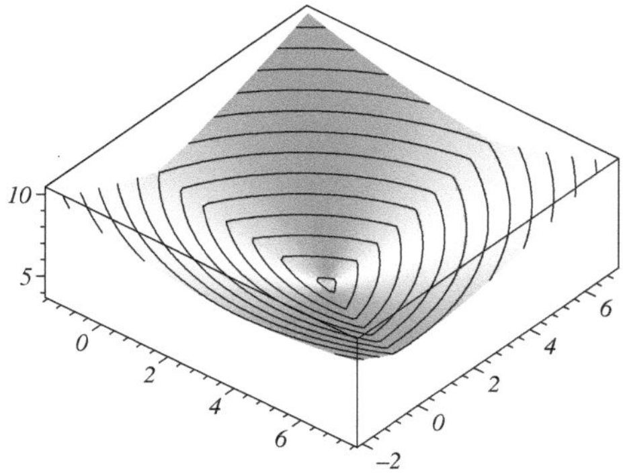

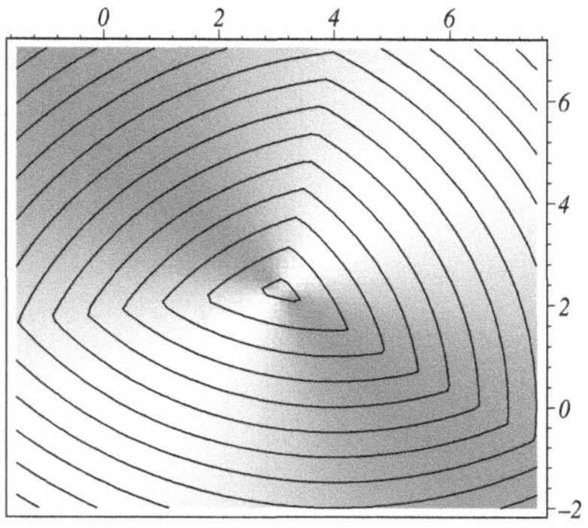

This minimax problem can be solved more easily if we transform it (with $r := d_{\max}$) to the following equivalent optimization problem:

$$\frac{1}{2}r^2 \longrightarrow \min_{x,\,y,\,r}$$

subject to the constraints

$$(x - x_\mu)^2 + (y - y_\mu)^2 \le r^2 \qquad (1 \le \mu \le m).$$

Substituting $\varrho := \frac{1}{2}(r^2 - x^2 - y^2)$, we get a simple *quadratic* optimization problem:

$$\frac{1}{2}(x^2 + y^2) + \varrho \longrightarrow \min_{x,y,\varrho}$$

subject to the (linear) constraints

$$x_\mu x + y_\mu y + \varrho \ge \frac{1}{2}(x_\mu^2 + y_\mu^2) \qquad (1 \le \mu \le m).$$

Later on, we will get to know solution algorithms for these types of problems which provide the following coordinates of the optimal helicopter location:

```
> M := [52/17, 39/17]: # coordinates of the optimal location
  r := evalf(subs({x[1]=M[1],x[2]=M[2]},d_max));
```

$$r := 3.8235$$

```
> p2 := pointplot(Points,symbol=circle,symbolsize=18,color=black):
  p3 := contourplot(d_max,r1,r2,levels=30,grid=[70,70],color=gray):
  p4 := plot([seq([z,M],z=Points)],color=blue):
  with(plottools): p5 := disk(M,0.1,color=black):
  p6 := circle(evalf(M),r,color=blue,thickness=3):
  display(p2,p3,p4,p5,p6,axes=frame,scaling=constrained);
```

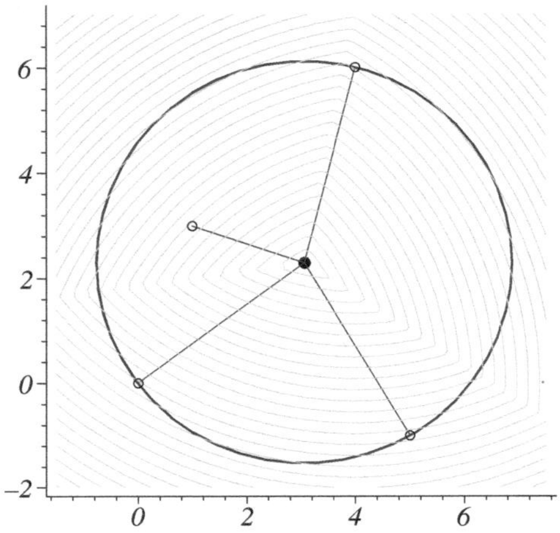

In addition, one sees, for example with the help of the KARUSH–KUHN–TUCKER conditions from chapter 2 (cf. exercise 14), that the location M is the optimum place if three points lie on the boundary of the blue (minimum) circumscribed circle. They are the vertices of an acute triangle. Naturally, it is possible that only two points lie on the circle. In this case, however, they must lie diametrically opposite to each other. ◁

Standard Form of Optimization Problems

The previous examples also served to demonstrate that optimization problems from quite different areas of application have common roots. Therefore, it is useful to classify them according to mathematical criteria. In this book, we consider optimization problems which are based on the following *standard form:*

$$f(x) \longrightarrow \min_{x \in \mathbb{R}^n}$$

subject to the constraints

$$g_i(x) \le 0 \quad \text{for} \quad i \in \mathcal{I} := \{1, \ldots, m\}$$

$$h_j(x) = 0 \quad \text{for} \quad j \in \mathcal{E} := \{1, \ldots, p\}.$$

Here '$\mathcal{I}$' stands for 'Inequality', '$\mathcal{E}$' for 'Equality'. As already denoted earlier, the function to be minimized is called the *objective function*. We write the *constraints*, respectively *restrictions*, in homogeneous form, and distinguish between inequality and equality constraints. The maximization of an objective function f is equivalent to the minimization of $-f$, and inequality restrictions of the type $g_i(x) \ge 0$ are equivalent to $-g_i(x) \le 0$. Generally, we assume that the objective function as well as the restriction functions are defined on a nonempty subset of $\mathbb{R}^n$ and are at least continuous there.

Unconstrained optimization problems play an important role. Our discussion of them in chapter 3 will occupy a big space because they come up in various applications and are consequently important as problems in their own right. There exists quite a range of methods to solve them. Numerical methods for unconstrained optimization problems usually only compute a stationary point, that is, the first-order necessary conditions are fulfilled. Only with the help of additional investigations like the second-order optimality condition is it possible to decide whether the stationary point is a local extreme point or a saddle point. *Global* minima can only be stated in special cases.

Minimax problems of the form

$$\max_{1 \le \mu \le m} f_\mu(x) \longrightarrow \min_{x \in \mathbb{R}^n}$$

are a special kind of unconstrained optimization problem with a nondifferentiable objective function even if every f_μ is differentiable. These are — similar to the CHEBYSHEV approximation — equivalent to a restricted problem:

$$\eta \longrightarrow \min_{x \in \mathbb{R}^n,\, \eta \in \mathbb{R}}$$

subject to the constraints

$$f_\mu(x) \le \eta \quad (\mu = 1, \ldots, m).$$

Optimization problems with equality constraints often occur as subproblems and can be reduced via elimination to unconstrained optimization problems as we will see later on.

With *restricted problems*, we are especially interested in the *linear* constraints $g_i(x) = a_i^T x - b_i$ for $i \in \mathcal{I}$ and $h_j(x) = \hat{a}_j^T x - \hat{b}_j$ for $j \in \mathcal{E}$.

In chapter 4 we will discuss *linear optimization* — here, we have $f(x) = c^T x$ — and *quadratic optimization*, that is, $f(x) = \frac{1}{2} x^T C x + c^T x$ with a symmetric positive semidefinite matrix $C \in \mathbb{R}^{n \times n}$, in detail. The above occur, for example, as subproblems in SQP[2] methods.

[2] The abbreviation stands for "Sequential Quadratic Programming".

Penalty Methods

We will conclude this introductory section with the presentation of the *classic* penalty method from the early days of numerical optimization. At that time, problems with nonlinear constraints could only be solved by reducing them to unconstrained problems by adding the constraints as 'penalty terms' to the objective function. Here, we will only consider the simple case of equality constraints and quadratic penalty terms. The optimization problem

$$f(x) \longrightarrow \min_{x \in \mathbb{R}^n}$$

subject to the constraints

$$h_j(x) = 0 \quad \text{for} \quad j = 1, \ldots, p$$

is transformed into the optimization problem

$$F(x) := f(x) + \frac{\lambda}{2} \sum_{j=1}^{p} h_j(x)^2 \longrightarrow \min_{x \in \mathbb{R}^n}$$

with a positive penalty parameter λ. Enlarging λ leads to a greater 'punishment' of nonfulfilled constraints. Thereby we hope that the 'unconstrained minimizer' $x^*(\lambda)$ will fulfill the constraints with growing accuracy and at the same time $f(x^*(\lambda))$ will better approximate the minimum we are looking for.

Example 5

Consider the problem

$$f(x_1, x_2) := x_1 x_2^2 \longrightarrow \min_{x \in \mathbb{R}^2}$$

subject to the constraint

$$h(x_1, x_2) := x_1^2 + x_2^2 - 2 = 0,$$

which has two global minima in $(x_1, x_2)^T$ with $x_1 = -\sqrt{2/3}$, $x_2 = \pm 2/\sqrt{3}$ and a local minimum in $x_1 = \sqrt{2}$, $x_2 = 0$. At the moment, we content ourselves with the following 'graphical inspection':

```
> restart: with(plots): f := x[1]*x[2]^2: h := x[1]^2+x[2]^2-2:
  F := subs({x[1]=r*cos(t),x[2]=r*sin(t)},[x[1],x[2],f]):
  plot(subs(r=sqrt(2),F[3]),t=0..2*Pi,color=blue,thickness=2);
  p1 := plot3d(F,t=0..2*Pi,r=0.1..sqrt(2),style=wireframe,grid=[25,25],
                color=blue): r := sqrt(2):
  p2 := spacecurve(F,t=0..2*Pi,color=blue,thickness=2):
  p3 := spacecurve([F[1],F[2],0],t=0..2*Pi,color=black,thickness=2):
  display(p1,p2,p3,orientation=[-144,61],axes=box);
```

Graph of the objective function

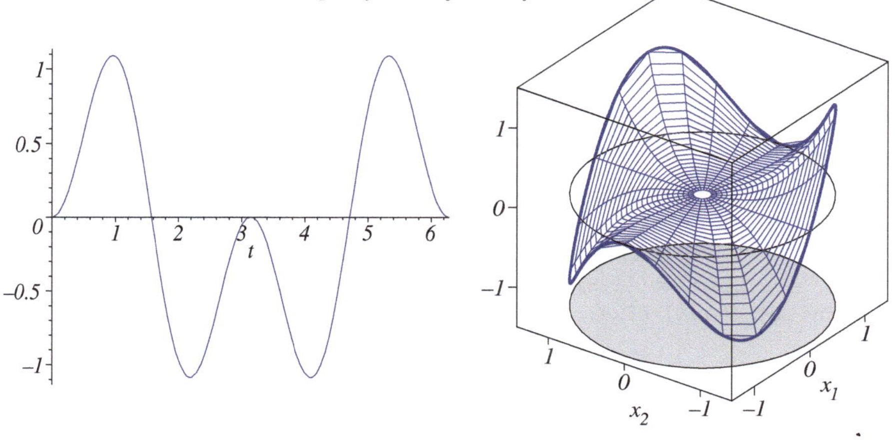

The figure on the right shows the constraint circle in the x_1x_2-plane, along with the surface $x_3 = f(x_1, x_2)$. The intersection of the cylinder $h(x_1, x_2) = 0$ and the surface $x_3 = f(x_1, x_2)$ gives the *space curve* shown by the thick blue line. Later on, we will be able to analyze and prove this in more detail with the help of the optimality conditions presented in chapter 2. Let us look at the penalty function:

```
> F := f+lambda/2*h^2;
```

$$F := x_1\,x_2^2 + \frac{1}{2}\,\lambda\,(x_1^2 + x_2^2 - 2)^2$$

We get the necessary optimality conditions:

```
> DF1 := diff(F,x[1]) = 0; DF2 := factor(diff(F,x[2])) = 0;
```

$$DF1 := x_2^2 + 2\,\lambda\,(x_1^2 + x_2^2 - 2)\,x_1 = 0$$
$$DF2 := 2\,x_2\,(\lambda\,x_2^2 + x_1 + \lambda\,x_1^2 - 2\,\lambda) = 0$$

If $x_2 = 0$, the first equation is simplified to:

```
> subs(x[2]=0,DF1); solve(%,x[1]): Sol := map(z->[z,0],[%]);
```

$$2\,\lambda\,(x_1^2 - 2)\,x_1 = 0, \quad Sol := \left[[0,\,0],\,[\sqrt{2},\,0],\,[-\sqrt{2},\,0]\right]$$

In these three points, F has the following Hessians:

```
> with(linalg): H := hessian(F,[x[1],x[2]]):
  seq(subs({x[1]=z[1],x[2]=z[2]},eval(H)),z=Sol);
```

$$\begin{bmatrix} -4\,\lambda & 0 \\ 0 & -4\,\lambda \end{bmatrix}, \begin{bmatrix} 8\,\lambda & 0 \\ 0 & 2\sqrt{2} \end{bmatrix}, \begin{bmatrix} 8\,\lambda & 0 \\ 0 & -2\sqrt{2} \end{bmatrix}$$

Obviously, the penalty function has a local minimum only in $\left(\sqrt{2},\,0\right)^T$. For large λ, the corresponding Hessian has the condition number $2\sqrt{2}\lambda$ and, therefore, is ill-conditioned.

Now consider the remaining case in which the second factor of *DF2* is equal to zero:

```
> DF2a := collect(op([1,3],DF2),lambda) = 0;
  expand(DF1-2*x[1]*DF2a): Eq[1] := x[2]^2 = solve(%,x[2]^2);
```

$$DF2a := \lambda\,(x_1^2 + x_2^2 - 2) + x_1 = 0\,, \quad Eq_1 := x_2^2 = 2\,x_1^2$$

Substituting Eq_1 into $DF2a$ leads to the following quadratic equation:

```
> Eq[2] := subs(Eq[1],DF2a);
  Eq[3] := x[1]^2 = solve(Eq[2],x[1]^2);
```

$$Eq_2 := \lambda\,(3\,x_1^2 - 2) + x_1 = 0\,, \quad Eq_3 := x_1^2 = \frac{1}{3}\,\frac{2\,\lambda - x_1}{\lambda}$$

Before looking at the solutions in more detail, we will prepare the Hessian for analyzing whether it is positive definite or not:

```
> diff(F,x[1]$2): subs(Eq[1],%): DF11 := factor(%);
  diff(F,x[2]$2): subs(Eq[1],%): DF22 := expand(%-2*lhs(Eq[2]));
```

$$DF11 := 2\,\lambda\,(5\,x_1^2 - 2)\,, \quad DF22 := 8\,\lambda\,x_1^2$$

```
> unprotect(Trace): Trace := expand(subs(Eq[3],DF11+DF22));
```

$$Trace := 8\,\lambda - 6\,x_1$$

```
> DF11*DF22-diff(F,x[1],x[2])^2: expand(%): subs(Eq[1],%): factor(%);
  Determinant := 8*x[1]^2*expand(subs(Eq[3],op(3,%)));
```

$$8\,x_1^2\,(6\,\lambda^2\,x_1^2 - 4\,\lambda^2 - 1 - 4\,\lambda\,x_1)\,, \quad Determinant := 8\,x_1^2\,(-6\,\lambda\,x_1 - 1)$$

```
> Qsol := solve(Eq[2],x[1]);
```

$$Qsol := \frac{1}{6}\,\frac{-1 + \sqrt{1 + 24\,\lambda^2}}{\lambda},\ \frac{1}{6}\,\frac{-1 - \sqrt{1 + 24\,\lambda^2}}{\lambda}$$

```
> subs(x[1]=Qsol[1],Determinant): 'Determinant' = factor(%);
```

$$Determinant = -\frac{2}{9}\,\frac{(-1 + \sqrt{1 + 24\,\lambda^2})^2\sqrt{1 + 24\,\lambda^2}}{\lambda^2}$$

```
> Qsol[2]^2-2/3: 'x[1]^2'-2/3 = simplify(%);
```

$$x_1^2 - \tfrac{2}{3} = \frac{1 + \sqrt{1 + 24\,\lambda^2}}{18\lambda^2}$$

Looking at the determinant, it is obvious that the first of the two solutions *Qsol* is not possible. For $x_1 = Qsol_2$, we have $x_1 < -\sqrt{2/3}$. Therefore, we get two minimizers, and because of $x_1^2 + x_2^2 = 3\,x_1^2 > 2$ these points do not lie within the circle. For $\lambda \longrightarrow \infty$, however, these minimizers converge to the global minimizer of the restricted optimization problem. For large λ their numerical calculation requires very good initial values because we will see at once that the Hessians of the penalty function are ill-conditioned in the (approximating) minimizers:

```
> subs(x[1]=Qsol[2],mu^2-Trace*mu+Determinant): Mu := solve(%,mu):
  mu[max]/mu[min] = asympt(eval(Mu[1]/Mu[2]),lambda,1);
```

$$\frac{\mu_{\max}}{\mu_{\min}} = \sqrt{6}\,\lambda + \mathrm{O}\Big(\frac{1}{\lambda}\Big)$$

In the following two graphics this fact becomes clear very quickly because the elliptic level curves around the local minimizers become more elongated for increasing penalty parameters:

penalty parameter = 3

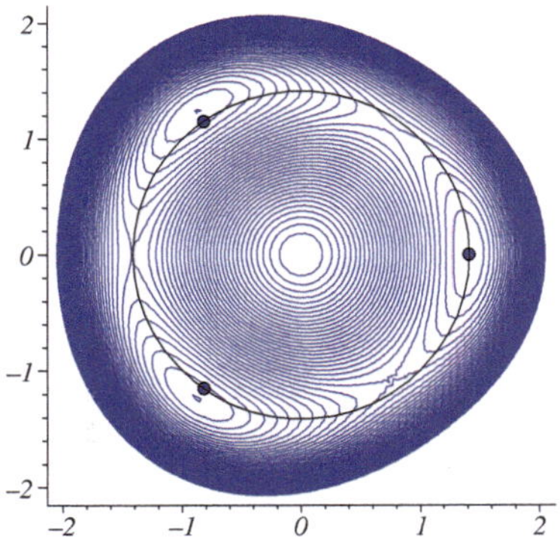

penalty parameter = 6

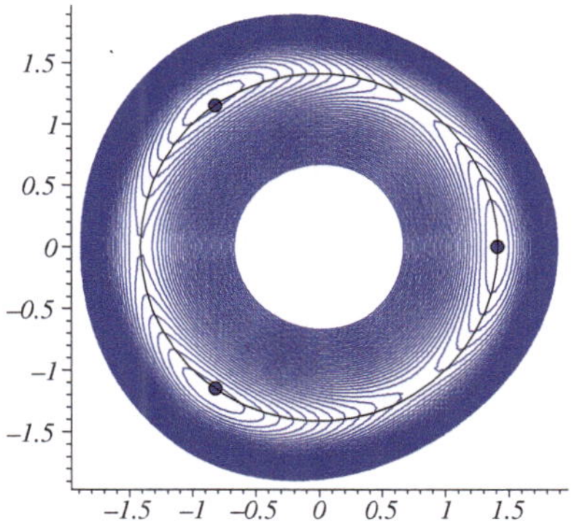

In comparison, we give a *Matlab*® script Penalty.m which calculates a minimizer of the penalty function for the parameters $\lambda = 1, 3, 6$ and 9 by means of the function fminsearch from the *Matlab*® Optimization Toolbox. Furthermore it plots the feasible region, the level curves of the objective function and the penalty functions and marks the exact and approximating minimizers.

```
% Script Penalty.m: Test of Penalty Method
clc; clear all; close all; N = 200;
X = linspace(-2,2,N); Y = X; T = linspace(0,2*pi,N);
```

```
Z = X.*Y.^2; [X,Y] = ndgrid(X,Y);
xx = sqrt(2).*cos(T); yy = sqrt(2).*sin(T);
x1 = -sqrt(2/3); y1 = sqrt(4/3);
f = @(x) x(1)*x(2)^2; h = @(x) x(1)^2+x(2)^2-2;
x = [2, 1]; % starting value of fminsearch
for lambda = [1, 3, 6, 9]
   Phi = Z+lambda/2*(X.^2+Y.^2-2).^2;
   phi =  @(x) f(x)+lambda/2*h(x)^2;
   [x,fval] = fminsearch(phi,x)
   % In the second iteration (lambda=3) fminsearch uses the result of
   % the first iteration (i.e. lambda=1) as starting value, and so on.
   figure, clf
   subplot(2,1,1);
   contour(X,Y,Z,40); colorbar;
   hold on
   P  = plot(xx,yy,'k'); set(P,'linewidth',2);
   PP = plot([sqrt(2),x1,x1],[0,y1,-y1],'ko');
   set(PP,'markersize',6,'markerfacecolor','k');
   axis([-2 2 -2 2]), axis equal, axis([-2 2 -2 2]), ylabel('y')
   title(['\lambda = ', num2str(lambda)]);
   subplot(2,1,2);
   contour(X,Y,Phi,160); colorbar;
   hold on
   P  = plot(xx,yy,'k'); set(P,'linewidth',2);
   PP = plot([sqrt(2),x1,x1],[0,y1,-y1],'ko');
   set(PP,'markersize',6,'markerfacecolor','k');
   % Plot the minimum of phi
   PP = plot(x(1),x(2),'r+');
   set(PP,'markersize',6,'linewidth',3);
   axis([-2 2 -2 2]), axis equal, axis([-2 2 -2 2])
   xlabel('x'), ylabel('y')
   waitforbuttonpress  % to wait for a click on a figure
end
```

◁

Penalty-based methods as presented above are at best suitable to find appropriate starting approximations for more efficient methods of optimization. Later on, however, we will see that a modified approach — it leads to the so-called 'exact penalty functions' — is of quite a practical interest.

1.2 Historical Overview

In this book we almost exclusively deal with continuous finite-dimensional optimization problems. This historical overview, however, does not only consider the finite-dimensional, but also the infinite-dimensional case (that is, *Calculus of Variations* and *Optimal Control*), because their developments have for large parts run in parallel and the theories are almost inextricably interweaved with each other.

Optimization Problems in Antiquity

The history of optimization begins, like so many other stories, with the 'ancient' Greeks. In his *Aeneid* VIRGIL recorded the legendary story of the Phoenician queen DIDO who — after her flight from Tyre — allegedly landed on the coast of Numidia in North Africa in 814 BC.

With seeming modesty she asked the ruler there for as much land as could be enclosed with a bull's hide. This bull's hide, however, she cut into many thin strips so that she was able to enclose the whole area around the harbor bay of what was later to become the city of Carthage.

Hence "DIDO's Problem" is to find the closed curve of fixed length which encloses the maximum area. In the language of modern mathematics it is the classical *isoperimetric problem*.

DIDO probably knew its solution, although not in today's strict sense, since the necessary terms and concepts were only devised over 2000 years later. This was probably also the reason why only simple extremal problems of geometrical nature were examined in antiquity.

ZENODORUS (200–140 BC), for example, dealt with a variant of the isoperimetric problem and proved that of all polygons with n vertices and equal perimeter the regular polygon has the greatest area. The following optimization problem came from the famous mathematician HERON OF ALEXANDRIA (second half of the first century AD). He postulated that beams of light always take the *shortest path*. Only in 1657 was it PIERRE DE FERMAT (1601–1665) who formulated the correct version of this law of nature stating that beams of light cross inhomogeneous media in *minimum time*. HERON's idea, however, was remarkable insofar as he was the first to formulate an extremal principle for a phenomenon of nature.

First Heroic Era: Development of Calculus

For a long time each extremal problem was solved individually using specific methods. In the 17th century people realized the necessity of developing more general methods in the form of a *calculus*. As the name suggests, this was the motivation for the creation of "Calculus". The necessary condition $f'(x) = 0$ for extrema was devised by FERMAT who discovered it in 1629 (allegedly only for polynomials). ISAAC NEWTON (1643–1727) also knew FERMAT's method — in a more general form. He dealt with determining maxima and minima in his work *Methods of Series and Fluxions* which had been completed in 1671 but was published only in 1736.

In 1684 GOTTFRIED WILHELM LEIBNIZ (1646–1716) published a paper in the *Acta Eruditorum* in which he used the necessary condition $f'(x) = 0$ as well as the second-order derivative to distinguish between maxima and minima.

Very surprisingly the obvious generalization to functions of several variables took a long time. It was only in 1755 that the necessary optimality condition $\nabla f(x) = 0$ was published by LEONHARD EULER (1707–1783) in his book *Institutiones Calculi Differentialis.* The LAGRANGE multiplier rule for problems with equality constraints appeared only in 1797 in LAGRANGE's *Théorie des fonctions analytiques.* Instead, the main focus was on problems in which an integral functional, like

$$J(y) = \int_a^b f(x, y, y')\, dx\,,$$

is to be maximized or minimized in a suitable function space. The isoperimetric problem is a problem of this kind. Further extremal problems — mainly motivated by physics — were studied by NEWTON, LEIBNIZ, JACOB BERNOULLI (1654–1705) and JOHN BERNOULLI (1667–1748) as well as EULER. In his book *Methodus Inveniendi Lineas Curvas Maximi Minimive Proprietate Gaudentes, sive Solutio Problematis Isoperimetrici Latissimo Sensu Accepti,* published in 1744, however, EULER developed first formulations of a general theory and for that used the EULER method named after him. From the necessary optimality conditions of the discretized problems he deduced the famous "EULER equation"

$$\frac{\partial f}{\partial y} = \frac{d}{dx}\frac{\partial f}{\partial y'}$$

via passage to the limit for an extremal function y. By doing this, he confirmed known results and was able to solve numerous more general problems.

JOSEPH-LOUIS LAGRANGE (1736–1813) greatly simplified EULER's method. Instead of the transition to the approximating open polygon, he embedded the extremal function $y(x)$ in a family of functions $y(x, \varepsilon) := y(x) + \varepsilon\eta(x)$ and demanded the disappearance of the "first variation"

$$\delta J := \frac{d}{d\varepsilon} J(y + \varepsilon\eta)\Big|_{\varepsilon=0}\,.$$

Following this expression, EULER coined the term *"Calculus of Variations"* for this new and important branch of Analysis.

In EULER's works there are already first formulations for variational problems with differential equation constraints. For these kinds of problems LAGRANGE devised the *multiplier method* named after him, of which the final version appeared in his *Méchanique analytique* in 1788. It took over 100 years until LAGRANGE's suggestion was 'properly' understood and until there was a mathematically 'clean' proof.

Calculus of Variations provided important impulse for the emergence of new branches of mathematics such as Functional Analysis and Convex Analysis. These in turn provided methods with which it was possible to also solve extremal problems from economics and engineering. First papers on the mathematical modeling of economic problems were published at the end of the thirties by the later winners of the Nobel Prize LEONID V. KANTOROVICH (1912–1986) and TJALLING C. KOOPMANS (1910–1985). In 1951 HAROLD W. KUHN (born 1925) and ALBERT W. TUCKER (1905–1995) extended the classical EULER–LAGRANGE multiplier concept to finite-dimensional extremal problems with inequality constraints. Later it became known that WILLIAM KARUSH (1917–1997) and FRITZ JOHN (1910–1994) had already achieved similar results in 1939 [Kar] and 1948 [John].

From the mid-fifties onwards there evolved the *Theory of Optimal Control* from the classical Calculus of Variations and other sources like dynamic programming, control and feedback control systems. Pioneering discoveries by LEV S. PONTRYAGIN (1908–1988) and his students V. G. BOLTYANSKII, R. V. GAMKRELIDZE and E. F. MISHCHENKO led to the *maximum principle*, and thus the theory had its breakthrough as an autonomous branch of mathematics.

The development of more powerful computers contributed to the fact that it was now possible to solve problems from applications which had so far been impossible to solve numerically because of their complex structure.

The most important and most widely used numerical methods of optimization were devised after the Second World War. Before that only some individual methods were known, about which sometimes only vague information is provided in the literature. The oldest method is probably the *least squares fit* which CARL FRIEDRICH GAUSS (1777–1855) allegedly already knew in 1794 [Gau] and which he used for astronomical and geodetic problems.

A particular success for him was the rediscovery of the planetoid CERES in late 1801. After the discovery during January 1801 by the Italian astronomer GIUSEPPE PIAZZI it had been observed only for a short period of time. From very few data GAUSS calculated its orbit using linear regression. FRANZ XAVER VON ZACH found CERES on the night of December 7, just where GAUSS predicted it would be.

In 1805 ADRIEN-MARIE LEGENDRE (1752–1833) presented a printed version of the least squares method for the first time. GAUSS, who published his results only in 1809, claimed priority for it.

An important predecessor of numerical optimization methods and prototype of a descent method with step size choice is the *gradient method*. It was proposed by AUGUSTIN-LOUIS CAUCHY (1789–1857) [Cau] in 1847 to solve systems of nonlinear equations by reducing them to linear regression problems of the form

$$F(x) = \frac{1}{2}\sum_{i=1}^{m} f_i(x)^2 \longrightarrow \min_{x\in\mathbb{R}^n}$$

as described in section 1.1. Here the iterations are carried out according to

$$x^{(k+1)} := x^{(k)} + \lambda_k d_k$$

— starting from an approximation $x^{(k)}$ and taking the direction of steepest descent $d_k := -\nabla F(x^{(k)})$; the *step size* λ_k is the smallest positive value λ which gives a local minimum of F on the half line $\{x^{(k)} + \lambda d_k \mid \lambda \geq 0\}$.

However, it seems that the gradient method has not gained general acceptance —probably because the rate of convergence slows down immensely after a few iterations. In a paper from 1944 LEVENBERG [Lev] regarded the GAUSS–NEWTON *method* as *the* standard method for solving nonlinear regression problems. This method is obtained by linearization

$$f(x) \approx f(x^{(k)}) + J_k(x - x^{(k)})$$

in a neighborhood $U(x^{(k)}, \Delta_k)$, where $J_k := f'(x^{(k)})$ is the Jacobian. The solution d_k to the linear regression problem

$$\|f(x^{(k)}) + J_k d\|_2 \longrightarrow \min_{d\in\mathbb{R}^n}$$

gives a new approximation $x^{(k+1)} := x^{(k)} + d_k$. Strictly speaking this only makes sense if $\|d_k\| \leq \Delta_k$ holds. In the case of $\|d_k\| > \Delta_k$ LEVENBERG proposed considering the *restricted* linear regression problem

$$\|f(x^{(k)}) + J_k d\|_2 \longrightarrow \min_{d\in\mathbb{R}^n,\, \|d\|_2 = \Delta_k}$$

instead. Then there exist exactly one LAGRANGE multiplier $\lambda_k > 0$ and one correction vector d_k with

$$(J_k^T J_k + \lambda_k I)\, d_k = -J_k^T f(x^{(k)})$$

and

$$\|d_k\|_2 = \Delta_k.$$

One can see that for $\lambda_k \approx 0$ a damped GAUSS–NEWTON correction and for large λ_k a damped gradient correction is carried out. Hence, this new method 'interpolates' between the GAUSS–NEWTON method and the gradient method by combining the usually very good local convergence properties of the GAUSS–NEWTON method with the descent properties of the gradient method. This flexibility is especially advantageous if the iteration runs through areas where the Hessian of the objective function — or its approximation — is not (yet) sufficiently positive definite.

LEVENBERG unfortunately does not make any suggestions for determining λ_k in a numerically efficient way and controlling the size of Δ_k. MARQUARDT [Mar] proposed first "ad hoc ideas" in a paper from 1963. Working independently from LEVENBERG, he had pursued a similar approach. This was only properly understood when the findings of LEVENBERG and MARQUARDT were studied in the context of *"trust region methods"* which were originally proposed by GOLDFELD, QUANDT and TROTTER [GQT] in 1966. In contrast to descent methods with choice of step size one chooses the step size — or an upper bound — *first, and then* the descent direction.

When the first computers appeared in the middle of the 20th century, nonlinear regression problems were of central importance. LEVENBERG's approach, however, remained unnoticed for a long time. Rather, at first the interest centered around methods with a gradient-like descent direction. Besides CAUCHY's classical gradient method these were iterative methods which minimized with alternating coordinates and which worked cyclically like the GAUSS–SEIDEL method or with relaxation control like the GAUSS–SOUTHWELL method [Sou]. In the latter case the minimization was done with respect to the coordinate with the steepest descent.

A *Quasi-*NEWTON *method* proposed by DAVIDON (1959) turned out to be revolutionary. It determined the descent directions $d_k = -H_k g_k$ from the gradient g_k using symmetric positive definite matrices H_k.

DAVIDON's paper on this topic remained unpublished at first. In 1963 a revised version of his algorithm was published by FLETCHER and POWELL. It is usually called the DAVIDON–FLETCHER–POWELL method (DFP method). DAVIDON's original work was published only 30 years later in [Dav].

DAVIDON himself called his method the "variable metric method" pointing out that the one-dimensional minimization was done in the direction of the minimum of an approximation of the objective function. For a quadratic objective function

$$f(x) = b^T x + \frac{1}{2} x^T A x$$

with a symmetric positive definite Hessian A the inner product $\langle u, v \rangle_A := v^T A u$ gives a norm $\|u\|_A := \sqrt{\langle u, u \rangle_A}$ and thereby a metric. If one chooses d_k as the steepest descent direction with respect to this norm, that means, as the solution to

$$\min_{d \neq 0} \frac{d^T g_k}{\|d\|_A},$$

it holds that $d_k = -\lambda A^{-1} g_k$ with a suitable $\lambda > 0$. If A^{-1} is known, the quadratic function can be minimized in one step. In the general case one works with changing positive definite matrices which are used as a tool to determine the new descent direction. Hence, the metric changes in each step — therefore the name of this method. In addition it is based on the idea of iteratively determining a good approximation of the inverse of the Hessian.

We thereby get an interesting connection with the *conjugate gradient method* by HESTENES and STIEFEL (1952) since in the quadratic case the following holds for the descent directions:

$$d_i^T A d_j = 0 \quad \text{for} \quad i \neq j\,.$$

This ensures that the optimality of the former search directions remains and that the method — at least in this special case — terminates at the minimal point after a finite number of iterations.

Besides these kinds of minimization methods there were various other methods which were commonly used. These were mostly solely based on the evaluation of functions and most of them did not have a sound theoretical basis. They were plain *search methods*. They were easy to implement, worked sufficiently reliably and fulfilled the expectations of their users in many applications. However, with higher-dimensional problems their rate of convergence was very slow — with quickly increasing costs. Therefore, due to their lack of efficiency most of these methods have gone out of use. However, very popular even today is the NELDER and MEAD *polytope method* from 1965 which will be presented in chapter 3.

Second Heroic Era: Discovery of Simplex Algorithm

The *simplex method* can be regarded as the first efficient method for solving a restricted optimization problem — here in the special case of a linear objective function and linear (inequality) constraints. This algorithm, developed by GEORGE DANTZIG (1914–2005) in 1947, greatly influenced the development of mathematical optimization. In problems of this kind the extrema are attained on the boundary of the feasible region — and in at least one 'vertex'. The simplex method utilizes this fact by iteratively switching from one vertex to the next and thereby — if we are minimizing — reducing the objective function until an optimal vertex is reached.

The representation of the simplex algorithm in tableau form, which had been dominant for a long time, as well as the fact that the problem was called 'linear program' indicate that it has its roots in the time of manual computation and mechanical calculating machines. More adequate forms of representation like the *active set method* only gradually gained acceptance. This also holds for first solution strategies for *quadratic optimization problems* which were devised by BEALE and WOLFE in 1959. These can be regarded as modifications of the simplex algorithm.

A survey by COLVILLE (1968) gives a very good overview of the algorithms available in the sixties to solve nonlinearly constrained optimization problems. Surprisingly, the *reduced gradient method* by ABADIE and CARPENTIER (1965, 1969) was among the best methods. First formulations of these kinds of *projected gradient methods* — at first for linear, later on also for nonlinear constraints — were devised by ROSEN (1960/61) and WOLFE (1962). The idea to linearize the nonlinear constraints does not only appear in the reduced gradient method. The *SLP method* (SLP = sequential linear programming) by GRIFFITH and STEWART (1961) solves nonlinear optimization problems by using a sequence of linear approximations — mostly using first-order TAYLOR series expansion. A disadvantage of this approach is that the solution to each LP-problem is attained in a vertex of the feasible region of the (linearized) constraints which is rather improbable for the nonlinear initial problem.

WILSON (1963) was the first to propose the use of quadratic subproblems — with a quadratic objective function and linear constraints. There one uses a second-order TAYLOR series expansion of the LAGRANGE function. One works with exact Hessians of the LAGRANGE function and uses the LAGRANGE multipliers of the preceding iteration step as its approximate values. If d_k is a solution of the quadratic subproblem, then $x^{(k+1)} := x^{(k)} + d_k$ gives the new approximative value; note that at first there is no 'line search'.

Similar to the classical NEWTON method one can prove *local quadratic convergence* for this method; in the literature it is called the LAGRANGE–NEWTON method. Beginning with HAN (1976), POWELL and FLETCHER WILSON's approach was taken up and developed further by numerous authors. The aim was to obtain a method with global convergence properties and — in analogy with Quasi-NEWTON methods — a simplified approximation of the Hessians. Furthermore, the (mostly nonfeasible) approximations $x^{(k)}$ were to reduce the objective function as well as their distance to the feasible region.

To measure and control the descent the so-called 'exact penalty functions' or 'merit functions' were introduced. This led to a modified step size control where the solutions of the quadratic subproblems were used as the descent directions of the penalty functions. The methods obtained in this way were called *SQP methods.*

Leonid Khachiyan's Breakthrough

The development of the last 30 years has been greatly influenced by the aftermath of a 'scientific earthquake' which was triggered in 1979 by the findings of the Russian mathematician KHACHIYAN (1952–2005) and in 1984 by those of the Indian-born mathematician KARMARKAR. The *New York Times*, which profiled KHACHIYAN's achievement in a November 1979 article entitled "Soviet Mathematician Is Obscure No More," called him "the mystery author of a new mathematical theorem that has rocked the world of computer analysis."

At first it *only* affected *linear* optimization and the up to that time unchallenged dominance of the simplex method. This method was seriously questioned for the first time ever when in 1972 KLEE and MINTY found examples in which the simplex algorithm ran through *all* vertices of the feasible region. This confirmed that the 'worst case complexity' depended *exponentially* on the dimension of the problem. Afterwards people began searching for LP-algorithms with *polynomial* complexity.

Based on SHOR's *ellipsoid method,* it was KHACHIYAN who found the first algorithm of this kind. When we speak of the 'ellipsoid method' today, we usually refer to the 'Russian algorithm' by KHACHIYAN. In many applications, however, it turned out to be less efficient than the simplex method. In 1984 KARMARKAR achieved the breakthrough when he *announced* a *polynomial* algorithm which he claimed to be fifty times faster than the simplex algorithm. This announcement was a bit of an exaggeration but it stimulated very fruitful research activities.

GILL, MURRAY, SAUNDERS and WRIGHT proved the equivalence between KARMARKAR's method and the classical *logarithmic barrier methods*, in particular when applied to linear optimization. Logarithmic barrier methods are methods which — unlike the example of an exterior penalty method from section 1.1 — solve restricted problems by transforming a penalty or barrier term into a parameterized family of unconstrained optimization problems the minimizers of which lie *in the interior of the feasible region.* First approaches to this method date back to FRISCH (1954). In the sixties FIACCO and MCCORMICK devised from that the so-called *interior-point methods.* Their book [Fi/Co] contains a detailed description of classical barrier methods and is regarded as the standard reference work. A disadvantage was that the Hessians of the barrier functions were ill-conditioned in the approximative minimizers. This is usually seen as the reason for large rounding errors. Probably this flaw was the reason why people lost interest in these methods. Now, due to the reawakened interest, the special problem structure was studied again and it was shown that the rounding errors are less problematic if the implementation is thorough enough. Efficient interior-point methods have in the meantime been applied to larger classes of nonlinear optimization problems and are still topics of current research.

Exercises

1. *In* GAUSS*'s Footsteps* (cf. [Hai])

A newly discovered planetoid orbiting the sun was seen at 10 different positions before it disappeared from view. The Cartesian coordinates (x_j, y_j), $j = 1, \ldots 10$, of these positions represented in a fitted coordinate system in the orbit plane are given in the following chart:

$$x_j: \quad -1.024940, -0.949898, -0.866114, -0.773392, -0.671372, \\ -0.559524, -0.437067, -0.302909, -0.155493, -0.007464$$

$$y_j: \quad -0.389269, -0.322894, -0.265256, -0.216557, -0.177152, \\ -0.147582, -0.128618, -0.121353, -0.127348, -0.148885$$

Our aim now is to determine the orbit of this object on the basis of these observations in order to be able to predict where it will be visible again. We assume the orbit to be defined by an ellipse of the form

$$x^2 = ay^2 + bxy + cx + dy + e.$$

This leads to an overdetermined system of linear equations with the unknown coefficients a, b, c, d and e, which is to be solved by means of the *method of least squares.* Do the same for the parabolic approach

$$x^2 = dy + e.$$

Which of the two trajectories is more likely?

2. GAUSS *Normal Equations*

Let $m, n \in \mathbb{N}$ with $m > n$, $A \in \mathbb{R}^{m \times n}$ and $b \in \mathbb{R}^m$. Consider the mapping φ given by

$$\varphi(x) := \|Ax - b\|_2 \quad \text{for } x \in \mathbb{R}^n$$

and show: A vector $u \in \mathbb{R}^n$ gives the minimum of φ if and only if

$$A^T A u = A^T b.$$

3. *Overdetermined System of Linear Equations* (cf. Example 1, page 3)

a) Formulate the linear $\| \ \|_1$-approximation problem $\|Ax - b\|_1 \to \min\limits_x$ as an equivalent linear optimization problem similar to the CHEBYSHEV *approximation problem.*

b) Calculate a solution to the above linear optimization problem with *Maple*® or in like manner.

c) Prove that the convex hull of the points $(3,2)$, $(2,1)$, $(\frac{3}{2},\frac{3}{2})$, $(1,3)$ yields solutions to the $\| \ \|_1$-approximation problem. Are there any other solutions?

4. *a)* The minimum of the function f defined by $f(x) := x_1^2+x_2^2-x_1x_2-3x_1$ for $x \in \mathbb{R}^2$ can instantly be calculated through differentiation. Another way is to use the *method of alternate directions:* The function f is firstly minimized with respect to x_1, then with respect to x_2, then again with respect to x_1 and so on. Use $x^{(0)} := (0,0)^T$ as the starting value and show that the sequence $\big(x^{(k)}\big)$ obtained with the method of alternate directions converges to the minimal point $x^* = (2,1)^T$.

b) Let

$$f(x) = \max\big\{|x_1 + 2x_2 - 7|, |2x_1 + x_2 - 5|\big\}.$$

- Visualize the function f by plotting its contour lines.
- Minimize f similar to the previous exercise, starting with $x^{(0)} := (0,\ 0)^T$. Why don't you get the minimizer $x^* = (1,3)^T$ with $f(x^*) = 0$?

5. *a)* Let $m \in \mathbb{N}$. Assume that the points $P_1, \ldots, P_m$ and $S_1, \ldots, S_m$ in $\mathbb{R}^2$ are given, and we are looking for a transformation consisting of a rotation and a translation that maps the scatterplot of $(P_1, \ldots, P_m)$ as closely as possible to the scatterplot of $(S_1, \ldots, S_m)$. That means: We are looking for $a, b \in \mathbb{R}$ and $\varphi \in [0, 2\pi)$ such that the value

$$d(a,b,\varphi) := \sum_{j=1}^{m} \|f_{a,b,\varphi}(P_j) - S_j\|_2^2$$

with

$$f_{a,b,\varphi}(x,y) := (a,b) + (x\cos\varphi - y\sin\varphi, x\sin\varphi + y\cos\varphi)$$

is minimized. Solve the problem analytically.

b) Calculate the values of x, y, φ for the special case $m = 4$ with

S_j	$(20,20)$	$(40,40)$	$(120,40)$	$(20,60)$
P_j	$(20,15)$	$(33,40)$	$(125,40)$	$(20,65)$

and

S_j	$(20,20)$	$(40,40)$	$(120,40)$	$(20,60)$
P_j	$(16,16)$	$(42,30)$	$(115,-23)$	$(42,58)$

.

6. Little LUCA has gotten 2 dollars from his grandfather. He takes the money and runs to the kiosk around the corner to buy sweets. His favorites are licorice twists (8 cents each) and jelly babies (5 cents each). LUCA wants to buy as many sweets as possible. From experience, however, he knows that he cannot eat more than 20 jelly babies. How many pieces of each type of candy should he buy in order to get as many pieces as possible for his 2 dollars, in view of the fact that not more than 20 jelly babies should be bought (as he does not want to buy more than he can consume immediately)?

a) Formulate the above situation as an *optimization problem.* Define an *objective function* f and maximize it subject to the *constraints* $g(s_1, s_2) \le 0$. (g can be a vector function; then the inequality is understood componentwise!) s_1 and s_2 denote the amount of licorice twists and jelly babies.

b) Visualize the *feasible region* $\mathcal{F}$, that is, the set of all pairs (s_1, s_2) such that $g(s_1, s_2) \le 0$ holds. Visualize $f\colon \mathcal{F} \longrightarrow \mathbb{N}$ and determine its maximum.

c) Is the optimal point unique?

7. The floor plan below shows the arrangement of the offices of a company with a staff of thirty people including the head of the company and two secretaries: On the plan the number in each office denotes the number of staff members working in this room.

The company has bought a copier and now the "sixty-four-dollar question" is: *Where should the copier go?*

The best spot is defined as the one that keeps *traffic in the hallways to a minimum.* The following facts are known about the "copying customs" of the staff members:

- Each staff member (except the head and the secretaries) uses the copier equally often. All the staff members do their own copying (that is, if there are four people in one office, each of them takes his own pile to the copier).
- Both secretaries copy 5 times as much as the other staff members; the head almost never uses the copier.
- There is nobody who walks from the server room or conference room to the copier.

Since the copier cannot be placed in any of the offices, it has to be put somewhere in the hallway, where the following restrictions have to be considered:

- The copier may not be placed directly at the main entrance.
- The copier should not be placed in the middle of the hallway. Consequently, one cannot put it between two opposing doors.
- The copier should not be placed directly in front of the glass wall of the conference room, so as not to disturb (customer) meetings.

Altogether we thus obtain four "taboo zones". Their exact position and layout can be seen in the *floorplan:*

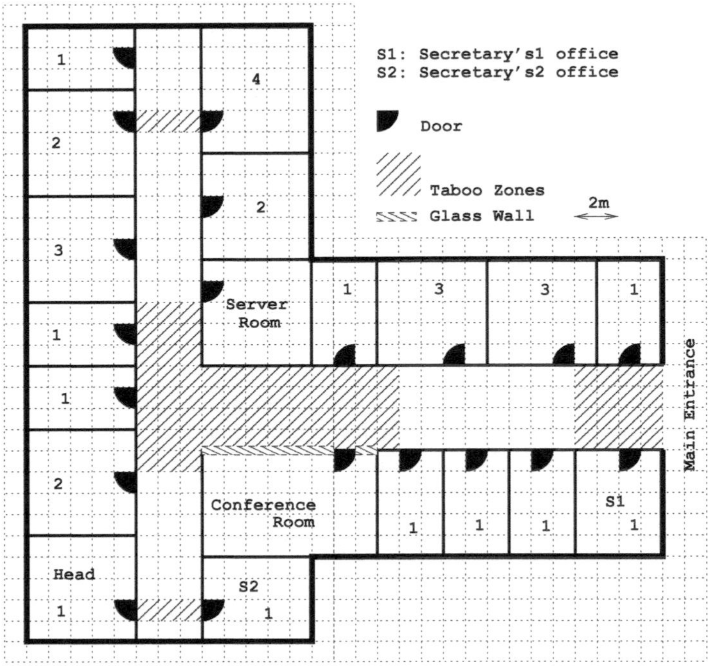

Hint: In practice it is almost impossible to solve this problem without any *simplifications.* In this case it holds that:

- It does not matter on which side of the hallway the copier is placed. Therefore the hallways may be treated as *one-dimensional* objects.
- The distance 'desk–copier' is measured from the door of the respective office and not from the actual desk in the office.
- You might need additional simplifications.

Implement the above problem and determine the optimal location for the copier. What would be the optimal solution if there were no "taboo zones"?

8. Everybody knows that in an advertisement a product can always do everything. Mathematical software is no exception to that. We want to investigate such promises in practice. The problem comes from the area of electric power supply:

A power plant has to provide power to meet the (estimated) power demand given in the following chart.

Expected Demand of Power

12 pm – 6 am	6 am – 9 am	9 am – 3 pm	3 pm – 6 pm	6 pm – 12 pm
15 GW	30 GW	25 GW	40 GW	27 GW

There are three types of generators available, 10 type 1, 10 type 2 and 12 type 3 generators. Each generator type has a minimal and maximal capacity; the production has to be somewhere in between (or else the generator has to be shut off). The running of a generator with minimal capacity costs a certain amount of money (dollars per hour). With each unit above minimal capacity there arise additional costs (dollars per hour) (cf. chart). Costs also arise every time a generator is switched on.

Technical information and costs for the different generator types

	m_i	M_i	e_i	c_i	f_i
Typ 1	850 MW	4000 MW	2000	4	4000
Typ 2	1250 MW	1750 MW	5200	2.6	2000
Typ 3	1500 MW	4000 MW	6000	6	1000

m_i, M_i : minimal and maximal capacity
e_i : costs per hour (minimal capacity)
c_i : costs per hour and per megawatt above minimal capacity
f_i : costs for switching on the generator

In addition to meeting the (estimated) power demands given in the chart an immediate increase by 15% must always be possible. This must be achieved without switching on any additional generators or exceeding the maximal capacity.

Let n_{ij} be the number of generators of type i which are in use in the j-th part of the day, $i = 1, 2, 3$ and $j = 1, 2, \ldots, 5$, and s_{ij} the number of generators that are switched on at the beginning of the j-th part of the day.

The total power supply of type i generators in the j-th part of the day is denoted by x_{ij}.

The costs can be described by the following function K:

$$K = \sum_{i=1}^{3}\sum_{j=1}^{5} c_i z_j (x_{ij} - m_i n_{ij}) + \sum_{i=1}^{3}\sum_{j=1}^{5} e_i z_j n_{ij} + \sum_{i=1}^{3}\sum_{j=1}^{5} f_i s_{ij}$$

x_{ij} are nonnegative *real numbers*, n_{ij} and s_{ij} are nonnegative *integers*. z_j denotes the number of hours in the j-th part of the day (which can be obtained from the above chart).

a) Which simplifications are "hidden" in the cost function?

b) Formulate the constraints!

c) Determine n_{ij}, s_{ij} and x_{ij} such that the total costs are as low as possible!

You might not find the global minimum but only a useful suggestion. You can try a software of your choice (for example the *Matlab*® Optimization Toolbox or *Maple*®). Your solution has to meet all the constraints.

9. *Portfolio optimization* (cf. [Bar1], p. 1ff)

We have a sum of money to split between three investment possibilities which offer rates of return r_1, r_2 and r_3. If x_1, x_2 and x_3 represent the portions of total investment, we expect an overall return

$$R = r_1 x_1 + r_2 x_2 + r_3 x_3 .$$

If the management charge associated with the j-th possibility is $c_j x_j$, then the total cost of our investment is $c_1 x_1 + c_2 x_2 + c_3 x_3$. For an aimed return R we want to pay the least charges to achieve this. Then we have to solve the *problem:*

$$\begin{aligned} & c_1 x_1 + c_2 x_2 + c_3 x_3 \longrightarrow \min \\ & x_1 + x_2 + x_3 = 1 \\ & r_1 x_1 + r_2 x_2 + r_3 x_3 = R \\ & x_1, x_2, x_3 \geq 0 \end{aligned}$$

Consider the corresponding quadratic *penalty problem*

$$\Phi_\lambda(x) := f(x) + \lambda P(x)$$

for $x = (x_1, x_2, x_3)^T \in \mathbb{R}^3$, $h_1(x) := x_1 + x_2 + x_3 - 1$, $h_2(x) := r_1 x_1 + r_2 x_2 + r_3 x_3 - R$ and $P(x) := h_1(x)^2 + h_2(x)^2 + \sum_{j=1}^{3} \psi(x_j)^2$ with

$$\psi(u) := \max(-u, 0) = \begin{cases} 0, & \text{if } u \geq 0 \\ -u, & \text{if } u < 0 \end{cases}$$

where negative investments x_j are penalized by a high management charge $\lambda \psi(x_j)^2$. Use the values $R = 1.25$ together with

$$c_1 = 10,\ c_2 = 9,\ c_3 = 14 \quad \text{and} \quad r_1 = 1.2,\ r_2 = 1.1,\ r_3 = 1.4 .$$

Calculate the minimum of Φ_λ for the parameter values $\lambda = 10^3, 10^6, 10^9$ for instance by means of the function *fminsearch* from the *Matlab*® Optimization Toolbox (cf. the m-file Penalty.m at the end of section 1.1).

How does the solution change in the cases where $c_3 = 12$ and $c_3 = 11$?

2

Optimality Conditions

Chapter 2

2.0 Introduction

In this chapter we will focus on necessary and sufficient optimality conditions for constrained problems.

As an introduction let us remind ourselves of the optimality conditions for *unconstrained* and *equality constrained* problems, which are commonly dealt with in basic Mathematics lectures.

We consider a real-valued function $f\colon D \longrightarrow \mathbb{R}$ with domain $D \subset \mathbb{R}^n$ and define, as usual, for a point $x_0 \in D$:

1) f has a *local minimum* in x_0

$$:\iff \exists\, U \in \mathbb{U}_{x_0}\ \forall\, x \in U \cap D\ \ f(x) \geq f(x_0)$$

W. Forst and D. Hoffmann, *Optimization—Theory and Practice*,
Springer Undergraduate Texts in Mathematics and Technology,
DOI 10.1007/978-0-387-78977-4_2, © Springer Science+Business Media, LLC 2010

2) f has a *strict local minimum* in x_0

$$:\iff \exists\, U \in \mathbb{U}_{x_0}\ \forall\, x \in U \cap D \setminus \{x_0\}\ \ f(x) > f(x_0)$$

3) f has a *global minimum* in x_0

$$:\iff \forall\, x \in D\ \ f(x) \geq f(x_0)$$

4) f has a *strict global minimum* in x_0

$$:\iff \forall\, x \in D \setminus \{x_0\}\ \ f(x) > f(x_0)$$

Here, $\mathbb{U}_{x_0}$ denotes the neighborhood system of x_0.

We often say "x_0 is a *local minimizer* of f" or "x_0 is a *local minimum point of* f" instead of "f has a *local minimum* in x_0" and so on. The *minimizer* is a point $x_0 \in D$, the *minimum* is the corresponding value $f(x_0)$.

Necessary Condition

Suppose that the function f has a local minimum in $x_0 \in \overset{\circ}{D}$, that is, in an interior point of D. Then:

a) If f is differentiable in x_0, then $\nabla f(x_0) = 0$ holds.

b) If f is twice continuously differentiable in a neighborhood of x_0, then the Hessian $H_f(x_0) = \nabla^2 f(x_0) = \left(\frac{\partial^2 f}{\partial x_\nu \partial x_\mu}(x_0)\right)$ is positive semidefinite.

We will use the notation $f'(x_0)$ (to denote the derivative of f at x_0; as we know, this is a linear map from $\mathbb{R}^n$ to $\mathbb{R}$, read as a *row vector*) as well as the corresponding transposed vector $\nabla f(x_0)$ (gradient, *column vector*).

Points $x \in \overset{\circ}{D}$ with $\nabla f(x) = 0$ are called *stationary points.* At a stationary point there can be a local minimum, a local maximum or a *saddlepoint.* To determine that there is a local minimum at a stationary point, we use the following:

Sufficient Condition

Suppose that the function f is twice continuously differentiable in a neighborhood of $x_0 \in D$; also suppose that the necessary optimality condition $\nabla f(x_0) = 0$ holds and that the Hessian $\nabla^2 f(x_0)$ is positive definite. Then f has a strict local minimum in x_0.

The proof of this proposition is based on the TAYLOR theorem and we regard it as known from Calculus. Let us recall that a symmetric (n,n)-matrix A is *positive definite* if and only if all principal subdeterminants

$$\det\begin{pmatrix} a_{11} & \dots & a_{1k} \\ \vdots & & \vdots \\ a_{k1} & \dots & a_{kk} \end{pmatrix} \qquad (k = 1, \dots, n)$$

are positive (cf. exercise 3).

Now let f be a real-valued function with domain $D \subset \mathbb{R}^n$ which we want to minimize subject to the *equality constraints*

$$h_j(x) = 0 \quad (j = 1, \ldots, p)$$

for $p < n$; here, let $h_1, \ldots, h_p$ also be defined on D. We are looking for local minimizers of f, that is, points $x_0 \in D$ which belong to the *feasible region*

$$\mathcal{F} := \big\{x \in D \mid h_j(x) = 0 \;\; (j = 1, \ldots, p)\big\}$$

and to which a neighborhood U exists with $f(x) \geq f(x_0)$ for all $x \in U \cap \mathcal{F}$.

Intuitively, it seems reasonable to solve the constraints for p of the n variables, and to eliminate these by inserting them into the objective function. For the *reduced objective function* we thereby get a nonrestricted problem for which under suitable assumptions the above necessary optimality condition holds.

After these preliminary remarks, we are now able to formulate the following *necessary optimality condition*: LAGRANGE **Multiplier Rule**

Let $D \subset \mathbb{R}^n$ be open and $f, h_1, \ldots, h_p$ continuously differentiable in D. Suppose that f has a local minimum in $x_0 \in \mathcal{F}$ subject to the constraints

$$h_j(x) = 0 \quad (j = 1, \ldots, p).$$

Let also the Jacobian $\left(\frac{\partial h_j}{\partial x_k}(x_0)\right)_{p,n}$ have rank p. Then there exist real numbers $\mu_1, \ldots, \mu_p$ — the so-called LAGRANGE multipliers *— with*

$$\nabla f(x_0) + \sum_{j=1}^{p} \mu_j \nabla h_j(x_0) = 0. \tag{1}$$

Corresponding to our preliminary remarks, a main tool in a *proof* would be the *Implicit Function Theorem.* We assume that interested readers are familiar with a proof from multidimensional analysis. In addition, the results will be generalized in theorem 2.2.5. Therefore we do not give a proof here, but instead illustrate the matter with the following simple problem, which was already introduced in chapter 1 (as example 5):

Example 1

With $f(x) := x_1 x_2^2$ and $h(x) := h_1(x) := x_1^2 + x_2^2 - 2$ for $x = (x_1, x_2)^T \in D := \mathbb{R}^2$ we consider the problem:

$$f(x) \longrightarrow \min \quad \text{subject to the constraint} \quad h(x) = 0\,.$$

We hence have $n = 2$ and $p = 1$.

Before we start, however, note that this problem can of course be solved very easily straight away: One inserts x_2^2 from the constraint $x_1^2 + x_2^2 - 2 = 0$ into $f(x)$ and thus gets a one-dimensional problem.

Points x meeting the constraint are different from 0 and thus also meet the rank condition. With $\mu := \mu_1$ the equation $\nabla f(x) + \mu \nabla h(x) = 0$ translates into

$$x_2^2 + \mu 2 x_1 = 0 \quad \text{and} \quad 2x_1x_2 + \mu 2 x_2 = 0\,.$$

Multiplication of the first equation by x_2 and the second by x_1 gives

$$x_2^3 + 2\mu x_1 x_2 = 0 \quad \text{and} \quad 2x_1^2 x_2 + 2\mu x_1 x_2 = 0$$

and thus

$$x_2^3 = 2x_1^2 x_2\,.$$

For $x_2 = 0$ the constraint yields $x_1 = \pm\sqrt{2}$. Of these two evidently only $x_1 = \sqrt{2}$ remains as a potential minimizer. If $x_2 \neq 0$, we have $x_2^2 = 2x_1^2$ and hence with the constraint $3x_1^2 = 2$, thus $x_1 = \pm\sqrt{2/3}$ and then $x_2 = \pm 2/\sqrt{3}$. In this case the distribution of the zeros and signs of f gives that only $x = (-\sqrt{2/3}, \pm 2/\sqrt{3})^T$ remain as potential minimizers. Since f is continuous on the compact set $\{x \in \mathbb{R}^2 \mid h(x) = 0\}$, we know that there exists a global minimizer. Altogether, we get: f attains its global minimum at $(-\sqrt{2/3}, \pm 2/\sqrt{3})^T$, the point $(\sqrt{2}, 0)^T$ yields a local minimum. The following picture illustrates the gradient condition very well:

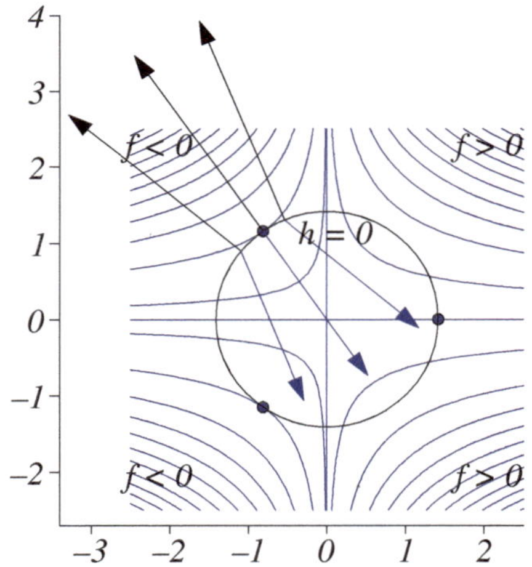

◁

The aim of our further investigations will be to generalize the Lagrange Multiplier Rule to *minimization problems with inequality constraints*:

$$(P) \qquad \begin{array}{l} f(x) \longrightarrow \min \quad \textit{subject to the constraints} \\ g_i(x) \le 0 \ \text{ for } \ i \in \mathcal{I} := \{1, \ldots, m\} \\ h_j(x) = 0 \ \text{ for } \ j \in \mathcal{E} := \{1, \ldots, p\} \end{array} .$$

With $m, p \in \mathbb{N}_0$ (hence, $\mathcal{E} = \emptyset$ or $\mathcal{I} = \emptyset$ are allowed), the functions $f, g_1, \ldots, g_m, h_1, \ldots, h_p$ are supposed to be continuously differentiable on an open subset D in $\mathbb{R}^n$ and $p \le n$. The set

$$\mathcal{F} := \left\{x \in D \mid g_i(x) \le 0 \ \text{ for } \ i \in \mathcal{I},\ h_j(x) = 0 \ \text{ for } \ j \in \mathcal{E}\right\}$$

— in analogy to the above — is called the *feasible region* or *set of feasible points of* (P).

In most cases we state the problem in the slightly shortened form

$$(P) \quad \begin{cases} f(x) \longrightarrow \min \\ g_i(x) \le 0 \ \text{ for } \ i \in \mathcal{I} \\ h_j(x) = 0 \ \text{ for } \ j \in \mathcal{E} \end{cases} .$$

The *optimal value* $v(P)$ to problem (P) is defined as

$$v(P) := \inf\{f(x) : x \in \mathcal{F}\}.$$

We allow $v(P)$ to attain the extended values $+\infty$ and $-\infty$. We follow the standard convention that the infimum of the empty set is ∞. If there are feasible points x_k with $f(x_k) \longrightarrow -\infty \quad (k \longrightarrow \infty)$, then $v(P) = -\infty$ and we say problem (P) — or the function f on $\mathcal{F}$ — is unbounded from below.

We say x_0 is a *minimal point* or a *minimizer* if x_0 is feasible and $f(x_0) = v(P)$.

In order to formulate optimality conditions for (P), we will need some simple tools from *Convex Analysis*. These will be provided in the following section.

2.1 Convex Sets, Inequalities

In the following consider the space $\mathbb{R}^n$ for $n \in \mathbb{N}$ with the euclidean norm and let C be a nonempty subset of $\mathbb{R}^n$. The standard *inner product* or *scalar product* on $\mathbb{R}^n$ is given by $\langle x, y\rangle := x^T y = \sum_{\nu=1}^{n} x_\nu y_\nu$ for $x, y \in \mathbb{R}^n$. The *euclidean norm* of a vector $x \in \mathbb{R}^n$ is defined by $\|x\| := \|x\|_2 := \sqrt{\langle x, x\rangle}$.

Definition

a) C is called *convex* $:\iff \forall\, x_1, x_2 \in C\ \forall\, \lambda \in (0,1)\ \ (1-\lambda)x_1 + \lambda x_2 \in C$

b) C is called a *cone* (with *apex* 0) $:\iff \forall\, x \in C\ \forall\, \lambda > 0\ \ \lambda x \in C$

Remark

C is a *convex cone* if and only if:
$\forall\, x_1, x_2 \in C \;\forall\, \lambda_1, \lambda_2 > 0 \;\; \lambda_1 x_1 + \lambda_2 x_2 \in C$

Proposition 2.1.1 (Separating Hyperplane Theorem)

Let C be closed and convex, and $b \in \mathbb{R}^n \setminus C$. Then there exist $p \in \mathbb{R}^n \setminus \{0\}$ and $\alpha \in \mathbb{R}$ such that $\langle p, x \rangle \geq \alpha > \langle p, b \rangle$ for all $x \in C$, that is, the hyperplane defined by $H := \{x \in \mathbb{R}^n \mid \langle p, x \rangle = \alpha\}$ strictly separates C and b. If furthermore C is a cone, we can choose $\alpha = 0$.

The following two little pictures show that none of the two assumptions that C is *convex* and *closed* can be dropped. The set C on the left is convex but not closed; on the right it is closed but not convex.

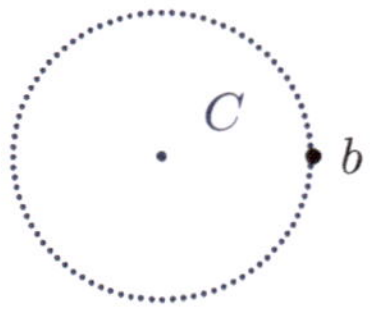

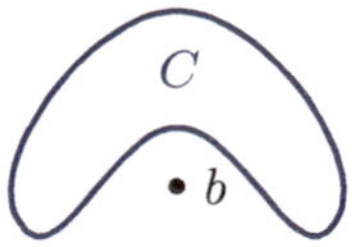

Proof: Since C is closed,

$$\delta := \delta(b, C) = \inf\big\{\|x - b\| : x \in C\big\}$$

is positive, and there exists a sequence (x_k) in C such that $\|x_k - b\| \longrightarrow \delta$. WLOG let $x_k \to q$ for a $q \in \mathbb{R}^n$ (otherwise use a suitable subsequence). Then q is in C with $\|p\| = \delta > 0$ for $p := q - b$.

For $x \in C$ and $0 < \tau < 1$ it holds that

$$\begin{aligned} \|p\|^2 &= \delta^2 \leq \|(1-\tau)q + \tau x - b\|^2 = \|q - b + \tau(x - q)\|^2 \\ &= \|p\|^2 + 2\tau\,\langle x - q, p\rangle + \tau^2\,\|x - q\|^2. \end{aligned}$$

From this we obtain

$$0 \leq 2\,\langle x - q, p\rangle + \tau\,\|x - q\|^2$$

and after passage to the limit $\tau \to 0$

$$0 \leq \langle x - q, p\rangle\,.$$

With $\alpha := \delta^2 + \langle b, p\rangle$ the first assertion $\langle p, x\rangle \geq \alpha > \langle p, b\rangle$ follows. If C is a *cone,* then for all $\lambda > 0$ and $x \in C$ the vectors $\frac{1}{\lambda}x$ and λx are also in C.

Therefore $\langle p, x \rangle = \lambda \langle p, \frac{1}{\lambda} x \rangle \geq \lambda \alpha$ holds and consequently $\langle p, x \rangle \geq 0$. $\lambda \langle p, x \rangle = \langle p, \lambda x \rangle \geq \alpha$ shows $0 \geq \alpha$, hence, $\langle p, b \rangle < \alpha \leq 0$. □

Definition

$$C^* := \left\{ y \in \mathbb{R}^n \mid \forall\, x \in C \ \langle y, x \rangle \geq 0 \right\}$$

is called the *dual cone* of C.

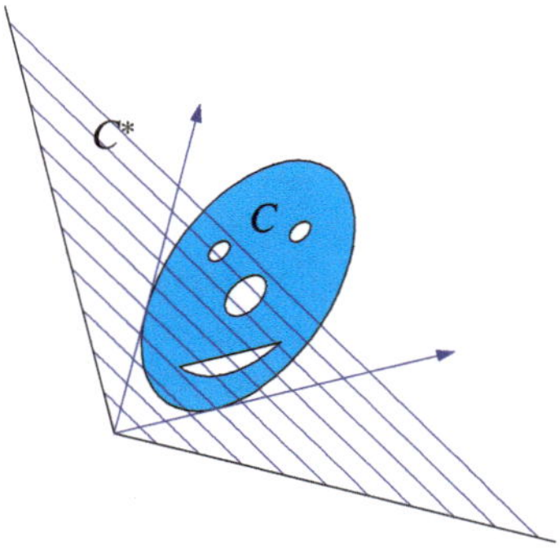

Remark C^* *is a closed, convex cone.*

We omit a *proof.* The statement is an immediate consequence of the definition of the dual cone.

As an *important application* let us now consider the following situation: Let $A = (a_1, \ldots, a_n) \in \mathbb{R}^{m \times n}$ be an (m, n)-matrix with columns $a_1, \ldots, a_n \in \mathbb{R}^m$.

Definition

$$\operatorname{cone}(A) := \operatorname{cone}(a_1, \ldots, a_n) := A\mathbb{R}^n_+ = \{Aw \mid w \in \mathbb{R}^n_+\}$$

is called the (positive) *conic hull* of $a_1, \ldots, a_n$.

Lemma 2.1.2

1) cone(A) *is a* closed, *convex cone.*

2) $(\operatorname{cone}(A))^* = \left\{ y \in \mathbb{R}^m \mid A^T y \geq 0 \right\}$

Proof:

1) It is obvious that $C_n := \operatorname{cone}(a_1, \ldots, a_n)$ is a convex cone. We will prove that it is *closed* by means of induction over n:

For $n = 1$ the cone $C_1 = \{\xi_1 a_1 \mid \xi_1 \geq 0\}$ is — in the nontrivial case — a closed half line. For the induction step from n to $n+1$ we assume that every conic hull generated by not more than n vectors is closed.

Firstly, consider the case that

$$-a_j \in \operatorname{cone}(a_1,\ldots,a_{j-1},a_{j+1},\ldots,a_{n+1}) \quad \text{for all } j = 1,\ldots,n+1 \,.$$

It follows that $C_{n+1} = \operatorname{span}\{a_1,\ldots,a_{n+1}\}$ and therefore obviously that C_{n+1} is closed:

The inclusion from left to right is trivial, and the other one follows, with $\xi_1,\ldots,\xi_{n+1} \in \mathbb{R}$ from

$$\sum_{j=1}^{n+1} \xi_j\, a_j \;=\; \sum_{j=1}^{n+1} |\xi_j| \operatorname{sign}(\xi_j)\, a_j \,.$$

Otherwise, assume WLOG $-a_{n+1} \notin \operatorname{cone}(a_1,\ldots,a_n) = C_n$; because of the induction hypothesis, C_n is closed and therefore $\delta := \delta(-a_{n+1}, C_n)$ is positive. Every $x \in C_{n+1}$ can be written in the form $x = \sum_{j=1}^{n+1} \xi_j\, a_j$ with $\xi_1,\ldots,\xi_{n+1} \in \mathbb{R}_+$. Then

$$\xi_{n+1} \le \frac{\|x\|}{\delta}$$

holds because in the nontrivial case $\xi_{n+1} > 0$ this follows directly from

$$\|x\| \;=\; \xi_{n+1} \Big\| - a_{n+1} - \underbrace{\sum_{j=1}^{n} \frac{\xi_j}{\xi_{n+1}} a_j}_{\in\, C_n} \Big\| \ge \xi_{n+1}\, \delta \,.$$

Let $(x^{(k)})$ be a sequence in C_{n+1} and $x \in \mathbb{R}^m$ with $x^{(k)} \longrightarrow x$ for $k \longrightarrow \infty$. We want to show $x \in C_{n+1}$: For $k \in \mathbb{N}$ there exist $\xi_1^{(k)},\ldots,\xi_{n+1}^{(k)} \in \mathbb{R}_+$ such that

$$x^{(k)} \;=\; \sum_{j=1}^{n+1} \xi_j^{(k)}\, a_j \,.$$

As $(x^{(k)})$ is a convergent sequence, there exists an $M > 0$ such that $\|x^{(k)}\| \le M$ for all $k \in \mathbb{N}$, and we get

$$0 \,\le\, \xi_{n+1}^{(k)} \,\le\, \frac{M}{\delta} \,.$$

WLOG let the sequence $\big(\xi_{n+1}^{(k)}\big)$ be convergent (otherwise, consider a suitable subsequence), and set $\xi_{n+1} := \lim \xi_{n+1}^{(k)}$. So we have

$$C_n \ni x^{(k)} - \xi_{n+1}^{(k)}\, a_{n+1} \longrightarrow x - \xi_{n+1}\, a_{n+1} \,.$$

By induction, C_n is closed, thus $x - \xi_{n+1} a_{n+1}$ is an element of C_n and consequently x is in C_{n+1}.

2) The definitions of cone(A) and of the dual cone give immediately:

$$\begin{aligned}\left(\operatorname{cone}(A)\right)^* &= \left\{y \in \mathbb{R}^m \mid \forall\, v \in \operatorname{cone}(A) \;\; \langle v, y\rangle \ge 0\right\}\\ &= \left\{y \in \mathbb{R}^m \mid \forall\, w \in \mathbb{R}^n_+ \;\; \langle Aw, y\rangle \ge 0\right\}\\ &= \left\{y \in \mathbb{R}^m \mid \forall\, w \in \mathbb{R}^n_+ \;\; \langle w, A^T y\rangle \ge 0\right\}\\ &\overset{\checkmark}{=} \left\{y \in \mathbb{R}^m \mid A^T y \ge 0\right\}\end{aligned}$$ □

A crucial tool for the following considerations is the

Theorem of the Alternative (FARKAS (1902))

For $A \in \mathbb{R}^{m\times n}$ and $b \in \mathbb{R}^m$ the following are strong alternatives:

1) $\exists\, x \in \mathbb{R}^n_+ \;\; Ax = b$

2) $\exists\, y \in \mathbb{R}^m \;\; A^T y \ge 0 \,\wedge\, b^T y < 0$

Proof: *1)* $\Longrightarrow \neg$ *2)*: For $x \in \mathbb{R}^n_+$ with $Ax = b$ and $y \in \mathbb{R}^m$ with $A^T y \ge 0$ we have $b^T y = x^T A^T y \ge 0$.

$\neg$ *1)* $\Longleftarrow$ *2)*: $C := \operatorname{cone}(A)$ is a closed convex cone which does not contain the vector b: Following the addendum in the Separating Hyperplane Theorem there exists a $y \in \mathbb{R}^m$ with $\langle y, x\rangle \ge 0 > \langle y, b\rangle$ for all $x \in C$, in particular $a_\nu^T y = \langle y, a_\nu\rangle \ge 0$, that is, $A^T y \ge 0$. □

If we *illustrate* the assertion, the theorem can be memorized easily: *1)* means nothing but $b \in \operatorname{cone}(A)$. With the open 'half space'

$$H_b := \left\{y \in \mathbb{R}^m \mid \langle y, b\rangle < 0\right\}$$

the condition *2)* states that $\left(\operatorname{cone}(A)\right)^*$ and H_b have a common point.

In the two-dimensional case, for example, we can illustrate the theorem with the following picture, which shows case *1):*

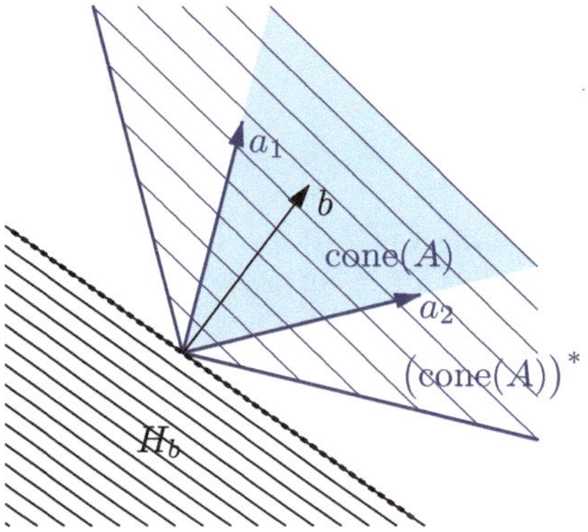

If you rotate the vector b out of cone(A), you get case *2).*

2.2 Local First-Order Optimality Conditions

We want to take up the minimization problem (P) from page 39 again and use the notation introduced there. For $x_0 \in \mathcal{F}$, the index set

$$\mathcal{A}(x_0) := \{i \in \mathcal{I} \mid g_i(x_0) = 0\}$$

describes the *inequality restrictions which are active at* x_0.

The active constraints have a special significance: They restrict feasible corrections around a feasible point. If a constraint is *inactive* ($g_i(x_0) < 0$) at the feasible point x_0, it is possible to move from x_0 a bit in any direction without violating this constraint.

Definition

Let $d \in \mathbb{R}^n$ and $x_0 \in \mathcal{F}$. Then d is called the *feasible direction of* $\mathcal{F}$ *at* x_0 $:\iff \exists\, \delta > 0\ \forall\, \tau \in [0, \delta]\ \ x_0 + \tau d \in \mathcal{F}$.

A 'small' movement from x_0 along such a direction gives feasible points.

The set of all feasible directions of $\mathcal{F}$ at x_0 is a *cone,* denoted by

$$\mathcal{C}_{fd}(x_0)\,.$$

Let d be a feasible direction of $\mathcal{F}$ at x_0. If we choose a δ according to the definition, then we have

$$\underbrace{g_i(x_0 + \tau d)}_{\leq 0} = \underbrace{g_i(x_0)}_{=0} + \tau g_i'(x_0)d + o(\tau)$$

for $i \in \mathcal{A}(x_0)$ and $0 < \tau \leq \delta$. Dividing by τ and passing to the limit as $\tau \to 0$ gives $g_i'(x_0)d \leq 0$. In the same way we get $h_j'(x_0)d = 0$ for all $j \in \mathcal{E}$.

Definition

For any $x_0 \in \mathcal{F}$

$$\mathcal{C}_\ell(P, x_0) := \left\{d \in \mathbb{R}^n \mid \forall\, i \in \mathcal{A}(x_0)\ g_i'(x_0)d \leq 0,\ \forall\, j \in \mathcal{E}\ h_j'(x_0)d = 0\right\}$$

is called the *linearizing cone* of (P) at x_0. Hence, $\mathcal{C}_\ell(x_0) := \mathcal{C}_\ell(P, x_0)$ *contains at least all feasible directions of* $\mathcal{F}$ *at* x_0:

$$\mathcal{C}_{fd}(x_0) \subset \mathcal{C}_\ell(x_0)$$

The linearizing cone is not only dependent on the *set* of feasible points $\mathcal{F}$ but also on the *representation* of $\mathcal{F}$ (compare Example 4). We therefore write more precisely $\mathcal{C}_\ell(P, x_0)$.

Definition

For any $x_0 \in D$

$$\mathcal{C}_{dd}(x_0) := \left\{d \in \mathbb{R}^n \mid f'(x_0)d < 0\right\}$$

is called the *cone of descent directions of f at x_0*.

Note that 0 is not in $\mathcal{C}_{dd}(x_0)$; also, for all $d \in \mathcal{C}_{dd}(x_0)$

$$f(x_0 + \tau d) = f(x_0) + \tau \underbrace{f'(x_0)d}_{<0} + o(\tau)$$

holds and therefore, $f(x_0 + \tau d) < f(x_0)$ for sufficiently small $\tau > 0$.

Thus, $d \in \mathcal{C}_{dd}(x_0)$ guarantees that the objective function f can be reduced along this direction. Hence, for a local minimizer x_0 of (P) it necessarily holds that $\mathcal{C}_{dd}(x_0) \cap \mathcal{C}_{fd}(x_0) = \emptyset$.

We will illustrate the above definitions with the following

Example 1

Let

$$\mathcal{F} := \left\{x = (x_1, x_2)^T \in \mathbb{R}^2 \mid x_1^2 + x_2^2 - 1 \leq 0,\ -x_1 \leq 0,\ -x_2 \leq 0\right\},$$

and f be defined by $f(x) := x_1 + x_2$. Hence, $\mathcal{F}$ is the part of the unit disk which lies in the first quadrant. The objective function f evidently attains a (strict, global) minimum at $(0,0)^T$.

In both of the following pictures $\mathcal{F}$ is colored in dark blue.

$x_0 := (0,0)^T$

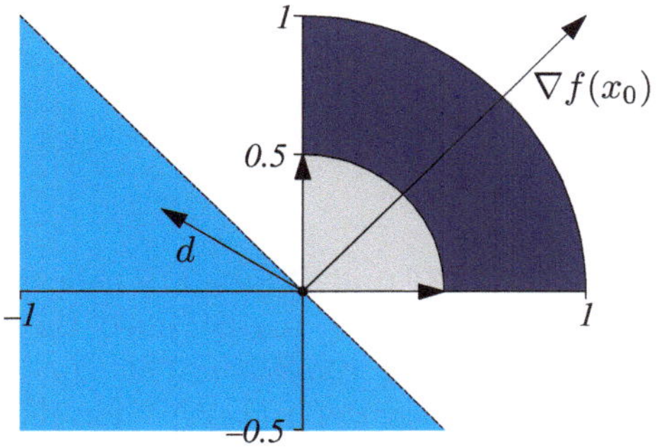

$x_0 := (1,0)^T$

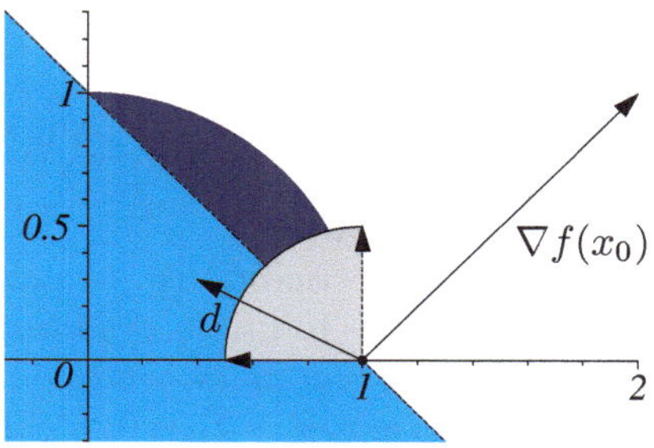

a) Let $x_0 := (0,0)^T$. $g_1(x) := x_1^2 + x_2^2 - 1$, $g_2(x) := -x_1$ and $g_3(x) := -x_2$ give $\mathcal{A}(x_0) = \{2,3\}$. A vector $d := (d_1, d_2)^T \in \mathbb{R}^2$ is a *feasible direction* of $\mathcal{F}$ at x_0 if and only if $d_1 \geq 0$ and $d_2 \geq 0$ hold. Hence, the set $\mathcal{C}_{fd}(x_0)$ of feasible directions is a convex cone, namely, the first quadrant, and it is represented in the left picture by the gray angular domain. $g_2'(x_0) = (-1, 0)$ and $g_3'(x_0) = (0, -1)$ produce

$$\mathcal{C}_\ell(x_0) = \left\{d \in \mathbb{R}^2 \mid -d_1 \leq 0,\ -d_2 \leq 0\right\}.$$

Hence, in this example, the *linearizing cone* and the *cone of feasible directions* are the same. Moreover, the *cone of descent directions* $\mathcal{C}_{dd}(x_0)$ — colored in light blue in the picture — is, because of $f'(x_0)d = (1,1)d = d_1 + d_2$, an open half space and disjoint to $\mathcal{C}_\ell(x_0)$.

b) If $x_0 := (1, 0)^T$, we have $\mathcal{A}(x_0) = \{1, 3\}$ and $d := (d_1, d_2)^T \in \mathbb{R}^2$ is a *feasible direction* of $\mathcal{F}$ at x_0 if and only if $d = (0, 0)^T$ or $d_1 < 0$ and $d_2 \geq 0$ hold. The set of feasible directions is again a convex cone. In the right picture it is depicted by the shifted gray angular domain. Because of $g_1'(x_0) = (2, 0)$ and $g_3'(x_0) = (0, -1)$, we get

$$\mathcal{C}_\ell(x_0) = \left\{d \in \mathbb{R}^2 \mid d_1 \leq 0,\ d_2 \geq 0\right\}.$$

As we can see, in this case the *linearizing cone* includes the cone of feasible directions properly as a subset. In the picture the *cone of descent directions* has also been moved to x_0. We can see that it contains feasible directions of $\mathcal{F}$ at x_0. Consequently, f does *not* have a local minimum in x_0. ◁

Proposition 2.2.1

For $x_0 \in \mathcal{F}$ it holds that $\mathcal{C}_\ell(x_0) \cap \mathcal{C}_{dd}(x_0) = \emptyset$ if and only if there exist $\lambda \in \mathbb{R}_+^m$ and $\mu \in \mathbb{R}^p$ such that

$$\nabla f(x_0) + \sum_{i=1}^{m} \lambda_i \nabla g_i(x_0) + \sum_{j=1}^{p} \mu_j \nabla h_j(x_0) = 0 \tag{2}$$

and

$$\lambda_i g_i(x_0) = 0 \text{ for all } i \in \mathcal{I}. \tag{3}$$

Together, these conditions — $x_0 \in \mathcal{F}$, $\lambda \geq 0$, (2) and (3) — are called KARUSH–KUHN–TUCKER *conditions*, or *KKT conditions.* (3) is called the *complementary slackness condition* or *complementarity condition.* This condition of course means $\lambda_i = 0$ or (in the nonexclusive sense) $g_i(x_0) = 0$ for all

$i \in \mathcal{I}$. A corresponding pair (λ, μ) or the scalars $\lambda_1, \ldots, \lambda_m, \mu_1, \ldots, \mu_p$ are called LAGRANGE *multipliers*. The function L defined by

$$L(x, \lambda, \mu) := f(x) + \sum_{i=1}^{m} \lambda_i g_i(x) + \sum_{j=1}^{p} \mu_j h_j(x) = f(x) + \lambda^T g(x) + \mu^T h(x)$$

for $x \in D$, $\lambda \in \mathbb{R}_+^m$ and $\mu \in \mathbb{R}^p$ is called the LAGRANGE *function* or *Lagrangian* of (P). Here we have combined the m functions g_i to a vector-valued function g and respectively the p functions h_j to a vector-valued function h.

Points $x_0 \in \mathcal{F}$ fulfilling (2) and (3) with a suitable $\lambda \in \mathbb{R}_+^m$ and $\mu \in \mathbb{R}^p$ play an important role. They are called KARUSH–KUHN–TUCKER *points*, or *KKT points*.

Owing to the complementarity condition (3), the multipliers λ_i corresponding to *inactive restrictions* at x_0 must be zero. So we can omit the terms for $i \in \mathcal{I} \setminus \mathcal{A}(x_0)$ from (2) and rewrite this condition as

$$\nabla f(x_0) + \sum_{i \in \mathcal{A}(x_0)} \lambda_i \nabla g_i(x_0) + \sum_{j=1}^{p} \mu_j \nabla h_j(x_0) = 0. \qquad (2')$$

Proof: By definition of $\mathcal{C}_\ell(x_0)$ and $\mathcal{C}_{dd}(x_0)$ it holds that:

$$d \in \mathcal{C}_\ell(x_0) \cap \mathcal{C}_{dd}(x_0) \iff \begin{cases} f'(x_0)d < 0 \\ \forall\, i \in \mathcal{A}(x_0) \;\; g_i'(x_0)d \le 0 \\ \forall\, j \in \mathcal{E} \;\; h_j'(x_0)d = 0 \end{cases}$$

$$\iff \begin{cases} f'(x_0)d < 0 \\ \forall\, i \in \mathcal{A}(x_0) \;\; -g_i'(x_0)d \ge 0 \\ \forall\, j \in \mathcal{E} \;\; -h_j'(x_0)d \ge 0 \\ \forall\, j \in \mathcal{E} \;\; h_j'(x_0)d \ge 0 \end{cases}$$

With that the Theorem of the Alternative from section 2.1 directly provides the following equivalence:
$\mathcal{C}_\ell(x_0) \cap \mathcal{C}_{dd}(x_0) = \emptyset$ if and only if there exist $\lambda_i \ge 0$ for $i \in \mathcal{A}(x_0)$ and $\mu_j' \ge 0$, $\mu_j'' \ge 0$ for $j \in \mathcal{E}$ such that

$$\nabla f(x_0) = \sum_{i \in \mathcal{A}(x_0)} \lambda_i(-\nabla g_i(x_0)) + \sum_{j=1}^{p} \mu_j'(-\nabla h_j(x_0)) + \sum_{j=1}^{p} \mu_j'' \nabla h_j(x_0).$$

If we now set $\lambda_i := 0$ for $i \in \mathcal{I} \setminus \mathcal{A}(x_0)$ and $\mu_j := \mu_j' - \mu_j''$ for $j \in \mathcal{E}$, the above is equivalent to: There exist $\lambda_i \ge 0$ for $i \in \mathcal{I}$ and $\mu_j \in \mathbb{R}$ for $j \in \mathcal{E}$ with

$$\nabla f(x_0) + \sum_{i=1}^{m} \lambda_i \nabla g_i(x_0) + \sum_{j=1}^{p} \mu_j \nabla h_j(x_0) = 0$$

and

$$\lambda_i g_i(x_0) = 0 \text{ for all } i \in \mathcal{I}. \qquad \square$$

So now the *question* arises whether not just $\mathcal{C}_{fd}(x_0) \cap \mathcal{C}_{dd}(x_0) = \emptyset$, but even $\mathcal{C}_\ell(x_0) \cap \mathcal{C}_{dd}(x_0) = \emptyset$ is true for any local minimizer $x_0 \in \mathcal{F}$. The following simple example gives a negative answer to this question:

Example 2 (KUHN–TUCKER (1951))

For $n = 2$ and $x = (x_1, x_2)^T \in \mathbb{R}^2 =: D$ let
$f(x) := -x_1$, $g_1(x) := x_2 + (x_1 - 1)^3$, $g_2(x) := -x_1$ and $g_3(x) := -x_2$.
For $x_0 := (1, 0)^T$, $m = 3$ and $p = 0$ we have:
$\nabla f(x_0) = (-1, 0)^T$, $\nabla g_1(x_0) = (0, 1)^T$, $\nabla g_2(x_0) = (-1, 0)^T$ and $\nabla g_3(x_0) = (0, -1)^T$.
Since $\mathcal{A}(x_0) = \{1, 3\}$, we get $\mathcal{C}_\ell(x_0) = \{(d_1, d_2)^T \in \mathbb{R}^2 \mid d_2 = 0\}$, as well as $\mathcal{C}_{dd}(x_0) = \{(d_1, d_2)^T \in \mathbb{R}^2 \mid d_1 > 0\}$; evidently, $\mathcal{C}_\ell(x_0) \cap \mathcal{C}_{dd}(x_0)$ is nonempty. However, the function f has a minimum at x_0 subject to the given constraints.

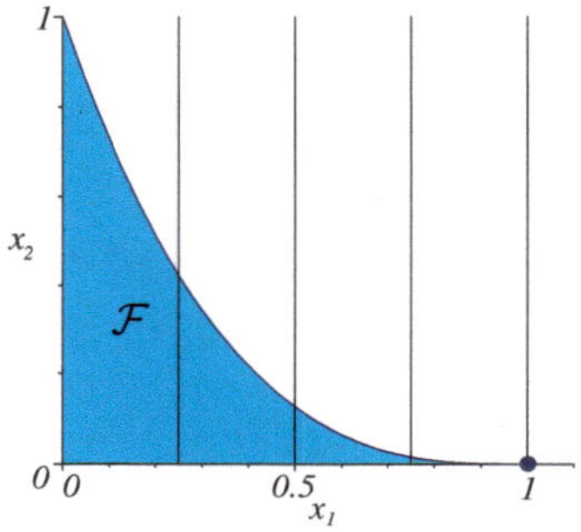

◁

Lemma 2.2.2

For $x_0 \in \mathcal{F}$ it holds that: $\mathcal{C}_\ell(x_0) \cap \mathcal{C}_{dd}(x_0) = \emptyset \iff \nabla f(x_0) \in \mathcal{C}_\ell(x_0)^*$

Proof:

$$\begin{aligned} \mathcal{C}_\ell(x_0) \cap \mathcal{C}_{dd}(x_0) = \emptyset &\iff \forall\, d \in \mathcal{C}_\ell(x_0) \ \ \langle \nabla f(x_0), d \rangle = f'(x_0)d \geq 0 \\ &\iff \nabla f(x_0) \in \mathcal{C}_\ell(x_0)^* \end{aligned} \qquad \square$$

The cone $\mathcal{C}_{fd}(x_0)$ of all feasible directions is too small to ensure general optimality conditions. Difficulties may occur due to the fact that *the boundary of $\mathcal{F}$ is curved.* Therefore, we have to consider a set which is less intuitive but bigger and with more suitable properties. To attain this goal, it is useful to state the *concept of being tangent to a set* more precisely:

Definition

A sequence (x_k) *converges in direction* d to x_0

$$:\iff x_k = x_0 + \alpha_k(d + r_k) \text{ with } \alpha_k \downarrow 0 \text{ and } r_k \to 0.$$

We will use the following *notation:* $x_k \xrightarrow{d} x_0$

$x_k \xrightarrow{d} x_0$ simply means: There exists a sequence of positive numbers (α_k) such that $\alpha_k \downarrow 0$ and

$$\frac{1}{\alpha_k}(x_k - x_0) \longrightarrow d \text{ for } k \longrightarrow \infty\,.$$

Definition

Let M be a nonempty subset of $\mathbb{R}^n$ and $x_0 \in M$. Then

$$\mathcal{C}_t\,(M, x_0) := \left\{d \in \mathbb{R}^n \mid \exists\ (x_k) \in M^{\mathbb{N}}\ \ x_k \xrightarrow{d} x_0\right\}$$

is called the *tangent cone* of M at x_0. The vectors of $\mathcal{C}_t\,(M, x_0)$ are called *tangents* or *tangent directions* of M at x_0.

Of main interest is the special case

$$\mathcal{C}_t(x_0) := \mathcal{C}_t\,(\mathcal{F}, x_0)\,.$$

Example 3

a) The following two figures illustrate the cone of tangents for

$$\mathcal{F} := \left\{x = (x_1, x_2)^T \in \mathbb{R}^2 \mid x_1 \geq 0,\, x_1^2 \geq x_2 \geq x_1^2\,(x_1 - 1)\right\}$$

and the points $x_0 \in \left\{(0,0)^T, (2,4)^T, (1,0)^T\right\}$. For convenience the origin is translated to x_0. The reader is invited to verify this:

$x_0 = (0,0)^T$ *and* $x_0 = (2,4)^T$ $\qquad\qquad$ $x_0 = (1,0)^T$

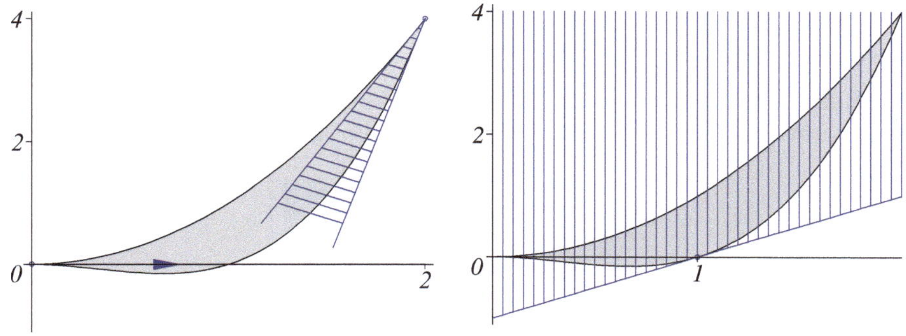

b) $\mathcal{F} := \left\{x \in \mathbb{R}^n \mid \|x\|_2 = 1\right\}$: $\quad \mathcal{C}_t(x_0) = \left\{d \in \mathbb{R}^n \mid \langle d, x_0\rangle = 0\right\}$

c) $\mathcal{F} := \left\{x \in \mathbb{R}^n \mid \|x\|_2 \leq 1\right\}$: Then $\mathcal{C}_t(x_0) = \mathbb{R}^n$ if $\|x_0\|_2 < 1$ holds, and $\mathcal{C}_t(x_0) = \left\{d \in \mathbb{R}^n \mid \langle d, x_0\rangle \leq 0\right\}$ if $\|x_0\|_2 = 1$.

These assertions have to be proven in exercise 10. $\triangleleft$

Lemma 2.2.3

1) $\mathcal{C}_t(x_0)$ *is a closed cone,* $0 \in \mathcal{C}_t(x_0)$.

2) $\overline{\mathcal{C}_{fd}(x_0)} \subset \mathcal{C}_t(x_0) \subset \mathcal{C}_\ell(x_0)$

Proof: The proof of *1)* is to be done in exercise 9.

2) First inclusion: As the tangent cone $\mathcal{C}_t(x_0)$ is closed, it is sufficient to show the inclusion $\mathcal{C}_{fd}(x_0) \subset \mathcal{C}_t(x_0)$. For $d \in \mathcal{C}_{fd}(x_0)$ and 'large' integers k it holds that $x_0 + \frac{1}{k} d \in \mathcal{F}$. With $\alpha_k := \frac{1}{k}$ and $r_k := 0$ this shows $d \in \mathcal{C}_t(x_0)$.

Second inclusion: Let $d \in \mathcal{C}_t(x_0)$ and $(x_k) \in \mathcal{F}^{\mathbb{N}}$ be a sequence with $x_k = x_0 + \alpha_k (d + r_k)$, $\alpha_k \downarrow 0$ and $r_k \to 0$. For $i \in \mathcal{A}(x_0)$

$$\underbrace{g_i(x_k)}_{\leq 0} = \underbrace{g_i(x_0)}_{=0} + \alpha_k \, g_i'(x_0)(d + r_k) + o(\alpha_k)$$

produces the inequality $g_i'(x_0) d \leq 0$. In the same way we get $h_j'(x_0) d = 0$ for $j \in \mathcal{E}$. □

Now the *question* arises whether $\mathcal{C}_t(x_0) = \mathcal{C}_\ell(x_0)$ always holds. The following example gives a negative answer:

Example 4

a) Consider $\mathcal{F} := \{x \in \mathbb{R}^2 \mid -x_1^3 + x_2 \leq 0 \, , \ -x_2 \leq 0\}$ and $x_0 := (0,0)^T$. In this case $\mathcal{A}(x_0) = \{1, 2\}$. This gives

$\mathcal{C}_\ell(x_0) = \{d \in \mathbb{R}^2 \mid d_2 = 0\}$ and $\mathcal{C}_t(x_0) = \{d \in \mathbb{R}^2 \mid d_1 \geq 0 \, , \ d_2 = 0\}$.

The last statement has to be shown in exercise 10.

b) Now let $\mathcal{F} := \{x \in \mathbb{R}^2 \mid -x_1^3 + x_2 \leq 0 \, , \ -x_1 \leq 0 \, , \ -x_2 \leq 0\}$ and $x_0 := (0,0)^T$. Then $\mathcal{A}(x_0) = \{1, 2, 3\}$ and therefore

$\mathcal{C}_\ell(x_0) = \{d \in \mathbb{R}^2 \mid d_1 \geq 0 \, , \ d_2 = 0\} = \mathcal{C}_t(x_0)$.

Hence, *the linearizing cone is dependent on the representation of the set of feasible points* $\mathcal{F}$ which is the same in both cases!

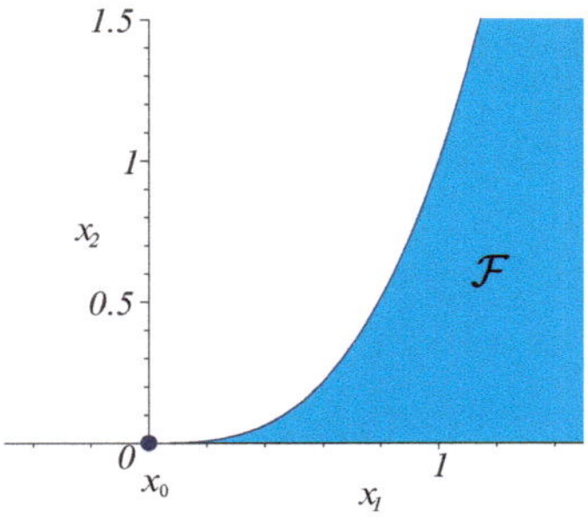

◁

Lemma 2.2.4

For a local minimizer x_0 of (P) it holds that $\nabla f(x_0) \in \mathcal{C}_t(x_0)^*$, hence $\mathcal{C}_{dd}(x_0) \cap \mathcal{C}_t(x_0) = \emptyset$.

Geometrically this condition states that for a local minimizer x_0 of (P) the angle between the gradient and any tangent direction, especially any feasible direction, does not exceed 90°.

Proof: Let $d \in \mathcal{C}_t(x_0)$. Then there exists a sequence $(x_k) \in \mathcal{F}^{\mathbb{N}}$ such that $x_k = x_0 + \alpha_k(d + r_k)$, $\alpha_k \downarrow 0$ and $r_k \longrightarrow 0$.

$$0 \le f(x_k) - f(x_0) = \alpha_k f'(x_0)(d + r_k) + o(\alpha_k)$$

gives the result $f'(x_0)d \ge 0$. □

The principal result in this section is the following:

Theorem 2.2.5 (Karush–Kuhn–Tucker)

Suppose that x_0 is a local minimizer of (P), and the constraint qualification[1] $\mathcal{C}_\ell(x_0)^* = \mathcal{C}_t(x_0)^*$ is fulfilled. Then there exist vectors $\lambda \in \mathbb{R}^m_+$ and $\mu \in \mathbb{R}^p$ such that

$$\nabla f(x_0) + \sum_{i=1}^{m} \lambda_i \nabla g_i(x_0) + \sum_{j=1}^{p} \mu_j \nabla h_j(x_0) = 0 \quad \textit{and}$$
$$\lambda_i g_i(x_0) = 0 \ \textit{for}\ i = 1, \dots, m.$$

Proof: If x_0 is a local minimizer of (P), it follows from lemma 2.2.4 with the help of the presupposed constraint qualification that

$$\nabla f(x_0) \in \mathcal{C}_t(x_0)^* = \mathcal{C}_\ell(x_0)^*;$$

lemma 2.2.2 yields $\mathcal{C}_\ell(x_0) \cap \mathcal{C}_{dd}(x_0) = \emptyset$ and the latter together with proposition 2.2.1 gives the result. □

In the presence of the presupposed constraint qualification $\mathcal{C}_t(x_0)^* = \mathcal{C}_\ell(x_0)^*$ the condition $\nabla f(x_0) \in \mathcal{C}_t(x_0)^*$ of lemma 2.2.4 transforms to $\nabla f(x_0) \in \mathcal{C}_\ell(x_0)^*$. This claim can be confirmed with the aid of a simple linear optimization problem:

Example 5 (Kleinmichel (1975))

For $x = (x_1, x_2)^T \in \mathbb{R}^2$ we consider the problem

$f(x) := x_1 + x_2 \longrightarrow \min$
$-x_1^3 + x_2 \le 1$
$x_1 \le 1, \ -x_2 \le 0$

[1] Guignard (1969)

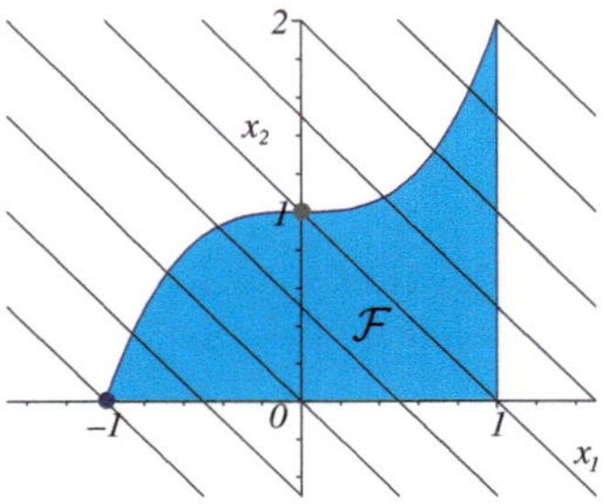

and ask whether the feasible points $x_0 := (-1,0)^T$ and $\widetilde{x_0} := (0,1)^T$ are local minimizers. (The examination of the picture shows immediately that this is not the case for $\widetilde{x_0}$, and that the objective function f attains a (strict, global) minimum at x_0. But we try to forget this for a while.) We have $\mathcal{A}(x_0) = \{1,3\}$. In order to show that $\nabla f(x_0) \in \mathcal{C}_\ell(x_0)^*$, hence, $f'(x_0)d \geq 0$ for all $d \in \mathcal{C}_\ell(x_0)$, we compute $\min\limits_{d \in \mathcal{C}_\ell(x_0)} f'(x_0)d$. So we have the following linear problem:

$$\begin{aligned} d_1 + d_2 &\longrightarrow \min \\ -3d_1 + d_2 &\leq 0 \\ -d_2 &\leq 0 \end{aligned}$$

Evidently it has the minimal value 0; lemma 2.2.2 gives that $\mathcal{C}_\ell(x_0) \cap \mathcal{C}_{dd}(x_0)$ is empty. Following proposition 2.2.1 there exist $\lambda_1, \lambda_3 \geq 0$ for x_0 satisfying

$$\begin{pmatrix}1\\1\end{pmatrix} + \lambda_1 \begin{pmatrix}-3\\1\end{pmatrix} + \lambda_3 \begin{pmatrix}0\\-1\end{pmatrix} = \begin{pmatrix}0\\0\end{pmatrix}.$$

The above yields $\lambda_1 = \frac{1}{3}$, $\lambda_3 = \frac{4}{3}$.

For $\widetilde{x_0}$ we have $\mathcal{A}(\widetilde{x_0}) = \{1\}$. In the same way as the above this leads to the subproblem

$$\begin{aligned} d_1 + d_2 &\longrightarrow \min \\ d_2 &\leq 0 \end{aligned}$$

whose objective function is unbounded; therefore $\mathcal{C}_\ell(\widetilde{x_0}) \cap \mathcal{C}_{dd}(\widetilde{x_0}) \neq \emptyset$.

So $\widetilde{x_0}$ is not a local minimizer, but the point x_0 remains as a candidate. $\lhd$

Convex Functions

Convexity plays a central role in optimization. We already had some simple results from Convex Analysis in section 2.1. Convex optimization problems — the functions f and g_i are supposed to be convex and the funcions h_j affinely linear — are by far easier to solve than general nonlinear problems. These assumptions ensure that the problems are well-behaved. They have two significant properties: *A local minimizer is always a global one. The KKT conditions are sufficient for optimality.* A special feature of *strictly* convex functions is that they have at most one minimal point. But convex functions also play an important role in problems that are not convex. Therefore a simple and short treatment of convex functions is given here:

Definition

Let $D \subset \mathbb{R}^n$ be nonempty and convex. A real-valued function f defined on at least D is called *convex* on D if and only if

$$f\big((1-\tau)x+\tau y\big) \leq (1-\tau)f(x)+\tau f(y)$$

holds for all $x, y \in D$ and $\tau \in (0,1)$. f is called *strictly convex* on D if and only if

$$f\big((1-\tau)x+\tau y\big) < (1-\tau)f(x)+\tau f(y)$$

for all $x, y \in D$ with $x \neq y$ and $\tau \in (0,1)$. The addition "on D" will be omitted, if D is the domain of definition. We say f is *concave* (on D) iff $-f$ is convex, and *strictly concave* (on D) iff $-f$ is strictly convex.

For a concave function the line segment joining two points on the graph is never above the graph.

Let $D \subset \mathbb{R}^n$ be nonempty and convex and $f\colon D \longrightarrow \mathbb{R}$ a *convex* function.

Properties

1) If f attains a local minimum at a point $x^* \in D$, then $f(x^*)$ is the global minimum.

2) f is continuous in $\overset{\circ}{D}$.

3) The function φ defined by $\varphi(\tau) := \frac{f(x+\tau h)-f(x)}{\tau}$ for $x \in \overset{\circ}{D}$, $h \in \mathbb{R}^n$ and sufficiently small, positive τ is isotone, that is, order-preserving.

4) For D open and a differentiable f it holds that $f(y)-f(x) \geq f'(x)(y-x)$ for all $x, y \in D$.

With the function f defined by $f(x) := 0$ for $x \in [0,1)$ and $f(1) := 1$ we can see that assertion *2)* cannot be extended to the whole of D.

Proof:

1) If there existed an $\overline{x} \in D$ such that $f(\overline{x}) < f(x^*)$, then we would have

$$f((1-\tau)x^* + \tau\overline{x}) \leq (1-\tau)f(x^*) + \tau f(\overline{x}) < f(x^*)$$

for $0 < \tau \leq 1$ and consequently a contradiction to the fact that f attains a local minimum at x^*.

2) For $x_0 \in \overset{\circ}{D}$ consider the function ψ defined by $\psi(h) := f(x_0+h)-f(x_0)$ for $h \in \mathbb{R}^n$ with a sufficiently small norm $\|h\|_\infty$: It is clear that the function ψ is convex. Let $\varrho > 0$ such that for

$$K := \{h \in \mathbb{R}^n \mid \|h\|_\infty \leq \varrho\}$$

it holds that $x_0+K \subset \overset{\circ}{D}$. Evidently, there exist $m \in \mathbb{N}$ and $a_1, \ldots, a_m \in \mathbb{R}^n$ with $K = \operatorname{conv}(a_1, \ldots, a_m)$ (convex hull). Every $h \in K$ may be represented as $h = \sum_{\mu=1}^{m} \gamma_\mu a_\mu$ with $\gamma_\mu \geq 0$ satisfying $\sum_{\mu=1}^{m} \gamma_\mu = 1$. With

$$\alpha := \begin{cases} \max\{|\psi(a_\mu)| \mid \mu = 1, \ldots, m\}, & \text{if positive} \\ 1 & , \text{ otherwise} \end{cases}$$

we have $\psi(h) \leq \sum_{\mu=1}^{m} \gamma_\mu \psi(a_\mu) \leq \alpha$. Now let $\varepsilon \in (0, \alpha]$. Then firstly for all $h \in \mathbb{R}^n$ with $\|h\|_\infty \leq \varepsilon \varrho / \alpha$ we have

$$\psi(h) = \psi\Big(\big(1 - \tfrac{\varepsilon}{\alpha}\big) 0 + \tfrac{\varepsilon}{\alpha}\big(\tfrac{\alpha}{\varepsilon} h\big)\Big) \leq \tfrac{\varepsilon}{\alpha} \psi\big(\tfrac{\alpha}{\varepsilon} h\big) \leq \varepsilon$$

and therefore with

$$0 = \psi(0) = \psi\Big(\frac{1}{2} h - \frac{1}{2} h\Big) \leq \frac{1}{2}\psi(h) + \frac{1}{2}\psi(-h)$$

$\psi(h) \geq -\psi(-h) \geq -\varepsilon$, hence, all together $|\psi(h)| \leq \varepsilon$.

3) Since f is convex, we have

$$\begin{aligned} f(x + \tau_0 h) &= f\big(\big(1 - \tfrac{\tau_0}{\tau_1}\big) x + \tfrac{\tau_0}{\tau_1}(x + \tau_1 h)\big) \\ &\leq \big(1 - \tfrac{\tau_0}{\tau_1}\big) f(x) + \tfrac{\tau_0}{\tau_1} f(x + \tau_1 h) \end{aligned}$$

for $0 < \tau_0 < \tau_1$. Transformation leads to

$$\frac{f(x + \tau_0 h) - f(x)}{\tau_0} \leq \frac{f(x + \tau_1 h) - f(x)}{\tau_1}.$$

4) This follows directly from *3)* (with $h = y - x$):

$$f'(x) h = \lim_{\tau \to 0+} \frac{f(x + \tau h) - f(x)}{\tau} \leq \frac{f(x + h) - f(x)}{1} \qquad \square$$

Constraint Qualifications

The condition $\mathcal{C}_\ell(x_0)^* = \mathcal{C}_t(x_0)^*$ is very abstract, extremely general, but not easily verifiable. Therefore, for practical problems, we will try to find regularity assumptions called *constraint qualifications* (CQ) which are more specific, easily verifiable, but also somewhat restrictive.

For the moment we will consider the case that we *only* have *inequality constraints*. Hence, $\boxed{\mathcal{E} = \emptyset}$ and $\mathcal{I} = \{1, \ldots, m\}$ with an $m \in \mathbb{N}_0$. Linear constraints pose fewer problems than nonlinear constraints. Therefore, we will assume the partition

$$\mathcal{I} = \mathcal{I}_1 \uplus \mathcal{I}_2.$$

If and only if $i \in \mathcal{I}_2$ let $g_i(x) = a_i^T x - b_i$ with suitable vectors a_i and b_i, that is, g_i is 'linear', more precisely affinely linear. Corresponding to this partition, we will also split up the set of active constraints $\mathcal{A}(x_0)$ for $x_0 \in \mathcal{F}$ into

$$\mathcal{A}_j(x_0) := \mathcal{I}_j \cap \mathcal{A}(x_0) \text{ for } j = 1, 2\,.$$

We will now focus on the following *Constraint Qualifications*:

(GCQ) GUIGNARD *Constraint Qualification*: $\mathcal{C}_\ell(x_0)^* = \mathcal{C}_t(x_0)^*$

(ACQ) ABADIE *Constraint Qualification*: $\mathcal{C}_\ell(x_0) = \mathcal{C}_t(x_0)$

(MFCQ) MANGASARIAN–FROMOVITZ *Constraint Qualification*:

$$\exists\, d \in \mathbb{R}^n \begin{cases} g_i'(x_0)d < 0 & \text{for } i \in \mathcal{A}_1(x_0) \\ g_i'(x_0)d \le 0 & \text{for } i \in \mathcal{A}_2(x_0) \end{cases}$$

(SCQ) SLATER *Constraint Qualification*:
The functions g_i are *convex* for all $i \in \mathcal{I}$ and
$\exists\, \tilde{x} \in \mathcal{F}\;\; g_i(\tilde{x}) < 0$ for $i \in \mathcal{I}_1$.

The conditions $g_i'(x_0)d < 0$ and $g_i'(x_0)d \le 0$ each define half spaces. (MFCQ) means nothing else but that the intersection of all of these half spaces is nonempty.

We will prove (SCQ) $\Longrightarrow$ (MFCQ) $\Longrightarrow$ (ACQ) .

The constraint qualification (GCQ) introduced in theorem 2.2.5 is a trivial consequence of (ACQ).

Proof: (SCQ) $\Longrightarrow$ (MFCQ): From the properties of convex and affinely linear functions and the definition of $\mathcal{A}(x_0)$ we get:

$$\begin{aligned} g_i'(x_0)(\tilde{x} - x_0) &\le g_i(\tilde{x}) - g_i(x_0) = g_i(\tilde{x}) < 0 \text{ for } i \in \mathcal{A}_1(x_0) \\ g_i'(x_0)(\tilde{x} - x_0) &= g_i(\tilde{x}) - g_i(x_0) = g_i(\tilde{x}) \le 0 \text{ for } i \in \mathcal{A}_2(x_0)\,. \end{aligned}$$

(MFCQ) $\Longrightarrow$ (ACQ): Lemma 2.2.3 gives that $\mathcal{C}_t(x_0) \subset \mathcal{C}_\ell(x_0)$ and $0 \in \mathcal{C}_t(x_0)$ always hold. Therefore it remains to prove that $\mathcal{C}_\ell(x_0) \setminus \{0\} \subset \mathcal{C}_t(x_0)$. So let $d_0 \in \mathcal{C}_\ell(x_0) \setminus \{0\}$. Take d as stated in (MFCQ). Then for a sufficiently small $\lambda > 0$ we have $d_0 + \lambda d \ne 0$. Since d_0 is in $\mathcal{C}_\ell(x_0)$, it follows that

$$\begin{aligned} g_i'(x_0)(d_0 + \lambda d) &< 0 \text{ for } i \in \mathcal{A}_1(x_0) \text{ and} \\ g_i'(x_0)(d_0 + \lambda d) &\le 0 \text{ for } i \in \mathcal{A}_2(x_0)\,. \end{aligned}$$

For the moment take a fixed λ. Setting $u := \frac{d_0 + \lambda d}{\|d_0 + \lambda d\|_2}$ produces

Chapter 2

$$\begin{aligned} g_i(x_0+tu) &= \underbrace{g_i(x_0)}_{=0} + t\,\underbrace{g_i'(x_0)u}_{<0} + o(t) \quad \text{for } i \in \mathcal{A}_1(x_0) \text{ and} \\ g_i(x_0+tu) &= \underbrace{g_i(x_0)}_{=0} + t\,\underbrace{g_i'(x_0)u}_{\le 0} \qquad\qquad \text{for } i \in \mathcal{A}_2(x_0). \end{aligned}$$

Thus, we have $g_i(x_0 + tu) \le 0$ for $i \in \mathcal{A}(x_0)$ and $t > 0$ sufficiently small. For the indices $i \in \mathcal{I} \setminus \mathcal{A}(x_0)$ this is obviously true. Hence, there exists a $t_0 > 0$ such that $x_0 + tu \in \mathcal{F}$ for $0 \le t \le t_0$. For the sequence (x_k) defined by $x_k := x_0 + \frac{t_0}{k} u$ it holds that $x_k \xrightarrow{u} x_0$. Therefore, $u \in \mathcal{C}_t(x_0)$ and consequently $d_0 + \lambda d \in \mathcal{C}_t(x_0)$. Passing to the limit as $\lambda \longrightarrow 0$ yields $d_0 \in \overline{\mathcal{C}_t(x_0)}$. Lemma 2.2.3 or respectively exercise 9 gives that $\mathcal{C}_t(x_0)$ is closed. Hence, $d_0 \in \mathcal{C}_t(x_0)$. □

Now we will consider the *general case*, where there may also occur *equality* constraints. In this context one often finds the following *linear independence constraint qualification* in the literature:

(LICQ) The vectors $\big(\nabla g_i(x_0) \mid i \in \mathcal{A}(x_0)\big)$ and $\big(\nabla h_j(x_0) \mid j \in \mathcal{E}\big)$ are linearly independent.

(LICQ) greatly reduces the number of active inequality constraints. Instead of (LICQ) we will now consider the following weaker constraint qualification which is a variant of (MFCQ), and is often cited as the ARROW–HURWITZ–UZAWA constraint qualification:

(AHUCQ) There exists a $d \in \mathbb{R}^n$ such that $\begin{cases} g_i'(x_0)d < 0 & \text{for } i \in \mathcal{A}(x_0), \\ h_j'(x_0)d = 0 & \text{for } j \in \mathcal{E}, \end{cases}$

and the vectors $\big(\nabla h_j(x_0) \mid j \in \mathcal{E}\big)$ are linearly independent.

We will show: (LICQ) $\Longrightarrow$ (AHUCQ) $\Longrightarrow$ (ACQ)

Proof: (LICQ) $\Longrightarrow$ (AHUCQ): (AHUCQ) follows, for example, directly from the solvability of the system of linear equations

$$\begin{aligned} g_i'(x_0)d &= -1 \quad \text{for } i \in \mathcal{A}(x_0), \\ h_j'(x_0)d &= 0 \quad\;\; \text{for } j \in \mathcal{E}. \end{aligned}$$

(AHUCQ) $\Longrightarrow$ (ACQ): Lemma 2.2.3 gives that again we only have to show $d_0 \in \mathcal{C}_t(x_0)$ for all $d_0 \in \mathcal{C}_\ell(x_0) \setminus \{0\}$. Take d as stated in (AHUCQ). Then we have $d_0 + \lambda d =: w \ne 0$ for a sufficiently small $\lambda > 0$ and thus

$$\begin{aligned} g_i'(x_0)w &< 0 \quad \text{for } i \in \mathcal{A}(x_0) \text{ and} \\ h_j'(x_0)w &= 0 \quad \text{for } j \in \mathcal{E}. \end{aligned}$$

Denote

$$A := \big(\nabla h_1(x_0), \ldots, \nabla h_p(x_0)\big) \in \mathbb{R}^{n \times p} .$$

For that $A^T A$ is regular because $\operatorname{rank}(A) = p$. Now consider the following system of linear equations dependent on $u \in \mathbb{R}^p$ and $t \in \mathbb{R}$:

$$\varphi_j(u,t) := h_j(x_0 + Au + tw) = 0 \quad (j = 1, \ldots, p)$$

For the corresponding vector-valued function φ we have $\varphi(0,0) = 0$, and because of

$$\tfrac{\partial \varphi_j}{\partial u_i}(u,t) = h_j'(x_0 + Au + tw)\nabla h_i(x_0) ,$$

we are able to solve $\varphi(u,t) = 0$ locally for u, that is, there exist a nullneighborhood $U_0 \subset \mathbb{R}$ and a continuously differentiable function $u \colon U_0 \longrightarrow \mathbb{R}^p$ satisfying

$$u(0) = 0 ,$$
$$h_j(\underbrace{x_0 + Au(t) + tw}_{=:\, x(t)}) = 0 \;\text{ for }\; t \in U_0 \quad (j = 1, \ldots, p).$$

Differentiation with respect to t at $t = 0$ leads to

$$h_j'(x_0)\big(Au'(0) + w\big) = 0 \quad (j = 1, \ldots, p)$$

and consequently — considering that $h_j'(x_0)w = 0$ and $A^T A$ is regular — to $u'(0) = 0$. Then for $i \in \mathcal{A}(x_0)$ it holds that

$$g_i(x(t)) = g_i(x_0) + t\, g_i'(x_0)x'(0) + \mathrm{o}(t) = t\, g_i'(x_0)\big(Au'(0) + w\big) + \mathrm{o}(t) .$$

With $u'(0) = 0$ we obtain

$$g_i(x(t)) = t\Big(g_i'(x_0)w + \frac{\mathrm{o}(t)}{t}\Big)$$

and the latter is negative for $t > 0$ sufficiently small.

Hence, there exists a $t_1 > 0$ with $x(t) \in \mathcal{F}$ for $0 \le t \le t_1$. From

$$x\left(\frac{t_1}{k}\right) = x_0 + \frac{t_1}{k}\Big(w + \underbrace{A\,\frac{u(t_1/k)}{t_1/k}}_{\longrightarrow\, 0 \ (k \to \infty)}\Big)$$

for $k \in \mathbb{N}$ we get $x\left(\frac{t_1}{k}\right) \xrightarrow{\;w\;} x_0$; this yields $w = d_0 + \lambda d \in \mathcal{C}_t(x_0)$ and also by passing to the limit as $\lambda \to 0$

$$d_0 \in \overline{\mathcal{C}_t(x_0)} = \mathcal{C}_t(x_0) . \qquad \Box$$

Convex Optimization Problems

Firstly suppose that $C \subset \mathbb{R}^n$ is nonempty and the functions $f, g_i \colon C \longrightarrow \mathbb{R}$ are *arbitrary* for $i \in \mathcal{I}$. We consider the *general* optimization problem

$$(P) \quad \begin{cases} f(x) \longrightarrow \min \\ g_i(x) \leq 0 \ \text{ for } \ i \in \mathcal{I} := \{1, \ldots, m\} \end{cases} .$$

In the following section the Lagrangian L to (P) defined by

$$L(x, \lambda) := f(x) + \sum_{i=1}^{m} \lambda_i g_i(x) = f(x) + \langle \lambda, g(x) \rangle \quad \text{for} \quad x \in C \ \text{ and } \ \lambda \in \mathbb{R}_+^m$$

will play an important role. As usual we have combined the m functions g_i to a vector-valued function g.

Definition

A pair $(x^*, \lambda^*) \in C \times \mathbb{R}_+^m$ is called a *saddlepoint* of L if and only if

$$L(x^*, \lambda) \leq L(x^*, \lambda^*) \leq L(x, \lambda^*)$$

holds for all $x \in C$ and $\lambda \in \mathbb{R}_+^m$, that is, x^* minimizes $L(\,\cdot\,, \lambda^*)$ and λ^* maximizes $L(x^*, \,\cdot\,)$.

Lemma 2.2.6

If (x^*, λ^*) is a saddlepoint of L, then it holds that:

- *x^* is a global minimizer of (P).*
- *$L(x^*, \lambda^*) = f(x^*)$*
- *$\lambda_i^* g_i(x^*) = 0$ for all $i \in \mathcal{I}$.*

Proof: Let $x \in C$ and $\lambda \in \mathbb{R}_+^m$. From

$$0 \geq L(x^*, \lambda) - L(x^*, \lambda^*) = \langle \lambda - \lambda^*, g(x^*) \rangle \tag{4}$$

we obtain for $\lambda := 0$

$$\langle \lambda^*, g(x^*) \rangle \geq 0 . \tag{5}$$

With $\lambda := \lambda^* + e_i$ we get — also from (4) —

$$g_i(x^*) \leq 0 \ \text{ for all } \ i \in \mathcal{I}, \ \text{that is,} \ g(x^*) \leq 0 . \tag{6}$$

Because of (6), it holds that $\langle \lambda^*, g(x^*) \rangle \leq 0$. Together with (5) this produces

$$\langle \lambda^*, g(x^*) \rangle = 0 \ \text{ and hence, } \ \lambda_i^* g_i(x^*) = 0 \ \text{ for all } \ i \in \mathcal{I} .$$

For $x \in \mathcal{F}$ it follows that

$$f(x^*) = L(x^*, \lambda^*) \le L(x, \lambda^*) = f(x) + \langle \lambda^*, \underbrace{g(x)}_{\le 0} \rangle \le f(x).$$

Therefore x^* is a global minimizer of (P). □

We assume now that C is open and convex and the functions $f, g_i \colon C \longrightarrow \mathbb{R}$ are continuously differentiable and *convex* for $i \in \mathcal{I}$. In this case we write more precisely (CP) instead of (P).

Theorem 2.2.7

If the SLATER *constraint qualification holds and x^* is a minimizer of (CP), then there exists a vector $\lambda^* \in \mathbb{R}^m_+$ such that (x^*, λ^*) is a saddlepoint of L.*

Proof: Taking into account our observations from page 55, theorem 2.2.5 gives that there exists a $\lambda^* \in \mathbb{R}^m_+$ such that

$$0 = L_x(x^*, \lambda^*) \quad \text{and} \quad \langle \lambda^*, g(x^*) \rangle = 0.$$

With that we get for $x \in C$[1]

$$L(x, \lambda^*) - L(x^*, \lambda^*) \ge L_x(x^*, \lambda^*)(x - x^*) = 0$$

and

$$L(x^*, \lambda^*) - L(x^*, \lambda) = -\langle \underbrace{\lambda}_{\ge 0}, \underbrace{g(x^*)}_{\le 0} \rangle \ge 0.$$

Hence, (x^*, λ^*) is a saddlepoint of L. □

The following example shows that the SLATER constraint qualification is essential in this theorem:

Example 6

With $n = 1$ and $m = 1$ we regard the *convex problem*

$$(P) \quad \begin{cases} f(x) := -x \longrightarrow \min \\ g(x) := x^2 \le 0 \end{cases}.$$

The only feasible point is $x^* = 0$ with value $f(0) = 0$. So 0 minimizes $f(x)$ subject to $g(x) \le 0$.

$L(x, \lambda) := -x + \lambda x^2$ for $\lambda \ge 0, x \in \mathbb{R}$. There is no $\lambda^* \in [0, \infty)$ such that (x^*, λ^*) is a saddlepoint of L. ◁

The following important observation shows that neither constraint qualifications nor second-order optimality conditions, which we will deal with in the

[1] By the convexity of f and g_i the function $L(\,\cdot\,, \lambda^*)$ is convex.

next section, are needed for a *sufficient* condition for general *convex optimization problems*:

Suppose that $f,\, g_i,\, h_j \colon \mathbb{R}^n \longrightarrow \mathbb{R}$ are continuously differentiable functions with f and g_i *convex* and h_j (affinely) *linear* $(i \in \mathcal{I}, j \in \mathcal{E})$, and consider the following *convex optimization problem*[2]

$$(CP) \quad \begin{cases} f(x) \longrightarrow \min \\ g_i(x) \le 0 \;\text{ for }\; i \in \mathcal{I} \\ h_j(x) = 0 \;\text{ for }\; j \in \mathcal{E} \end{cases}.$$

We will show that *for this special kind of problem every KKT point already gives a (global) minimum*:

Theorem 2.2.8

Suppose $x_0 \in \mathcal{F}$ and there exist vectors $\lambda \in \mathbb{R}^m_+$ and $\mu \in \mathbb{R}^p$ such that

$$\nabla f(x_0) + \sum_{i=1}^{m} \lambda_i \nabla g_i(x_0) + \sum_{j=1}^{p} \mu_j \nabla h_j(x_0) = 0 \;\; \textit{and}$$
$$\lambda_i g_i(x_0) = 0 \;\textit{for}\; i = 1, \ldots, m,$$

then (CP) attains its global minimum at x_0.

The *Proof* of this theorem is surprisingly simple:

Taking into account *4)* on page 53, we get for $x \in \mathcal{F}$:

$$\begin{aligned} f(x) - f(x_0) &\underset{f \text{ convex}}{\ge} f'(x_0)(x - x_0) \\ &= -\sum_{i=1}^{m} \lambda_i g_i'(x_0)(x - x_0) - \sum_{j=1}^{p} \mu_j \underbrace{h_j'(x_0)(x - x_0)}_{= h_j(x) - h_j(x_0) = 0} \\ &\underset{g_i \text{ convex}}{\ge} -\sum_{i=1}^{m} \lambda_i (g_i(x) - g_i(x_0)) = -\sum_{i=1}^{m} \lambda_i g_i(x) \ge 0 \end{aligned}$$ □

The following example shows that *even if we have convex problems the KKT conditions are not necessary for minimal points:*

Example 7

With $n = 2$, $m = 2$ and $x = (x_1, x_2)^T \in D := \mathbb{R}^2$ we consider:

[2] Since the functions h_j are assumed to be (affinely) linear, exercise 6 gives that this problem can be written in the form from page 58 by substituting the two inequalities $h_j(x) \le 0$ and $-h_j(x) \le 0$ for every equation $h_j(x) = 0$.

$$(P) \quad \begin{cases} f(x) := x_1 \longrightarrow \min \\ g_1(x) := x_1^2 + (x_2-1)^2 - 1 \le 0 \\ g_2(x) := x_1^2 + (x_2+1)^2 - 1 \le 0 \end{cases}$$

Obviously, only the point $x_0 := (0,0)^T$ is feasible. Hence, x_0 is the (global) minimal point. Since $\nabla f(x_0) = (1,0)^T$, $\nabla g_1(x_0) = (0,-2)^T$ and $\nabla g_2(x_0) = (0,2)^T$, the gradient condition of the KKT conditions is *not* met. f is *linear,* the functions g_ν are *convex*. Evidently, however, the SLATER condition is not fulfilled.

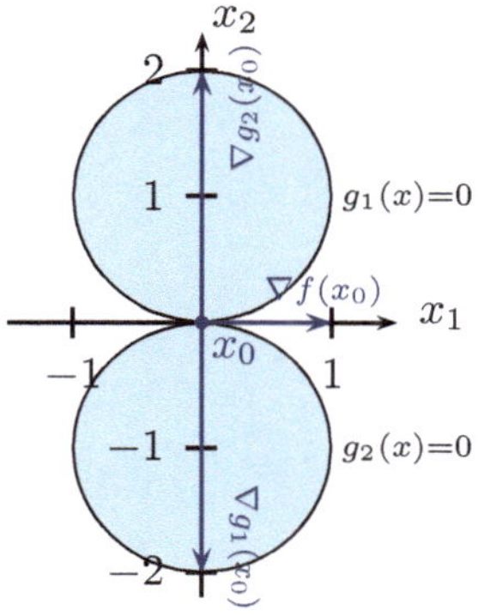

Of course, one could also argue from proposition 2.2.1: The cones

$$\mathcal{C}_{dd}(x_0) = \{d \in \mathbb{R}^2 \mid f'(x_0)d < 0\} = \{d \in \mathbb{R}^2 \mid d_1 < 0\}$$

and

$$\mathcal{C}_\ell(x_0) = \{d \in \mathbb{R}^2 \mid \forall\, i \in \mathcal{A}(x_0)\ g_i'(x_0)d \le 0\} = \{d \in \mathbb{R}^2 \mid d_2 = 0\}$$

are clearly not disjoint. ◁

2.3 Local Second-Order Optimality Conditions

To get a finer characterization, it is natural to examine the effects of second-order terms near a given point too. The following second-order results take the 'curvature' of the feasible region in a neighborhood of a 'candidate' for a minimizer into account. The necessary second-order condition $s^T H s \ge 0$ and the sufficient second-order condition $s^T H s > 0$ for the Hessian H of the Lagrangian with respect to x regard only certain subsets of vectors s.

Suppose that the functions f, g_i and h_j are twice continuously differentiable.

Chapter 2

Theorem 2.3.1 (Necessary second-order condition)

Suppose $x_0 \in \mathcal{F}$ and there exist $\lambda \in \mathbb{R}^m_+$ and $\mu \in \mathbb{R}^p$ such that

$$\nabla f(x_0) + \sum_{i=1}^{m} \lambda_i \nabla g_i(x_0) + \sum_{j=1}^{p} \mu_j \nabla h_j(x_0) = 0 \quad \textit{and}$$
$$\lambda_i \, g_i(x_0) = 0 \quad \textit{for all} \quad i \in \mathcal{I}.$$

If (P) has a local minimum at x_0, then

$$s^T \Big(\nabla^2 f(x_0) + \sum_{i=1}^{m} \lambda_i \nabla^2 g_i(x_0) + \sum_{j=1}^{p} \mu_j \nabla^2 h_j(x_0) \Big) s \geq 0$$

holds for all $s \in C_{t+}(x_0)$, where

$$\mathcal{F}_+ := \mathcal{F}_+(x_0) := \big\{ x \in \mathcal{F} \mid g_i(x) = 0 \;\; \textit{for all} \;\; i \in \mathcal{A}_+(x_0) \big\} \quad \textit{with}$$
$$\mathcal{A}_+(x_0) := \big\{ i \in \mathcal{A}(x_0) \mid \lambda_i > 0 \big\} \quad \textit{and}$$
$$C_{t+}(x_0) := C_t(\mathcal{F}_+, x_0) = \big\{ d \in \mathbb{R}^n \mid \exists \; (x_k) \in \mathcal{F}_+^{\mathbb{N}} \;\; x_k \xrightarrow{d} x_0 \big\}.$$

With the help of the Lagrangian L the second and fifth lines can be written more clearly

$$\nabla_x L(x_0, \lambda, \mu) = 0 \,,$$

respectively

$$s^T \nabla^2_{x\,x} L(x_0, \lambda, \mu) \, s \geq 0 \,.$$

Proof: It holds that

$$\lambda_i \, g_i(x) = 0 \;\; \text{for all} \;\; x \in \mathcal{F}_+$$

because we have $\lambda_i = 0$ for $i \in \mathcal{I} \setminus \mathcal{A}_+(x_0)$ and $g_i(x) = 0$ for $i \in \mathcal{A}_+(x_0)$, respectively.

With the function φ defined by

$$\varphi(x) := f(x) + \sum_{i=1}^{m} \lambda_i \, g_i(x) + \sum_{j=1}^{p} \mu_j \, h_j(x) = L(x, \lambda, \mu)$$

for $x \in D$ this leads to the following relation:

$$\varphi(x) = f(x) \;\; \text{for} \;\; x \in \mathcal{F}_+ \,.$$

x_0 gives a local minimum of f on $\mathcal{F}$, therefore one of φ on $\mathcal{F}_+$.

Now let $s \in C_{t+}(x_0)$. Then by definition of the tangent cone there exists a sequence $(x^{(k)})$ in $\mathcal{F}_+$, such that $x^{(k)} = x_0 + \alpha_k (s + r_k)$, $\alpha_k \downarrow 0$ and $r_k \to 0$.

By assumption $\nabla\varphi(x_0) = 0$. With the TAYLOR theorem we get

$$\varphi(x_0) \le \varphi(x^{(k)}) = \varphi(x_0) + \alpha_k \underbrace{\varphi'(x_0)}_{=0}(s + r_k) + \frac{1}{2}\alpha_k^2 (s + r_k)^T \nabla^2\varphi\big(x_0 + \tau_k(x^{(k)} - x_0)\big)(s + r_k)$$

for all sufficiently large k and a suitable $\tau_k \in (0, 1)$.

Dividing by $\alpha_k^2/2$ and passing to the limit as $k \to \infty$ gives the result

$$s^T \nabla^2\varphi(x_0) s \ge 0. \qquad \square$$

In the following example we will see that $x_0 := (0,0,0)^T$ is a stationary point. With the help of theorem 2.3.1 we want to show that the necessary condition for a minimum is *not* met.

Example 8

$$\begin{aligned} f(x) &:= x_3 - \frac{1}{2}x_1^2 \longrightarrow \min \\ g_1(x) &:= -x_1^2 - x_2 - x_3 \le 0 \\ g_2(x) &:= -x_1^2 + x_2 - x_3 \le 0 \\ g_3(x) &:= -x_3 \le 0 \end{aligned}$$

For the point $x_0 := (0,0,0)^T$ we have $f'(x_0) = (0,0,1)$, $\mathcal{A}(x_0) = \{1,2,3\}$ and $g_1'(x_0) = (0,-1,-1)$, $g_2'(x_0) = (0,1,-1)$, $g_3'(x_0) = (0,0,-1)$.
We start with the gradient condition:

$$\nabla_x L(x_0,\lambda) = \begin{pmatrix} 0 \\ 0 \\ 1 \end{pmatrix} + \lambda_1 \begin{pmatrix} 0 \\ -1 \\ -1 \end{pmatrix} + \lambda_2 \begin{pmatrix} 0 \\ 1 \\ -1 \end{pmatrix} + \lambda_3 \begin{pmatrix} 0 \\ 0 \\ -1 \end{pmatrix} = \begin{pmatrix} 0 \\ 0 \\ 0 \end{pmatrix}$$

$$\iff \begin{cases} -\lambda_1 + \lambda_2 = 0 \\ -\lambda_1 - \lambda_2 - \lambda_3 = -1 \end{cases}$$

$$\iff \lambda_2 = \lambda_1, \ \lambda_3 = 1 - 2\lambda_1$$

For $\lambda_1 := 1/2$ we obtain $\lambda = (1/2, 1/2, 0)^T \in \mathbb{R}_+^3$ and $\lambda_i g_i(x_0) = 0$ for $i \in \mathcal{I}$. Hence, we get $\mathcal{A}_+(x_0) = \{1,2\}$,

$$\mathcal{F}_+ = \{x \in \mathbb{R}^3 \mid g_1(x) = g_2(x) = 0, \ g_3(x) \le 0\} = \{(0,0,0)^T\}$$

and therefore $\mathcal{C}_{t+}(x_0) = \{(0,0,0)^T\}$. *In this way* no decision can be made!

Setting $\lambda_1 := 0$ we obtain respectively $\lambda = e_3$, $\mathcal{A}_+(x_0) = \{3\}$, $\mathcal{F}_+ = \{x \in \mathcal{F} \mid x_3 = 0\}$, $\mathcal{C}_{t+}(x_0) = \{\alpha e_1 \mid \alpha \in \mathbb{R}\}$ and

$$H := \nabla^2 f(x_0) + \nabla^2 g_3(x_0) = \begin{pmatrix} -1 & 0 & 0 \\ 0 & 0 & 0 \\ 0 & 0 & 0 \end{pmatrix}.$$

H is negative definite on $\mathcal{C}_{t+}(x_0)$. Consequently there is *no* local minimum of (P) at $x_0 = 0$. ◁

In order to expand the second-order necessary condition to a sufficient condition, we will now have to make stronger assumptions.

Before we do that, let us recall that there will remain a 'gap' between these two conditions. This fact is well-known (even for real-valued functions of *one* variable) and is usually demonstrated by the functions f_2, f_3 and f_4 defined by

$$f_k(x) := x^k \text{ for } x \in \mathbb{R}, k = 2, 3, 4,$$

at the point $x_0 = 0$.

The following **Remark** can be proven in the same way as *2)* in lemma 2.2.3:

$$\mathcal{C}_{t+}(x_0) \subset \mathcal{C}_{\ell+}(x_0) := \left\{ s \in \mathbb{R}^n \;\middle|\; \begin{array}{l} g_i'(x_0)s = 0 \text{ for } i \in \mathcal{A}_+(x_0) \\ g_i'(x_0)s \leq 0 \text{ for } i \in \mathcal{A}(x_0) \setminus \mathcal{A}_+(x_0) \\ h_j'(x_0)s = 0 \text{ for } j \in \mathcal{E} \end{array} \right\}$$

Theorem 2.3.2 (Sufficient second-order condition)

Suppose $x_0 \in \mathcal{F}$ and there exist vectors $\lambda \in \mathbb{R}^m_+$ and $\mu \in \mathbb{R}^p$ such that

$$\nabla_x L(x_0, \lambda, \mu) = 0 \text{ and } \lambda^T g(x_0) = 0.$$

Furthermore, suppose that

$$s^T \nabla^2_{xx} L(x_0, \lambda, \mu)\, s > 0$$

for all $s \in \mathcal{C}_{\ell+}(x_0) \setminus \{0\}$. Then (P) attains a strict local minimum at x_0.

Proof (indirect): If f does *not* have a strict local minimum at x_0, then there exists a sequence $(x^{(k)})$ in $\mathcal{F} \setminus \{x_0\}$ with $x^{(k)} \longrightarrow x_0$ and $f(x^{(k)}) \leq f(x_0)$. For $s_k := \frac{x^{(k)} - x_0}{\|x^{(k)} - x_0\|_2}$ it holds that $\|s_k\|_2 = 1$. Hence, there exists a convergent subsequence. WLOG suppose $s_k \longrightarrow s$ for an $s \in \mathbb{R}^n$. With $\alpha_k := \|x^{(k)} - x_0\|_2$ we have $x^{(k)} = x_0 + \alpha_k s_k$ and WLOG $\alpha_k \downarrow 0$. From

$$f(x_0) \geq f(x^{(k)}) = f(x_0) + \alpha_k f'(x_0) s_k + o(\alpha_k)$$

it follows that

$$f'(x_0)s \leq 0.$$

For $i \in \mathcal{A}(x_0)$ and $j \in \mathcal{E}$ we get in the same way:

$$\underbrace{g_i(x^{(k)})}_{\leq 0} = \underbrace{g_i(x_0)}_{=0} + \alpha_k g_i'(x_0)s_k + o(\alpha_k) \implies g_i'(x_0)s \leq 0$$

$$\underbrace{h_j(x^{(k)})}_{=0} = \underbrace{h_j(x_0)}_{=0} + \alpha_k h_j'(x_0)s_k + o(\alpha_k) \implies h_j'(x_0)s = 0$$

With the assumption $\nabla_x L(x_0, \lambda, \mu) = 0$ it follows that

$$\underbrace{f'(x_0)s}_{\leq 0} + \underbrace{\sum_{i=1}^{m} \lambda_i g_i'(x_0)s} + \sum_{j=1}^{p} \mu_j \underbrace{h_j'(x_0)s}_{=0} = 0$$

$$= \sum_{i \in \mathcal{A}_+(x_0)} \lambda_i \underbrace{g_i'(x_0)s}_{\leq 0}$$

and from that $g_i'(x_0)s = 0$ for all $i \in \mathcal{A}_+(x_0)$.

Since $\|s\|_2 = 1$, we get $s \in \mathcal{C}_{\ell+}(x_0) \setminus \{0\}$. For the function φ defined by

$$\varphi(x) := f(x) + \sum_{i=1}^{m} \lambda_i g_i(x) + \sum_{j=1}^{p} \mu_j h_j(x) = L(x, \lambda, \mu)$$

it holds by assumption that $\nabla \varphi(x_0) = 0$.

$$\varphi(x^{(k)}) = \underbrace{f(x^{(k)})}_{\leq f(x_0)} + \sum_{i=1}^{m} \lambda_i \underbrace{g_i(x^{(k)})}_{\leq 0} + \sum_{j=1}^{p} \mu_j \underbrace{h_j(x^{(k)})}_{=0} \leq f(x_0) = \varphi(x_0)$$

The TAYLOR theorem yields

$$\varphi(x^{(k)}) = \varphi(x_0) + \alpha_k \underbrace{\varphi'(x_0)}_{=0} s_k + \frac{1}{2} \alpha_k^2 s_k^T \nabla^2 \varphi\big(x_0 + \tau_k(x^{(k)} - x_0)\big) s_k$$

with a suitable $\tau_k \in (0, 1)$. From this we deduce, as usual, $s^T \nabla^2 \varphi(x_0) s \leq 0$. With $s \in \mathcal{C}_{\ell+}(x_0) \setminus \{0\}$ we get a contradiction to our assumption. □

The following example gives a simple illustration of the necessary and sufficient second-order conditions of theorems 2.3.1 and 2.3.2:

Example 9 (FIACCO and MCCORMICK (1968))

$f(x) := (x_1 - 1)^2 + x_2^2 \longrightarrow \min$

$g_1(x) := x_1 - \varrho x_2^2 \leq 0$

We are looking for a $\varrho > 0$ such that $x_0 := (0,0)^T$ is a local minimizer of the problem: With $\nabla f(x_0) = (-2,0)^T, \nabla g_1(x_0) = (1,0)^T$ the condition $\nabla_x L(x_0, \lambda, \mu) = 0$ firstly yields $\lambda_1 = 2$.

In this case (MFCQ) is fulfilled with $\mathcal{A}_1(x_0) = \mathcal{A}(x_0) = \{1\} = \mathcal{A}_+(x_0)$. We have

$$\mathcal{C}_{\ell+}(x_0) = \{d \in \mathbb{R}^2 \mid d_1 = 0\} = \mathcal{C}_{t+}(x_0)\,.$$

The matrix

$$\nabla^2 f(x_0) + 2\nabla^2 g_1(x_0) = \begin{pmatrix} 2 & 0 \\ 0 & 2 \end{pmatrix} + 2\begin{pmatrix} 0 & 0 \\ 0 & -2\varrho \end{pmatrix} = 2\begin{pmatrix} 1 & 0 \\ 0 & 1-2\varrho \end{pmatrix}$$

is negative definite on $\mathcal{C}_{t+}(x_0)$ for $\varrho > 1/2$. Thus the second-order necessary condition of theorem 2.3.1 is violated and so there is no local minimum at x_0. For $\varrho < 1/2$ the Hessian is positive definite on $\mathcal{C}_{\ell+}(x_0)$. Hence, the sufficient conditions of theorem 2.3.2 are fulfilled and thus there is a strict local minimum at x_0. When $\varrho = 1/2$, this result is not determined by the second-order conditions; but we can confirm it in the following simple way: $f(x) = (x_1-1)^2 + x_2^2 = x_1^2 + 1 + (x_2^2 - 2x_1)$. Because of $x_2^2 - 2x_1 \geq 0$ this yields $f(x) \geq 1$ and $f(x) = 1$ only for $x_1 = 0$ and $x_2^2 - 2x_1 = 0$. Hence, there is a strict local minimum at x_0.

$\varrho = 1/4$

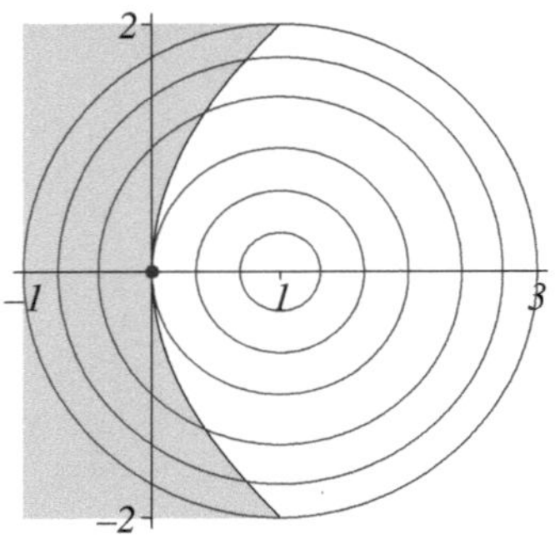

$\varrho = 1$

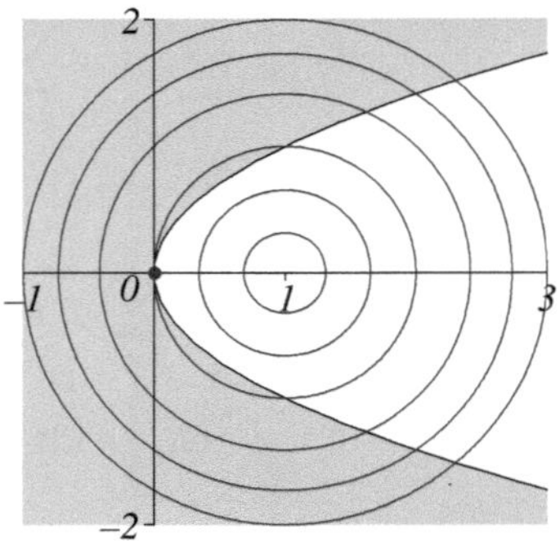

◁

2.4 Duality

Duality plays a crucial role in the theory of optimization and in the development of corresponding computational algorithms. It gives insight from a theoretical point of view but is also significant for computational purposes and economic interpretations, for example shadow prices. We shall concentrate on some of the more basic results and limit ourselves to a particular duality — Lagrange duality — which is the most popular and useful one for many purposes.

Given an arbitrary optimization problem, called *primal problem*, we consider a problem that is closely related to it, called the Lagrange *dual problem*. Several properties of this dual problem are demonstrated in this section. They help to provide strategies for solving the primal and the dual problem. The Lagrange dual problem of large classes of important nonconvex optimization problems can be formulated as an easier problem than the original one.

Lagrange Dual Problem

With $n \in \mathbb{N}$, $m, p \in \mathbb{N}_0$, $\emptyset \neq C \subset \mathbb{R}^n$, functions $f\colon C \longrightarrow \mathbb{R}$, $g = (g_1, \ldots, g_m)^T\colon C \longrightarrow \mathbb{R}^m$, $h = (h_1, \ldots, h_p)^T\colon C \longrightarrow \mathbb{R}^p$ and the feasible region

$$\mathcal{F} := \big\{x \in C \mid g(x) \leq 0,\ h(x) = 0\big\}$$

we regard the *primal problem* in standard form:

$$(P) \quad \begin{cases} f(x) \longrightarrow \min \\ x \in \mathcal{F} \end{cases}$$

There is a certain flexibility in defining a given problem: Some of the constraints $g_i(x) \leq 0$ or $h_j(x) = 0$ can be included in the definition of the set C.

Substituting the two inequalities $h_j(x) \leq 0$ and $-h_j(x) \leq 0$ for every equation $h_j(x) = 0$ we can assume WLOG $p = 0$. Then we have

$$\mathcal{F} = \big\{x \in C \mid g(x) \leq 0\big\}.$$

The *Lagrangian function* L is defined as a weighted sum of the objective function and the constraint functions, defined by

$$L(x, \lambda) := f(x) + \lambda^T g(x) = f(x) + \langle \lambda, g(x) \rangle = f(x) + \sum_{i=1}^{m} \lambda_i g_i(x)$$

for $x \in C$ and $\lambda = (\lambda_1, \ldots, \lambda_m)^T \in \mathbb{R}^m_+$.

The vector λ is called the *dual variable* or *multiplier* associated with the problem. For $i = 1, \ldots, m$ we refer to λ_i as the *dual variable* or *multiplier* associated with the inequality constraint $g_i(x) \leq 0$.

The LAGRANGE *dual function,* or *dual function,* φ is defined by

$$\varphi(\lambda) := \inf_{x \in C} L(x, \lambda)$$

on the *effective domain* of φ

$$\mathcal{F}_D := \Big\{\lambda \in \mathbb{R}^m_+ \mid \inf_{x \in C} L(x, \lambda) > -\infty\Big\}.$$

The LAGRANGE *dual problem,* or *dual problem,* then is defined by

$$(D) \quad \begin{cases} \varphi(\lambda) \longrightarrow \max \\ \lambda \in \mathcal{F}_D \end{cases}.$$

In the general case, the dual problem may not have a solution, even if the primal problem has one; conversely, the primal problem may not have a solution, even if the dual problem has one:

Example 10

For both examples let $C := \mathbb{R}, m := 1$ and $p := 0$:

a)
$$(P) \quad \begin{cases} f(x) := x + 2010 \longrightarrow \min \\ g(x) := \frac{1}{2}x^2 \leq 0 \end{cases}$$

1. $x^* := 0$ is the only feasible point. Thus $\inf\{f(x) \mid x \in \mathcal{F}\} = f(0) = 2010$.
2. $L(x, \lambda) := f(x) + \lambda g(x) = x + 2010 + \frac{\lambda}{2}x^2 \quad (\lambda \geq 0,\, x \in \mathbb{R})$
   $\mathcal{F}_D = \mathbb{R}_{++}$ (for $\lambda > 0$: parabola opening upwards; for $\lambda = 0$: unbounded from below): $\varphi(\lambda) = 2010 - \frac{1}{2\lambda}$

b)
$$(P) \quad \begin{cases} f(x) := \exp(-x) \longrightarrow \min \\ g(x) := -x \leq 0 \end{cases}$$

1. We have $\inf\{f(x) \mid x \in \mathcal{F}\} = \inf\{\exp(-x) \mid x \geq 0\} = 0$, but there exists no $x \in \mathcal{F} = \mathbb{R}_+$ with $f(x) = 0$.
2. $L(x, \lambda) := f(x) + \lambda g(x) = \exp(-x) - \lambda x \quad (\lambda \geq 0)$ shows $\mathcal{F}_D = \{0\}$ with $\varphi(0) = 0$. So we have $\sup\{\varphi(\lambda) \mid \lambda \in \mathcal{F}_D\} = 0 = \varphi(0)$. ◁

The dual objective function φ — as the pointwise infimum of a family of affinely linear functions — is always a *concave function*, even if the initial problem is not convex. Hence the dual problem can always be written ($\varphi \mapsto -\varphi$) as a *convex minimum problem:*

Remark *The set $\mathcal{F}_D$ is convex, and φ is a concave function on $\mathcal{F}_D$.*

Proof: Let $x \in C$, $\alpha \in [0, 1]$ and $\lambda, \mu \in \mathcal{F}_D$:

$$\begin{aligned} L(x, \alpha\lambda + (1-\alpha)\mu) &= f(x) + \langle \alpha\lambda + (1-\alpha)\mu, g(x)\rangle \\ &= \alpha\big(f(x) + \langle \lambda, g(x)\rangle\big) + (1-\alpha)\big(f(x) + \langle \mu, g(x)\rangle\big) \\ &= \alpha L(x, \lambda) + (1-\alpha) L(x, \mu) \\ &\geq \alpha\varphi(\lambda) + (1-\alpha)\varphi(\mu) \end{aligned}$$

This inequality has two implications: $\alpha\lambda + (1-\alpha)\mu \in \mathcal{F}_D$, and further, $\varphi(\alpha\lambda + (1-\alpha)\mu) \geq \alpha\varphi(\lambda) + (1-\alpha)\varphi(\mu)$. □

As we shall see below, the dual function yields lower bounds on the optimal value

$$p^* := v(P) := \inf(P) := \inf\{f(x) : x \in \mathcal{F}\}$$

of the primal problem (P). The optimal value of the dual problem (D) is defined by

$$d^* := v(D) := \sup(D) := \sup\{\varphi(\lambda) : \lambda \in \mathcal{F}_D\}.$$

We allow $v(P)$ and $v(D)$ to attain the extended values $+\infty$ and $-\infty$ and follow the standard convention that the infimum of the empty set is ∞ and the supremum of the empty set is $-\infty$. If there are feasible points x_k with $f(x_k) \to -\infty$ $(k \to \infty)$, then $v(P) = -\infty$ and we say problem (P) — or the function f on $\mathcal{F}$ — is unbounded from below. If there are feasible points λ_k with $\varphi(\lambda_k) \to \infty$ $(k \to \infty)$, then $v(D) = \infty$ and we say problem (D) — or the function φ on $\mathcal{F}_D$ — is unbounded from above. The problems (P) and (D) always have optimal values — possibly ∞ or $-\infty$. The question is whether or not they have *optimizers*, that is, there exist feasible points achieving these values. If there exists a feasible point achieving $\inf(P)$, we sometimes write $\min(P)$ instead of $\inf(P)$, accordingly $\max(D)$ instead of $\sup(D)$ if there is a feasible point achieving $\sup(D)$. In example 10, *a)* we had $\min(P) = \sup(D)$, in example 10, *b)* we got $\inf(P) = \max(D)$.

What is *the relationship between* d^* *and* p^*? The following theorem gives a first answer:

Weak Duality Theorem

If x is feasible to the primal problem (P) and λ is feasible to the dual problem (D), then we have $\varphi(\lambda) \le f(x)$. *In particular*

$$d^* \le p^* .$$

Proof: Let $x \in \mathcal{F}$ and $\lambda \in \mathcal{F}_D$:

$$\varphi(\lambda) \le L(x, \lambda) = f(x) + \underbrace{\lambda^T}_{\ge 0} \underbrace{g(x)}_{\le 0} \le f(x)$$

This implies immediately $d^* \le p^*$. □

Although very easy to show, the weak duality result has useful implications: For instance, it implies that the primal problem has no feasible points if the optimal value of (D) is ∞. Conversely, if the primal problem is unbounded from below, the dual problem has no feasible points. Any feasible point λ to the dual problem provides a lower bound $\varphi(\lambda)$ on the optimal value p^* of problem (P), and any feasible point x to the primal problem (P) provides an upper bound $f(x)$ on the optimal value d^* of problem (D). One aim is to generate *good* bounds. This can help to get termination criteria for algorithms: If one has a feasible point x to (P) and a feasible point λ to (D), whose values are close together, then these values must be close to the optima in both problems.

Corollary

If $f(x^*) = \varphi(\lambda^*)$ for some $x^* \in \mathcal{F}$ and $\lambda^* \in \mathcal{F}_D$, then x^* is a minimizer to the primal problem (P) and λ^* is a maximizer to the dual problem (D).

Proof:

$$\varphi(\lambda^*) \le \sup\{\varphi(\lambda) \mid \lambda \in \mathcal{F}_D\} \le \inf\{f(x) \mid x \in \mathcal{F}\} \le f(x^*) = \varphi(\lambda^*)$$

Hence, equality holds everywhere, in particular
$f(x^*) = \inf\{f(x) \mid x \in \mathcal{F}\}$ and $\varphi(\lambda^*) = \sup\{\varphi(\lambda) \mid \lambda \in \mathcal{F}_D\}$. □

The difference $p^* - d^*$ is called the *duality gap*. If this duality gap is zero, that is, $p^* = d^*$, then we say that *strong duality* holds. We will see later on: If the functions f and g are convex (on the convex set C) and a certain constraint qualification holds, then one has strong duality. In nonconvex cases, however, a duality gap

$$p^* - d^* > 0$$

has to be expected. The following examples illustrate the necessity of making more demands on f, g and C to get a close relation between the problems (P) and (D):

Example 11 With $n := 1$, $m := 1$:

a) $d^* = -\infty$, $p^* = \infty$ $\quad C := \mathbb{R}_+$, $f(x) := -x$, $g(x) := \pi$ $\quad (x \in C)$:

$L(x, \lambda) = -x + \lambda\pi \quad (x \in C, \lambda \in \mathbb{R}_+)$
$\mathcal{F} = \emptyset$, $p^* = \infty$; $\inf_{x \in C} L(x, \lambda) = -\infty$, $\mathcal{F}_D = \emptyset$, $d^* = -\infty$

b) $d^* = 0$, $p^* = \infty$ $\quad C := \mathbb{R}_{++}$, $f(x) := x$, $g(x) := x$ $\quad (x \in C)$:

$L(x, \lambda) = x + \lambda x = (1 + \lambda)x$
$\mathcal{F} = \emptyset$, $p^* = \infty$; $\mathcal{F}_D = \mathbb{R}_+$, $\varphi(\lambda) = 0$, $d^* = 0$

c) $-\infty = d^* < p^* = 0$

$C := \mathbb{R}$, $f(x) := x^3$, $g(x) := -x \quad (x \in \mathbb{R})$:
$\mathcal{F} = \mathbb{R}_+$, $p^* = \min(P) = 0$
$L(x, \lambda) = x^3 - \lambda x \quad \big(x \in \mathbb{R},\ \lambda \ge 0\big)$
$\mathcal{F}_D = \emptyset$, $d^* = -\infty$

d) $d^* = \max(D) < \min(P) = p^*$

$C := [0, 1]$, $f(x) := -x^2$, $g(x) := 2x - 1 \quad (x \in C)$:
$\mathcal{F} = [0, 1/2]$, $p^* = \min(P) = f(1/2) = -1/4$
$L(x, \lambda) = -x^2 + \lambda(2x - 1) \quad \big(x \in [0, 1],\ \lambda \ge 0\big)$
For $\lambda \in \mathcal{F}_D = \mathbb{R}_+$ we get

$$\varphi(\lambda) = \min\big(L(0, \lambda), L(1, \lambda)\big) = \min\big(-\lambda, \lambda - 1\big) = \begin{cases} -\lambda\ , & \lambda \ge 1/2 \\ \lambda - 1\ , & \lambda < 1/2 \end{cases}$$

and hence, $d^* = \max(D) = \varphi(1/2) = -1/2$. ◁

With $m, n \in \mathbb{N}$, a real (m,n)-matrix A, vectors $b \in \mathbb{R}^m$ and $c \in \mathbb{R}^n$ we consider a *linear problem* in standard form, that is,

$$(P) \quad \begin{cases} c^T x \longrightarrow \min \\ Ax = b, \ x \geq 0 \end{cases} .$$

The LAGRANGE dual problem of this linear problem is given by

$$(D) \quad \begin{cases} b^T \mu \longrightarrow \max \\ A^T \mu \leq c \end{cases} .$$

Proof: With $f(x) := c^T x$, $h(x) := b - Ax \quad (x \in \mathbb{R}^n_+ =: C)$ we have

$$L(x,\mu) = \langle c, x \rangle + \langle \mu, b - Ax \rangle = \langle \mu, b \rangle + \langle x, c - A^T \mu \rangle \quad (\mu \in \mathbb{R}^m).$$

$$\inf_{x \in C} \{\langle \mu, b \rangle + \langle x, c - A^T \mu \rangle\} \overset{\checkmark}{=} \begin{cases} b^T \mu \,, & \text{if } \ A^T \mu \leq c \\ -\infty \,, & \text{else} \end{cases}$$ □

It is easy to verify that the LAGRANGE dual problem of (D) — transformed into standard form — is again the primal problem (cf. exercise 18).

Geometric Interpretation

We give a geometric interpretation of the dual problem that helps to find and understand examples which illustrate the various possible relations that can occur between the primal and the dual problem. This visualization can give insight in theoretical results. For the sake of simplicity, we consider only the case $m = 1$, that is, only *one* inequality constraint:

We look at the image of C under the map (g, f), that is,

$$B := \{ (g(x), f(x)) \mid x \in C \}.$$

In the *primal problem* we have to find a pair $(v, w) \in B$ with minimal ordinate w in the (v, w)-plane, that is, the point (v, w) in B which minimizes w subject to $v \leq 0$. It is the point (v^*, w^*) — the image under (g, f) of the minimizer x^* to problem (P) — in the following figure, which illustrates a typical case for $n = 2$:

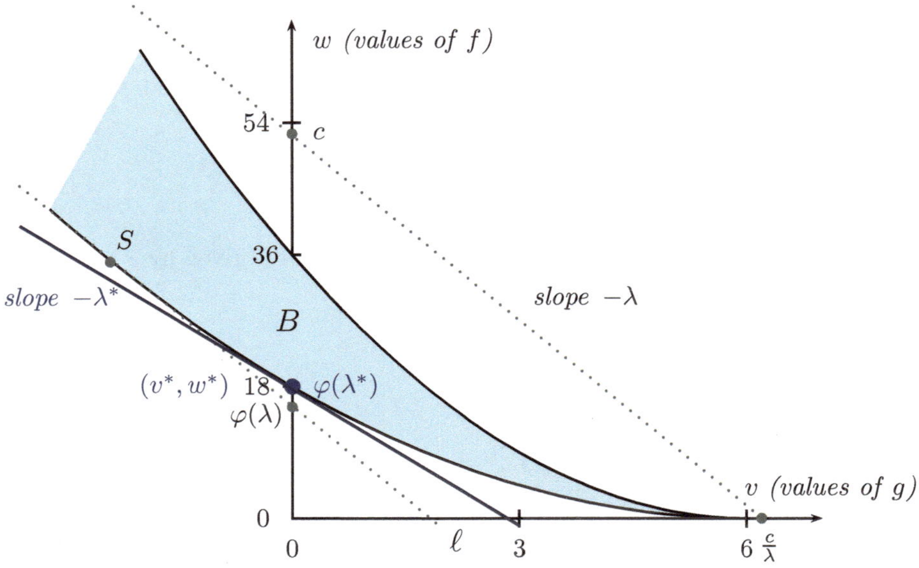

To get $\varphi(\lambda)$ for a fixed $\lambda \geq 0$, we have to minimize $L(x, \lambda) = f(x) + \lambda g(x)$ over $x \in C$, that is, $w + \lambda v$ over $(v, w) \in B$.

For any constant $c \in \mathbb{R}$, the equation $w + \lambda v = c$ describes a straight line with slope $-\lambda$ and intercept c on the w-axis. Hence we have to find the lowest line with slope $-\lambda$ which intersects the region B (move the line $w + \lambda v = c$ parallel to itself as far down as possible while it touches B). This leads to the line ℓ tangent to B at the point S in the figure. (The region B has to lie above the line and to touch it.) Then the intercept on the w-axis gives $\varphi(\lambda)$.

The geometric description of the *dual problem* (D) is now clear: Find the value λ^* which defines the slope of a tangent to B intersecting the ordinate at the highest possible point.

Example 12

Let $n := 2$, $m := 1$, $C := \mathbb{R}^2_+$ and $x = (x_1, x_2)^T \in C$:

$$(P) \quad \begin{cases} f(x) := x_1^2 + x_2^2 \longrightarrow \min \\ g(x) := 6 - x_1 - x_2 \leq 0 \end{cases}$$

$g(x) \leq 0$ implies $6 \leq x_1 + x_2$. The equality $6 = x_1 + x_2$ gives
$f(x) = x_1^2 + (6 - x_1)^2 = 2\left((x_1 - 3)^2 + 9\right)$.

The minimum is attained at $x^* = (3, 3)$ with $f(x^*) = 18$: $\min(P) = 18$

$$\begin{aligned} L(x, \lambda) &= x_1^2 + x_2^2 + \lambda(-x_1 - x_2 + 6) \quad (\lambda \geq 0,\, x \in C) \\ &= (x_1 - \lambda/2)^2 + (x_2 - \lambda/2)^2 + 6\lambda - \lambda^2/2 \end{aligned}$$

So we get the minimum for $x_1 = x_2 = \lambda/2$ with value $6\lambda - \lambda^2/2$.

$\varphi(\lambda) = 6\lambda - \lambda^2/2$ describes a parabola, therefore we get the maximum at $\lambda = 6$ with value $\varphi(\lambda) = 18$: $\max(D) = 18$

To get the region $B := \{(g(x), f(x)) : x \in C\}$, we proceed as follows:
For $x \in C$ we have $v := g(x) \leq 6$. The equation $-x_1 - x_2 + 6 = v$ gives $x_2 = -x_1 + 6 - v$ and further

$$\begin{aligned} f(x) &= x_1^2 + x_2^2 = x_1^2 + (x_1 + (v-6))^2 \\ &= 2x_1^2 + 2(v-6)x_1 + (v-6)^2 \\ &= 2\big(x_1 + (v-6)/2\big)^2 + (v-6)^2/2 \geq (v-6)^2/2 \end{aligned}$$

with equality for $x_1 = -(v-6)/2$.

$$f(x) = 2x_1 \underbrace{(x_1 + v - 6)}_{\leq 0} + (v-6)^2 \leq (v-6)^2$$

with equality for $x_1 = 0$. So we have

$$B = \big\{(v, w) \mid v \leq 6,\ (v-6)^2/2 \leq w \leq (v-6)^2\big\}. \qquad \triangleleft$$

The attentive reader will have noticed that this example corresponds to the foregoing figure.

Example 13 We look once more at example 11, *d)*:
$B := \{(g(x), f(x)) \mid x \in C\} = \big\{(2x-1, -x^2) \mid 0 \leq x \leq 1\big\}$
$v := g(x) = 2x - 1 \in [-1, 1]$ gives $x = (1+v)/2$, hence,
$w := f(x) = -(1+v)^2/4$.

Duality Gap

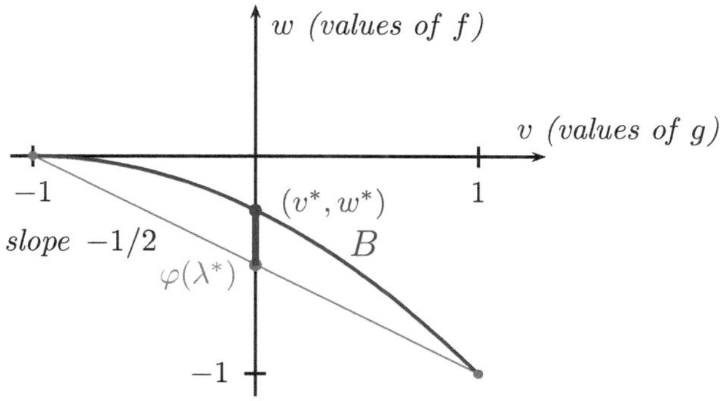

$\triangleleft$

Chapter 2

Saddlepoints and Duality

For the following *characterization of strong duality* neither convexity nor differentiability is needed:

Theorem 2.4.1

Let x^* be a point in C and $\lambda^* \in \mathbb{R}^m_+$. Then the following statements are equivalent:

a) (x^*, λ^*) *is a saddlepoint of the* LAGRANGE *function* L.

b) x^* *is a minimizer to problem* (P) *and* λ^* *is a maximizer to problem* (D) *with*

$$f(x^*) = L(x^*, \lambda^*) = \varphi(\lambda^*) .$$

In other words: *A saddlepoint of the Lagrangian L exists if and only if the problems (P) and (D) have the same value and admit optimizers, that is,*

$$\min(P) = \max(D) .$$

Proof: First, we show that *a)* implies *b)*:

$$\begin{aligned} L(x^*, \lambda^*) &= \inf_{x \in C} L(x, \lambda^*) \le \sup_{\lambda \in \mathbb{R}^m_+} \inf_{x \in C} L(x, \lambda) \\ &\overset{\checkmark}{\le} \inf_{x \in C} \sup_{\lambda \in \mathbb{R}^m_+} L(x, \lambda) \le \sup_{\lambda \in \mathbb{R}^m_+} L(x^*, \lambda) = L(x^*, \lambda^*) \end{aligned}$$

Consequently, $\infty > \varphi(\lambda^*) = \inf_{x \in C} L(x, \lambda^*) = \sup_{\lambda \in \mathbb{R}^m_+} L(x^*, \lambda) = L(x^*, \lambda^*)$.
By lemma 2.2.6 we know already: x^* is a minimizer of (P) with $f(x^*) = L(x^*, \lambda^*)$. *b)* now follows by the corollary to the weak duality theorem.

Conversely, suppose now that *b)* holds true:

$$\begin{aligned} \varphi(\lambda^*) &= \inf \{L(x, \lambda^*) \mid x \in C\} \le L(x^*, \lambda^*) \\ &= f(x^*) + \langle \lambda^*, g(x^*) \rangle \le f(x^*) \end{aligned} \tag{7}$$

We have $\varphi(\lambda^*) = f(x^*)$, by assumption. Therefore, equality holds everywhere in (7), especially, $\langle \lambda^*, g(x^*) \rangle = 0$. This leads to
$L(x^*, \lambda^*) = f(x^*) \le L(x, \lambda^*)$ for $x \in C$ and
$L(x^*, \lambda) = f(x^*) + \langle \lambda, g(x^*) \rangle \le f(x^*) = L(x^*, \lambda^*)$ for $\lambda \in \mathbb{R}^m_+$. □

Perturbation and Sensitivity Analysis

In this subsection, we discuss how changes in parameters affect the solution of the primal problem. This is called *sensitivity analysis*. How sensitive are the minimizer and its value to 'small' perturbations in the data of the problem? If parameters change, sensitivity analysis often helps to avoid having to solve a problem again.

For $u \in \mathbb{R}^m$ we consider the 'perturbed' optimization problem

$$(P_u) \quad \begin{cases} f(x) \longrightarrow \min \\ x \in \mathcal{F}_u \end{cases}$$

with the feasible region

$$\mathcal{F}_u := \{x \in C \mid g(x) \leq u\}.$$

The vector u is called the *'perturbation vector'*. Obviously we have $(P_0) = (P)$.

If a variable u_i is positive, this means that we 'relax' the i-th constraint $g_i(x) \leq 0$ to $g_i(x) \leq u_i$; if u_i is negative we tighten this constraint.

We define the *perturbation* or *sensitivity function*

$$p\colon \mathbb{R}^m \longrightarrow \mathbb{R} \cup \{-\infty, \infty\}$$

associated with the problem (P) by

$$p(u) := \inf\{f(x) \mid x \in \mathcal{F}_u\} = \inf\{f(x) \mid x \in C,\ g(x) \leq u\} \quad \text{for } u \in \mathbb{R}^m$$

(with $\inf \emptyset := \infty$). Obviously we have $p(0) = p^*$.

The function p gives the minimal value of the problem (P_u) as a function of 'perturbations' of the right-hand side of the constraint $g(x) \leq 0$.

Its *effective domain* is given by the set

$$\operatorname{dom}(p) := \{u \in \mathbb{R}^m \mid p(u) < \infty\} \overset{\checkmark}{=} \{u \in \mathbb{R}^m \mid \exists\, x \in C\ g(x) \leq u\}.$$

Obviously the function p *is antitone*, that is, order-reversing: If the vector u increases, the feasible region $\mathcal{F}_u$ increases and so p decreases (in the weak sense).

Remark

If the original problem (P) *is convex, then the effective domain* $\operatorname{dom}(p)$ *is convex and the perturbation function* p *is convex on it.*

Since $-\infty$ is possible as a value for p on $\operatorname{dom}(p)$, *convexity* here means the convexity of the *epigraph*[3]

$$\operatorname{epi}(p) := \{(u, z) \in \mathbb{R}^m \times \mathbb{R} \mid u \in \operatorname{dom}(p),\ p(u) \leq z\}$$

[3] The prefix 'epi' means 'above'. A *real-valued* function p is convex if and only if the set $\operatorname{epi}(p)$ is convex (cf. exercise 8).

Proof: The convexity of $\text{dom}(p)$ and p is given immediately by the convexity of the set C and the convexity of the function g:

Let $u, v \in \text{dom}(p)$ and $\varrho \in (0,1)$. For $\alpha, \beta \in \mathbb{R}$ with $p(u) < \alpha$ and $p(v) < \beta$ there exist vectors $x, y \in C$ with $g(x) \le u$, $g(y) \le v$ and $f(x) < \alpha$, $f(y) < \beta$. The vector $\widetilde{x} := \varrho x + (1-\varrho) y$ belongs to C with

$$g(\widetilde{x}) \le \varrho g(x) + (1-\varrho) g(y) \le \varrho u + (1-\varrho) v =: \widetilde{u}$$

and

$$f(\widetilde{x}) \le \varrho f(x) + (1-\varrho) f(y) < \varrho\alpha + (1-\varrho)\beta .$$

This shows $p(\widetilde{u}) \le f(\widetilde{x}) < \varrho\alpha + (1-\varrho)\beta$, hence, $p(\widetilde{u}) \le \varrho p(u) + (1-\varrho) p(v)$. □

Remark

We assume that strong duality holds and that the dual optimal value is attained. Let λ^* be a maximizer to the dual problem (D). Then we have

$$p(u) \ge p(0) - \langle \lambda^*, u \rangle \quad \textit{for all } u \in \mathbb{R}^m .$$

Proof: For a given $u \in \mathbb{R}^m$ and any feasible point x to the problem (P_u), that is, $x \in \mathcal{F}_u$, we have

$$p(0) = p^* = d^* = \varphi(\lambda^*) \le f(x) + \langle \lambda^*, g(x) \rangle \le f(x) + \langle \lambda^*, u \rangle .$$

From this follows $p(0) \le p(u) + \langle \lambda^*, u \rangle$. □

This inequality gives a lower bound on the optimal value of the perturbed problem (P_u). The hyperplane given by $z = p(0) - \langle \lambda^*, u \rangle$ 'supports' the epigraph of the function p at the point $(0, p(0))$. For a problem with only one inequality constraint the inequality shows that the affinely linear function $u \mapsto p^* - \lambda^* u$ $(u \in \mathbb{R})$ lies below the graph of p and is tangent to it at the point $(0, p^*)$.

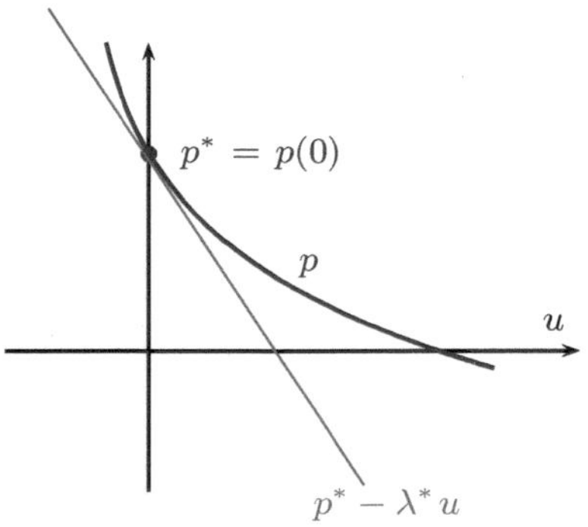

We get the following rough sensitivity results:

If λ_i^* is 'small', relaxing the i-th constraint causes a small decrease of the optimal value $p(u)$. Conversely, if λ_i^* is 'large', tightening the i-th constraint causes a large increase of the optimal value $p(u)$.

Under the assumptions of the foregoing remark we have:

Remark

If the function p is differentiable[4] *at the point $u = 0$, then the maximizer λ^* of the dual problem (D) is related to the gradient of p at $u = 0$:*

$$\nabla p(0) = -\lambda^*$$

Here the LAGRANGE multipliers λ_i^* are exactly the *local sensitivities* of the function p with respect to perturbations of the constraints.

Proof: The differentiability at the point $u = 0$ gives:
$p(u) = p(0) + \langle \nabla p(0), u \rangle + r(u)\,\|u\|$ with $r(u) \to 0$ for $\mathbb{R}^m \ni u \to 0$.
Hence we obtain $-\langle \nabla p(0) + \lambda^*, u \rangle \le r(u)\,\|u\|$. We set $u := -t[\nabla p(0) + \lambda^*]$ for $t > 0$ and get $t\,\|\nabla p(0) + \lambda^*\|^2 \le t\,\|\nabla p(0) + \lambda^*\|\; r\big(-t[\nabla p(0) + \lambda^*]\big)$. This shows: $\|\nabla p(0) + \lambda^*\| \le r\big(-t[\nabla p(0) + \lambda^*]\big)$. Passage to the limit $t \to 0$ yields $\nabla p(0) + \lambda^* = 0$. □

For the rest of this section we consider only the special case of a *convex optimization problem*, where the functions f and g are *convex* and continuously differentiable and the set C is convex.

Economic Interpretation of Duality

The equation

$$\nabla p(0) = -\lambda^*$$

or

$$-\frac{\partial p}{\partial u_i}(0) = \lambda_i^* \quad \text{for } i = 1, \ldots, m$$

leads to the following interpretation of dual variables in economics:

The components λ_i^* of the LAGRANGE multiplier λ^* are often called *shadow prices* or *attribute costs*. They represent the *'marginal' rate of change* of the optimal value

$$p^* = v(P) = \inf(P)$$

[4] *Subgradients* generalize the concept of gradient and are helpful if the function p is *not* differentiable at the point $u = 0$. We do not pursue this aspect and its relation to the concept of *stability*.

of the primal problem (P) with respect to changes in the constraints. They describe the incremental change in the value p^* per unit increase in the right-hand side of the constraint.

If, for example, the variable $x \in \mathbb{R}^n$ determines how an enterprise 'operates', the objective function f describes the cost for some production process, and the constraint $g_i(x) \leq 0$ gives a bound on a special resource, for example labor, material or space, then $p^*(u)$ shows us how much the costs (and with it the profit) change when the resource changes. λ_i^* determines approximately how much fewer costs the enterprise would have, for a 'small' increase in availability of the i-th resource. Under these circumstances λ_i^* has the dimension of dollars (or euros) per unit of capacity of the i-th resource and can therefore be regarded as a value per unit resource. So we get the maximum price we should pay for an additional unit of u_i.

Strong Duality

Below we will see: If the SLATER constraint qualification holds and the original problem is convex, then we have strong duality, that is, $p^* = d^*$. We see once more: The class of convex programs is a class of 'well-behaved' optimization problems. Convex optimization is relatively 'easy'.

We need a slightly different separation theorem (compared to proposition 2.1.1). We quote it without proof (for a proof see, for example: [Fra], p. 49f):

Separation Theorem

Given two disjoint nonempty convex sets $\mathcal{V}$ and $\mathcal{W}$ in $\mathbb{R}^k$, there exist a real α and a vector $p \in \mathbb{R}^k \setminus \{0\}$ with

$$\langle p, v\rangle \geq \alpha \text{ for all } v \in \mathcal{V} \quad \text{and} \quad \langle p, w\rangle \leq \alpha \text{ for all } w \in \mathcal{W}.$$

In other words: *The hyperplane* $\{x \in \mathbb{R}^k \mid \langle p, x\rangle = \alpha\}$ *separates* $\mathcal{V}$ *and* $\mathcal{W}$.

The **example**

$$\mathcal{V} := \left\{x = (x_1, x_2)^T \in \mathbb{R}^2 \mid x_1 \leq 0\right\} \quad \text{and}$$
$$\mathcal{W} := \left\{x = (x_1, x_2)^T \in \mathbb{R}^2 \mid x_1 > 0,\, x_1 x_2 \geq 1\right\}$$

(with separating 'line' $x_1 = 0$) shows that the sets cannot be 'strictly' separated.

Strong Duality Theorem

Suppose that the SLATER *constraint qualification*

$$\exists\, \tilde{x} \in \mathcal{F} \;\; g_i(\tilde{x}) < 0 \;\text{ for all } i \in \mathcal{I}_1$$

holds for the convex problem (P). Then we have strong duality, and the value of the dual problem (D) is attained if $p^* > -\infty$.

In order to simplify the *proof,* we verify the theorem under the slightly stronger condition

$$\exists\, \tilde{x} \in \mathcal{F} \;\; g_i(\tilde{x}) < 0 \;\text{ for all } i \in \mathcal{I}\,.$$

For an extension of the proof to the (refined) SLATER *constraint qualification* see for example [Rock], p. 277.

Proof: There exists a feasible point, hence we have $p^* < \infty$. If $p^* = -\infty$, then we get $d^* = -\infty$ by the weak duality theorem. Hence, we can suppose that p^* *is finite.* The two sets

$$\mathcal{V} := \{(v, w) \in \mathbb{R}^m \times \mathbb{R} \mid \exists\, x \in C \;\; g(x) \le v \text{ and } f(x) \le w\}$$

$$\mathcal{W} := \{(0, w) \in \mathbb{R}^m \times \mathbb{R} \mid w < p^*\}$$

are nonempty and *convex.* By the definition of p^* they are *disjoint:* Let (v, w) be in $\mathcal{W} \cap \mathcal{V}$: $(v, w) \in \mathcal{W}$ shows $v = 0$ and $w < p^*$. For $(v, w) \in \mathcal{V}$ there exists an $x \in C$ with $g(x) \le v = 0$ and $f(x) \le w < p^*$, which is a contradiction to the definition of p^*.

The quoted separation theorem gives the existence of a pair

$(\lambda, \mu) \in \mathbb{R}^m \times \mathbb{R} \setminus \{(0, 0)\}$ and an $\alpha \in \mathbb{R}$ such that:

$$\langle \lambda, v\rangle + \mu w \ge \alpha \text{ for all } (v, w) \in \mathcal{V} \quad \text{and} \tag{8}$$

$$\langle \lambda, v\rangle + \mu w \le \alpha \text{ for all } (v, w) \in \mathcal{W} \tag{9}$$

From (8) we get $\lambda \ge 0$ and $\mu \ge 0$. (9) means that $\mu w \le \alpha$ for all $w < p^*$, hence $\mu p^* \le \alpha$. (8) and the definition of $\mathcal{V}$ give for any $x \in C$:

$$\langle \lambda, g(x)\rangle + \mu f(x) \ge \alpha \ge \mu p^* \tag{10}$$

For $\boxed{\mu = 0}$ we get from (10) that $\langle \lambda, g(x)\rangle \ge 0$ for any $x \in C$, especially $\langle \lambda, g(\tilde{x})\rangle \ge 0$ for a point $\tilde{x} \in C$ with $g_i(\tilde{x}) < 0$ for all $i \in \mathcal{I}$. This shows $\lambda = 0$ arriving at a *contradiction* to $(\lambda, \mu) \ne (0, 0)$. So we have $\boxed{\mu > 0}$: We divide the inequality (10) by μ and obtain

$$L\big(x, \lambda/\mu\big) \ge p^* \text{ for any } x \in C\,.$$

From this follows $\varphi\big(\lambda/\mu\big) \ge p^*$. By the weak duality theorem we have $\varphi\big(\lambda/\mu\big) \le d^* \le p^*$. This shows strong duality and that the dual value is attained. □

Strong duality can be obtained for some special nonconvex problems too: It holds for any optimization problem with quadratic objective function and one quadratic inequality constraint, provided SLATER's constraint qualification holds. See for example [Bo/Va], Appendix B.

Exercises

1. *Orthogonal Distance Line Fitting*

Consider the following *approximation problem* arising from quality control in manufacturing using coordinate measurement techniques [Ga/Hr]. Let

$$M := \{(x_1, y_1), (x_2, y_2), \ldots, (x_m, y_m)\}$$

be a set of $m \in \mathbb{N}$ given points in $\mathbb{R}^2$. The task is to find a line L

$$L(c, n_1, n_2) := \{(x, y) \in \mathbb{R}^2 \mid c + n_1 x + n_2 y = 0\}$$

in Hessian normal form with $n_1^2 + n_2^2 = 1$ which *best approximates* the point set M such that the *sum of squares of the distances of the points from the straight line* becomes minimal. If we calculate $r_j := c + n_1 x_j + n_2 y_j$ for a point (x_j, y_j), then $|r_j|$ is its distance to L.

a) Formulate the above problem as a constrained optimization problem.

b) Show the existence of a solution and determine the optimal parameters c, n_1 and n_2 by means of the LAGRANGE multiplier rule. Explicate when and in which sense these parameters are uniquely defined.

c) Find a (minimal) example which consists of three points and has infinitely many optimizers.

d) Solve the optimization problem with *Matlab*® or *Maple*® and test your program with the following data (cf. [Ga/Hr]):

x_j	1.0	2.0	3.0	4.0	5.0	6.0	7.0	8.0	9.0	10.0
y_j	0.2	1.0	2.6	3.6	4.9	5.3	6.5	7.8	8.0	9.0

2. *a)* Solve the optimization problem

$$f(x_1, x_2) := 2x_1 + 3x_2 \longrightarrow \max$$
$$\sqrt{x_1} + \sqrt{x_2} = 5$$

using LAGRANGE multipliers (cf. [Br/Ti]).

b) Visualize the contour lines of f as well as the set of feasible points, and mark the solution. Explain the result!

3. Let $n \in \mathbb{N}$ and $A = (a_{\nu,\mu})$ be a real symmetric (n,n)-matrix with the submatrices A_k

$$A_k := \begin{pmatrix} a_{11} & a_{12} & \ldots & a_{1k} \\ a_{21} & a_{22} & \ldots & a_{2k} \\ \vdots & \vdots & \vdots & \vdots \\ a_{k1} & a_{k2} & \ldots & a_{kk} \end{pmatrix} \quad \text{for} \quad k \in \{1, \ldots, n\}.$$

Then the following statements are equivalent:

a) A is positive definite.

b) $\exists\, \delta > 0\ \forall\, x \in \mathbb{R}^n\ x^T A x \ge \delta \|x\|^2$

c) $\forall\, k \in \{1, ..., n\}\ \det A_k > 0$

4. Consider a function $f\colon \mathbb{R}^n \longrightarrow \mathbb{R}$.

a) If f is differentiable, then the following holds:
f convex $\iff \forall x, y \in \mathbb{R}^n\ f(y) - f(x) \ge f'(x)(y - x)$

b) If f is twice continuously differentiable, then:
f convex $\iff \forall x \in \mathbb{R}^n\ \nabla^2 f(x)$ positive semidefinite

c) What do the corresponding characterizations of strictly convex functions look like?

5. In the "colloquial speech" of mathematicians one can sometimes hear the following statement: *"Strictly convex functions always have exactly one minimizer."*

However, is it really right to use this term so carelessly? Consider two typical representatives $f_i\colon \mathbb{R}^2 \longrightarrow \mathbb{R}$, $i \in \{1, 2\}$:

$$f_1(x, y) = x^2 + y^2$$
$$f_2(x, y) = x^2 - y^2$$

Visualize these functions and plot their contour lines. Which function is convex? Show this analytically as well. Is the above statement correct?

Let $D_j \subset \mathbb{R}^2$ for $j \in \{1, 2, 3, 4, 5\}$ be a region in $\mathbb{R}^2$ with

$$\begin{aligned} D_1 &:= \{(x, y) \in \mathbb{R}^2 \,:\, x_1^2 + x_2^2 \le 0.04\} \\ D_2 &:= \{(x, y) \in \mathbb{R}^2 \,:\, (x_1 - 0.55)^2 + (x_2 - 0.7)^2 \le 0.04\} \\ D_3 &:= \{(x, y) \in \mathbb{R}^2 \,:\, (x_1 - 0.55)^2 + x_2^2 \le 0.04\}. \end{aligned}$$

The outer boundary of the regions D_4 and D_5 is defined by

$$\begin{aligned} x &= 0.5(0.5 + 0.2\cos(6\vartheta))\cos\vartheta + x_c \\ y &= 0.5(0.5 + 0.2\cos(6\vartheta))\sin\vartheta + y_c \end{aligned}\ ,\quad \vartheta \in [0, 2\pi)\,,$$

where $(x_c, y_c) = (0, 0)$ for D_4 and $(x_c, y_c) = (0, -0.7)$ for D_5.

If we now restrict the above functions f_i to D_j ($i \in \{1, 2\}$, $j \in \{1, 2, 3, 4, 5\}$), does the statement about the uniqueness of the minimizers still hold? Find all the minimal points, where possible! Where do they lie? Which role does the convexity of the region and the function play?

6. Show that a function $f\colon \mathbb{R}^n \longrightarrow \mathbb{R}$ is affinely linear if and only if it is convex as well as concave.

7. Let X be a real vector space. For $m \in \mathbb{N}$ and $x_1, \ldots, x_m \in X$ let

$$\operatorname{conv}(x_1, \ldots, x_m) := \left\{ \sum_{i=1}^{m} \lambda_i \, x_i \mid \lambda_1, \ldots, \lambda_m > 0, \sum_{i=1}^{m} \lambda_i = 1 \right\}.$$

Verify that the following assertions hold for a nonempty subset $A \subset X$:

a) A convex $\Longleftrightarrow$ $\forall\, m \in \mathbb{N}\ \forall\, a_1, \ldots, a_m \in A \quad \operatorname{conv}(a_1, \ldots, a_m) \subset A$

b) Let A be convex and $f\colon A \longrightarrow \mathbb{R}$ a convex function. For $x_1, x_2, \ldots, x_m \in A$ and $x \in \operatorname{conv}(x_1, \ldots, x_m)$ in a representation as given above, it then holds that

$$f\left(\sum_{i=1}^{m} \lambda_i x_i\right) \le \sum_{i=1}^{m} \lambda_i f(x_i).$$

c) The intersection of an arbitrary number of convex sets is convex. Consequently there exists the smallest convex superset $\operatorname{conv}(A)$ of A, called the *convex hull* of A.

d) It holds that $\operatorname{conv}(A) = \bigcup_{\substack{m \in \mathbb{N} \\ a_1, \ldots, a_m \in A}} \operatorname{conv}(a_1, \ldots, a_m)$.

e) CARATHÉODORY's lemma:

For $X = \mathbb{R}^n$ it holds that $\operatorname{conv}(A) = \bigcup_{\substack{m \le n+1 \\ a_1, \ldots, a_m \in A}} \operatorname{conv}(a_1, \ldots, a_m)$.

f) In which way does this lemma have to be modified for $X = \mathbb{C}^n$?

g) For $X \in \{\mathbb{R}^n, \mathbb{C}^n\}$ and A compact the convex hull $\operatorname{conv}(A)$ is also compact.

8. For a nonempty subset $D \subset \mathbb{R}^n$ and a function $f\colon D \longrightarrow \mathbb{R}$ let

$$\operatorname{epi}(f) := \{(x, y) \in D \times \mathbb{R} : f(x) \le y\}$$

be the *epigraph* of f. Show that for a convex set D we have

$$f \text{ convex} \iff \operatorname{epi}(f) \text{ convex}.$$

9. Prove part *1)* of lemma 2.2.3 and additionally show the following assertions for $\mathcal{F}$ *convex* and $x_0 \in \mathcal{F}$:

a) $\mathcal{C}_{fd}(x_0) = \{\mu(x - x_0) \mid \mu > 0, x \in \mathcal{F}\}$

b) $\mathcal{C}_t(x_0) = \overline{\mathcal{C}_{fd}(x_0)}$

c) $\mathcal{C}_t(x_0)$ is convex.

10. Prove for the *tangent cones* of the following sets

$$\begin{aligned}
\mathcal{F}_1 &:= \{x \in \mathbb{R}^n \mid \|x\|_2 = 1\}, \\
\mathcal{F}_2 &:= \{x \in \mathbb{R}^n \mid \|x\|_2 \le 1\}, \\
\mathcal{F}_3 &:= \{x \in \mathbb{R}^2 \mid -x_1^3 + x_2 \le 0, -x_2 \le 0\}:
\end{aligned}$$

a) For $x_0 \in \mathcal{F}_1$ it holds that $\mathcal{C}_t(x_0) = \{d \in \mathbb{R}^n \mid \langle d, x_0 \rangle = 0\}$.

b) For $x_0 \in \mathcal{F}_2$ we have $\mathcal{C}_t(x_0) = \begin{cases} \mathbb{R}^n, & \|x_0\|_2 < 1, \\ \{d \in \mathbb{R}^n \mid \langle d, x_0 \rangle \le 0\}, & \|x_0\|_2 = 1. \end{cases}$

c) For $x_0 := (0,0)^T \in \mathcal{F}_3$ it holds that $\mathcal{C}_t(x_0) = \{d \in \mathbb{R}^2 \mid d_1 \ge 0, d_2 = 0\}$.

11. With $f(x) := x_1^2 + x_2^2$ for $x \in \mathbb{R}^2$ consider

$$(P) \quad \begin{cases} f(x) \longrightarrow \min \\ -x_2 \le 0 \\ x_1^3 - x_2 \le 0 \\ x_1^3(x_2 - x_1^3) \le 0 \end{cases}$$

and determine the linearizing cone, the tangent cone and the respective dual cones at the (strict global) minimal point $x_0 := (0,0)^T$.

12. Let x_0 be a feasible point of the optimization problem (P). According to page 56 *it holds that* (LICQ) $\Longrightarrow$ (AHUCQ) $\Longrightarrow$ (ACQ).
Show by means of the following examples (with $n = m = 2$ and $p = 0$) that these two implications do not hold in the other direction:

a) $f(x) := x_1^2 + (x_2+1)^2$, $g_1(x) := -x_1^3 - x_2$, $g_2(x) := -x_2$, $x_0 := (0,\, 0)^T$

b) $f(x) := x_1^2 + (x_2+1)^2$, $g_1(x) := x_2 - x_1^2$, $g_2(x) := -x_2$, $x_0 := (0,\, 0)^T$

13. Let the following optimization problem be given:

$$\begin{aligned} & f(x) \longrightarrow \min, \quad x \in \mathbb{R}^2 \\ & g_1(x_1, x_2) := 3(x_1 - 1)^3 - 2x_2 + 2 \le 0 \\ & g_2(x_1, x_2) := (x_1 - 1)^3 + 2x_2 - 2 \le 0 \\ & g_3(x_1, x_2) := -x_1 \le 0 \\ & g_4(x_1, x_2) := -x_2 \le 0 \end{aligned}$$

a) Plot the feasible region.

b) Solve the optimization problem for the following objective functions:

(i) $f(x_1, x_2) := (x_1 - 1)^2 + (x_2 - \frac{3}{2})^2$

(ii) $f(x_1, x_2) := (x_1 - 1)^2 + (x_2 - 4)^2$
Regard the objective function on the 'upper boundary' of $\mathcal{F}$.

(iii) $f(x_1, x_2) := (x_1 - \frac{5}{4})^2 + (x_2 - \frac{5}{4})^2$
Do the KKT conditions hold at the optimal point?

Hint: In addition illustrate these problems graphically.

14. *Optimal Location of a Rescue Helicopter* (see example 4 of chapter 1) *a)* Formulate the minimax problem

$$d_{\max}(x, y) := \max_{1 \le j \le m} \sqrt{(x - x_j)^2 + (y - y_j)^2}$$

as a quadratic optimization problem

$$\begin{cases} f(x, y, \varrho) \longrightarrow \min \\ g_j(x, y, \varrho) \le 0 \qquad (j = 1, \ldots, m) \end{cases}$$

(with f quadratic, g_j linear). You can find some *hints* on page 13.

b) Visualize the function $d_{\max}$ by plotting its contour lines for the points $(0, 0)$, $(5, -1)$, $(4, 6)$, $(1, 3)$.

c) Give the corresponding Lagrangian. Solve the problem by means of the KARUSH–KUHN–TUCKER conditions.

15. Determine a triangle with minimal area containing two disjoint disks with radius 1. WLOG let $(0, 0)$, $(x_1, 0)$ and (x_2, x_3) with $x_1, x_3 \ge 0$ be the vertices of the triangle; (x_4, x_5) and (x_6, x_7) denote the centers of the disks.

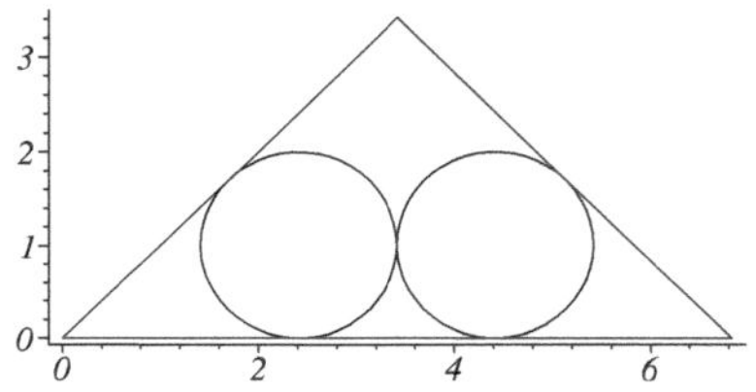

a) Formulate this problem as a minimization problem in terms of seven variables and nine constraints (see [Pow 1]).

b) $x^* = \left(4 + 2\sqrt{2}, 2 + \sqrt{2}, 2 + \sqrt{2}, 1 + \sqrt{2}, 1, 3 + \sqrt{2}, 1\right)^T$ is a solution of this problem; calculate the corresponding LAGRANGE multipliers λ^*, such that the KARUSH–KUHN–TUCKER conditions are fulfilled.

c) Check the sufficient second-order optimality conditions for (x^*, λ^*).

16. Find the point $x \in \mathbb{R}^2$ that lies closest to the point $p := (2, 3)$ under the constraints $g_1(x) := x_1 + x_2 \le 0$ and $g_2(x) := x_1^2 - 4 \le 0$.

a) Illustrate the problem graphically.

b) Verify that the problem is convex and fulfills (SCQ).

c) Determine the KKT points by differentiating between three cases: none is active, exactly the first one is active, exactly the second one is active.

d) Now conclude with theorem 2.2.8.

The problem can of course be solved elementarily. We, however, want to practice the theory with simple examples.

17. In a small power network the power r runs through two different channels. Let x_i be the power running through channel i for $i = 1, 2$. The total loss is given by the function $f\colon \mathbb{R}^2 \longrightarrow \mathbb{R}$ with

$$f(x_1, x_2) := x_1 + \frac{1}{2}\left(x_1^2 + x_2^2\right).$$

Determine the current flow such that the total loss stays minimal. The constraints are given by $x_1 + x_2 = r\,,\ \ x_1 \geq 0\,, x_2 \geq 0\,.$

18. Verify in the *linear case* that the LAGRANGE dual problem of (D) (cf. p. 71) — transformed into standard form — is again the primal problem.

19. Consider the optimization problem (cf. [Erik]):

$$\begin{cases} f(x) \;:=\; \sum\limits_{i=1}^{n} x_i \log(\frac{x_i}{p_i}) \longrightarrow \min\,, \quad x \in \mathbb{R}^n \\ A^T x \;=\; b\,,\ x \geq 0 \end{cases}$$

where $A \in \mathbb{R}^{n\times m}, b \in \mathbb{R}^m$ and $p_1, p_2, \dots, p_n \in \mathbb{R}_{++}$ are given. Let further $0 \ln 0$ be defined as 0. Prove:

a) The dual problem is given by

$$\varphi(\lambda) := b^T\lambda - \sum_{i=1}^{n} p_i \exp(e_i^T A\lambda - 1) \longrightarrow \max\,, \quad \lambda \in \mathbb{R}^m\,.$$

b) $\nabla\varphi(\lambda) \;=\; b - A^T x$ with $x_i = p_i \exp(e_i^T A\lambda - 1)\,.$

c) $\nabla^2\varphi(\lambda) \;=\; -A^T X A$, where $X = \operatorname{Diag}(x)$ with x from *b)*.

20. *Support Vector Machines* (cf. [Cr/Sh])

Support vector machines have been extensively used in *machine learning* and *data mining applications* such as classification and regression, *text categorization* as well as *medical applications*, for example *breast cancer diagnosis*. Let two classes of patterns be given, i. e., samples of observable characteristics which are represented by points x_i in $\mathbb{R}^n$. The patterns are given in the form (x_i, y_i), $i = 1, \dots, m$, with $y_i \in \{1, -1\}$. $y_i = 1$ means that x_i belongs to class 1; otherwise x_i belongs to class 2. In the simplest case we are looking for a *separating hyperplane* described by $\langle w\,, x\rangle + \beta = 0$ with $\langle w\,, x_i\rangle + \beta \geq 1$ if $y_i = 1$ and $\langle w\,, x_i\rangle + \beta \leq -1$ if $y_i = -1$. These conditions can be written as $y_i\big(\langle w\,, x_i\rangle + \beta\big) \geq 1$ $(i = 1, \dots, m)$. We aim to maximize the 'margin' (distance) $2/\sqrt{\langle w\,, w\rangle}$ between the two hyperplanes $\langle w\,, x\rangle + \beta = 1$ and $\langle w\,, x\rangle + \beta = -1$. This gives a linearly constrained convex quadratic minimization problem

$$\begin{cases} \frac{1}{2}\langle w\,, w\rangle \longrightarrow \min \\ y_i\big(\langle w\,, x_i\rangle + \beta\big) \geq 1 \quad (i = 1, \dots, m)\,. \end{cases} \tag{11}$$

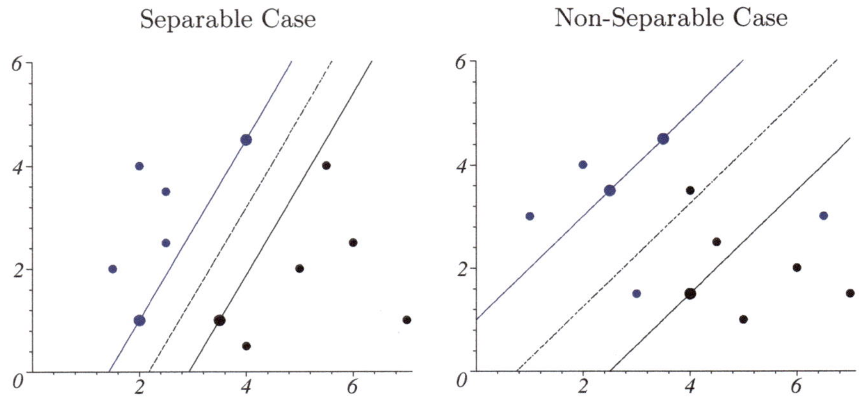

In the case that the two classes are *not* linearly separable (by a hyperplane), we introduce nonnegative penalties ξ_i for the 'misclassification' of x_i and minimize both $\langle w, w\rangle$ and $\sum_{i=1}^m \xi_i$. We solve this optimization problem in the following way with *soft margins*

$$(P) \qquad \begin{cases} \frac{1}{2}\langle w, w\rangle + C\sum_{i=1}^m \xi_i \longrightarrow \min \\ y_i\big(\langle w, x_i\rangle + \beta\big) \ge 1 - \xi_i,\ \xi_i \ge 0 \quad (i = 1, \ldots, m). \end{cases} \tag{12}$$

Here, C is a weight parameter of the penalty term.

a) Introducing the dual variables $\lambda \in \mathbb{R}_+^m$, derive the LAGRANGE dual problem to (P):

$$(D) \qquad \begin{cases} -\frac{1}{2}\sum_{i,j=1}^m y_i y_j \langle x_i, x_j\rangle \lambda_i\lambda_j + \sum_{i=1}^m \lambda_i \longrightarrow \max \\ \sum_{i=1}^m y_i\lambda_i = 0, \quad 0 \le \lambda_i \le C \quad (i = 1, \ldots, m) \end{cases} \tag{13}$$

Compute the coefficients $w \in \mathbb{R}^n$ and $\beta \in \mathbb{R}$ of the separating hyperplane by means of the dual solution λ and show

$w = \sum_{j=1}^m y_j\lambda_j x_j, \quad \beta = y_j - \langle w, x_j\rangle \quad$ if $0 < \lambda_j < C$.

Vectors x_j with $\lambda_j > 0$ are called *support vectors.*

b) Calculate a support vector 'machine' for *breast cancer diagnosis* using the file wisconsin-breast-cancer.data from the *Breast Cancer Wisconsin Data Set* (cf. http://archive.ics.uci.edu/ml/). The file wisconsin-breast-cancer.names gives information on the data set: It contains 699 instances consisting of 11 attributes. The first attribute gives the sample code number. Attributes 2 through 10 describe the medical status and give a 9-dimensional vector x_i. The last attribute is the class attribute ("2" for *benign,* "4" for *malignant).* Sixteen samples have a missing attribute, denoted by "?". Remove these samples from the data set. Now split the data into two portions: The first 120 instances are used as training data. Take software of your choice to solve the quadratic problem (P), using the penalty parameter $C = 1000$. The remaining instances are used to evaluate the 'performance' of the *classifier* or *decision function* given by $f(x) := \operatorname{sgn}\big\{\langle w, x\rangle + \beta\big\}$.

3

Unconstrained Optimization Problems

Chapter 3

Unconstrained optimization methods seek a local minimum (or a local maximum) in the absence of restrictions, that is,

$$f(x) \longrightarrow \min \quad (x \in D)$$

for a real-valued function $f\colon D \longrightarrow \mathbb{R}$ defined on a nonempty subset D of $\mathbb{R}^n$ for a given $n \in \mathbb{N}$. Unconstrained optimization involves the theoretical study of optimality criteria and above all algorithmic methods for a wide variety of problems. In section 2.0 we have repeated — as essential basics — the well-known (first- and second-order) optimality conditions for smooth real-valued functions. Often constraints complicate a given task but in some cases they simplify it. Even though most optimization problems in 'real life' have restrictions to be satisfied, the study

W. Forst and D. Hoffmann, *Optimization—Theory and Practice*,
Springer Undergraduate Texts in Mathematics and Technology,
DOI 10.1007/978-0-387-78977-4_3,

of unconstrained problems is useful for two reasons: Firstly, they occur directly in some applications, so they are important in their own right. Secondly, unconstrained problems often originate as a result of transformations of constrained optimization problems. Some methods, for example, solve a general problem by converting it into a sequence of unconstrained problems.

In section 3.1 *elementary search and localization methods* like the NELDER–MEAD polytope method and SHOR'S ellipsoid method are treated. The first method is widely used in applications, as the effort is small. The results, however, are in general rather poor. SHOR's ellipsoid method has attracted great attention, mainly because of its applications to *linear optimization* in the context of *interior-point methods.* Often methods proceed by finding a suitable direction and then minimizing along this direction (*"line search"*). This is treated in section 3.2. *Trust region methods,* which we are going to cover in section 3.3, start with a given step size or an upper bound for it and then determine a suitable search direction. In section 3.4 the concept of *conjugate directions* is introduced. *If* the objective function is quadratic, the resulting method terminates after a finite number of steps. The extension to *any* differentiable function $f\colon \mathbb{R}^n \longrightarrow \mathbb{R}$ goes back to FLETCHER/REEVES (1964). Quasi-NEWTON methods in section 3.5 are based on the *"least change secant update principle"* by C. G. BROYDEN and are thus well motivated.

Due to the multitude of methods presented, this chapter might at times read a bit like a 'cookbook'. To enliven it, we will illustrate the most important methods with insightful examples, whose results will be given in tabular form. In order to achieve comparability, we will choose *one* framework example which picks up on our considerations from exercise 20 in chapter 2 *(classification, support vector machines).* To facilitate understanding, we content ourselves, at first, with the discussion of an example with only twelve given points for the different methods. At the end of the chapter we will consider a more serious problem, *breast cancer diagnosis,* and compile important results of the different methods in a table for comparison.

3.0 Logistic Regression

As the *framework example* we consider a specific method for *binary classification,* called *logistic regression:* Besides support vector machines, this method arises in many applications, for example in medicine, natural language processing and supervised learning (see [Bo/Va]). The aim is, given a 'training set', to get a function that is a 'good' classifier. In our discussion we follow the notation of [LWK], in which the interested reader may also find supplementary considerations.

We consider — in a general form — the *logistic regression function* given by

$$f(w,\beta) := \sum_{\mu=1}^{m} \log\left(1+\exp(-y_\mu(\langle w\,, x_\mu\rangle + \beta))\right).$$

Given m training instances $x_1,\ldots,x_m \in \mathbb{R}^n$ and 'labels' $y_1,\ldots,y_m \in \{-1,1\}$, one 'estimates' $(w,\beta)\in\mathbb{R}^n\times\mathbb{R}$ by minimizing $f(w,\beta)$.

A professional will recognize this immediately as a *log-likelihood model* since the probability is the product of the individual probabilities

$$P(y_\mu \,|\, x_\mu; w, \beta) := \frac{1}{1 + \exp\big(-y_\mu(\langle w, x_\mu\rangle + \beta)\big)}$$

and therefore the logarithm of the probability is nothing else but the sum of the logarithms of the individual probabilities. This knowledge, however, is *not* necessary to understand the examples given here. These very brief remarks only serve to illustrate that the above is an important type of example. By taking the logarithm of the original function we obtain a *convex problem* (see below).

To simplify the calculation as well as the implementation, we transform

$$\begin{pmatrix} x_\mu \\ 1 \end{pmatrix} \longmapsto x_\mu \quad \text{and} \quad \begin{pmatrix} w \\ \beta \end{pmatrix} \longmapsto w$$

and, in doing so, get — after relabeling the variables and inserting a regularization term $\lambda \cdot \frac{1}{2}\langle w, w\rangle$ — the simple form of a *regularized logistic regression function*

$$f(w) := \lambda \cdot \frac{1}{2}\langle w, w\rangle + \sum_{\mu=1}^{m} \log\big(1 + \exp(-y_\mu \langle w, x_\mu\rangle)\big),$$

where $\lambda \geq 0$ is a suitable parameter. The resulting optimization methods are iterative processes, generating a sequence $\big(w^{(k)}\big)$ which is hoped to converge to a minimizer.

For the *gradient* and the *Hessian* of f we get

$$\nabla f(w) = \lambda w - X\big(y \,.\!*(1-h)\big) \quad \text{and} \quad \nabla^2 f(w) = \lambda I + X^T \operatorname{Diag}\big(h \,.\!*(1-h)\big)X$$

with the notations

$$X := \begin{pmatrix} x_1^T \\ \vdots \\ x_m^T \end{pmatrix}, \quad y := \begin{pmatrix} y_1 \\ \vdots \\ y_m \end{pmatrix} \quad \text{and} \quad h := 1 \,./\Big(1 + \exp\big(-y \,.* (Xw)\big)\Big),$$

where — following the *Matlab*® convention — the operations .* and ./ denote the coordinatewise multiplication and division, respectively.

From that the *convexity* of f can be directly deduced.

In the following we will each time at first consider $m = 12$ and the data:

μ	1	2	3	4	5	6	7	8	9	10	11	12
x_μ	1.0	2.0	2.5	3.0	3.5	4.0	4.0	4.5	5.0	6.0	6.5	7.0
	3.0	4.0	3.5	1.5	4.5	1.5	3.5	2.5	1.0	2.0	3.0	1.5
y_μ	1	1	1	1	1	−1	−1	−1	−1	−1	1	−1

3.1 Elementary Search and Localization Methods

The Nelder and Mead Polytope Method (1965)

The Nelder and Mead method is a *direct search method* which works "moderately well". It is based on the evaluation of functions at the vertices of a polytope[1] which is modified iteratively by replacing 'old' vertices with better ones. The method is *widely* used in applications. Therefore one should be familiar with it. However, we will not need *any* of the theoretical considerations we have discussed so far! General assertions about the convergence of this method are not known. Even for convex problems of two variables the method does not necessarily converge to a stationary point — and *if* it converges, then often very slowly. Since we do not need any derivatives, the method can also be applied to problems with a nondifferentiable objective function, or to problems where the computation of the derivatives is laborious. To sum up the method, we can say: Small effort but rather poor results.

A *polytope* is the convex hull of a finite number of vectors $x_1, \ldots, x_m$ in $\mathbb{R}^n$. If these vectors are affinely independent, we call it a *simplex* with the *vertices* $x_1, \ldots, x_m$. However, we will use the term "polytope" in this section as a synonym for "simplex", always assuming the affine independence of the 'generating' points.

Suppose that $f \colon \mathbb{R}^n \longrightarrow \mathbb{R}$ is a given function, and that we have an n-dimensional polytope with the $n+1$ vertices $x_1, x_2, \ldots, x_{n+1}$. Let these be arranged in such a way that

$$f_1 \le f_2 \le \cdots \le f_{n+1}$$

holds for the function values $f_j := f(x_j)$. In each iteration step the currently worst vertex x_{n+1} is replaced by a new vertex. For that, denote by

$$x^c := \frac{1}{n} \sum_{j=1}^{n} x_j$$

the barycenter (centroid) of the best n vertices.

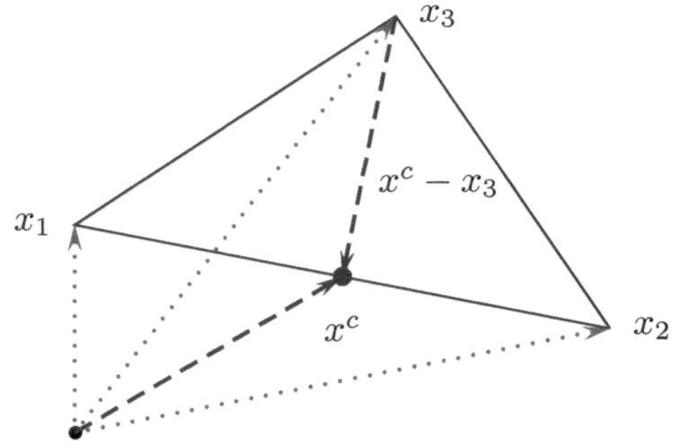

[1] We will not use the common term "simplex" in the name of this method, in order to avoid confusion with the *Simplex Algorithm*.

The polytope will be modified using the operations

Reflection — *Expansion* — *Contraction* — *Shrinking*.

At the start of each iteration we try a *reflection* of the polytope and for that we compute

$$x^r := x^c + \alpha(x^c - x_{n+1}), \quad f^r := f(x^r)$$

with a fixed constant $\alpha > 0$ (often $\alpha = 1$). α is called the reflection coefficient, and x^r is called the reflected point.

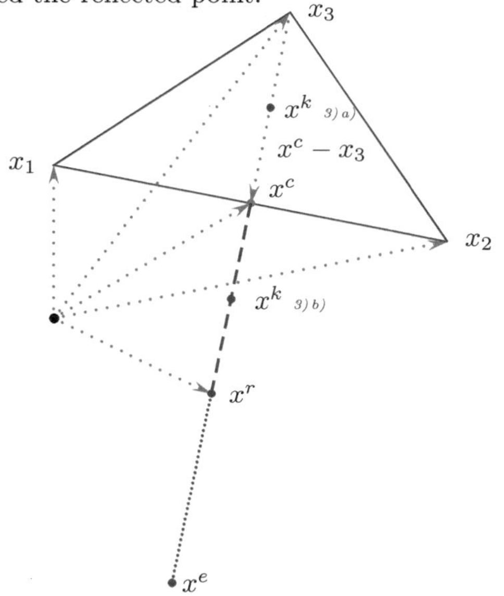

Then we consider the following three cases:

1) $f_1 \le f^r < f_n$: In this case we replace x_{n+1} with x^r.

2) $f^r < f_1$: The 'direction of the reflection' seems 'worthy' of further exploration. Therefore we *expand* the polytope and, in order to do that, we compute

$$x^e := x^c + \beta(x^r - x^c), \; f^e := f(x^e)$$

with a fixed constant $\beta > 1$ (often $\beta = 2$). β is called the expansion coefficient, and x^e is called the extrapolated point.

If $f^e < f^r$: Replace x_{n+1} with x^e.
Else: Replace x_{n+1} with x^r.

3) $f^r \ge f_n$: The polytope seems to be 'too big'. Therefore we try a (partial) *contraction* of the polytope:

a) If $f^r \ge f_{n+1}$, we compute

$$x^k := x^c + \gamma(x_{n+1} - x^c), \; f^k := f(x^k)$$

with a fixed constant γ, $0 < \gamma < 1$ (often $\gamma = \frac{1}{2}$). γ is called the contraction coefficient, and x^k is called the contracted point.

If $f^k < f_{n+1}$, we replace x_{n+1} with x^k; otherwise we *shrink* the whole polytope:

$$\widetilde{x_j} := \frac{1}{2}(x_j + x_1) \quad (j = 1, \ldots, n+1)$$

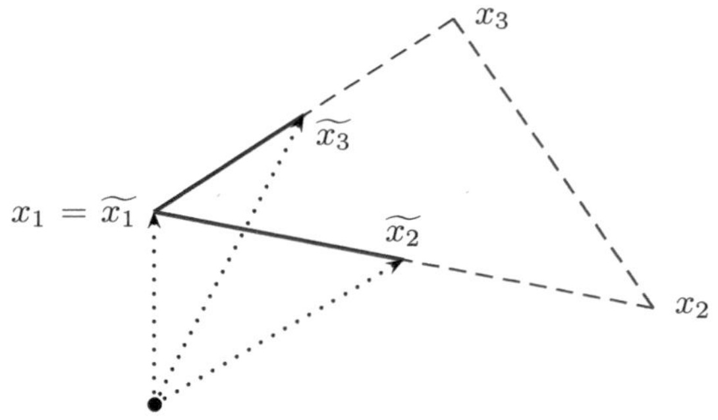

b) If $f^r < f_{n+1}$, we compute

$$x^k := x^c + \gamma(x^r - x^c).$$

If $f^k \leq f^r$, we replace x_{n+1} with x^k; otherwise we shrink the polytope as well.

In each iteration step there is a rearrangement and we choose the notation such that $f_1 \leq f_2 \leq \cdots \leq f_{n+1}$ holds again.

Termination Condition:

$$\frac{1}{n+1} \sum_{j=1}^{n+1} |f_j - f(x^c)|^2 < \varepsilon^2$$

to a given $\varepsilon > 0$. This condition states: The function is nearly constant in the given $n+1$ points.

A variant of this method with information on its convergence can be found in [Kel].

Example 1

In order to apply the Nelder and Mead method to our framework example (cf. section 3.0), we utilize the function `fminsearch` from the *Matlab®* *Optimization Toolbox.* The option `optimset('Display', 'iter')` gives additional information on the operation executed in each step (reflection, expansion, contraction). For the starting point $w^{(0)} = (0,0,0)^T$, the regularization parameter $\lambda = 1$ and the tolerance of 10^{-7} we need 168 iteration steps, of which we will obviously only list a few (rounded to eight decimal places):

k	$w^{(k)}$			$f(w^{(k)})$
0	0	0	0	8.31776617
15	−0.11716667	0.06591667	0.03558333	7.57510570
30	−0.31787421	0.17843116	0.09626648	7.17602741
45	−0.46867136	0.56801005	0.01999832	5.90496742
60	−0.48864968	0.77331672	−0.05325103	5.68849106
75	−0.49057101	0.77096765	−0.05059686	5.68818353
90	−0.50062724	0.76632985	0.00262412	5.68256597
110	−0.50501507	0.75304960	0.07187413	5.67970021
135	−0.50485529	0.75325130	0.07065917	5.67969938
168	−0.50485551	0.75324905	0.07066912	5.67969938

Our numerical tests were run on an *Apple iMac G5* with *Matlab*® 7.0.

The given distribution of points and the obtained 'separating' line

$$w_1 x_1 + w_2 x_2 + w_3 = 0$$

for $w := w^{(168)}$ look like this:

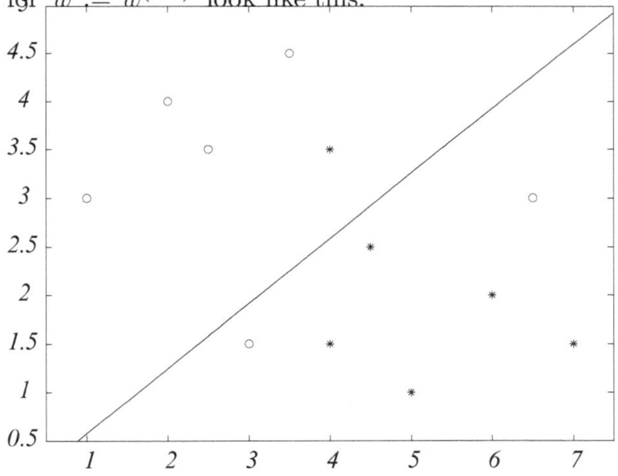

The Shor Ellipsoid Method[2]

SHOR's ellipsoid method is a *localization method* which has attracted great attention mainly because of its applications to *linear optimization*. However,

[2] Cf. [Shor], ch. 8.5, pp. 86–91.

we will not discuss the corresponding algorithm published by KHACHIYAN in 1979.

Starting with an ellipsoid $\mathcal{E}^{(0)}$ containing 'the' minimal point we are looking for, we will consider a segment of it containing the point and then the corresponding enclosing ellipsoid of minimal volume in each iteration step.

Since this method uses *ellipsoids*, we will give some preliminary remarks about them:

Let $n \in \mathbb{N}$ be fixed. For $r > 0$ and $x_0 \in \mathbb{R}^n$

$$B_{x_0}^r := \left\{x \in \mathbb{R}^n : \|x - x_0\|_2 \le r\right\} = \left\{x \in \mathbb{R}^n : (x - x_0)^T(x - x_0) \le r^2\right\}$$

is the *ball* with center x_0 and radius $r > 0$.

For a symmetric positive definite matrix $P \ \left(\in \mathbb{R}^{n \times n}\right)$

$$\mathcal{E} := \mathcal{E}(P, x_0) := \left\{x \in \mathbb{R}^n \mid (x - x_0)^T P^{-1}(x - x_0) \le 1\right\}$$

describes an *ellipsoid* with *center* x_0.

The special case $P := r^2 I$ gives $\mathcal{E} = B_{x_0}^r$.

For $a, b > 0$ and $\xi_0, \eta_0 \in \mathbb{R}$ we get, for example,

$$\left(\frac{\xi - \xi_0}{a}\right)^2 + \left(\frac{\eta - \eta_0}{b}\right)^2 \le 1$$

with

$$P := \begin{pmatrix} a^2 & 0 \\ 0 & b^2 \end{pmatrix}, \text{ hence, } P^{-1} = \begin{pmatrix} a^{-2} & 0 \\ 0 & b^{-2} \end{pmatrix},$$

in the form

$$(\xi - \xi_0, \eta - \eta_0)\, P^{-1} \begin{pmatrix} \xi - \xi_0 \\ \eta - \eta_0 \end{pmatrix} \le 1\,.$$

$\mathcal{E}$ is the image of the unit ball

$$\mathrm{UB} := B_0^1 = \left\{u \in \mathbb{R}^n : \|u\|_2 \le 1\right\}$$

under the mapping h defined by

$$h(u) := P^{1/2}u + x_0 \text{ for } u \in \mathbb{R}^n,$$

since for $u \in \mathrm{UB}$ and $x := h(u) = P^{1/2}u + x_0$ it holds that

$$\begin{aligned}(x - x_0)^T P^{-1}(x - x_0) &= (P^{1/2}u)^T P^{-1} P^{1/2} u = \left\langle u, P^{1/2} P^{-1} P^{1/2} u\right\rangle \\ &= \langle u, u\rangle \le 1\end{aligned}$$

and back. We obtain the *volume* $\mathrm{vol}(\mathcal{E}(P,x_0))$ of the above ellipsoid with the help of the Transformation Theorem:

$$\mathrm{vol}(\mathcal{E}(P,x_0)) = \int\limits_{\mathcal{E}} dx = \det P^{1/2} \int\limits_{\mathrm{UB}} du = \omega_n \sqrt{\det P},$$

where ω_n denotes the volume of the n-dimensional unit ball (see, for example, [Co/Jo], p. 459).

In addition we will need the following two tools:

Remark 3.1.1

For $\alpha \in \mathbb{R}$ and $u \in \mathbb{R}^n \setminus \{0\}$ it holds that

$$\det\left(I - \alpha u u^T\right) = 1 - \alpha u^T u.$$

Proof: $(I - \alpha u u^T)u = (1 - \alpha u^T u)u$ and $(I - \alpha u u^T)v = v$ hold for $v \in \mathbb{R}^n$ with $v \perp u$. With that the matrix $I - \alpha u u^T$ has the eigenvalue 1 of multiplicity $(n-1)$ and the simple eigenvalue $1 - \alpha u^T u$. □

Remark 3.1.2

For $n \geq 2$ it holds that $\left(\frac{n}{n+1}\right)^{n+1}\left(\frac{n}{n-1}\right)^{n-1} < \exp\left(-\frac{1}{n}\right)$.

Proof: The assertion is equivalent to

$$\left(\frac{n+1}{n}\right)^{n+1}\left(\frac{n-1}{n}\right)^{n-1} > \exp\left(\frac{1}{n}\right)$$

and hence to

$$(n+1)\ln(1+1/n) + (n-1)\ln(1-1/n) > 1/n.$$

Using the power series expansion $\ln(1+x) = \sum_{k=1}^{\infty}(-1)^{k+1}\frac{1}{k}x^k$ for $-1 < x \leq 1$, we get

$$\begin{aligned}
\ell hs &= \sum_{k=1}^{\infty}(-1)^{k+1}\frac{n+1}{k\,n^k} - \sum_{k=1}^{\infty}\frac{n-1}{k\,n^k} \\
&= -\sum_{\kappa=1}^{\infty}\frac{2n}{2\kappa n^{2\kappa}} + \sum_{\kappa=1}^{\infty}\frac{2}{(2\kappa-1)\,n^{2\kappa-1}} \\
&= -\sum_{\kappa=1}^{\infty}\frac{1}{\kappa n^{2\kappa-1}} + \sum_{\kappa=1}^{\infty}\frac{2}{(2\kappa-1)\,n^{2\kappa-1}} \\
&= \sum_{\kappa=1}^{\infty}\frac{1}{(2\kappa-1)\kappa}\,\frac{1}{n^{2\kappa-1}} > \frac{1}{n}.
\end{aligned}$$

□

Suppose that $f\colon \mathbb{R}^n \longrightarrow \mathbb{R}$ is a given *differentiable convex* function with a minimal point $x^* \in \mathbb{R}^n$, hence, $f(x) \ge f(x^*)$ for all $x \in \mathbb{R}^n$. As is generally known,

$$f(x) \ge f(x_0) + f'(x_0)(x - x_0)$$

holds for all $x, x_0 \in \mathbb{R}^n$ (see chapter 2). From $f'(x_0)(x - x_0) > 0$ it follows that $f(x) > f(x_0)$. Therefore, each time x^* lies in the half space

$$\big\{x \in \mathbb{R}^n \mid f'(x_0)(x - x_0) \le 0\big\}.$$

The following *method* is based on this observation:

Locate x^* in an ellipsoid $\mathcal{E}^{(0)}$. If we have an ellipsoid $\mathcal{E}^{(k)}$ for $k \in \mathbb{N}_0$ containing x^*, with center $x^{(k)}$, $x^* \in \mathcal{E}^{(k)} \cap \big\{x \in \mathbb{R}^n \mid g_k^T\big(x - x^{(k)}\big) \le 0\big\} =: \mathcal{S}_k$ holds for $g_k := \nabla f\big(x^{(k)}\big)$ — as stated above. (Other classes of so-called *cutting-plane methods* will be considered in chapter 8.)

Now choose $\mathcal{E}^{(k+1)}$ as the *hull ellipsoid*, that is, the ellipsoid of minimal volume containing $\mathcal{S}_k$.[3]

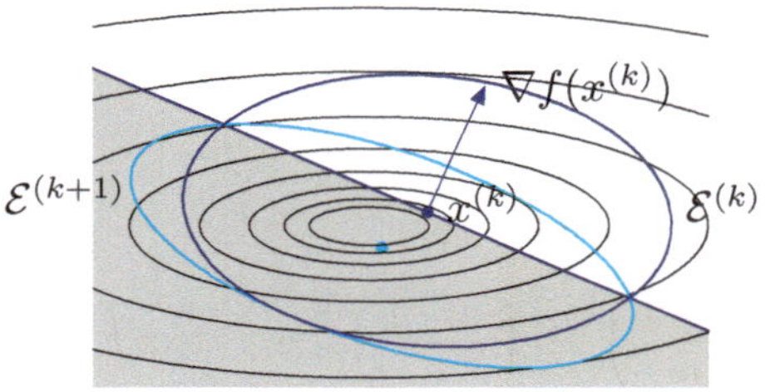

Each of these ellipsoids can be written in the form

$$\mathcal{E}^{(k)} = \Big\{x \in \mathbb{R}^n \mid (x - x^{(k)})^T A_k^{-1}(x - x^{(k)}) \le 1\Big\}$$

with a symmetric positive definite matrix A_k. With

$$\widehat{g_k} := \frac{g_k}{\sqrt{g_k^T A_k\, g_k}} \quad \text{and} \quad b_k := A_k\, \widehat{g_k}$$

for a nonzero gradient $\nabla f(x^{(k)})$ we note the following *updating formulae* for the ellipsoid $\mathcal{E}^{(k+1)}$ (without proof[4]):

$$\begin{aligned} x^{(k+1)} &:= x^{(k)} - \frac{1}{n+1}\, b_k \\ A_{k+1} &:= \frac{n^2}{n^2-1}\Big(A_k - \frac{2}{n+1}\, b_k b_k^T\Big) \end{aligned}$$

[3] The unique ellipsoid of minimal volume containing a given convex body is called a LÖWNER–JOHN ellipsoid.

[4] cf. [John]

For the determinant we get from the above (considering remark 3.1.1):

$$\det A_{k+1} = \left(\frac{n^2}{n^2-1}\right)^n \det A_k \left(1 - \frac{2}{n+1} b_k^T A_k^{-1} b_k\right)$$

By definition of b_k it follows that $b_k^T A_k^{-1} b_k = (A_k^{-1/2} b_k)^T A_k^{-1/2} b_k = 1$ and hence with remark 3.1.2

$$\frac{\det A_{k+1}}{\det A_k} = \left(\frac{n^2}{n^2-1}\right)^n \frac{n-1}{n+1} = \left(\frac{n}{n+1}\right)^{n+1} \left(\frac{n}{n-1}\right)^{n-1} < \exp\left(-\frac{1}{n}\right).$$

Properties of the Ellipsoid Method:

- If $n = 1$, it is identical to the *bisection method.*
- If $\mathcal{E}^{(k)}$ and g_k are given, we obtain $\mathcal{E}^{(k+1)}$ with very simple updating formulae. However, the *direct* execution of this *rank 1-update* can pose numerical problems, for example, we might lose the positive definiteness. We will get to know later on how to execute this *numerically stable.*
- vol $(\mathcal{E}^{(k+1)}) \le \exp(-\frac{1}{2n})$ vol$(\mathcal{E}^{(k)})$
  $n = 2:$ 0.7788
  $n = 3:$ 0.8465
- *Note:* The ellipsoid method is *not* a descent method.
- The ellipsoid method can be generalized to nondifferentiable convex objective functions and can also be applied to *problems with convex constraints.*

With the help of the estimate

$$\begin{aligned} f(x^*) &\ge f(x^{(k)}) + f'(x^{(k)})(x^* - x^{(k)}) \\ &\ge f(x^{(k)}) + \inf_{x \in \mathcal{E}^{(k)}} f'(x^{(k)})(x - x^{(k)}) \\ &= f(x^{(k)}) - \sqrt{g_k^T A_k g_k} \end{aligned}$$

we obtain the following *stop-criteria*:

- $\sqrt{g_k^T A_k g_k} < \varepsilon\left(|f(x^{(k)})| + 1\right)$
- Or: $U_k - L_k < \varepsilon\left(|U_k| + 1\right)$, where
  $U_k := \min_{j \le k} f(x^{(j)})$ and $L_k := \max_{j \le k} \left(f(x^{(j)}) - \sqrt{g_j^T A_j g_j}\right)$.

3.2 Descent Methods with Line Search

In the following we will look at minimization methods which, beginning with a starting point $x^{(0)}$, iteratively construct a sequence of approximations $(x^{(k)})$ such that

$$f(x^{(k)}) > f(x^{(k+1)}) \text{ for } k \in \mathbb{N}_0 .$$

For differentiable objective functions they are all based on the idea of choosing a *descent direction* d_k, that is, $f'(x^{(k)})\, d_k < 0$, at the current point $x^{(k)}$ and then determining a corresponding *step size* $\lambda_k > 0$ such that

$$f\big(x^{(k)} + \lambda_k d_k\big) = \min_{\lambda \geq 0} f\big(x^{(k)} + \lambda d_k\big) \tag{1}$$

or weaker — in a sense to be stated more precisely —

$$f\big(x^{(k)} + \lambda_k d_k\big) \approx \min_{\lambda \geq 0} f\big(x^{(k)} + \lambda d_k\big).$$

$x^{(k+1)} := x^{(k)} + \lambda_k d_k$ yields a better approximation. In the case of (1) it holds that

$$0 = \frac{d}{d\lambda} f(x^{(k)} + \lambda d_k)\big|_{\lambda=\lambda_k} = f'(x^{(k+1)})\, d_k .$$

This *one-dimensional minimization* is called *'line search'*. It has turned out that putting a lot of effort into the computation of the exact solution of (1) is not worthwhile. The exact search is often too laborious and costly. Therefore, a more tolerant search strategy for *efficient step sizes* is useful. The exact step size is only used for theoretical considerations and in the special case of quadratic optimization problems.

Before addressing specific aspects of the so-called 'inexact line search', we will discuss the *choice of suitable descent directions.*

The following example shows that — even in the simplest special case — the step sizes have to be chosen with care:

Example 2

$f(x) := x^2$ for $x \in \mathbb{R}$, $x^{(0)} := 1$, $d_k := -1$ and $\lambda_k := 2^{-k-2}$ for $k \in \mathbb{N}_0$. We get

$$x^{(k+1)} := x^{(k)} + \lambda_k d_k = x^{(k)} - \lambda_k = x^{(0)} - \sum_{\kappa=0}^{k} \lambda_\kappa$$

and then

$$x^{(k+1)} = 1 - \frac{1}{4} \sum_{\kappa=0}^{k} \Big(\frac{1}{2}\Big)^{\kappa} = \frac{1}{2} + \Big(\frac{1}{2}\Big)^{k+2}.$$

In this way it follows that

$$0 < x^{(k+1)} < x^{(k)}, \quad \text{hence,} \quad f(x^{(k+1)}) < f(x^{(k)}),$$

with $x^{(k)} \longrightarrow 1/2$, $f(x^{(k)}) \longrightarrow 1/4$. However, the minimum lies at $x^* = 0$ and has the value 0.

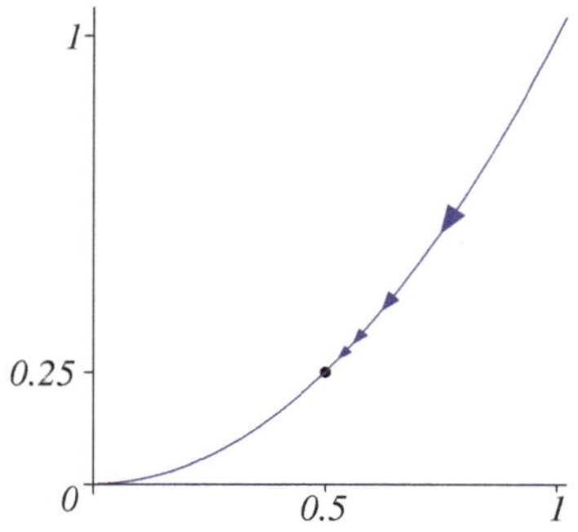

Here the step sizes have been chosen too small!

The choice $d_k := (-1)^{k+1}$ with $\lambda_k := 1+3/2^{k+2}$ gives $x^{(k)} = \frac{1}{2}(-1)^k(1+\frac{1}{2^k})$.

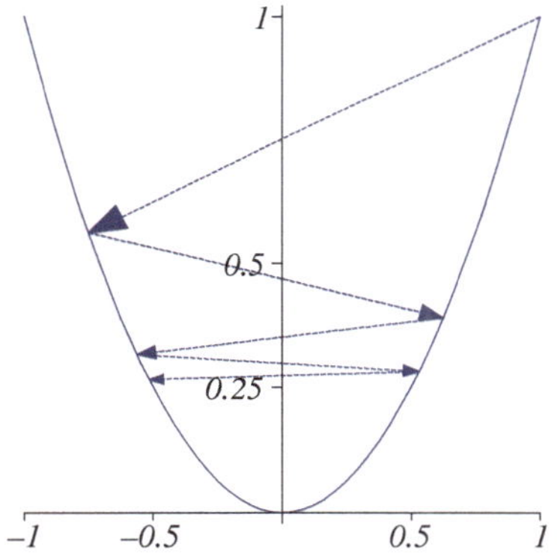

Here the step sizes have been chosen too large! ◁

Coordinatewise Descent Methods

Cyclic Control: We cyclically go through the n coordinate axes $x_1, \ldots, x_n$. In the k-th iteration of a cycle we fix all except the k-th variable and minimize the objective function. Then we repeat the cycle (Gauss–Seidel method).

Relaxation Control: In this variant of the *alternating variable method* we choose the number j of the coordinate axes such that $\left|\frac{\partial f}{\partial x_j}(x^{(k)})\right|$ becomes maximal (Southwell).

Gradient Method

The *gradient method* — also called *steepest descent method* — was already proposed by CAUCHY in 1847 (see section 1.2). The choice of the negative gradient as the descent direction is based on the observation that

$$\frac{d}{d\lambda} f(x^{(k)} + \lambda d)\big|_{\lambda=0} = f'(x^{(k)})d$$

holds and that the minimization problem

$$\min_{\|d\|_2=1} f'(x^{(k)})d$$

for a nonzero gradient $\nabla f(x^{(k)})$ has the solution

$$d_k := -\frac{1}{\|\nabla f(x^{(k)})\|_2} \nabla f(x^{(k)})$$

— for example, because of the CAUCHY–SCHWARZ inequality. This *locally optimal choice* of d will turn out not to be *globally* very advantageous.

The way the gradient method works can very easily be examined in the special case of a *convex quadratic objective function* f defined by

$$f(x) := \frac{1}{2} x^T A x + b^T x$$

with a symmetric positive definite matrix A and a vector $b \in \mathbb{R}^n$. For the gradient it holds that

$$\nabla f(x) = Ax + b =: g(x).$$

In this case the step size can easily be determined by means of 'exact line search':

$$\begin{aligned} \frac{d}{d\lambda} f(x^{(k)} + \lambda(-g_k)) &= (A(x^{(k)} + \lambda(-g_k)) + b)^T(-g_k) \\ &= -(g_k - \lambda A g_k)^T g_k = 0 \iff \lambda = \frac{g_k^T g_k}{g_k^T A g_k} > 0 \end{aligned}$$

Here, $g_k := g(x^{(k)})$ denotes the gradient of f at the point $x^{(k)}$. As an improved approximation we obtain

$$x^{(k+1)} = x^{(k)} - \frac{g_k^T g_k}{g_k^T A g_k} g_k.$$

Example 3

We want to minimize the convex quadratic objective function f given by

$$f(x_1, x_2) := \frac{1}{2}(x_1^2 + 9x_2^2) = \frac{1}{2} x^T A x \text{ with } A := \begin{pmatrix} 1 & 0 \\ 0 & 9 \end{pmatrix}.$$

For the gradient it holds that $\nabla f(x) = Ax$. Choosing $x^{(0)} := (9,1)^T$ as the starting vector and using exact 'line search', we prove for the sequence $(x^{(k)})$:

$$x^{(k)} = 0.8^k \begin{pmatrix} 9 \\ (-1)^k \end{pmatrix}$$

This is clear for $k = 0$. For the induction step from k to $k+1$ we obtain

$$g_k = 0.8^k \cdot 9 \begin{pmatrix} 1 \\ (-1)^k \end{pmatrix}, \quad Ag_k = 0.8^k \cdot 9 \begin{pmatrix} 1 \\ 9\,(-1)^k \end{pmatrix}$$

as well as

$$\lambda_k = \frac{g_k^T g_k}{g_k^T A g_k} = \frac{2}{10} \,.$$

Together with

$$x^{(k)} - \frac{1}{5} g_k = 0.8^k \begin{pmatrix} 9 - \frac{9}{5} \\ (-1)^k (1 - \frac{9}{5}) \end{pmatrix} = 0.8^{k+1} \begin{pmatrix} 9 \\ (-1)^{k+1} \end{pmatrix} = x^{(k+1)}$$

this gives the result.

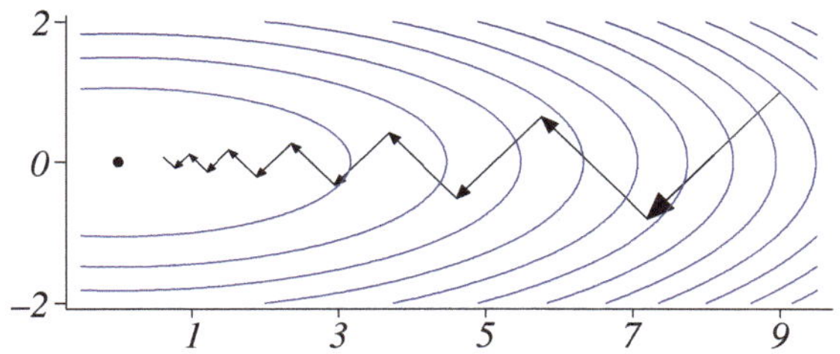

◁

Kantorovich's Inequality

Let A be a symmetric positive definite matrix with the eigenvalues $0 < \alpha_1 \le \cdots \le \alpha_n$. Then for any $x \in \mathbb{R}^n \setminus \{0\}$ it holds that

$$\frac{x^T A x}{x^T x} \cdot \frac{x^T A^{-1} x}{x^T x} \le \frac{(\alpha_1 + \alpha_n)^2}{4 \alpha_1 \alpha_n} \,.$$

Proof: With the corresponding orthonormal eigenvectors $z_1, \ldots, z_n$ we presume the representation $x = \sum_{\nu=1}^{n} \xi_\nu z_\nu$ and obtain

$$\frac{x^T A x}{x^T x} \cdot \frac{x^T A^{-1} x}{x^T x} = \frac{\sum_{\nu=1}^{n} \alpha_\nu \xi_\nu^2}{\sum_{\mu=1}^{n} \xi_\mu^2} \cdot \frac{\sum_{\nu=1}^{n} \frac{1}{\alpha_\nu} \xi_\nu^2}{\sum_{\mu=1}^{n} \xi_\mu^2}$$

$$= \Big(\sum_{\nu=1}^{n} u_\nu \alpha_\nu \Big) \Big(\sum_{\nu=1}^{n} \frac{u_\nu}{\alpha_\nu} \Big)$$

with $u_\nu := \xi_\nu^2 / \sum_{\mu=1}^{n} \xi_\mu^2$. Here $u_\nu \ge 0$ holds with $\sum_{\nu=1}^{n} u_\nu = 1$. For the weighted sum $\alpha := \sum_{\nu=1}^{n} u_\nu \alpha_\nu$ we have $\alpha_1 \le \alpha \le \alpha_n$, and from the convexity of the function $\mathbb{R}_{++} \ni t \mapsto 1/t$ the inequality

$$\frac{1}{t} \le \frac{1}{\alpha_1} + \frac{1}{\alpha_n} - \frac{t}{\alpha_1 \alpha_n} \quad \text{for } \alpha_1 \le t \le \alpha_n$$

follows. Hence,

$$\sum_{\nu=1}^{n} \frac{u_\nu}{\alpha_\nu} \le \sum_{\nu=1}^{n} u_\nu \left(\frac{1}{\alpha_1} + \frac{1}{\alpha_n} - \frac{\alpha_\nu}{\alpha_1 \alpha_n} \right) = \frac{1}{\alpha_1} + \frac{1}{\alpha_n} - \frac{\alpha}{\alpha_1 \alpha_n} .$$

This gives

$$\begin{aligned} \left(\sum_{\nu=1}^{n} u_\nu \alpha_\nu \right) \left(\sum_{\nu=1}^{n} \frac{u_\nu}{\alpha_\nu} \right) &\le \alpha \left(\frac{1}{\alpha_1} + \frac{1}{\alpha_n} - \frac{\alpha}{\alpha_1 \alpha_n} \right) \\ &\le \max_{\alpha_1 \le t \le \alpha_n} t \left(\frac{1}{\alpha_1} + \frac{1}{\alpha_n} - \frac{t}{\alpha_1 \alpha_n} \right) \overset{\checkmark}{=} \frac{(\alpha_1 + \alpha_n)^2}{4 \alpha_1 \alpha_n} . \qquad \square \end{aligned}$$

We revert to the above convex quadratic objective function

$$f(x) = \frac{1}{2} x^T A x + b^T x$$

with the minimal point x^*, that is, $\nabla f(x^*) = Ax^* + b = 0$, and introduce the *error function*

$$E(x) := \frac{1}{2} (x - x^*)^T A (x - x^*) .$$

Since $E(x) = f(x) + \frac{1}{2} x^{*T} A x^* = f(x) - f(x^*)$, the functions E and f differ only in a constant. Using the notation introduced above

$$x^{(k+1)} := x^{(k)} - \lambda_k g_k , \; \lambda_k := \frac{g_k^T g_k}{g_k^T A g_k} ,$$

we obtain:

Lemma 3.2.1

$$E\big(x^{(k+1)}\big) = \left(1 - \frac{(g_k^T g_k)^2}{(g_k^T A g_k)(g_k^T A^{-1} g_k)} \right) E\big(x^{(k)}\big) \le \left(\frac{c(A) - 1}{c(A) + 1} \right)^2 E\big(x^{(k)}\big)$$

Hence, with regard to the norm[5] $\| \; \|_A$ *defined by* $\|x\|_A := \sqrt{\langle x, Ax \rangle}$ *it holds that*

[5] cf. exercise 4

$$\|x^{(k+1)} - x^*\|_A \leq \frac{c(A)-1}{c(A)+1} \cdot \|x^{(k)} - x^*\|_A .$$

Proof: Here, for the condition number $c(A) := \|A\|_2 \cdot \|A^{-1}\|_2$ of A it holds that $c(A) = \alpha_n/\alpha_1$. Setting $y_k := x^{(k)} - x^*$, we obtain

$$y_{k+1} = y_k - \lambda_k g_k \quad \text{and} \quad Ay_k = A(x^{(k)} - x^*) = g_k ;$$

with that the first part follows

$$\frac{E(x^{(k)}) - E(x^{(k+1)})}{E(x^{(k)})} = \frac{\lambda_k g_k^T A y_k - \frac{1}{2}\lambda_k^2 g_k^T A g_k}{\frac{1}{2} y_k^T A y_k} = \frac{\left(g_k^T g_k\right)^2}{(g_k^T A g_k)(g_k^T A^{-1} g_k)} .$$

The second part follows with the Kantorovich inequality:

$$1 - \frac{\left(g_k^T g_k\right)^2}{(g_k^T A g_k)(g_k^T A^{-1} g_k)} \leq 1 - \frac{4\alpha_1 \alpha_n}{(\alpha_1 + \alpha_n)^2} = \left(\frac{\alpha_n - \alpha_1}{\alpha_n + \alpha_1}\right)^2 \qquad \square$$

In conclusion we can note the following significant *disadvantages of the gradient method*:

- Slow convergence in the case of strong eccentricity (ill condition)
- For quadratic objective functions the method is not necessarily finite (cf. example 3 on page 100 f).

Example 4

In order to also apply the *steepest descent method* to our framework example (see section 3.0), we utilize the function fminunc — with the option optimset('HessUpdate', 'steepdesc') — from the *Matlab® Optimization Toolbox.* For the starting point $w^{(0)} = (0,0,0)^T$ and the regularization parameter $\lambda = 1$ we need 66 iteration steps to reach the tolerance of 10^{-7}. Again, we will only list a few of them (rounded to eight decimal places):

k	$w^{(k)}$			$f(w^{(k)})$	$\|\nabla f(w^{(k)})\|_\infty$
0	0	0	0	8.31776617	6.00000000
2	−0.22881147	0.30776833	0.03037985	6.45331017	2.54283988
4	−0.54704228	0.81975751	0.08597370	5.69424031	0.29568040
6	−0.50630235	0.75340273	0.07669618	5.67972415	0.00699260
14	−0.50508694	0.75290067	0.07254714	5.67970141	0.00194409
24	−0.50491650	0.75315790	0.07116240	5.67969952	0.00051080
34	−0.50487153	0.75322512	0.07079865	5.67969939	0.00013416
44	−0.50485972	0.75324278	0.07070312	5.67969938	0.00003524
54	−0.50485662	0.75324742	0.07067802	5.67969938	0.00000925
66	−0.50485573	0.75324874	0.07067088	5.67969938	0.00000186

Chapter 3

Requirements on the Step Size Selection

When choosing a suitable step size λ_k , we try to find a happy medium between the maximal demand of exact line search — compare (1) — and the minimum demand $f(x^{(k)} + \lambda_k d_k) < f(x^{(k)})$. A λ_k fulfilling (1) is sometimes called *ray minimizing.* As a weakening the smallest local minimal point or even the smallest stationary point on the ray $\{x^{(k)} + \lambda d_k \mid \lambda \geq 0\}$ are often considered.

More tolerant conditions are based on, for example, GOLDSTEIN (1965), WOLFE (1963), ARMIJO (1966) and POWELL (1976). For the following discussions suppose that in general

$$-g_k^T d_k \geq \gamma \|d_k\|_2 \cdot \|g_k\|_2 \tag{2}$$

holds for $k \in \mathbb{N}_0$ with a fixed $0 < \gamma \leq 1$. Geometrically speaking this means: The descent direction d_k must not be 'almost orthogonal' to the gradient $g_k := \nabla f(x^{(k)})$ (less than the right angle, uniformly in k).

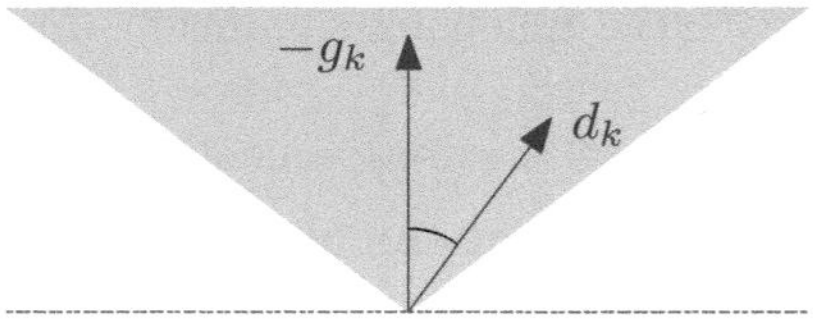

GOLDSTEIN **Conditions:** λ_k as a positive solution of two inequalities:

Suppose that

$$f(x^{(k)} + \lambda_k d_k) \leq f(x^{(k)}) + \alpha \lambda_k \underbrace{g_k^T d_k}_{<0} \tag{3}$$

holds with an $\alpha \in (0, 1)$.

Hence, the step size λ_k should only guarantee a *sufficient descent* (sufficiently large decrease of the value of the objective function), that is, $x^{(k+1)}$ should not be *too* far from $x^{(k)}$, hence, not be too close to the 'right edge' of the descent region. Furthermore suppose that

$$f(x^{(k)} + \lambda_k d_k) \geq f(x^{(k)}) + \beta \lambda_k g_k^T d_k \tag{4}$$

holds with a $\beta \in (\alpha, 1)$. This should guarantee a minimum size of the step size, that is, $x^{(k+1)}$ should not be close to the 'left edge'. (Often we choose $\alpha = 1/4$ and $\beta = 3/4$.)

In this kind of problem $k \in \mathbb{N}_0$, $x^{(k)}$ and d_k are fixed. Therefore, we consider the real-valued function φ of one real-valued variable defined by

$$\varphi(\lambda) := \varphi_k(\lambda) := f\big(x^{(k)} + \lambda d_k\big) \text{ for } \lambda \in \mathbb{R}_+ ,$$

taking into account the observations

$$\varphi(0) = f\big(x^{(k)}\big), \quad \varphi'(\lambda) = f'\big(x^{(k)} + \lambda d_k\big)\, d_k ,$$

$$\text{in particular } \varphi'(0) = f'\big(x^{(k)}\big)\, d_k = g_k^T d_k < 0 .$$

With that, conditions (3) and (4) can be written more clearly:

$$\varphi(\lambda_k) \le \varphi(0) + \alpha\lambda_k\varphi'(0) \quad \text{and} \quad \varphi(\lambda_k) \ge \varphi(0) + \beta\lambda_k\varphi'(0) .$$

The idea to demand (3) and (4) suggests itself: If we consider the difference quotient defined by

$$q(\lambda) := \frac{\varphi(\lambda) - \varphi(0)}{\lambda - 0}$$

for $\lambda > 0$, we have

$$q(\lambda) \longrightarrow a := \varphi'(0) < 0 \quad (\lambda \longrightarrow 0) .$$

The condition $\beta a \le q(\lambda_k) \le \alpha a$ precisely means $\varphi(\lambda_k) \le \varphi(0) + \alpha\lambda_k\varphi'(0)$ and $\varphi(\lambda_k) \ge \varphi(0) + \beta\lambda_k\varphi'(0)$.

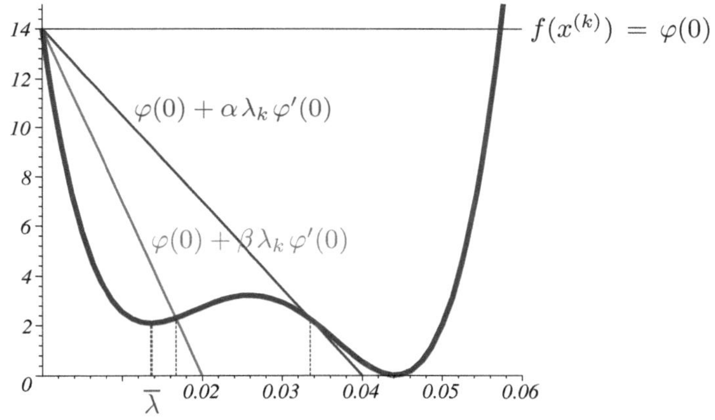

Disadvantage of (4):

In the plotted example the 'first' minimizing $\overline{\lambda}$ lies outside the interval in which conditions (3) and (4) are met.

$$f'(x^{(k+1)})\, d_k \approx \frac{f(x^{(k+1)}) - f(x^{(k)})}{\lambda_k} \overset{!}{\ge} \beta f'(x^{(k)})\, d_k$$

suggests a modification of (4): With a $\beta \in (\alpha, 1)$ suppose that

$$f'(x^{(k+1)})\, d_k \ge \beta g_k^T d_k \text{ respectively } \varphi'(\lambda_k) \ge \beta\varphi'(0) \qquad (5)$$

holds. (3), (5) are called WOLFE–POWELL *conditions.* (5) gives that the derivative grows sufficiently. The step size does not become too small $\big(x^{(k+1)}$ does not lie too close to $x^{(k)}\big)$. According to [Bonn], these conditions are the *"most intelligent in the current state of the art."* Now the lower estimate will always include the minimizing $\overline{\lambda}$.

With a sharpening of (5) we can force λ_k to lie closer to a local minimum $\overline{\lambda}$:

$$\Big|f'(x^{(k+1)})\,d_k\Big| \;\le\; -\beta\, g_k^T d_k \tag{6}$$

For example, for $\beta = 0.9$ we obtain an 'inexact line search'; for $\beta = 0.1$, however, we conduct an almost 'exact line search'.

We will now examine in which case the WOLFE–POWELL conditions always have a solution:

Lemma 3.2.2

Suppose that $f \in C^2(\mathbb{R}^n)$ *and* $0 < \alpha \le \beta < 1$, $0 < \gamma \le 1$. *For* $x \in \mathbb{R}^n$ *with* $g := \nabla f(x) \neq 0$ *and* $d \in \mathbb{R}^n \setminus \{0\}$ *let also*

$$-g^T d \ge \gamma\,\|g\|_2 \|d\|_2 \quad \textit{and} \quad \inf\{f(x+\lambda d) \mid \lambda \ge 0\} > -\infty\,.$$

Then it holds that:

1) There exists a $\lambda > 0$ *such that*

$$f(x+\lambda d) \;\le\; f(x) + \alpha\lambda g^T d \tag{7}$$

$$\textit{and} \quad f'(x+\lambda d)\,d \;\ge\; \beta g^T d\,. \tag{8}$$

2) Let $\overline{\lambda}$ *be the smallest of all numbers* $\lambda > 0$ *fulfilling (7) and (8). Then we have*

$f'(x+td)d \,<\, f'(x+\overline{\lambda}d)\,d$ *for all* $0 \le t < \overline{\lambda}$, *and* $t \longmapsto f(x+td)$ *is strictly antitone (order-reversing) on the interval* $[0, \overline{\lambda}]$.

Setting $L := \max_{0 \le t \le \overline{\lambda}} \|\nabla^2 f(x+td)\|_2$, *it holds for all positive* λ *fulfilling (7) and (8) that*

$$f(x+\lambda d) \;\le\; f(x) - \frac{\alpha(1-\beta)\gamma^2}{L}\,\|g\|_2^2\,.$$

Here, the corresponding matrix norm is denoted by the same symbol $\|\ \ \|_2$.

Proof: 1) The function $\varphi\colon \mathbb{R} \longrightarrow \mathbb{R}$ given by $\varphi(t) := f(x+td)$ is twice continuously differentiable and, because of $\varphi'(t) = f'(x+td)\,d$, it holds that

$\varphi'(0) = g^T d \le -\gamma \|g\|_2 \|d\|_2 < 0$.

First of all, we will show that there exists a $\lambda > 0$ which fulfills (8). The latter is equivalent to $\varphi'(\lambda) \ge \beta\varphi'(0)$. If $\varphi'(\lambda) < \beta\varphi'(0)\,(< 0)$ for all $\lambda > 0$, then

$$\varphi(t) = \varphi(0) + \int_0^t \varphi'(\lambda)\,d\lambda \le \underbrace{\varphi(0) + t\beta\varphi'(0)}_{\longrightarrow\, -\infty\ (t\to\infty)},$$

which would be a contradiction to the assumption that φ is bounded from below in $\mathbb{R}_+$. Because of

$$g^T d = \varphi'(0) < \beta\varphi'(0) \le \varphi'(\lambda),$$

the continuity of φ' results in a minimal $\overline{\lambda} > 0$ such that $\varphi'(\overline{\lambda}) = \beta\varphi'(0) < 0$, hence,

$$\varphi'(t) < \beta\varphi'(0) \quad \text{for} \quad 0 \le t < \overline{\lambda}.$$

$\overline{\lambda}$ then also fulfills (7):

$$\begin{aligned} \varphi(\overline{\lambda}) &= \varphi(0) + \varphi'(\tau)\overline{\lambda} \quad \text{with a suitable } \tau \in (0, \overline{\lambda}) \\ &\le \varphi(0) + \beta\overline{\lambda}\underbrace{\varphi'(0)}_{<0} \le \varphi(0) + \alpha\overline{\lambda}\varphi'(0). \end{aligned}$$

2) The first two of the assertions that still have to be proven are a direct result of the above observations. Only the following estimate remains to prove: From $\varphi''(t) = d^T \nabla^2 f(x + td)\,d$ follows

$$|\varphi''(t)| \le \|\nabla^2 f(x + td)\|_2 \|d\|_2^2 \le L\,\|d\|_2^2 \ \text{ for } 0 \le t \le \overline{\lambda}.$$

Since φ' is not constant, L has to be positive. Then it holds that

$$(\beta - 1)\,\varphi'(0) = \varphi'(\overline{\lambda}) - \varphi'(0) = \int_0^{\overline{\lambda}} \varphi''(t)\,dt \le \overline{\lambda} L\,\|d\|_2^2.$$

This yields a lower bound for $\overline{\lambda}$:

$$\overline{\lambda} \ge -\frac{(1-\beta)\,\varphi'(0)}{L\,\|d\|_2^2} = -\frac{(1-\beta)\,g^T d}{L\,\|d\|_2^2}$$

If λ meets conditions (7) and (8), it is obvious that $\lambda \ge \overline{\lambda}$; consequently, using $\varphi'(0)^2 = (-g^T d)^2 \ge \gamma^2 \|g\|_2^2\,\|d\|_2^2$, we get

$$\begin{aligned} \varphi(\lambda) - \varphi(0) &\underset{(7)}{\le} \alpha\lambda\underbrace{\varphi'(0)}_{<0} \le \alpha\overline{\lambda}\varphi'(0) \\ &\le -\frac{\alpha(1-\beta)\,\varphi'(0)^2}{L\,\|d\|_2^2} \le -\frac{\alpha(1-\beta)\gamma^2}{L}\,\|g\|_2^2. \end{aligned}$$ □

Chapter 3

These observations are the basis for the following

Algorithm

Let $0 < \alpha \le \beta < 1$, $0 < \gamma \le 1$ and a starting vector $x^{(0)} \in \mathbb{R}^n$ be given.

Iteration Step for $k \in \mathbb{N}_0$:

If $g_k = 0$: STOP; $x^{(k)}$ is a stationary point of f. *Else:*

a) Choose a *descent direction* $d_k \in \mathbb{R}^n$ such that

$$-g_k^T d_k \ge \gamma \|g_k\|_2 \|d_k\|_2 .$$

b) Calculate a *step size* $\lambda_k > 0$ such that

$$f(x^{(k+1)}) \le f(x^{(k)}) + \alpha \lambda_k g_k^T d_k \quad (9)$$

$$g_{k+1}^T d_k \ge \beta g_k^T d_k \quad (10)$$

holds for $x^{(k+1)} := x^{(k)} + \lambda_k d_k$.

Proposition 3.2.3

Suppose that $f \in C^2(\mathbb{R}^n)$, $x^{(0)} \in \mathbb{R}^n$ *and that the level set*

$$N := \{x \in \mathbb{R}^n \mid f(x) \le f(x^{(0)})\}$$

is compact. Then the above algorithm can be carried out. It will either terminate after a finite number of steps or generate a sequence $(x^{(k)})$ *such that:*

1) $f(x^{(k+1)}) < f(x^{(k)})$ *for* $k \in \mathbb{N}_0$.

2) $(x^{(k)})$ *has at least one accumulation point* $x^* \in N$.

3) Each of these accumulation points x^* *is a stationary point of f, that is,* $\nabla f(x^*) = 0$.

Proof: If there exists a $k \in \mathbb{N}_0$ such that $g_k = 0$, then $x^{(k)}$ is a stationary point. Otherwise the set $\{x^{(k)} \mid k \in \mathbb{N}_0\}$ is infinite. *1)* holds because of (9), and *2)* results from the fact that all $x^{(k)}$ are in the compact set N. In order to prove *3)*, we set $L := \max_{x \in N} \|\nabla^2 f(x)\|_2$; lemma 3.2.2 gives

$$f(x^{(k+1)}) \le f(x^{(k)}) - \frac{\alpha(1-\beta)\gamma^2}{L} \|g_k\|_2^2 \quad \text{for } k \in \mathbb{N}_0. \quad (11)$$

By construction the sequence $\big(f(x^{(k)})\big)$ is strictly antitone. Because of

$$f(x^{(k)}) \ge \min_{x \in N} f(x) > -\infty$$

it is bounded from below and hence convergent. In particular, $f(x^{(k)}) - f(x^{(k+1)}) \longrightarrow 0$ holds for $k \to \infty$. Consequently, (11) shows that (g_k) is a null sequence. The continuity of ∇f yields $\nabla f(x^*) = 0$ for each accumulation point x^* of $(x^{(k)})$. □

For remarks on the limited utility of this proposition see, for example, [Ja/St], p. 144.

Addendum:

The assertion of proposition 3.2.3 also holds if we replace conditions (9) and (10) in the algorithm with the following rule which is easy to implement:

ARMIJO **Step Size Rule**:

With a fixed $\sigma > 0$ *independent of* k *firstly choose* $\overline{\lambda}_0 \geq \sigma \|g_k\|_2 / \|d_k\|_2$. *Then calculate the smallest* $j \in \mathbb{N}_0$ *such that*

$$f(x^{(k)} + \overline{\lambda}_j d_k) \leq f(x^{(k)}) + \alpha \overline{\lambda}_j g_k^T d_k$$

holds for $\overline{\lambda}_j := \overline{\lambda}_0 / 2^j$ *and set* $\lambda_k := \overline{\lambda}_j$.

Proof: Consider once more

$$\varphi(\lambda) := f(x^{(k)} + \lambda d_k)$$

for $\lambda \in \mathbb{R}$ and $k \in \mathbb{N}_0$.

1) This iterative halving of the step size terminates after a finite number of steps; since, because of

$$\frac{\varphi(\lambda) - \varphi(0)}{\lambda} \longrightarrow \varphi'(0) < 0 \text{ for } \lambda \longrightarrow 0,$$

$\varphi(\lambda) - \varphi(0) \leq \lambda \alpha \varphi'(0)$ holds for a sufficiently small $\lambda > 0$, that is,

$$f(x^{(k)} + \lambda d_k) \leq f(x^{(k)}) + \alpha \lambda g_k^T d_k .$$

2) We will prove: With

$$L := \max_{x \in N} \left\|\nabla^2 f(x)\right\|_2 \text{ and } c := \min\left\{\frac{\alpha(1-\beta)\gamma^2}{2L}, \alpha\gamma\sigma\right\} > 0$$

it holds that $f(x^{(k)} + \overline{\lambda}_j d_k) \leq f(x^{(k)}) - c\,\|g_k\|_2^2$:

We distinguish two cases: If $j = 0$,

$$\varphi(\overline{\lambda}_0) \leq \varphi(0) + \alpha \overline{\lambda}_0 g_k^T d_k$$

holds at first. Together with the assumptions $\overline{\lambda}_0 \geq \sigma \|g_k\|_2/\|d_k\|_2$ and $g_k^T d_k \leq -\gamma \|g_k\|_2 \|d_k\|_2$ we get

$$\varphi(\overline{\lambda}_0) \leq \varphi(0) + \alpha\sigma \frac{\|g_k\|_2}{\|d_k\|_2}\Big(-\gamma\|g_k\|_2\|d_k\|_2\Big) = \varphi(0) - \alpha\gamma\sigma\|g_k\|_2^2 .$$

If $j > 0$, then the step size rule gives

$$\varphi(\overline{\lambda}_{j-1}) > \varphi(0) + \alpha\overline{\lambda}_{j-1}\varphi'(0),$$

since $j-1$ does not yet meet the condition. It follows that $2\overline{\lambda}_j = \overline{\lambda}_{j-1} \geq \overline{\lambda}$ with $\overline{\lambda}$ from lemma 3.2.2 (cf. proof of *1)*).

Hence, it holds that: $\overline{\lambda}_j \geq \frac{\overline{\lambda}}{2} \underset{\text{(p. 107)}}{\geq} -\frac{(1-\beta)\varphi'(0)}{2L\|d_k\|_2^2}$ and with that

$$\begin{aligned}\varphi(\overline{\lambda}_j) &\leq \varphi(0) + \alpha\overline{\lambda}_j\varphi'(0)\\ &\leq \varphi(0) - \frac{\alpha(1-\beta)\varphi'(0)^2}{2L\|d_k\|_2^2} \leq \varphi(0) - \frac{\alpha(1-\beta)\gamma^2}{2L}\|g_k\|_2^2 .\end{aligned}$$

□

3.3 Trust Region Methods

In contrast to line search methods, where we are looking for an appropriate step size to a given descent direction, trust region methods have a given step size (or an upper bound for it) and we have to determine an appropriate search direction.

For this method we use a 'local model' φ_k of a given twice continuously differentiable function f on the ball with center $x^{(k)}$ and radius Δ_k

$$B_k := B_{x^{(k)}}^{\Delta_k}$$

for a fixed $k \in \mathbb{N}_0$ with $f_k := f(x^{(k)})$, $g_k := \nabla f(x^{(k)}) \neq 0$ and a symmetric matrix H_k, for example, H_k as the Hessian $\nabla^2 f(x^{(k)})$,

$$f(x) \approx \varphi_k(x) := f_k + g_k^T(x - x^{(k)}) + \frac{1}{2}(x - x^{(k)})^T H_k(x - x^{(k)}) .$$

We calculate the search direction d_k as the *global* minimizer of the following problem:

$$\begin{aligned} &f_k + g_k^T d + \frac{1}{2} d^T H_k d \longrightarrow \min \\ &\|d\|_2 \leq \Delta_k \text{ respectively } \tfrac{1}{2}(d^T d - \Delta_k^2) \leq 0 . \end{aligned} \tag{12}$$

Since the function value and the value of the first derivative of f and φ_k correspond in $x^{(k)}$, we can expect a solution of (12) to give a good approximation of a minimum of f on the region B_k for sufficiently small radii Δ_k.

The idea behind this method is to determine the radius Δ_k such that we can 'trust' the model φ_k in the neighborhood B_k. Then we speak of a 'trust region'.

In order to do this, we calculate the quotient between the descent of the function f and the descent of the modeling function φ_k for a solution d_k of (12). The closer the quotient is to 1, the more 'trustworthy' the model seems to be. If the correspondence is bad, we reduce the radius Δ_k.

With the help of the KARUSH–KUHN–TUCKER conditions it follows that

$$\begin{aligned} \exists\, \lambda_k \in \mathbb{R}_+ \ (H_k + \lambda_k I)\, d_k + g_k &= 0 \\ \|d_k\|_2 \le \Delta_k\,,\ \lambda_k \big(\Delta_k - \|d_k\|_2\big) &= 0\,. \end{aligned} \tag{13}$$

A simple transformation where we insert the KKT conditions and take into account that $\Delta_k = \|d_k\|_2$, hence, $\Delta_k^2 = \langle d_k\,, d_k\rangle$, for $\lambda_k \neq 0$, yields for all $d \in \mathbb{R}^n$ with $\|d\|_2 \le \Delta_k$

$$\begin{aligned} \varphi_k(x^{(k)} + d_k) \le \varphi_k(x^{(k)} + d) &= f_k + \langle g_k\,, d\rangle + \frac{1}{2}\,\langle d\,, H_k d\rangle \\ &= \varphi_k(x^{(k)} + d_k) + \tfrac{1}{2}(d - d_k)^T (H_k + \lambda_k I)(d - d_k) + \frac{1}{2}\lambda_k\,(\Delta_k^2 - d^T d)\,. \end{aligned}$$

For $d \in \mathbb{R}^n$ with $\|d\|_2 = \Delta_k$ we hence get

$$(d - d_k)^T (H_k + \lambda_k I)(d - d_k) \ge 0\,.$$

This gives that the matrix $H_k + \lambda_k I$ is *positive semidefinite.*

Proof: If $\|d_k\|_2 < \Delta_k$, this is clear right away, since $\lambda_k = 0$, and H_k is positive semidefinite, because of the necessary optimality condition for unconstrained problems (cf. page 36).

If $\|d_k\|_2 = \Delta_k$, we proceed as follows: For $\lambda := -2\,\langle d_k\,, x\rangle / \|x\|_2^2$ we have

$$\|d_k + \lambda x\|_2^2 = \|d_k\|_2^2 + 2\lambda\,\langle d_k\,, x\rangle + \lambda^2\,\|x\|_2^2 = \|d_k\|_2^2 = \Delta_k^2\,.$$

Hence, for any $x \in \mathbb{R}^n$ with $\langle d_k\,, x\rangle \neq 0$ there exists a $\lambda \neq 0$ such that $\|d_k + \lambda x\|_2 = \Delta_k$.

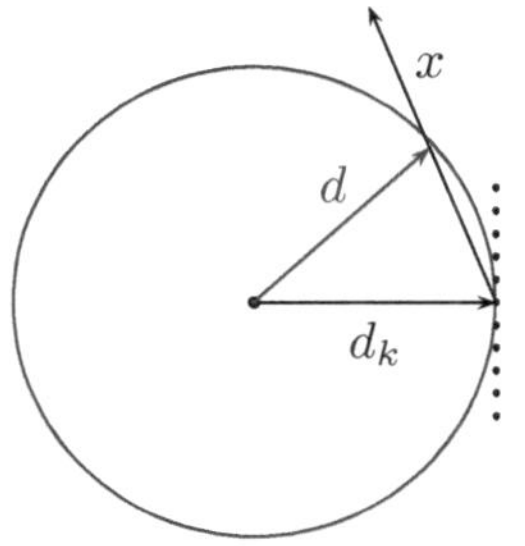

For $d := d_k + \lambda x$

$$0 \leq (d - d_k)^T (H_k + \lambda_k I)(d - d_k) = \lambda^2 x^T (H_k + \lambda_k I) x$$

holds and consequently $x^T(H_k + \lambda_k I)x \geq 0$. However, if $\langle d_k, x \rangle = 0$, $\langle d_k, x + t d_k \rangle = t \|d_k\|_2^2 > 0$ holds for any positive t, therefore firstly $\langle x + t d_k, (H_k + \lambda_k I)(x + t d_k) \rangle \geq 0$ and after passing to the limit $\langle x, (H_k + \lambda_k I)x \rangle \geq 0$. □

In the following we will **assume** that the matrix $H_k + \lambda_k I$ is even *positive definite*. Then, because of (13), it holds that:

a) $-g_k^T d_k = d_k^T (H_k + \lambda_k I) d_k > 0$, *that means* d_k *is a descent direction.*

b) d_k *yields a strict global minimum of (12).*

The vector d_k is called the LEVENBERG–MARQUARDT *direction* to the parameter value λ_k. For $\lambda_k = 0$, $d_k = -H_k^{-1} g_k$ corresponds to a (quasi-)NEWTON direction (cf. section 3.5) and for a large λ_k to the direction of steepest descent, because of $d_k \approx -\frac{1}{\lambda_k} g_k$.

Remark

$-g_k^T d_k > 0$ *also holds if the matrix* $H_k + \lambda_k I$ *is only positive semidefinite;* otherwise we would have a stationary point, because of

$$-g_k^T d_k = d_k^T (H_k + \lambda_k I) d_k = 0 \iff (H_k + \lambda_k I) d_k = 0 \iff g_k = 0.$$

This would be a contradiction to the assumption (cf. page 110).

The equation

$$(H_k + \lambda I) d(\lambda) + g_k = 0$$

defines the so-called LEVENBERG–MARQUARDT *trajectory* $d(\lambda)$. *We are looking for* a parameter value $\lambda \geq 0$ such that $\|d(\lambda)\|_2 = \Delta_k$. *Then with* $\lambda_k = \lambda$ *and* $d_k = d(\lambda)$ *condition (13) is met.*

Suppose that the matrix H_k has the eigenvalues $\alpha_1 \geq \alpha_2 \geq \cdots \geq \alpha_n$ with the corresponding orthonormal eigenvectors $v_1, v_2, \ldots, v_n$. Starting from the representation $g_k = \sum_{\nu=1}^{n} \beta_\nu v_\nu$, we obtain the LEVENBERG–MARQUARDT trajectory $d(\lambda)$ for $\lambda \notin \{-\alpha_1, \ldots, -\alpha_n\}$:

$$d(\lambda) = -\sum_{\nu=1}^{n} \frac{\beta_\nu}{\alpha_\nu + \lambda} v_\nu, \quad \text{consequently} \quad \|d(\lambda)\|_2^2 = \sum_{\nu=1}^{n} \frac{\beta_\nu^2}{(\alpha_\nu + \lambda)^2}.$$

Hence, $\|d(\cdot)\|_2$ is *strictly antitone* in $(-\alpha_n, \infty)$ with

$$\|d(\lambda)\|_2 \longrightarrow 0 \text{ for } \lambda \longrightarrow \infty$$

and, if $\beta_n \neq 0$,

$$\|d(\lambda)\|_2 \longrightarrow \infty \text{ for } \lambda \downarrow -\alpha_n .$$

If $\beta_n \neq 0$, there exists therefore exactly one $\lambda \in (-\alpha_n, \infty)$ such that $\|d(\lambda)\|_2 = \Delta_k$.

Closer Case Differentiation:

In general there are three cases:

i) $\alpha_n > 0$: Then $d(0)$ is defined, since 0 is not an eigenvalue of H_k. If $\|d(0)\|_2 \leq \Delta_k$, then $\lambda_k := 0$ and $d_k := d(0)$ solve conditions (13). In the remaining case $\|d(0)\|_2 > \Delta_k$ there exists exactly one $\lambda_k > 0$ such that $\|d(\lambda_k)\|_2 = \Delta_k$.

We consider $J := \{\nu \in \{1, \ldots, n\} \mid \alpha_\nu = \alpha_n\}$.

ii) $\alpha_n \leq 0$ and suppose that there exists a $\nu_0 \in J$ with $\beta_{\nu_0} \neq 0$:
$\|d(\lambda)\|_2^2 \geq \dfrac{\beta_{\nu_0}^2}{(\alpha_n + \lambda)^2} \uparrow \infty$ for $\lambda \downarrow -\alpha_n \geq 0$. Then there exists exactly one λ_k in the interval $(-\alpha_n, \infty)$ such that $\|d(\lambda_k)\|_2 = \Delta_k$.

iii) $\alpha_n \leq 0$ and suppose $\beta_\nu = 0$ for all $\nu \in J$:

$$d(\lambda) = -\sum_{\nu \notin J} \frac{\beta_\nu}{\alpha_\nu + \lambda} v_\nu \longrightarrow -\sum_{\nu \notin J} \frac{\beta_\nu}{\alpha_\nu - \alpha_n} v_\nu =: \widehat{d} \text{ for } \lambda \downarrow -\alpha_n .$$

If $\|\widehat{d}\,\|_2 \leq \Delta_k$ we set $\lambda_k := -\alpha_n$. Every vector d of the form $\widehat{d} + \sum_{\nu \in J} \gamma_\nu v_\nu$ ($\gamma_\nu \in \mathbb{R}$ for $\nu \in J$) with $\|d\|_2 = \Delta_k$ solves (13).
If $\Delta_k < \|\widehat{d}\,\|_2$, then there exists exactly one $\lambda_k \in (-\alpha_n, \infty)$ such that $\|d(\lambda_k)\|_2 = \Delta_k$.

Solution of the Equation $\|d(\lambda)\|_2 = \Delta_k$:

By definition of $d(\lambda)$ it holds that

$$d(\lambda) = -(H_k + \lambda I)^{-1} g_k \text{ respectively } (H_k + \lambda I)\, d(\lambda) = -g_k .$$

Differentiation gives

$$d(\lambda) + (H_k + \lambda I)\, d'(\lambda) = 0 \text{ and}$$

$$\frac{d}{d\lambda} \langle d(\lambda), d(\lambda) \rangle = 2 \langle d(\lambda), d'(\lambda) \rangle = 2 \left\langle d(\lambda), -(H_k + \lambda I)^{-1} d(\lambda) \right\rangle .$$

$N := \|d(\cdot)\|_2$ has a pole in $-\alpha_n$. Therefore, we consider

$$\psi(\lambda) := \frac{1}{N(\lambda)} - \frac{1}{\Delta_k}$$

and solve the equation $\psi(\lambda) = 0$ using NEWTON's method

$$\lambda_k^{(r+1)} = \lambda_k^{(r)} - \frac{\psi(\lambda_k^{(r)})}{\psi'(\lambda_k^{(r)})}.$$

With $\psi'(\lambda) = -N'(\lambda)/N(\lambda)^2$ we obtain

$$\begin{aligned}
\lambda_k^{(r+1)} &= \lambda_k^{(r)} + \left(\frac{1}{N(\lambda_k^{(r)})} - \frac{1}{\Delta_k}\right) \frac{N(\lambda_k^{(r)})^2}{N'(\lambda_k^{(r)})} \\
&= \lambda_k^{(r)} + \left(1 - \frac{N(\lambda_k^{(r)})}{\Delta_k}\right) \frac{N(\lambda_k^{(r)})}{N'(\lambda_k^{(r)})} \\
&= \lambda_k^{(r)} + \left(1 - \frac{\|d(\lambda_k^{(r)})\|_2}{\Delta_k}\right) \frac{\|d(\lambda_k^{(r)})\|_2^2}{\left\langle d(\lambda_k^{(r)}), d'(\lambda_k^{(r)}) \right\rangle}.
\end{aligned}$$

$d(\lambda_k^{(r)})$ and $\left\langle d(\lambda_k^{(r)}), d'(\lambda_k^{(r)}) \right\rangle$ can be calculated as follows:

If $H_k + \lambda_k^{(r)} I$ is positive definite, the CHOLESKY decomposition

$$H_k + \lambda_k^{(r)} I = L L^T \qquad (L \text{ lower triangular matrix})$$

gives $d(\lambda_k^{(r)})$ as the solution of

$$L L^T u = -g_k .$$

Furthermore it holds that

$$\begin{aligned}
\left\langle d(\lambda_k^{(r)}), d'(\lambda_k^{(r)}) \right\rangle &= \left\langle d(\lambda_k^{(r)}), -(H_k + \lambda_k^{(r)} I)^{-1} d(\lambda_k^{(r)}) \right\rangle \\
&= -\left\langle d(\lambda_k^{(r)}), (L L^T)^{-1} d(\lambda_k^{(r)}) \right\rangle \\
&= -\left\langle d(\lambda_k^{(r)}), (L^T)^{-1} \underbrace{L^{-1} d(\lambda_k^{(r)})}_{=:\,w} \right\rangle = -\langle w, w \rangle .
\end{aligned}$$

Remark

The equation $\|d(\lambda)\|_2 = \Delta_k$ or its variants have only to be solved with low accuracy. We demand, for example,

$$\|d(\lambda)\|_2 \in \left[0.75\,\Delta_k,\, 1.5\,\Delta_k\right].$$

Then normally a few steps of the (scalar) NEWTON method suffice.

Notation:

$$r_k := \frac{f(x^{(k)}) - f(x^{(k)} + d_k)}{f(x^{(k)}) - \varphi_k(x^{(k)} + d_k)}$$

The numerator gives the reduction of the function f, the denominator that of the modeling function φ_k.

The above considerations lead to the following

Algorithm

0) Let $x^{(k)}$ and Δ_k be given; calculate corresponding g_k and H_k.

1) Solve approximately: $\varphi_k(x^{(k)} + d) \longrightarrow \min$
subject to the constraint $\|d\|_2 \le \Delta_k$.

2) Calculate r_k.

3) If $r_k < 1/4$: $\Delta_{k+1} := \frac{1}{2}\Delta_k$ and $x^{(k+1)} := x^{(k)}$;
else, that is, $r_k \ge 1/4$:
If $r_k > 3/4$ and $\|d_k\|_2 = \Delta_k$: $\Delta_{k+1} := 2\Delta_k$;
else: $\Delta_{k+1} = \Delta_k$;
$x^{(k+1)} := x^{(k)} + d_k$.

In *0)* we assume that it has been set how to calculate H_k each time.

Example 5

$$f(x) := x_1^4 + x_1^2 + x_2^2 \qquad (x = (x_1, x_2)^T \in \mathbb{R}^2)$$

$$\nabla f(x) = \begin{pmatrix} 4x_1^3 + 2x_1 \\ 2x_2 \end{pmatrix}, \quad \nabla^2 f(x) = \begin{pmatrix} 12x_1^2 + 2 & 0 \\ 0 & 2 \end{pmatrix}$$

$$x^{(0)} := \begin{pmatrix} 1 \\ 1 \end{pmatrix}, \; H_0 := \nabla^2 f(x^{(0)}) = \begin{pmatrix} 14 & 0 \\ 0 & 2 \end{pmatrix}, \; \Delta_0 := \tfrac{1}{2}, \; \lambda_0^{(0)} := 0$$

$$\varphi_0(x^{(0)} + d) := f_0 + g_0^T d + \tfrac{1}{2} d^T H_0 d$$

$$g_0 = \begin{pmatrix} 6 \\ 2 \end{pmatrix}, \; L = \begin{pmatrix} \sqrt{14} & 0 \\ 0 & \sqrt{2} \end{pmatrix}, \; d(\lambda) = -(H_0 + \lambda I)^{-1} g_0$$

$$d(\lambda_0^{(0)}) = -H_0^{-1} g_0 = \begin{pmatrix} -\frac{3}{7} \\ -1 \end{pmatrix}, \; \left\| d(\lambda_0^{(0)}) \right\|_2 = 1.0880$$

$$L w = d(\lambda_0^{(0)}), \; w = \begin{pmatrix} -\frac{3}{7 \cdot \sqrt{14}} \\ -\frac{1}{\sqrt{2}} \end{pmatrix}, \; \langle w, w \rangle = \tfrac{352}{14 \cdot 49} = 0.5131$$

$$\lambda_0^{(1)} = 0 + \left(1 - \tfrac{1.0880}{0.5}\right)\left(-\tfrac{1.0880^2}{0.5131}\right) = 2.713$$

$$d(\lambda_0^{(1)}) = \begin{pmatrix} -0.3590 \\ -0.4244 \end{pmatrix}, \ \left\| d(\lambda_0^{(1)}) \right\|_2 = 0.556 \in [0.375\,,\, 0.75]$$

Hence, the value is already within the demanded interval $\left[0.75\,\Delta_0\,,\, 1.5\,\Delta_0\right]$. Therefore, we actually do not need the computation in the next three lines. They only serve to show the quality of the next step.

$$w = \begin{pmatrix} -0.0878 \\ -0.1955 \end{pmatrix}, \ \langle w\,, w\rangle = 0.0459$$

$$\lambda_0^{(2)} = 2.713 + \left(1 - \tfrac{0.556}{0.5}\right)\left(-\tfrac{0.556^2}{0.0459}\right) = 3.467$$

$$d(\lambda_0^{(2)}) = \begin{pmatrix} -0.3435 \\ -0.3658 \end{pmatrix}, \ \left\| d(\lambda_0^{(2)}) \right\|_2 = 0.5018$$

$$x^{(1)} = x^{(0)} + \underbrace{d(\lambda_0^{(1)})}_{=:\, d_0} = \begin{pmatrix} 0.6410 \\ 0.5756 \end{pmatrix}$$

$$f(x^{(0)}) = 3, \ f(x^{(0)} + d_0) = 0.9110$$

$$\varphi_0(x^{(0)} + d_0) = 1.0795$$

$$r_0 = \tfrac{2.089}{1.9205}\,, \ \Delta_1 := 1$$

Modeling function φ_0

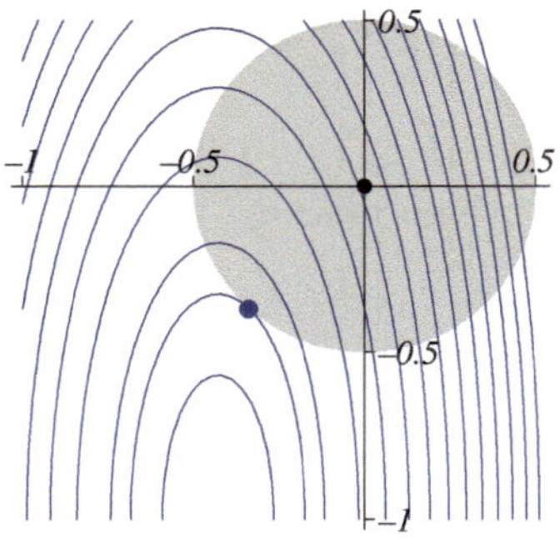

◁

To conclude these discussions, we want to introduce POWELL's *dogleg*[6] *trajectory*. We will see that it can be regarded as an approximation of the LEVENBERG–MARQUARDT trajectory replacing it with a path consisting of two line segments.

To solve the nonlinear equation $(H_k + \lambda I)d(\lambda) = -g_k$ just in order to get a new direction d_k for the actual value Δ_k seems wasting one's breath. The *dogleg method* fixes a new direction more directly.

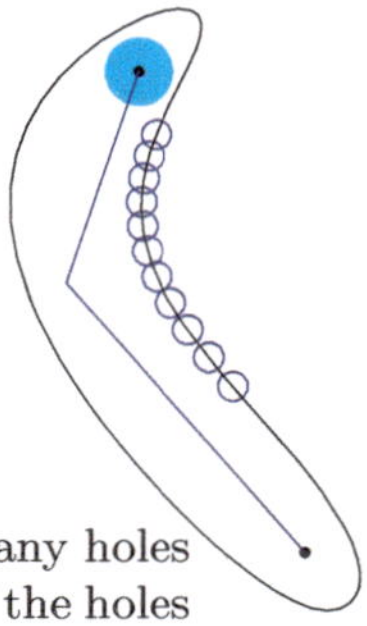

Here again we are dealing with the problem

$$\varphi_k(x^{(k)} + d) = f_k + g_k^T d + \tfrac{1}{2} d^T H_k\, d \longrightarrow \min$$
$$\|d\|_2 \le \Delta_k\,.$$

We assume that the matrix H_k is positive definite. Because of

[6] In golf, *"dogleg"* is a reference for the direction of a golf hole. While many holes are designed in a straight line from the tee-off point to the green, some of the holes may bend somewhat to the right or left. This is called a *"dogleg"*, referencing the partial bend at the knee of a dog's leg. On rare occasions, a hole's direction can bend twice. This is called a *"double dogleg"*.

$$\frac{d}{d\lambda}\,\varphi_k\Big(x^{(k)} - \lambda g_k\Big) = 0 \iff \lambda = \frac{\|g_k\|_2^2}{\langle g_k\,, H_k\, g_k\rangle}\,,$$

we obtain *the minimum point*

$$x^S := x^{(k)} + d^S \;\text{ with }\; d^S := -\frac{\|g_k\|_2^2}{\langle g_k\,, H_k\, g_k\rangle}\, g_k$$

in the direction of steepest descent and as the *unconstrained minimum point*

$$x^N := x^{(k)} + d^N \;\text{ with }\; d^N := -H_k^{-1}\, g_k\,.$$

We will prove:

For the steepest descent direction d^S *and the quasi-*NEWTON *direction* d^N *it holds that* $\|d^S\|_2 \le \|d^N\|_2$.

This follows from the estimate

$$\begin{aligned}\|d^S\|_2 = \frac{\|g_k\|_2^3}{\langle g_k\,, H_k\, g_k\rangle} &\le \frac{\|g_k\|_2^3}{\langle g_k\,, H_k\, g_k\rangle}\cdot\frac{\|g_k\|_2\,\|H_k^{-1}g_k\|_2}{\langle g_k\,, H_k^{-1}\, g_k\rangle}\\ &= \underbrace{\frac{\|g_k\|_2^2}{\langle g_k\,, H_k\, g_k\rangle}\cdot\frac{\|g_k\|_2^2}{\langle g_k\,, H_k^{-1}\, g_k\rangle}}_{\le 1}\,\|d^N\|_2\,.\end{aligned}$$

The latter is, for example, a result of the *proof* of KANTOROVICH's inequality (cf. page 101), if we take into account the inequality

$$\left(\sum_{\nu=1}^{n} u_\nu\, \alpha_\nu\right)\left(\sum_{\nu=1}^{n} \frac{u_\nu}{\alpha_\nu}\right) \ge 1$$

for the weighted arithmetic and harmonic mean (cf. e.g. [HLP], pp. 12–18).

Following POWELL we choose $x^{(k+1)}$ such that:

1) $\|d^N\|_2 \le \Delta_k$: $\quad x^{(k+1)} := x^N$

2) $\|d^S\|_2 \ge \Delta_k$: $\quad x^{(k+1)} := x^{(k)} - \dfrac{\Delta_k}{\|g_k\|_2}\, g_k$

3) $\|d^S\|_2 < \Delta_k < \|d^N\|_2$: There exists exactly one $\lambda \in (0,1)$ such that $\|d^S + \lambda(d^N - d^S)\|_2 = \Delta_k$. Set $x^{(k+1)} := (1-\lambda)x^S + \lambda x^N$.

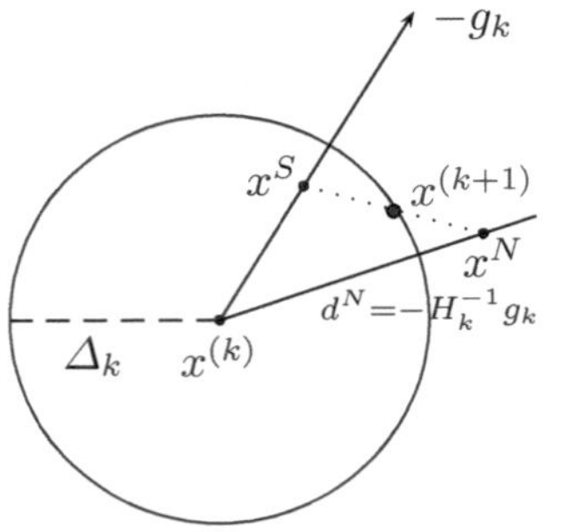

Example 6

In order to apply the *trust region method* to the framework example, we utilize the function fminunc — with the option optimset('GradObj', 'on') — from the *Matlab*® *Optimization Toolbox*. For the starting point $w^{(0)} = (0,0,0)^T$ and the regularization parameter $\lambda = 1$, six iteration steps suffice to reach the tolerance of 10^{-7}. The approximate solution of the quadratic subproblem (cf. step 1 of the trust region algorithm on p. 115) is restricted to a two-dimensional subspace which is determined with the aid of a (preconditioned) conjugate gradient process (cf. section 3.4).

k	$w^{(k)}$			$f(w^{(k)})$	$\lVert\nabla f(w^{(k)})\rVert_\infty$
0	0	0	0	8.31776617	6.00000000
1	−0.40705741	0.62664555	0	5.75041913	0.79192755
2	−0.49570677	0.74745305	0.04755343	5.68031561	0.05564070
3	−0.50487306	0.75569907	0.06543416	5.67972329	0.01257655
4	−0.50447655	0.75299924	0.06972656	5.67970043	0.00221882
5	−0.50485565	0.75335803	0.07043384	5.67969942	0.00056029
6	−0.50483824	0.75323767	0.07062618	5.67969938	0.00010095

Least Squares Problems

In applications there often occur problems in which the function to be minimized has the following special form

$$F(x) = \frac{1}{2}\sum_{\mu=1}^{m} f_\mu(x)^2$$

with an $m \in \mathbb{N}$. For that suppose that twice continuously differentiable functions $f_\mu \colon \mathbb{R}^n \longrightarrow \mathbb{R}$ are given. With $f(x) := (f_1(x), \ldots, f_m(x))^T$ for $x \in \mathbb{R}^n$ we then get

$$F(x) = \frac{1}{2}\langle f(x), f(x)\rangle = \frac{1}{2}\,\|f(x)\|_2^2\,.$$

Parameter estimates via the least squares method as well as systems of nonlinear equations are examples of this kind of problem. In the following we will discuss methods which make use of the special form of such objective functions. With the help of the Jacobian of f

$$J(x) := \left(\frac{\partial f_\mu}{\partial x_\nu}(x)\right) \in \mathbb{R}^{m\times n}$$

we obtain the gradient

$$g(x) := \nabla F(x) = J(x)^T f(x),$$

and the Hessian is

$$H(x) := \nabla^2 F(x) = J(x)^T J(x) + R(x) \tag{14}$$

with

$$R(x) := \sum_{\mu=1}^{m} f_\mu(x)\, H_\mu(x) \quad\text{and}\quad H_\mu := \nabla^2 f_\mu\,.$$

From (14) we can deduce that $J(x)^T J(x)$ — here, there occur only first-order derivatives — is an essential part of the Hessian in the vicinity of a minimizer x^* if, for example, $F(x^*) = 0$, or if all f_μ are affinely linear. In the latter case $H_\mu = 0$ holds.

Starting from the current approximation $x^{(k)}$ of the minimum point x^* with

$$f_k := f\big(x^{(k)}\big)\,,\; J_k := J\big(x^{(k)}\big) \;\text{ and }\; R_k := R\big(x^{(k)}\big)\,,$$

$x^{(k+1)} = x^{(k)} + d_k$ can be calculated as follows:

a) NEWTON–RAPHSON *Method* (cf. (14) with $H_k := H\big(x^{(k)}\big)$)

$$\big(J_k^T J_k + R_k\big)\, d_k = -J_k^T f_k$$

Advantage: The algorithm is — with standard assumptions — (locally) *quadratically* convergent.

Disadvantage: Second-order partial derivatives are needed explicitly.

b) GAUSS–NEWTON *Method*

$$\big(J_k^T J_k\big)\, d_k = -J_k^T f_k$$

In this case there occur only first-order partial derivatives of f. The method is *at the most* quadratically convergent. Slow convergence can occur in particular if $J(x^*)^T J(x^*)$ is nondefinite.

c) LEVENBERG–MARQUARDT *Method*

$$\left(J_k^T J_k + \lambda_k I\right) d_k = -J_k^T f_k \,, \; \lambda_k \geq 0$$

MARQUARDT (1963) provided first 'ad hoc' formulations for the choice of λ_k. A more specific control of λ_k is possible if we regard the LEVENBERG–MARQUARDT method as a trust region method.

There are many modifications of this method of which we only want to mention the method introduced by DENNIS, GAY and WELSCH in 1981 (cf. [DGW]). It takes account of the term R_k in the Hessian ignored by the GAUSS–NEWTON method and uses quasi-NEWTON updates to approximate it. We do not go into details because this is beyond the scope of our presentation.

3.4 Conjugate Gradient Method

The conjugate gradient (CG) method was introduced in 1952 by HESTENES and STIEFEL to solve *large* systems of linear equations with a *sparse* symmetric and positive definite coefficient matrix. For that they made use of the equivalence

$$Ax = -b \iff f(x) = \min_{z \in \mathbb{R}^n} f(z)$$

to the minimization problem for the convex quadratic objective function f defined by $f(z) := \frac{1}{2}\langle z, Az\rangle + \langle b, z\rangle$, whose well-known gradient is $\nabla f(z) = Az + b$. This iterative method has proved to be particularly efficient since the main effort per iteration step lies in the calculation of a matrix-vector product and which is therefore very small.

Originally the CG-method was not designed as a minimization method — at most to minimize quadratic functions. Only when FLETCHER and REEVES (1964) suggested it, was the method used to minimize *'any kind' of function.*

For the remainder of this section let $n \in \mathbb{N}$, A a real-valued symmetric and positive definite (n,n)-matrix, $b \in \mathbb{R}^n$ and — corresponding to that — the quadratic objective function f be defined by

$$f(x) := \frac{1}{2}\langle x, Ax\rangle + \langle b, x\rangle \,.$$

Definition

For $m \in \mathbb{N}$ the vectors $d_1, \ldots, d_m \in \mathbb{R}^n \setminus \{0\}$ are called A-conjugate if and only if $\langle d_\nu, Ad_\mu\rangle = 0$ for all $1 \leq \nu < \mu \leq m$.

Remark *Such $d_1, \ldots, d_m$ are linearly independent.*

In particular $m \leq n$ holds.

Using the definiteness of A, this can be proved immediately by applying the definition.

Proposition 3.4.1

To A-conjugate vectors $d_0, \ldots, d_{n-1}$ and any starting vector $x^{(0)} \in \mathbb{R}^n$ let $x^{(1)}, \ldots, x^{(n)}$ be recursively defined by

$$x^{(k+1)} := x^{(k)} + \lambda_k d_k \quad \textit{with} \quad \lambda_k := -\frac{\langle g_k , d_k \rangle}{\langle d_k , A d_k \rangle}.$$

Then it holds that $f\big(x^{(k+1)}\big) = \min_{\lambda \in \mathbb{R}} f\big(x^{(k)} + \lambda d_k\big)$, $\langle g_{k+1} , d_k \rangle = 0$ *and*

$$f\big(x^{(n)}\big) = \min_{x \in \mathbb{R}^n} f(x).$$

Proof: The simple derivation of the relations $\langle g_{k+1} , d_k \rangle = 0$ and $f\big(x^{(k+1)}\big) = \min\big\{f\big(x^{(k)} + \lambda d_k\big) \mid \lambda \in \mathbb{R}\big\}$ is familiar to us from earlier discussions of descent methods. However, we will give a short explanation:

$$0 = \frac{d}{d\lambda} f\big(x^{(k)} + \lambda d_k\big)|_{\lambda=\lambda_k} = f'\big(x^{(k+1)}\big) d_k = \langle g_{k+1} , d_k \rangle$$

and

$$\langle g_{k+1} , d_k \rangle = \big\langle A x^{(k)} + \lambda_k A d_k + b, d_k \big\rangle = \langle g_k , d_k \rangle + \lambda_k \langle A d_k , d_k \rangle.$$

The vectors $d_0, \ldots, d_{n-1}$ are linearly independent since we have assumed them to be A-conjugate. Hence, they form a basis of the $\mathbb{R}^n$. From

$$x^{(n)} = x^{(n-1)} + \lambda_{n-1} d_{n-1} = \cdots = x^{(j+1)} + \sum_{\nu=j+1}^{n-1} \lambda_\nu d_\nu$$

for $0 \le j \le n-1$ follows[7]

$$g_n = g_{j+1} + \sum_{\nu=j+1}^{n-1} \lambda_\nu A d_\nu$$

and from that

$$\langle g_n , d_j \rangle = \underbrace{\langle g_{j+1} , d_j \rangle}_{=0} + \sum_{\nu=j+1}^{n-1} \lambda_\nu \underbrace{\langle A d_\nu , d_j \rangle}_{=0} = 0,$$

hence, $g_n = 0$. This gives $f\big(x^{(n)}\big) = \min_{x \in \mathbb{R}^n} f(x)$. □

[7] It is not necessary to remind ourselves of $\nabla f(x) = Ax + b$, is it?

Chapter 3

In order to obtain an algorithm from proposition 3.4.1, we need A-conjugate vectors, one talks of *A-conjugate directions.* First of all, we will give the

Remark

There exists an orthonormal basis of eigenvectors to A. Its vectors are evidently A-conjugate.

This observation is more of a theoretical nature. The following method, in which the conjugate directions d_k are calculated at the same time as the $x^{(k)}$, is more elegant.

Generation of A-Conjugate Directions

HESTENES–STIEFEL **CG-Algorithm**

0) Let $g_0 := \nabla f\big(x^{(0)}\big) = Ax^{(0)} + b$ and $d_0 := -g_0$ to a given $x^{(0)} \in \mathbb{R}^n$.
(Starting step: steepest descent)

For $k = 0, 1, \ldots, n-1$:

If $g_k = 0$: STOP; $x^{(k)}$ is the minimal point of f. *Else:*

1) $x^{(k+1)} := x^{(k)} + \lambda_k d_k$, where $\lambda_k := -\dfrac{\langle g_k, d_k\rangle}{\langle A d_k, d_k\rangle}$.

Hence, we have $f(x^{(k+1)}) = \min_{\lambda \in \mathbb{R}} f(x^{(k)} + \lambda d_k)$ (exact line search).

2) $g_{k+1} := \nabla f\big(x^{(k+1)}\big)$; $d_{k+1} := -g_{k+1} + \gamma_{k+1} d_k$ with

$$\gamma_{k+1} := \frac{\|g_{k+1}\|_2^2}{\|g_k\|_2^2} = \frac{\langle g_{k+1}, g_{k+1}\rangle}{\langle g_k, g_k\rangle}.$$

In order to calculate d_{k+1}, we only use the preceding vector d_k — besides g_k and g_{k+1}.

Before making some more basic comments about this algorithm — especially that it is *well-defined,* we will give an example:

Example 7 (cf. page 100f)

$$f(x) := \frac{1}{2}\big(x_1^2 + 9x_2^2\big),\ A := \begin{pmatrix} 1 & 0 \\ 0 & 9 \end{pmatrix},\ b := 0$$

$$\nabla f(x) = \begin{pmatrix} x_1 \\ 9x_2 \end{pmatrix}$$

$$x^{(0)} := \begin{pmatrix} 9 \\ 1 \end{pmatrix},\ g_0 = 9\begin{pmatrix} 1 \\ 1 \end{pmatrix},\ d_0 = -9\begin{pmatrix} 1 \\ 1 \end{pmatrix},\ \lambda_0 = \frac{2}{10} = \frac{1}{5}$$

$$
\begin{aligned}
x^{(1)} &= \tfrac{4}{5}\begin{pmatrix} 9 \\ -1 \end{pmatrix}, \quad g_1 = \frac{36}{5}\begin{pmatrix} 1 \\ -1 \end{pmatrix} \\
d_1 &= -g_1 + \frac{\langle g_1, g_1 \rangle}{\langle g_0, g_0 \rangle}\, d_0 \\
&= -\frac{36}{5}\begin{pmatrix} 1 \\ -1 \end{pmatrix} - \frac{\left(\frac{36}{5}\right)^2 \cdot 2}{81 \cdot 2} \cdot 9 \begin{pmatrix} 1 \\ 1 \end{pmatrix} = \frac{36}{25}\begin{pmatrix} -9 \\ 1 \end{pmatrix} \\
\lambda_1 &= \frac{50}{90} = \frac{5}{9} \\
x^{(2)} &= x^{(1)} + \lambda_1 d_1 = 0
\end{aligned}
$$

◁

Important properties of this method are contained in the following

Proposition 3.4.2

There exists a smallest $m \in \{0, \ldots, n\}$ *such that* $d_m = 0$. *With that we have* $g_m = 0$ *and for all* $\ell \in \{0, \ldots, m\}$, $0 \le \nu < k \le \ell$ *and* $0 \le r \le \ell$:

$$
\begin{aligned}
&a) \quad \langle d_\nu , g_k \rangle &&= 0 \\
&b) \quad \langle g_\nu , g_k \rangle &&= 0 \\
&c) \quad \langle d_\nu , A d_k \rangle &&= 0 \\
&d) \quad \langle g_r , d_r \rangle &&= -\langle g_r , g_r \rangle .
\end{aligned}
$$

$x^{(m)}$ *then gives the minimum of* f.

If $g_m = 0$ for an $m \in \{1, \ldots, n\}$, then *2)* shows $d_m = 0$. The other way round it follows from $d_m = 0$ via

$$
\|d_m\|_2^2 = \|g_m\|_2^2 - 2\gamma_m \underbrace{\langle g_m , d_{m-1} \rangle}_{=0} + \underbrace{\gamma_m^2 \|d_{m-1}\|_2^2}_{\ge 0},
$$

that $\|d_m\|_2 \ge \|g_m\|_2$, therefore $g_m = 0$.

The CG-method is in particular well-defined, since — because A is positive definite — γ_{r+1} and λ_r are defined for $0 \le r \le m-1$. Here, following *d)*, λ_r is positive. In addition, by *c)* the vectors $d_0, \ldots, d_{m-1}$ are A-conjugate, hence linearly independent. This proves $m \le n$.

The name 'conjugate gradient method' is unfortunately chosen, since not the gradients but the directions d_k are A-conjugate.

Proof: Denote by $\mathcal{A}(\ell)$ the validity of the assertions *a)* to *d)* for all $0 \le \nu < k \le \ell$ and $0 \le r \le \ell$. Hence, we have to prove $\mathcal{A}(\ell)$ for $\ell \in \{0, \ldots, m\}$:

$\mathcal{A}(0)$: Assertions *a)* to *c)* are empty and consequently true. *d)* holds by definition of d_0.

For $0 \le \ell < m$ we will show $\mathcal{A}(\ell) \Longrightarrow \mathcal{A}(\ell+1)$. (In particular $g_\ell \ne 0$ and $d_\ell \ne 0$ hold, since $\ell < m$.)

Chapter 3

a) From $g_{\ell+1} = \nabla f\bigl(x^{(\ell+1)}\bigr) = Ax^{(\ell+1)} + b = g_\ell + \lambda_\ell A d_\ell$ it follows for $\nu < \ell+1$ that

$$\langle d_\nu , g_{\ell+1} \rangle = \begin{cases} 0 \ \text{ for } \ \nu = \ell \ \text{ (by definition of } \lambda_\ell) \\ \underbrace{\langle d_\nu , g_\ell \rangle}_{=0} + \lambda_\ell \underbrace{\langle d_\nu , A d_\ell \rangle}_{=0} = 0 \ \text{ for } \ \nu < \ell . \end{cases}$$

b) We have to prove $\langle g_\nu , g_{\ell+1} \rangle = 0$ for $\nu < \ell+1$:

$$\begin{aligned} \langle g_\nu , g_{\ell+1} \rangle &= \langle -d_\nu + \gamma_\nu d_{\nu-1} , g_{\ell+1} \rangle \quad (\text{with } d_{-1} := 0 \text{ and } \gamma_0 := 0) \\ &= -\underbrace{\langle d_\nu , g_{\ell+1} \rangle}_{\underset{a)}{=}0} + \gamma_\nu \underbrace{\langle d_{\nu-1} , g_{\ell+1} \rangle}_{\underset{a)}{=}0} = 0 . \end{aligned}$$

d) $\langle g_{\ell+1} , d_{\ell+1} \rangle = \langle g_{\ell+1} , -g_{\ell+1} + \gamma_{\ell+1} d_\ell \rangle \underset{a)}{=} -\|g_{\ell+1}\|_2^2$

c) We have to prove $\langle d_\nu , A d_{\ell+1} \rangle = 0$ for $\nu < \ell+1$:

Since $g_{\nu+1} - g_\nu = \lambda_\nu A d_\nu$ it holds that

$$\begin{aligned} \langle d_\nu , A d_{\ell+1} \rangle &= \langle d_\nu , A(-g_{\ell+1} + \gamma_{\ell+1} d_\ell) \rangle \\ &= -\langle A d_\nu , g_{\ell+1} \rangle + \gamma_{\ell+1} \langle d_\nu , A d_\ell \rangle \\ &= \begin{cases} \frac{1}{\lambda_\nu} \underbrace{\langle g_\nu - g_{\nu+1} , g_{\ell+1} \rangle}_{\underset{b)}{=}0} + \gamma_{\ell+1} \underbrace{\langle d_\nu , A d_\ell \rangle}_{=0} = 0 & \text{for } \ \nu < \ell \\ -\frac{1}{\lambda_\ell} \|g_{\ell+1}\|_2^2 + \gamma_{\ell+1} \langle d_\ell , A d_\ell \rangle = 0 & \text{for } \ \nu = \ell . \end{cases} \end{aligned}$$

The latter follows with

$$\gamma_{\ell+1} = \frac{\|g_{\ell+1}\|_2^2}{\|g_\ell\|_2^2} \quad \text{and} \quad \lambda_\ell = \frac{-\langle g_\ell , d_\ell \rangle}{\langle d_\ell , A d_\ell \rangle} \underset{d)}{=} \frac{\|g_\ell\|_2^2}{\langle d_\ell , A d_\ell \rangle} . \qquad \square$$

Hence, in an *exact calculation* at the latest $x^{(n)}$ will give the minimum of f. In practice, however, g_n is different from zero most of the time, because of the propagation of rounding errors. For the numerical aspects of this method see, for example, [No/Wr], p. 112 ff.

Corollary *For $\ell < m$ it holds that:*

a) $S_\ell := \operatorname{span}\{g_0, \ldots, g_\ell\} = \operatorname{span}\{d_0, \ldots, d_\ell\} = \operatorname{span}\{g_0, A g_0, \ldots, A^\ell g_0\}$

b) $f\bigl(x^{(\ell+1)}\bigr) = \min_{u \in S_\ell} f\bigl(x^{(\ell)} + u\bigr) = \min_{u \in S_\ell} f\bigl(x^{(0)} + u\bigr)$

The spaces S_ℓ are called "KRYLOV *spaces*".

Proof: *a)*

α) $\operatorname{span}\{g_0, \ldots, g_\ell\} = \operatorname{span}\{d_0, \ldots, d_\ell\}$ can be deduced inductively from $d_{\ell+1} = -g_{\ell+1} + \gamma_{\ell+1} d_\ell$ for $\ell + 1 < m$ with $d_0 = -g_0$.

β) The second equation is (with α)) trivial for $\ell = 0$.

$g_{\ell+1} = g_\ell + \lambda_\ell A d_\ell$ for $\ell + 1 < m$ with α) gives the induction assertion $\mathrm{span}\{d_0, \ldots, d_{\ell+1}\} \subset \mathrm{span}\{g_0, A g_0, \ldots, A^{\ell+1} g_0\}$. For the other inclusion we only have to show $A^{\ell+1} g_0 \in \mathrm{span}\{d_0, \ldots, d_{\ell+1}\}$ in the induction step for $\ell + 1 < m$: From the induction hypothesis we have

$$A^\ell g_0 = \sum_{\nu=0}^{\ell} \mu_\nu d_\nu \quad \text{for suitable } \mu_0, \ldots, \mu_\ell \in \mathbb{R},$$

hence, $A^{\ell+1} g_0 = \sum_{\nu=0}^{\ell} \mu_\nu A d_\nu$. The already familiar relation $\lambda_\nu A d_\nu = g_{\nu+1} - g_\nu$ for $\nu = 0, \ldots, \ell$ gives the result (with α) and taking into account $\lambda_\nu \neq 0$).

b) By *a)* any $u \in S_\ell$ is of the form

$$\sum_{\nu=0}^{\ell} \tau_\nu d_\nu \quad \text{with suitable } \tau_0, \ldots, \tau_\ell \in \mathbb{R}.$$

For $j \in \{0, \ldots, \ell\}$ it holds that

$$\begin{aligned} \frac{d}{d\tau_j} f\Big(x^{(\ell)} + \sum_{\nu=0}^{\ell} \tau_\nu d_\nu\Big) &= f'\big(x^{(\ell)} + \sum_{\nu=0}^{\ell} \tau_\nu d_\nu\big) d_j = \Big(g_\ell + \sum_{\nu=0}^{\ell} \tau_\nu A d_\nu\Big)^T d_j \\ &= \underbrace{\langle g_\ell, d_j \rangle}_{=0 \text{ for } j<\ell} + \tau_j \underbrace{\langle d_j, A d_j \rangle}_{>0} . \end{aligned}$$

Hence, we have

$$\tau_j = \begin{cases} 0 & , \text{ if } j < \ell \\ \lambda_\ell & , \text{ if } j = \ell \end{cases}$$

for the minimal point on $x^{(\ell)} + S_\ell$. From that follows

$$x^{(\ell)} + \sum_{\nu=0}^{\ell} \tau_\nu d_\nu = x^{(\ell)} + \lambda_\ell d_\ell = x^{(\ell+1)} .$$

The second partial assertion directly results from

$$x^{(\ell)} = x^{(0)} + \sum_{\nu=0}^{\ell-1} \lambda_\nu d_\nu . \qquad \square$$

Rate of Convergence

Let the real-valued symmetric and positive definite matrix A have the eigenvalues $0 < \alpha_1 \leq \cdots \leq \alpha_n$ with corresponding orthonormal eigenvectors

Chapter 3

$v_1, \ldots, v_n$. Let x^* be the minimal point of f; hence, $Ax^* + b = 0$. 'TAYLOR series expansion' around x^* yields

$$f(x) = f(x^*) + \frac{1}{2}\langle x - x^*, A(x - x^*)\rangle = f(x^*) + \frac{1}{2}\|x - x^*\|_A^2 .$$

Consequently,

$$E(x) := f(x) - f(x^*) = \frac{1}{2}\langle x - x^*, A(x - x^*)\rangle .$$

Starting from the representation $h := x^{(0)} - x^* = \sum_{\nu=1}^{n} \xi_\nu v_\nu$ with $\xi_\nu \in \mathbb{R}$, we obtain

$$E\big(x^{(0)}\big) = \frac{1}{2}\sum_{\nu=1}^{n} \alpha_\nu \xi_\nu^2 .$$

For $k \in \{1, \ldots, m\}$ with suitable $\gamma_\kappa \in \mathbb{R}$ the vectors $u \in x^{(0)} + S_{k-1}$ can be written in the form

$$u = x^{(0)} + \sum_{\kappa=0}^{k-1} \gamma_\kappa A^\kappa g_0 = x^{(0)} + p(A)\, g_0$$

with the polynomial $p \in \Pi_{k-1}$ defined by $p(\tau) := \sum_{\kappa=0}^{k-1} \gamma_\kappa \tau^\kappa$. Via

$$g_0 := \nabla f\big(x^{(0)}\big) = Ax^{(0)} + b = Ax^{(0)} - Ax^* = Ah$$

due to

$$u - x^* = x^{(0)} + p(A)\, g_0 - x^* = h + p(A)\, Ah = (I + p(A)\, A)\, h ,$$

we obtain

$$\begin{aligned} E(u) &= \frac{1}{2}\langle u - x^*, A(u - x^*)\rangle \\ &= \frac{1}{2}\langle (I + p(A)\, A)\, h, A(I + p(A)\, A)\, h\rangle = \frac{1}{2}\langle h, A(I + p(A)\, A)^2 h\rangle \\ &= \frac{1}{2}\sum_{\nu=1}^{n} \alpha_\nu \big(1 + \alpha_\nu p(\alpha_\nu)\big)^2 \xi_\nu^2 \\ &\le \max_{1 \le \nu \le n} (1 + \alpha_\nu p(\alpha_\nu))^2\, E\big(x^{(0)}\big) . \end{aligned}$$

With $\widehat{\Pi}_k := \big\{q \in \Pi_k \mid q(0) = 1\big\}$ it follows that

$$E\big(x^{(k)}\big) \le \min_{q \in \widehat{\Pi}_k} \max_{1 \le \nu \le n} |q(\alpha_\nu)|^2\, E\big(x^{(0)}\big) \le \Big(\min_{q \in \widehat{\Pi}_k} \max_{\alpha_1 \le \alpha \le \alpha_n} |q(\alpha)|\Big)^2 E\big(x^{(0)}\big) .$$

This leads to an extremal problem whose solution can be stated with the help of the CHEBYSHEV polynomials:

For $t \in [-1, 1]$ and $k \in \mathbb{N}_0$ we consider

$$T_k(t) := \cos(k \arccos t) .$$

Evidently

$$\max\{T_k(t) : t \in [-1, 1]\} = 1 . \tag{15}$$

The relation

$$\cos(k+1)\vartheta + \cos(k-1)\vartheta = 2\cos\vartheta \cos k\vartheta$$

for $\vartheta := \arccos t$ and $k \in \mathbb{N}$, which follows directly from the addition theorem of the cosine, shows the *recursion:*

$$T_0(t) = 1, \; T_1(t) = t \; \text{ and } \; T_{k+1}(t) + T_{k-1}(t) = 2tT_k(t) .$$

In this way we can see inductively for $k \in \mathbb{N}$: T_k is the restriction of a polynomial[8] of degree k with highest coefficient 2^{k-1} on $[-1, 1]$. The first CHEBYSHEV polynomials are given by

$$T_0(t) = 1, T_1(t) = t, T_2(t) = 2t^2 - 1, T_3(t) = 4t^3 - 3t, T_4(t) = 8t^4 - 8t^2 + 1 .$$

A short *Maple*® program yields a plot of $T_1, \ldots, T_4$:

```
> restart: with(plots): with(orthopoly):
  p := plot([seq(T(n,t),n=1..4)],t=-1..1,color=[blue$2,black$2],
            labels=[' ',' '],linestyle=[1,2]):
  display(p);
```

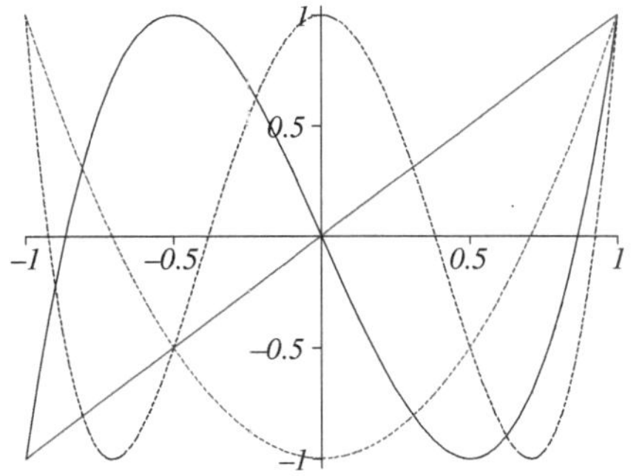

For real-valued t with $|t| \geq 1$ the addition theorem of cosh yields respectively:

$$T_k(t) = \cosh(k \operatorname{arcosh} t) ,$$

with that

[8] In the notation we do not make a distinction between the polynomials and their restrictions to $[-1, 1]$.

$$T_k(t) = \frac{1}{2}\left(\left(t+\sqrt{t^2-1}\right)^k + \left(t-\sqrt{t^2-1}\right)^k\right)$$

and thus

$$T_k(t) \geq \frac{1}{2}\left(t+\sqrt{t^2-1}\right)^k \quad \text{for } t \geq 1. \tag{16}$$

The function τ defined by

$$\alpha \longmapsto (\alpha_n + \alpha_1 - 2\alpha)/(\alpha_n - \alpha_1) =: \tau(\alpha)$$

is a one-to-one mapping of the interval $[\alpha_1, \alpha_n]$ to $[-1, 1]$. Hence, by (15) it holds for the polynomial $q^* \in \widehat{\Pi}_k$ defined by

$$q^*(\alpha) := \frac{T_k(\tau(\alpha))}{T_k(\tau(0))} = \frac{T_k((\alpha_n + \alpha_1 - 2\alpha)/(\alpha_n - \alpha_1))}{T_k((\alpha_n + \alpha_1)/(\alpha_n - \alpha_1))}$$

that

$$\max\left\{|q^*(\alpha)| \,:\, \alpha \in [\alpha_1, \alpha_n]\right\} = \left(T_k(\tau(0))\right)^{-1} = \left(T_k\left(\frac{\kappa+1}{\kappa-1}\right)\right)^{-1}$$

with the condition number $\kappa := c(A) = \alpha_n/\alpha_1$ of A (cf. exercise 16).

$$\begin{aligned} T_k\left(\tfrac{\kappa+1}{\kappa-1}\right) &\underset{(16)}{\geq} \tfrac{1}{2}\left(\tfrac{\kappa+1}{\kappa-1} + \sqrt{\tfrac{4\kappa}{(\kappa-1)^2}}\right)^k \\ &= \tfrac{1}{2}\left(\tfrac{(\sqrt{\kappa}+1)^2}{\kappa-1}\right)^k = \tfrac{1}{2}\left(\tfrac{\sqrt{\kappa}+1}{\sqrt{\kappa}-1}\right)^k \end{aligned}$$

gives the following estimate

$$\max\left\{|q^*(\alpha)| \,:\, \alpha \in [\alpha_1, \alpha_n]\right\} = \left(T_k\left(\frac{\kappa+1}{\kappa-1}\right)\right)^{-1} \leq 2\left(\frac{\sqrt{\kappa}-1}{\sqrt{\kappa}+1}\right)^k.$$

With the above considerations we thus obtain

$$E(x^{(k)}) \leq 4\left(\frac{\sqrt{\kappa}-1}{\sqrt{\kappa}+1}\right)^{2k} E(x^{(0)}).$$

Compared to the gradient method (cf. lemma 3.2.1) this is a much better estimate, because here we have replaced the condition number κ of A with $\sqrt{\kappa}$.

Hence, a better condition number of A ensures a quicker convergence of the CG-method. We will use this fact in the technique of preconditioning to speed up the CG-method.

Preconditioning

We carry out a substitution $x = M\widehat{x}$ with an invertible matrix M in the objective function f, and get

$$\begin{aligned}\widehat{f}(\widehat{x}) := f(x) = f(M\widehat{x}) &= \tfrac{1}{2}\widehat{x}^T M^T A M \widehat{x} + b^T M \widehat{x}\\ &= \tfrac{1}{2}\widehat{x}^T \widehat{A}\widehat{x} + \widehat{b}^T\widehat{x}\end{aligned}$$

with

$$\widehat{A} := M^T A M \quad \text{and} \quad \widehat{b} := M^T b\,.$$

The gradient $g := \nabla f$ transforms as follows:

$$\widehat{g}(\widehat{x}) := \nabla\widehat{f}(\widehat{x}) = \widehat{A}\widehat{x} + \widehat{b} = M^T(AM\widehat{x} + b) = M^T g(x)$$

We now want to choose M such that $c(M^TAM) \ll c(A)$ holds. If we know the CHOLESKY decomposition $A = LL^T$, the choice $M := L^{-T}$ gives

$$\widehat{A} = M^T A M = L^{-1}AL^{-T} = I\,.$$

If the matrix A is sparse, L usually has mostly nonzero entries (problem of *"fill in"*). Therefore, this suggestion for the choice of M is only useful for an approximative $\overline{L}$ which is sparse (*incomplete* CHOLESKY *decomposition*[9]):

$$\begin{aligned} A &= \overline{L}\,\overline{L}^T + E \quad \text{(E stands for "error")}\\ M &:= \overline{L}^{-T}: \quad M^T A M = I + \underbrace{\overline{L}^{-1}E\overline{L}^{-T}}_{\text{"perturbation"}}\,. \end{aligned}$$

The occurring inverse will of course not actually be calculated. We will only solve the corresponding systems of linear equations.

The CG-Method in the Nonquadratic Case

The CG-method can be rephrased in such a way that it can also be applied to *any* differentiable function $f\colon \mathbb{R}^n \longrightarrow \mathbb{R}$. A first version goes back to FLETCHER/REEVES (1964). A variant was devised by POLAK/RIBIÈRE (1971). They proposed to take for γ_{k+1}

$$\frac{\langle g_{k+1}, g_{k+1} - g_k\rangle}{\langle g_k, g_k\rangle} \quad \text{instead of} \quad \frac{\langle g_{k+1}, g_{k+1}\rangle}{\langle g_k, g_k\rangle}\,.$$

This modification is said to be more robust and efficient, even though it will theoretically give the same result in special cases. In the literature on the topic it is also reported in many instances that this variant has much better convergence properties.

We consider $g := \nabla f$ and are looking for a stationary point, hence, a zero of the gradient g:

[9] Also see [Me/Vo].

FLETCHER **and** REEVES **Algorithm with Restart:**

0) To a given $x^{(0)} \in \mathbb{R}^n$ let $g_0 := g(x^{(0)})$ and $d_0 := -g_0$.

For $k = 0, 1, 2 \ldots$: If $g_k = 0$: $x^{(k)}$ is a stationary point: STOP; else:

1) Calculate — using exact or inexact line search — $\lambda_k > 0$ such that

$$f(x^{(k)} + \lambda_k d_k) \approx \min\{f(x^{(k)} + \lambda d_k) \mid \lambda \geq 0\}$$

and set $x^{(k+1)} := x^{(k)} + \lambda_k d_k$ and $g_{k+1} := g(x^{(k+1)})$.

2) $d_{k+1} := -g_{k+1} + \gamma_{k+1} d_k$, where

$$\gamma_{k+1} := \begin{cases} 0 & \text{, if } k+1 \equiv 0 \mod n \qquad \text{(Restart)} \\ \dfrac{\langle g_{k+1}, g_{k+1} - g_k \rangle}{\langle g_k, g_k \rangle} & \text{, else.} \end{cases}$$

When inexact line search is used, it is assumed that a specific method has been chosen.

Here — in contrast to the special case — λ_k cannot be noted down directly in the iteration step but the calculation requires a minimization in a given direction.

In this more general case d_{k+1} is also always a descent direction when we use exact line search, because

$$\langle g_{k+1}, d_{k+1} \rangle = -\|g_{k+1}\|_2^2 + \gamma_{k+1} \langle g_{k+1}, d_k \rangle = -\|g_{k+1}\|_2^2 < 0 .$$

This property can likely be lost in inexact line search. We often have to reach a compromise between great effort (exact line search) and possibly unfavorable properties (inexact line search). In practice we would of course replace the condition $g_k = 0$, for example, by $\|g_k\|_2 \leq \varepsilon$ or $\|g_k\|_2 \leq \varepsilon \|g_0\|_2$ for a given $\varepsilon > 0$.

For generalizations, additions and further numerical aspects see, for example, [No/Wr], pp. 122ff.

3.5 Quasi-Newton Methods

Another important class of methods for constructing descent directions are Quasi-NEWTON methods which we are going to look at in this section.

Suppose that $f: \mathbb{R}^n \longrightarrow \mathbb{R}$ is twice continuously differentiable. As usual we consider the minimization problem

$$f(x) \longrightarrow \min_{x \in \mathbb{R}^n}$$

by determining *stationary points*, that is, by solving the system of nonlinear equations $\nabla f(x) = 0$. In order to do that, we can, for example, use the following *approximation methods* starting from an $x^{(0)} \in \mathbb{R}^n$ and with the abbreviations $f_k := f(x^{(k)})$, $g_k := \nabla f(x^{(k)})$ for $k \in \mathbb{N}_0$:

1) NEWTON's *method*: $x^{(k+1)} := x^{(k)} \underbrace{-\nabla^2 f(x^{(k)})^{-1} g_k}_{=: d_k}$, assuming that the Hessian $\nabla^2 f(x^{(k)})$ is invertible.

2) *Damped* NEWTON *method*: $x^{(k+1)} := x^{(k)} + \lambda_k d_k$ for a $\lambda_k \in [0, 1]$ such that
$$f(x^{(k+1)}) = \min_{0 \le \lambda \le 1} f(x^{(k)} + \lambda d_k).$$

3) Modification of the descent direction: $d_k := -B_k^{-1} g_k$ with a suitable symmetric positive definite matrix $B_k \approx \nabla^2 f(x^{(k)})$.[10]

Chapter 3

Example 8

A simple *Matlab*® program for NEWTON's *method* applied to our framework example — with the starting point $w^{(0)} = (0, 0, 0)^T$, the regularization parameter $\lambda = 1$ and the tolerance of 10^{-7} — gives after only five iterations:

k	$w^{(k)}$			$f(w^{(k)})$	$\lVert \nabla f(w^{(k)}) \rVert_\infty$
0	0	0	0	8.31776617	6.00000000
1	−0.41218838	0.61903615	0.04359951	5.74848374	0.79309457
2	−0.49782537	0.74273470	0.06860137	5.68008777	0.04949655
3	−0.50481114	0.75317925	0.07065585	5.67969939	0.00031030
4	−0.50485551	0.75324907	0.07066909	5.67969938	0.00000002
5	−0.50485551	0.75324907	0.07066909	5.67969938	0.00000000

The idea underlying many quasi-NEWTON methods is: If the Hessian is cumbersome or too expensive or time-consuming to compute, it — or its inverse — is approximated by a suitable (easy to compute) matrix.

We consider for $x \in \mathbb{R}^n$ with

$$\Phi_{k+1}(x) := f_{k+1} + g_{k+1}^T (x - x^{(k+1)}) + \frac{1}{2}(x - x^{(k+1)})^T B_{k+1}(x - x^{(k+1)})$$

[10] Approximate matrices to $\nabla^2 f(x^{(k)})$ will be denoted by B_k, those to $\left(\nabla^2 f(x^{(k)})\right)^{-1}$ by H_k. This is surely a bit confusing but seems to be standard in the literature.

the corresponding quadratic approximation Φ_{k+1} of f in $x^{(k+1)}$. For that it holds that

$$\nabla\Phi_{k+1}(x) = B_{k+1}(x - x^{(k+1)}) + g_{k+1}\,.$$

We demand from B_{k+1} that $\nabla\Phi_{k+1}$ has to match with the gradient of f at $x^{(k)}$ and $x^{(k+1)}$:

$$\begin{aligned}\nabla\Phi_{k+1}(x^{(k+1)}) &= g_{k+1}\\ \nabla\Phi_{k+1}(x^{(k)}) &= B_{k+1}(x^{(k)} - x^{(k+1)}) + g_{k+1} \stackrel{!}{=} g_k\end{aligned}$$

The relation

$$B_{k+1}\,p_k = q_k$$

results from the above with $p_k := x^{(k+1)} - x^{(k)}$ and $q_k := g_{k+1} - g_k$. It is called the *secant relation* or Quasi-NEWTON condition. We are looking for suitable *updating formulae for B_k which meet the Quasi-*NEWTON *condition.* The first method of this kind was devised by DAVIDON (1959) and later developed further by FLETCHER and POWELL (1963) (cf. [Dav], [Fl/Po]).

Our presentation follows the discussions of J. GREENSTADT and D. GOLDFARB (1970) which themselves are based on the *"least change secant update" principle* by C. G. BROYDEN.

Least Change Principle

Since we do not want the discussion with its numerous formulae to become confused by too many indices, we will temporarily introduce — for a fixed $k \in \mathbb{N}_0$ — a *different notation*:

$x^{(k)} \longmapsto x\,,\ x^{(k+1)} \longmapsto x'$

$p_k \longmapsto p\,,\quad q_k \longmapsto q = \nabla f(x') - \nabla f(x)$

$B_k \longmapsto B\,,\ B_{k+1} \longmapsto B'$

The Quasi-NEWTON condition then reads $B'p = q$.

Denote by $\mathbb{M}_n$ the set of all real-valued (n, n)-matrices.

Definition

For a fixed symmetric and positive definite matrix $W \in \mathbb{M}_n$ let

$$\|A\|_W := \|W^{1/2} A\, W^{1/2}\|_F\,,$$

where $\|\ \|_F$ is the well-known FROBENIUS norm with

$$\|A\|_F^2 := \sum_{i,j=1}^{n} a_{i,j}^2 = \operatorname{trace}\,(A^T A)$$

for any matrix $A \in \mathbb{M}_n$.

Remark

The norm $\| \ \|_W$ *is strictly convex in the following weak sense: For* different $A_1, A_2 \in \mathbb{M}_n$ *with* $\|A_1\|_W = \|A_2\|_W = 1$ *it holds that* $\left\|\frac{1}{2}(A_1 + A_2)\right\|_W < 1$.

Proof: This follows immediately from the parallelogram identity in the HILBERT space $\left(\mathbb{M}_n, \| \ \|_F\right)$. □

This remark shows that the following approximation problem will have *at most* one (global) minimizer:

Proposition 3.5.1

Suppose $W \in \mathbb{M}_n$ *symmetric and positive definite,* $p, q \in \mathbb{R}^n$ *with* $p \neq 0$, $c := W^{-1}p$ *and* $B \in \mathbb{M}_n$ *symmetric. Then we obtain the following "rank 2-update of* B*", that is, a matrix of (at most) rank two is added to* B,

$$B' = B + \frac{(q - Bp)c^T + c(q - Bp)^T}{\langle c, p\rangle} - \frac{\langle q - Bp, p\rangle}{\langle c, p\rangle^2} cc^T$$

as the unique solution of the convex optimization problem

$$\min\left\{\|A - B\|_W : A \in \mathbb{M}_n \text{ with } A^T = A \text{ and } Ap = q\right\}.$$

This is called the "principle of least change".

Proof: The *existence* of a minimizer A of this problem follows by a routine compactness argument.

a) For the moment we only consider the case $W = I$:
For $A \in \mathbb{M}_n$ we have to minimize the term $1/2\|A - B\|_F^2$ subject to the constraints $Ap = q$ and $A = A^T$. With the corresponding Lagrangian

$$L(A, \varrho, \sigma) := \frac{1}{2} \sum_{i,j=1}^{n} (a_{ij} - b_{ij})^2 + \sum_{i=1}^{n} \varrho_i \Big(\sum_{j=1}^{n} a_{ij}\, p_j - q_i\Big) + \sum_{i,j=1}^{n} \sigma_{ij}\, (a_{ij} - a_{ji})$$

($\varrho \in \mathbb{R}^n$ and $A, \sigma \in \mathbb{M}_n$) it holds for the minimizer A and $1 \leq i, j \leq n$ that

$$\frac{\partial L}{\partial a_{ij}} = a_{ij} - b_{ij} + \varrho_i p_j + \sigma_{ij} - \sigma_{ji} = 0 \tag{17}$$

$$\sum_{j=1}^{n} a_{ij}\, p_j = q_i \tag{18}$$

$$a_{ij} = a_{ji}\,. \tag{19}$$

Exchanging the indices in (17) yields

$$a_{ji} - b_{ji} + \varrho_j p_i + \sigma_{ji} - \sigma_{ij} = 0;$$

addition to (17) and consideration of (19) leads to the relation

$$2a_{ij} - 2b_{ij} + \varrho_i p_j + p_i \varrho_j = 0$$

respectively

$$A = B - \tfrac{1}{2}\big(\varrho p^T + p\varrho^T\big)\,. \tag{20}$$

It follows from (20) that

$$q = Ap = Bp - \tfrac{1}{2}\big(\varrho p^T p + p\varrho^T p\big) = Bp - \tfrac{1}{2}\big(\varrho\,\langle p,p\rangle + p\,\langle \varrho,p\rangle\big); \tag{21}$$

setting $w := q - Bp$ we obtain from that:

$$\langle p,w\rangle = -\langle p,\varrho\rangle\,\langle p,p\rangle \quad\text{or}\quad \langle p,\varrho\rangle = -\frac{\langle p,w\rangle}{\langle p,p\rangle}\,.$$

With (21) we get

$$\langle p,p\rangle\,\varrho = -2w - \langle \varrho,p\rangle\,p \quad\text{or}\quad \varrho = -\frac{2}{\langle p,p\rangle}\,w + \frac{\langle p,w\rangle}{\langle p,p\rangle^2}\,p\,.$$

Hence,

$$\begin{aligned} A &= B - \frac{1}{2}\left\{\varrho p^T + p\varrho^T\right\} \\ &= B - \frac{1}{2}\left\{\Big(-\frac{2}{\langle p,p\rangle}\,w + \frac{\langle p,w\rangle}{\langle p,p\rangle^2}\,p\Big)p^T + p\Big(-\frac{2}{\langle p,p\rangle}\,w^T + \frac{\langle p,w\rangle}{\langle p,p\rangle^2}\,p^T\Big)\right\} \\ &= B + \frac{wp^T + pw^T}{\langle p,p\rangle} - \frac{\langle p,w\rangle}{\langle p,p\rangle^2}\,pp^T\,. \end{aligned}$$

This gives the result in this case (taking into account $p = c$).

b) For arbitrary W we set $M := W^{1/2}$ and obtain the desired result by considering

$$M(A-B)M = \underbrace{MAM}_{=:\,\widetilde{A}} - \underbrace{MBM}_{=:\,\widetilde{B}}$$

and

$$Ap = q \iff (MAM)\,\underbrace{M^{-1}p}_{=:\,\widetilde{p}} = \underbrace{Mq}_{=:\,\widetilde{q}} \iff \widetilde{A}\widetilde{p} = \widetilde{q}\,.$$

The optimization problem is equivalent to

$$\min\Big\{\|\widetilde{A} - \widetilde{B}\|_F \,:\, \widetilde{A} \in \mathbb{M}_n \text{ with } \widetilde{A}^T = \widetilde{A} \text{ and } \widetilde{A}\widetilde{p} = \widetilde{q}\Big\}.$$

Following *a)* we obtain with $\widetilde{w} := \widetilde{q} - \widetilde{B}\widetilde{p}$ the solution

$$\widetilde{A} = \widetilde{B} + \frac{\widetilde{w}\widetilde{p}^T + \widetilde{p}\widetilde{w}^T}{\langle \widetilde{p}, \widetilde{p} \rangle} - \frac{\langle \widetilde{p}, \widetilde{w} \rangle}{\langle \widetilde{p}, \widetilde{p} \rangle^2} \widetilde{p}\widetilde{p}^T .$$

Because of

$$\begin{aligned} &\widetilde{p} = Mc, \ \widetilde{w} = Mw, \\ &\langle \widetilde{p}, \widetilde{p} \rangle = \langle M^{-1}p, M^{-1}p \rangle = \langle p, M^{-2}p \rangle = \langle p, W^{-1}p \rangle = \langle p, c \rangle , \\ &\langle \widetilde{p}, \widetilde{w} \rangle = \langle M^{-1}p, Mw \rangle = \langle MM^{-1}p, w \rangle = \langle p, w \rangle \end{aligned}$$

we have — as claimed —

$$A = B + \frac{wc^T + cw^T}{\langle c, p \rangle} - \frac{\langle p, w \rangle}{\langle c, p \rangle^2} cc^T . \qquad \square$$

Chapter 3

Special Cases

1) The case $W = I$, which was firstly discussed in the proof, yielded (with $c = p$) the POWELL-*Symmetric*-BROYDEN *Update*

$$B'_{\text{PSB}} = B + \frac{(q - Bp)p^T + p(q - Bp)^T}{\langle p, p \rangle} - \frac{\langle q - Bp, p \rangle}{\langle p, p \rangle^2} pp^T$$

of the so-called PSB *method.*

2) If $\langle p, q \rangle > 0$, then there exists a symmetric positive definite matrix W with $Wq = p$: We can obtain such a matrix, for example, by setting

$$W := I - \frac{qq^T}{\langle q, q \rangle} + \frac{pp^T}{\langle p, q \rangle} :$$

Obviously $Wq = p$, and for $x \in \mathbb{R}^n$ it follows with the CAUCHY–SCHWARZ inequality (specifically with the characterization of the equality) that

$$\langle x, Wx \rangle = \langle x, x \rangle - \underbrace{\frac{\langle q, x \rangle^2}{\langle q, q \rangle}}_{\leq \langle x, x \rangle} + \frac{\langle p, x \rangle^2}{\langle p, q \rangle} \geq 0 \qquad \text{as well as}$$

$$\langle x, Wx \rangle = 0 \iff x = \lambda q \text{ with a suitable } \lambda \in \mathbb{R} \text{ and } \langle p, x \rangle = 0 .$$

Hence $x = \lambda q$ follows from $\langle x, Wx \rangle = 0$ and further $0 = \langle p, x \rangle = \lambda \langle p, q \rangle$ and from that $\lambda = 0$, hence, $x = 0$.

With $c = q$ we obtain the *updating formula of* DAVIDON–FLETCHER–POWELL

$$\begin{aligned} B'_{\text{DFP}} &= B + \frac{(q-Bp)q^T + q(q-Bp)^T}{\langle p,q\rangle} - \frac{\langle q-Bp,p\rangle}{\langle p,q\rangle^2}\, qq^T \\ &= \Big(I - \frac{qp^T}{\langle p,q\rangle}\Big) B \underbrace{\Big(I - \frac{pq^T}{\langle p,q\rangle}\Big)}_{=:Q} + \frac{qq^T}{\langle p,q\rangle} \end{aligned}$$

of the so-called DFP *method*, also often referred to as the *variable metric method*. Since $Q^2 = Q$ the matrix Q is a *projection* with $Qp = 0$.

Chapter 3

When updating a symmetric positive definite $B := B_k$ in a quasi-NEWTON method, it is desirable that $B' := B_{k+1}$ is symmetric positive definite too:

Proposition 3.5.2

If B is symmetric positive definite, then — with the assumption $\langle p,q\rangle > 0$ — B'_{DFP} is also symmetric positive definite. For $H'_{DFP} := (B'_{DFP})^{-1}$ and $H := B^{-1}$ it holds that

$$H'_{DFP} = H + \frac{pp^T}{\langle p,q\rangle} - \frac{Hqq^TH}{\langle q,Hq\rangle}\,, \tag{22}$$

and we get the original DFP update (cf. [Fl/Po]).

Proof: For $x \in \mathbb{R}^n$ we have

$$\langle x, B'_{\text{DFP}}x\rangle = \Big\langle x, Q^TBQx + \frac{qq^T}{\langle p,q\rangle}x\Big\rangle = \underbrace{\langle Qx, B(Qx)\rangle}_{\geq 0} + \underbrace{\frac{\langle q,x\rangle^2}{\langle p,q\rangle}}_{\geq 0} \geq 0\,.$$

From $\langle x, B'_{\text{DFP}}x\rangle = 0$ it thus follows $\langle Qx, B(Qx)\rangle = 0$ and $\langle q,x\rangle = 0$, hence, $Qx = 0$ and $\langle q,x\rangle = 0$, and finally $x = 0$. If we denote the right-hand side of (22) by $\widetilde{H}'$, $B'_{\text{DFP}}\widetilde{H}' = I$ follows after some transformations from

$$B'_{\text{DFP}}\widetilde{H}' = \Big(Q^TBQ + \frac{qq^T}{\langle p,q\rangle}\Big)\Big(H + \frac{pp^T}{\langle p,q\rangle} - \frac{Hqq^TH}{\langle q,Hq\rangle}\Big)\,,$$

hence, $H'_{\text{DFP}} = \widetilde{H}'$. □

Remark

Suppose that in the damped NEWTON method $x^{(k+1)} := x^{(k)} - \lambda_k \nabla^2 f\big(x^{(k)}\big)^{-1} g_k$ we do not approximate $\nabla^2 f\big(x^{(k)}\big)$ with B_k, but $\nabla^2 f\big(x^{(k)}\big)^{-1}$ with H_k. Then we get the following new kind of Quasi-NEWTON method

$$x^{(k+1)} := x^{(k)} - \lambda_k H_k g_k\,.$$

The Quasi-NEWTON condition transforms into $H_{k+1}q_k = p_k$.

If we exchange q and p as well as B and H, it follows with proposition 3.5.1:

Proposition 3.5.1'

Suppose that $W \in \mathbb{M}_n$ is symmetric and positive definite, $p, q \in \mathbb{R}^n$ with $q \neq 0$, $d := W^{-1}q$ and $H \in \mathbb{M}_n$ symmetric. Then the unambiguously determined solution of

$$\min \{\|G - H\|_W : G \in \mathbb{M}_n \text{ with } G^T = G \text{ and } Gq = p\}$$

is given by

$$H' = H + \frac{(p - Hq)d^T + d(p - Hq)^T}{\langle d, q\rangle} - \frac{\langle p - Hq, q\rangle}{\langle d, q\rangle^2} dd^T .$$

Special Cases

1) $W = I$: In this case we have $d = q$, and with

$$H'_G = H + \frac{(p - Hq)q^T + q(p - Hq)^T}{\langle q, q\rangle} - \frac{\langle p - Hq, q\rangle}{\langle q, q\rangle^2} qq^T$$

we obtain the GREENSTADT *update*.

2) If $\langle p, q\rangle > 0$, there exists a symmetric and positive definite matrix W such that $Wp = q$. Consequently, $d = p$, and we obtain the BROYDEN–FLETCHER–GOLDFARB–SHANNO *updating formula* (BFGS formula)

$$\begin{aligned} H'_{\text{BFGS}} &= H + \frac{(p - Hq)p^T + p(p - Hq)^T}{\langle p, q\rangle} - \frac{\langle p - Hq, q\rangle}{\langle p, q\rangle^2} pp^T \\ &= \left(I - \frac{pq^T}{\langle p, q\rangle}\right) H \left(I - \frac{qp^T}{\langle p, q\rangle}\right) + \frac{pp^T}{\langle p, q\rangle} . \end{aligned}$$

Remark

This updating formula is the one most commonly used. Due to empirical evidence the BFGS method seems to be the best for general purposes, since good convergence properties remain valid even with inexact line search. Besides that, it has effective self-correcting properties.

Proposition 3.5.2'

If $\langle p, q\rangle > 0$ and H symmetric positive definite, then H'_{BFGS} is also symmetric positive definite. For $B'_{BFGS} := \left(H'_{BFGS}\right)^{-1}$ and $B := H^{-1}$ the formula

$$B'_{BFGS} = B + \frac{qq^T}{\langle p, q\rangle} - \frac{Bpp^TB}{\langle p, Bp\rangle}$$

holds which is the dual of the DFP formula (22).

Chapter 3

More General Quasi-Newton Methods

Many of the properties of the DFP and BFGS formulae extend to more general classes which can dominate these in special cases, especially for certain nonquadratic problems with inexact line search.

a) BROYDEN's class (1970)

This updating formula, which is also called the *SSVM technique (self-scaling variable metric)*, is a convex combination of those of the DFP and BFGS methods; hence, for $\vartheta \in [0, 1]$ we have

$$H'_\vartheta := \vartheta H'_{\text{BFGS}} + (1-\vartheta) H'_{\text{DFP}} =$$
$$H + \left(1 + \vartheta \frac{\langle q, Hq\rangle}{\langle p, q\rangle}\right) \frac{pp^T}{\langle p, q\rangle} - (1-\vartheta) \frac{Hqq^T H}{\langle q, Hq\rangle} - \frac{\vartheta}{\langle p, q\rangle}\left(pq^T H + Hqp^T\right) .$$

If H is positive definite, $\vartheta \in [0, 1]$ and $\langle p, q\rangle > 0$, then the matrix H'_ϑ is positive definite. Clearly $\vartheta = 0$ yields the DFP and $\vartheta = 1$ the BFGS update. A BROYDEN *method* is a quasi-NEWTON method in which a BROYDEN update is used in each step, possibly with varying parameters ϑ.

b) If we replace the matrix H by γH with a positive γ as an additional scaling factor, we obtain for $\vartheta \in [0, 1]$ the OREN–LUENBERGER class

$$H \longmapsto \gamma H \longmapsto H'_{\gamma,\vartheta} := \vartheta H'_{\text{BFGS}}(\gamma H) + (1-\vartheta) H'_{\text{DFP}}(\gamma H).$$

An OREN–LUENBERGER method with exact line search terminates after at most n iterations for quadratic functions:

Proposition 3.5.3

Suppose that f is given by

$$f(x) = \frac{1}{2}\langle x, Ax\rangle + \langle b, x\rangle$$

with $b \in \mathbb{R}^n$ and a symmetric positive definite matrix $A \in \mathbb{M}_n$. In addition let $x^{(0)} \in \mathbb{R}^n$ and a symmetric positive definite matrix $H_0 \in \mathbb{M}_n$ be given. Then each method of the OREN–LUENBERGER *class yields, starting from $\left(x^{(0)}, H_0\right)$, with exact line search*

$$x^{(k+1)} = x^{(k)} - \lambda_k H_k g_k, \quad \min_{\lambda \geq 0} f\left(x^{(k)} - \lambda H_k g_k\right) = f\left(x^{(k+1)}\right),$$

sequences $x^{(k)}$, H_k, $p_k := x^{(k+1)} - x^{(k)}$ and $q_k := g_{k+1} - g_k$ such that:

a) There exists a smallest $m \in \{0, \ldots, n\}$ such that $g_m = 0$. Then $x^{(m)} = -A^{-1}b$ is the minimal point of f.

b) The following assertions hold:

1) $\langle p_i, q_k \rangle = \langle p_i, Ap_k \rangle = 0$ *for* $0 \le i < k \le m-1$

$\langle p_i, q_i \rangle > 0$ *for* $0 \le i \le m-1$

H_i *is symmetric positive definite for* $0 \le i \le m$.

2) $\langle p_i, g_k \rangle = 0$ *for* $0 \le i < k \le m$.

3) $H_k q_i = \gamma_{i,k} p_i$ *for* $0 \le i < k \le m$, *where*

$$\gamma_{i,k} := \begin{cases} \gamma_{i+1} \cdot \gamma_{i+2} \cdot \cdots \cdot \gamma_{k-1} & \text{for } i < k-1 \\ 1 & \text{for } i = k-1. \end{cases}$$

c) If $m = n$, *it holds in addition:*

$$H_n = PDP^{-1}A^{-1}$$

with

$$D := \operatorname{Diag}(\gamma_{0,n}, \gamma_{1,n}, \ldots, \gamma_{n-1,n}),$$
$$P := (p_0, p_1, \ldots, p_{n-1}).$$

Hence, for BROYDEN*'s class (that is,* $\gamma_k = 1$ *for all k)* $H_n = A^{-1}$ *holds.*

Proof: See [St/Bu], theorem 5.11.10, p. 320ff.

This result is somewhat idealized because an exact line search cannot be guaranteed in practice for arbitrary functions f.

Quasi-Newton Methods in the Nonquadratic Case

We content ourselves with presenting the *framework* of general DFP and BFGS methods:

DFP method:

0) Starting with $x^{(0)} \in \mathbb{R}^n$ *and* $H_0 = I$ *for* $k \in \mathbb{N}_0$ *let:*

1) $d_k := -H_k g_k \quad \left(H_k \approx \nabla^2 f(x^{(k)})^{-1}\right)$

If $g_k \ne 0$: d_k *is a descent direction of* f *in* $x^{(k)}$, *since for*

$\varphi(t) := f\big(x^{(k)} + t\,d_k\big)$ *it holds that*

$\varphi'(0) = g_k^T d_k = -g_k^T H_k g_k < 0$.

2) Calculate a $\lambda_k > 0$ *with* $f\big(x^{(k)} + \lambda_k d_k\big) = \min_{\lambda \ge 0} f\big(x^{(k)} + \lambda d\big)$ *or via suitable inexact line search.*

$x^{(k+1)} := x^{(k)} + \lambda_k d_k$

3) Update with a given $\gamma \in (0,1)$:

$$H_{k+1} := \begin{cases} H_k + \dfrac{p_k p_k^T}{\langle p_k, q_k \rangle} - \dfrac{H_k q_k q_k^T H_k}{\langle q_k, H_k q_k \rangle}, & \text{if } \langle p_k, q_k \rangle \geq \gamma \|p_k\|_2 \|q_k\|_2 \\ H_0 & \text{, else restart.} \end{cases}$$

BFGS method:

$d_k := B_k^{-1} g_k \qquad \left(B_k \approx \nabla^2 f(x^{(k)})\right)$

Updating formula:

$$B_{k+1} := \begin{cases} B_k + \dfrac{q_k q_k^T}{\langle p_k, q_k \rangle} - \dfrac{B_k p_k p_k^T B_k}{\langle p_k, B_k p_k \rangle}, & \text{if } \langle p_k, q_k \rangle \geq \gamma \|p_k\|_2 \|q_k\|_2 \\ B_0 & \text{, else restart.} \end{cases}$$

The literature on quasi-NEWTON methods is quite extensive. We did not intend to cover the whole range in detail. This is beyond the scope of our presentation and it might be confusing for a student.

Convergence properties of these methods are somewhat difficult to prove. We refer the interested reader for example to [No/Wr], section 6.4 or [De/Sc] where a comprehensive treatment of quasi-NEWTON methods can be found.

Example 9

In order to finally apply the DFP and BFGS methods to our framework example, we again utilize the function fminunc — here with the options optimset('HessUpdate', 'dfp') and optimset('HessUpdate', 'bfgs') . For the starting point $w^{(0)} = (0,0,0)^T$ and the regularization parameter $\lambda = 1$ we need 62 (DFP) and 15 (BFGS) iteration steps to reach the tolerance of 10^{-7}. As in the preceding examples, we only list a few of them:

DFP

k	$w^{(k)}$			$f(w^{(k)})$	$\|\nabla f(w^{(k)})\|_\infty$
0	0	0	0	8.31776617	6.00000000
2	−0.32703997	0.28060590	0.01405140	6.67865969	5.81620726
4	−0.48379776	0.71909320	0.07610771	5.68355186	0.19258622
6	−0.50560601	0.75191947	0.07912688	5.67974239	0.01290898
14	−0.50582098	0.75201407	0.07398917	5.67972332	0.03566679
23	−0.50463179	0.75347752	0.06704473	5.67971060	0.01623329
33	−0.50452876	0.75367743	0.06994114	5.67970303	0.01552518
43	−0.50502368	0.75304119	0.07161968	5.67970002	0.00302611
52	−0.50482358	0.75328767	0.07044751	5.67969941	0.00022112
62	−0.50485551	0.75324907	0.07066907	5.67969938	0.00000005

BFGS

k	$w^{(k)}$			$f(w^{(k)})$	$\|\nabla f(w^{(k)})\|_\infty$
0	0	0	0	8.31776617	6.00000000
1	−0.30805489	0.19253431	0	7.20420404	7.47605903
2	−0.38153345	0.43209082	0.03570334	6.10456800	3.59263698
3	−0.46878531	0.67521531	0.07011404	5.69897036	0.57610147
4	−0.49706979	0.73581744	0.07772590	5.68053511	0.09150746
6	−0.50578467	0.75171902	0.07917689	5.67974092	0.01002935
8	−0.50627420	0.75246799	0.07785948	5.67973141	0.01278409
10	−0.50590101	0.75350315	0.07263553	5.67970764	0.02001700
12	−0.50489391	0.75333394	0.07043112	5.67969945	0.00175692
15	−0.50485548	0.75324905	0.07066907	5.67969938	0.00000059

Lastly, we have compiled — for our framework example — some important data from the different methods for comparison:

	$\lambda = 0$		$\lambda = 1$		$\lambda = 2$	
	iter	*sec*	*iter*	*sec*	*iter*	*sec*
Nelder–Mead	205	0.20	168	0.17	158	0.16
Steepest Descent	524	1.26	66	0.40	52	0.38
Trust Region	10	0.47	6	0.44	5	0.44
Newton	6	0.014	5	0.013	5	0.013
DFP	633	0.67	62	0.32	38	0.31
BFGS	20	0.30	15	0.29	14	0.29
$c(H^*)$	522.63		48.26		28.16	

The table shows that the regularization parameter λ may have great influence on the number of iterations *(iter),* the time needed and the condition of the corresponding Hessian H^* in the respective minimizer. One should, however, heed the fact that these are the results of one particular problem and that no conclusions can be drawn from them to other problems. Moreover, it is well possible that individual methods cannot be used, because necessary prerequisites are not met (for example, smoothness or convexity). Also, it may be dangerous to only look at the number of iteration steps as the costs per iteration can differ greatly!

Example 10

As already mentioned at the beginning of this chapter, with *breast cancer diagnosis* we now want to turn to a more serious 'real-life' problem for which we will again assemble important results from the different methods in tabular form:

	$\lambda = 0$		$\lambda = 1$		$\lambda = 2$	
	iter	*sec*	*iter*	*sec*	*iter*	*sec*
Nelder–Mead	6 623	2.42	5 892	2.17	5 779	2.25
Steepest Descent	29 857	85.10	2 097	6.58	1 553	5.12
Trust Region	38	0.61	7	0.44	7	0.45
Newton	11	0.03	8	0.02	7	0.02
DFP	10 855	8.66	2 180	1.83	1 606	1.78
BFGS	104	0.39	79	0.36	76	0.36
$c(H^*)$	$\approx 1.6\,e4$		371.45		273.58	

We have split the data (wisconsin-breast-cancer.data) into two portions: The first 120 instances are used as training data. The remaining 563 instances are used as test data to evaluate the 'performance' of the *classifier* or *decision function*. The *exact figures* are displayed in the following table:

	$\lambda = 0$		$\lambda = 1$		$\lambda = 2$	
	M	*B*	*M*	*B*	*M*	*B*
P	362	3	366	5	362	5
N	17	181	13	179	17	179

Here 'P' stands for a positive test, 'N' for a negative test (from a medical point of view!). A malignant tumor is denoted by 'M' and a benign tumor by 'B'.

Accuracy, sensitivity and *specificity* are measures of the performance of a binary classification method, in this case of a medical test. *Sensitivity* measures the probability of a positive test among patients with disease. The higher the sensitivity, the fewer real cases of breast cancer (in our example) go undetected. The *specificity* measures the probability of a negative test among patients without disease (healthy people who are identified as not having breast cancer). *Accuracy* measures the proportion of all test persons correctly identified.

Accuracy	$\frac{362+181}{563} = 0.9645$	$\frac{366+179}{563} = 0.9680$	$\frac{362+179}{563} = 0.9609$
Sensitivity	$\frac{362}{379} = 0.9551$	$\frac{366}{379} = 0.9657$	$\frac{362}{379} = 0.9551$
Specificity	$\frac{181}{184} = 0.9837$	$\frac{179}{184} = 0.9728$	$\frac{179}{184} = 0.9728$

A comparison with the results for the *support vector machine method* shows that their accuracy values are almost the same: 95.74 % (SVM) versus 96.80 % (logistic regression with $\lambda = 1$).

Exercises

1. The ROSENBROCK function $f: \mathbb{R}^2 \longrightarrow \mathbb{R}$, $x \mapsto 100\left(x_2 - x_1^2\right)^2 + (1 - x_1)^2$, also compare `http://en.wikipedia.org/wiki/Rosenbrock_function`, is frequently utilized to test optimization methods.

   a) The absolute minimum of f, at the point $(1,1)^T$, can be seen without any calculations. Show that it is the only extremal point.

   b) Graph the function f on $[-1.5, 1.5] \times [-0.5, 2]$ as a 3D plot or as a level curve plot to understand why f is also referred to as the *banana function* and the like.

   c) Implement the NELDER–MEAD method. Visualize the level curves of the given function together with the polytope for each iteration. Finally visualize the trajectory of the centers of gravity of the polytopes!

   d) Test the program with the starting polytope given by the vertices $(-1,1)^T, (0,1)^T, (-0.5,2)^T$ and the parameters $(\alpha, \beta, \gamma) := (1, 2, 0.5)$ and $\varepsilon := 10^{-4}$ using the ROSENBROCK function. How many iterations are needed? What is the distance between the calculated solution and the exact minimizer $(1,1)^T$?

   e) Find $(\alpha, \beta, \gamma) \in [0.9, 1.1] \times [1.9, 2.1] \times [0.4, 0.6]$ such that with $\varepsilon := 10^{-4}$ the algorithm terminates after as few iterations as possible. What is the distance between the solution and the minimizer $(1,1)^T$ in this case?

   f) If the distance in *e)* was greater than in *d)*, reduce ε until the algorithm gives a result — with the (α, β, γ) found in *e)* — which is not farther away from $(1,1)^T$ and at the same time needs fewer iterations than the solution in *d)*.

2. Implement SHOR's *ellipsoid method* and use it to search for the minimizers of the functions $f_1, f_2: \mathbb{R}^2 \longrightarrow \mathbb{R}$ defined by $f_1(x) := 3x_1^2 + x_2^2 + 3x_1$ and $f_2(x) := x_1^2 + \cos(\pi(x_1 + x_2)) + \frac{1}{2}(x_2 - 1)^2$ as well as the ROSENBROCK function. Let the starting ellipsoid $\mathcal{E}^{(0)}$ be given by

$$x^{(0)} := (0, -1)^T \quad \text{and} \quad A_0 := \begin{pmatrix} 5 & 0 \\ 0 & 5 \end{pmatrix}.$$

   For $k = 1, 2, \ldots$ visualize the level curves of each function together with the corresponding ellipsoids $\mathcal{E}^{(k)}$ of the kth iteration.

   What is your observation when you are trying to find the minimum of the concave function f given by $f(x_1, x_2) := -x_1^2 - x_2^2$?

3. *Modified* ARMIJO *step size rule*
   Let a differentiable function $\varphi: \mathbb{R}_+ \longrightarrow \mathbb{R}$ with $\varphi'(0) < 0$ be given, in

Chapter 3

addition $0 < \alpha < 1$, an initial step size λ_0 and a step size factor $\varrho > 1$. Determine the maximal step size λ in $S := \{\lambda_0 \varrho^k \mid k \in \mathbb{Z}\}$ for which $\varphi(\lambda) \leq \varphi(0) + \alpha \lambda \varphi'(0)$ holds.

a) Implement this ARMIJO step size rule with *Matlab®* or *Maple®* and test it for different φ, α and ϱ. Observe the distance from the exact minimizer of φ. Study the examples

$$\begin{aligned} \varphi(t) &:= 1 - \frac{1}{1 - t + 2t^2} \\ \varphi(t) &:= 2 - 4t + \exp(t) \\ \varphi(t) &:= 1 - t\exp(-t^2) \end{aligned}$$

with $\alpha \in \{0.1, 0.9\}$, $\varrho \in \{1.5, 2\}$ and $\lambda_0 := 0.1$.

b) Implement the *gradient method* with this step size rule. (In each iteration step the function φ is defined by $\varphi(t) := f(x^{(k)} - t\, g_k)$, where f is the objective function to be minimized.) Compare the results of this *inexact line search* with those of the *exact line search.* Test it with the function f given by

$$f(x_1, x_2) := \frac{1}{2}x_1^2 + \frac{9}{2}x_2^2 ,$$

and the *weakened banana-shaped valley function* of ROSENBROCK defined by

$$f(x_1, x_2) := 10\,(x_2 - x_1^2)^2 + (1 - x_1)^2 .$$

The number of iterations needed to obtain a euclidean distance from the minimizer of less than $\varepsilon = 10^{-k}$ $(k = 1, \ldots, 6)$ can serve as a criterion for comparison.

4. Let $f\colon \mathbb{R}^n \longrightarrow \mathbb{R}$ be a continuously differentiable function and $H \in \mathbb{M}_n$ a symmetric positive definite matrix. Show:

a) $\langle x, y \rangle_H := \langle x, Hy \rangle$ defines an inner product on $\mathbb{R}^n$. The thereby induced norm is denoted by $\|\ \|_H$.

b) The direction of steepest descent of f in a point x with $\nabla f(x) \neq 0$ with respect to the norm $\|\ \|_H$, that is, the solution to the optimization problem

$$\min \nabla f(x)^T d \quad \text{such that} \quad \|d\|_H = 1,$$

is given by

$$d = -\frac{H^{-1}\nabla f(x)}{\|H^{-1}\nabla f(x)\|_H} .$$

c) Let $(x_k) \in (\mathbb{R}^n)^{\mathbb{N}}$ and $(d_k) \in (\mathbb{R}^n)^{\mathbb{N}}$ with $d_k := -H^{-1}\nabla f(x_k) \neq 0$. Then the sequence (d_k) meets the *angle condition,* that is, there exists a constant $c > 0$ such that for all $k \in \mathbb{N}$

$$-\frac{\langle \nabla f(x_k), d_k \rangle}{\|\nabla f(x_k)\|\ \|d_k\|} \geq c.$$

5. *Golden Section Search*

The one-dimensional 'line search' is the basis of many multidimensional optimization methods. We will have a closer look at the following simple variant for the minimization of a continuous function $f\colon \mathbb{R} \longrightarrow \mathbb{R}$.

Algorithm:

Initialize $t_1^{(0)}, t_2^{(0)}$ such that $t^* \in \left[t_1^{(0)}, t_2^{(0)}\right]$ holds for an optimizer t^*. Set $j = 0$. Calculate

$$t_3^{(j)} := \alpha\, t_1^{(j)} + (1-\alpha)\, t_2^{(j)}, \quad t_4^{(j)} := \alpha\, t_2^{(j)} + (1-\alpha)\, t_1^{(j)},$$

where $\alpha := \frac{1}{2}(\sqrt{5}-1)$.

- If $f\big(t_3^{(j)}\big) > f\big(t_4^{(j)}\big)$, set $t_1^{(j+1)} = t_3^{(j)}$ and $t_2^{(j+1)} = t_2^{(j)}$.
- Else, set $t_2^{(j+1)} = t_4^{(j)}$ and $t_1^{(j+1)} = t_1^{(j)}$.

Set $j = j+1$ and repeat the above calculations until convergence is reached, that is, $t_2^{(j)} - t_1^{(j)} \le \varepsilon$, where ε is a given tolerance.

a) Show that this algorithm *always* converges (*where to* is a different question), hence $t_2^{(j)} - t_1^{(j)} \longrightarrow 0$ as $j \longrightarrow \infty$. How many iterations are necessary to go below a given tolerance ε?

b) Implement this algorithm and determine the minimum of the functions $f_1, f_2, f_3\colon \mathbb{R} \longrightarrow \mathbb{R}$ with

$$f_1(t) := |t|, \quad f_2(t) := t^2 + 5\,|\sin(t)|, \quad f_3(t) := t^2 + 5\,|\sin(3t)|$$

and the starting interval $\big(t_1^{(0)}, t_2^{(0)}\big) = (-5, 5)$ numerically. What can be observed?

c) Show that for convex functions the algorithm converges to a minimizer.

6. Let $f\colon \mathbb{R}^2 \longrightarrow \mathbb{R}$ with $f(x) = \frac{1}{2}\langle x, Ax\rangle + \langle b, x\rangle$ and A symmetric, positive definite. Implement the *gradient method* to minimize f (cf. p. 100f).

a) Try to select the matrix A, the vector b and the starting vector x_0 such that the trajectory (the line connecting the approximations x_0, $x_1, x_2, \ldots$) forms a zigzag line similar to the following picture:

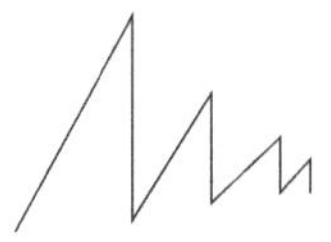

b) Try to choose the matrix such that the angle between consecutive directions becomes as small as possible ('wild zigzagging').

Make a conjecture as to what the smallest possible angle in this zigzag line is. Prove this!

Chapter 3

c) Which conditions does the matrix A have to meet to get a 'spiral' trajectory (see below)?

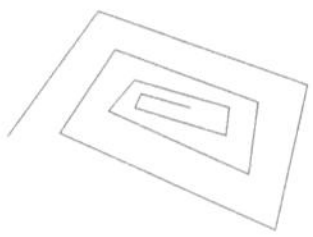

7. Let $f\colon \mathbb{R}^2 \longrightarrow \mathbb{R}$. Implement the *gradient method* to solve

$$f(x) \longrightarrow \min\,, \quad x \in \mathbb{R}^2.$$

Use the Golden Section Search algorithm (GSS) from exercise 5 to determine the step size. Visualize the level curves of f together with the approximations of the solutions x_k, $k = 0, 1, 2, \ldots$.

a) Do the first 20 iteration steps with the gradient method. Choose

$$f(x) := x_1^4 + 20x_2^2 \quad \text{and} \quad x_0 := (1,\, 0.5)^T$$

as the test function and the starting point. Let ε be the termination tolerance for the one-dimensional minimization.

b) Deactivate the visualization part in your program (since with smaller problems the visualization is often the most costly operation). Then repeat *a)* and complete the following table:

ε	$\lVert x - x_* \rVert_2$	total runtime	percentage needed for GSS
10^{-1}			
10^{-2}			
10^{-4}			
10^{-6}			
10^{-8}			
10^{-10}			
10^{-12}			

What conclusions can you draw? What strategy would you recommend for the selection of ε?

8. We are going to use the gradient method to minimize the function $f\colon \mathbb{R}^n \longrightarrow \mathbb{R}$ defined by $f(x) := \frac{1}{2}\langle x, Ax\rangle$ with a positive definite matrix A. When choosing the step size, however, assume that we allow there to be a certain tolerance $\delta\widehat{\lambda}_k$, that is, assume

$$x^{(k+1)} = x^{(k)} - \lambda_k\, g_k$$

with $|\lambda_k - \widehat{\lambda}_k| \le \delta\widehat{\lambda}_k$, where $\widehat{\lambda}_k$ minimizes the term $f(x^{(k)} - \lambda g_k)$ with reference to $\lambda \in \mathbb{R}$.

a) Express the speed of convergence of this algorithm with the help of the condition number $c(A)$ and the tolerance δ.

b) What is the largest value δ for which the algorithm still converges? Explain this geometrically!

9. The (stationary) temperature distribution T in a long bar can be expressed by means of the differential equation

$$-T''(x) = f(x)\ , \quad x \in (0,\, 1)\,. \tag{23}$$

$f(x)$ is the intensity of the heat source. The temperature at both ends is given by

$$T(0) = T(1) = 0\,.$$

For twice continuously differentiable functions $T\colon\ [0,\, 1] \longrightarrow \mathbb{R}$ the solution to (23) is also the *minimizer* of the following variational problem

$$\frac{1}{2}a(T,T) - \ell(T) \longrightarrow \min\,,\ \ T(0) = T(1) = 0\,, \tag{24}$$

where a and ℓ are given by $a(T,P) = \int_0^1 T'(x)P'(x)\,dx$ and $\ell(P) = \int_0^1 f(x)P(x)\,dx$ (cf. [St/Bu], p. 540ff).

If we discretize (24), we obtain a 'classic' optimization problem. Let $x_i = (i-1)h$, $i = 1,\, 2,\, \dots, N$, be equidistant points in the interval $[0,1]$ with $h = 1/(N-1)$ and $x_{i+\frac{1}{2}} = x_i + \frac{h}{2}$, $i = 1,\, 2,\, \dots,\, N-1$. Let furthermore $g_i := g(x_i)$ and $g_{i+\frac{1}{2}} := g(x_{i+\frac{1}{2}})$ for a function $g\colon\ [0,1] \longrightarrow \mathbb{R}$. We will use the following approximations

$$\begin{aligned} \frac{dg}{dx}(x_{i+\frac{1}{2}}) &\approx \frac{1}{h}(g_{i+1} - g_i) \\ g_{i+\frac{1}{2}} &\approx \frac{1}{2}(g_{i+1} + g_i) \\ \int_{x_i}^{x_{i+1}} g(x)\,dx &\approx h g_{i+\frac{1}{2}}\,. \end{aligned}$$

a) Formulate the approximations of (24) as an unconstrained optimization problem!

b) Show that this problem has a unique solution!

c) Find an approximate solution to (24) via

(i) cyclic control (relaxation control),

(ii) the gradient method.

Use the ARMIJO step size rule. Visualize the temperature distribution in each iteration step. What is your observation?

Chapter 3

d) Replace the ARMIJO step size rule by something else. What happens, for example, if you set $\lambda_k = 0.1$ (0.01), $\lambda_k = 0.2$ or $\lambda_k = 1/k$?

Hint: In the special case $f(x) \equiv c$ in $(0,1)$ the solution to (23) with the constraints $T(0) = T(1) = 0$ can be determined analytically. It is

$$T(x) = \frac{1}{2} c\, x\,(1-x)\,.$$

The vector $\big(T(x_1), \ldots, T(x_{N-1})\big)^T$ also yields the minimum of the approximating variational problem at the same time. This fact can be used to test the program.

10. Calculate the LEVENBERG–MARQUARDT trajectory to

a) $f(x) := 2x_1^2 - 2x_1x_2 + x_2^2 + 2x_1 - 2x_2$ at the point $x^{(0)} := (0,\, 0)^T$,

b) $f(x) := x_1x_2$ at the point $x^{(0)} := (1,\, 0.5)^T$,

c) $f(x) := x_1x_2$ at the point $x^{(0)} := (1,\, 1)^T$.

Treat the examples according to the case differentiations on page 113.

11. Do one iteration step of the *trust region method* for the example $f(x) := x_1^4 + x_1^2 + x_2^2$. Choose $x^{(0)} := (1,\, 1)^T$, $H_0 := \nabla^2 f(x^{(0)})$ and $\Delta_0 := \frac{3}{4}$ and determine d_0 with the help of POWELL's *dogleg trajectory.*

12. We again consider the weakened ROSENBROCK function $f\colon \mathbb{R}^2 \longrightarrow \mathbb{R}$ defined by $f(x) := 10\left(x_2 - x_1^2\right)^2 + (1-x_1)^2$.

a) Plot the level curves of f and the level curves of the quadratic approximation of f at the point $x_0 := (0, -1)^T$.

b) Calculate the solutions to the quadratic minimization problem

$$\begin{cases} \varphi(x_0 + d) := f(x_0) + \nabla f(x_0)^T d + \frac{1}{2} d^T \nabla^2 f(x_0) d \longrightarrow \min \\ \|d\| \le \Delta_0 \end{cases}$$

for the trust regions with the radii $\Delta_0 := 0.25, 0.75, 1.25$.

c) Repeat *a)* and *b)* for $x_0 := (0, 0.5)^T$.

13. Consider the quadratic function $f\colon \mathbb{R}^3 \longrightarrow \mathbb{R}$ defined by

$$f(x) := x_1^2 - x_1x_2 + x_2^2 - x_2x_3 + x_3^2.$$

First write f in the form

$$f(x) = \tfrac{1}{2}\,\langle x, Ax\rangle + \langle b, x\rangle$$

with a symmetric positive definite matrix $A \in \mathbb{M}_3$ and $b \in \mathbb{R}^3$. Calculate the minimum of this function by means of the *conjugate gradient method.* Take $x^{(0)} := (0, 1, 2)^T$ as a starting vector and apply exact line search.

14. Calculate the minimum of the quadratic function f in the preceding exercise by means of the DFP method. Start at $x^{(0)} := (0,1,2)^T$ with $H_0 := I$ and apply exact line search. Give the inverse of the Hessian of f.

15. Let $f\colon \mathbb{R}^2 \longrightarrow \mathbb{R}$ be defined by $f(x) := -12x_2+4x_1^2+4x_2^2-4x_1x_2$. Write f as

$$f(x) = \tfrac{1}{2}\langle x, Ax\rangle + \langle b, x\rangle$$

with a positive definite symmetric matrix $A \in \mathbb{M}_2$ and $b \in \mathbb{R}^2$. To $d_1 := (1,0)^T$ find all the vectors $d_2 \in \mathbb{R}^2$ such that the pair (d_1, d_2) is A-conjugate.

Minimize f starting from $x^{(0)} := (-\frac{1}{2}, 1)^T$ in the direction of d_1 and from the thus obtained point in the direction of $d_2 := (1,2)^T$.

Sketch the situation (level curves; $x^{(0)}, x^{(1)}, x^{(2)}$).

16. Let $n \in \mathbb{N}$ and T_n the n-th CHEBYSHEV *polynomial*, cf. p. 126ff. Verify:

a) $T_n(t) = 2^{n-1} \prod_{\nu=0}^{n-1} \left(t - \cos\left(\frac{2\nu+1}{2n}\pi\right)\right)$

b) Among all polynomials of degree n with leading coefficient 2^{n-1} on the interval $[-1,1]$ T_n has the smallest maximum norm, which is 1.

17. Let $f\colon \mathbb{R}^2 \longrightarrow \mathbb{R}$ be given by the function in exercise 12. Starting from $x^{(0)} := (0,-1)^T$ carry out some steps of the FLETCHER and REEVES algorithm using 'inexact line search' by hand and sketch the matter.

18. Solve the least squares problem

$$F(x) := \frac{1}{2}\sum_{i=1}^{5} f_i(x)^2 \longrightarrow \min$$

with $f_i(x) := x_1 e^{x_2 t_i} - y_i$ and

t_i	1	2	4	5	8
y_i	3	4	6	11	20

Use $x^{(0)} := (3, 0.5)^T$ as a starting point (cf. [GNS], p. 743f (available in appendix D at the book web site http://www.siam.org/books/ot108) or p. 409f of the first edition).

a) We apply the routine `fminunc` of the *Matlab*® Optimization Toolbox which realizes several minimization methods: steepest descent, bfgs, dfp, trust region method. By default we have `GradObj = off`. In this case the gradient will be calculated numerically by finite differences. First implement the objective function (without gradient) as described in `help fminunc` or `doc fminunc` beginning with the headline

Chapter 3

```
function F = myfun(x).
```

Now test this implementation. Further apply `fminunc` to minimize F. Use `optimset('Display','iter','TolX',1e-6)` to get more information about the iterative process and to raise the precision to 10^{-6}. Next implement the gradient (cf. `help fminunc`). Activate the gradient for the solution algorithm by means of `optimset('GradObj','on')` and repeat the computations. For which of the methods is the gradient absolutely necessary?
By means of `optimset('HessUpdate','steepdesc')`, for instance, you can activate the gradient method. Compare the numerical results of the different methods.

b) Carry out similar experiments with `lsqnonlin` and `lsqcurvefit` and try to activate the LEVENBERG–MARQUARDT method as well as the GAUSS–NEWTON method.

c) Visualize the results!

19. *Practical Realization of the BFGS Update*

Let us assume $B = LL^T$ with a lower triangular matrix $L \in \mathbb{M}_n$ and $p, q \in \mathbb{R}^n$ with $\langle p, q\rangle > 0$. Try to calculate — in an efficient way — the CHOLESKY decomposition $B' = L'(L')^T$ of the matrix B' defined by $B' := B + \dfrac{qq^T}{\langle p, q\rangle} - \dfrac{Bpp^TB}{\langle p, Bp\rangle}$. Show that this needs at most $O(n^2)$ operations.

Hints:

a) Define $L^* := L + uv^T$ with $u := q/\sqrt{\langle p, q\rangle} - Lv$ and $v := L^Tp/\|L^Tp\|_2$. Then it holds that $L^*(L^*)^T = B'$.

b) Choose GIVENS rotations $G_{n-1,n}, \ldots, G_{1,2}$ such that

$$G_{1,2} \cdots G_{n-1,n} v = e_1 \,;$$

then for $Q := G_{n-1,n}^T \cdots G_{1,2}^T$ it holds that $\overline{L} := L^*Q$ has lower HESSENBERG form.

c) Eliminate the upper diagonal of $\overline{L}$ using appropriate GIVENS rotations.

4

Linearly Constrained Optimization Problems

Chapter 4

The problems we consider in this chapter have general objective functions but the constraints are linear. Section 4.1 gives a short introduction to *linear optimization* (LO) — also referred to as *linear programming*, which is the historically entrenched term. LO is the simplest type of constrained optimization: the objective function *and* all constraints are linear. The classical, and still well usable algorithm to solve linear programs is the *Simplex Method.* Quadratic problems which we treat in section 4.2 are linearly constrained optimization problems with a quadratic objective function. Quadratic optimization is often considered to be an essential field in its own right. More important, however, it forms the basis of several algorithms for general nonlinearly constrained problems. In section 4.3 we give a concise outline of *projection methods*, in particular the feasible direction methods of ZOUTENDIJK, ROSEN and WOLFE. They are extensions of the steepest descent method and are closely related to the simplex algorithm and the active set method. Then we will discuss some basic ideas of *SQP methods* — more generally treated in section 5.2 — which have proven to be very efficient for wide classes of problems.

W. Forst and D. Hoffmann, *Optimization—Theory and Practice*,
Springer Undergraduate Texts in Mathematics and Technology,
DOI 10.1007/978-0-387-78977-4_4,

We write — as usual — for $k \in \mathbb{N}$ and vectors $z, y \in \mathbb{R}^k$

$$z \leq y \;:\Longleftrightarrow\; y \geq z \;:\Longleftrightarrow\; \forall \kappa \in \{1, \ldots, k\} \quad y_\kappa \geq z_\kappa$$

and

$$z < y \;:\Longleftrightarrow\; y > z \;:\Longleftrightarrow\; \forall \kappa \in \{1, \ldots, k\} \quad y_\kappa > z_\kappa,$$

in particular

$$0 \leq y \;:\Longleftrightarrow\; y \geq 0 \;:\Longleftrightarrow\; \forall \kappa \in \{1, \ldots, k\} \quad y_\kappa \geq 0$$

and

$$0 < y \;:\Longleftrightarrow\; y > 0 \;:\Longleftrightarrow\; \forall \kappa \in \{1, \ldots, k\} \quad y_\kappa > 0.$$

For $j_1, \ldots, j_k \in \mathbb{R}$ and $J := (j_1, \ldots, j_k)$ we denote by $S(J)$ the set of the components of J, that is, $S(J) := \{j_1, \ldots, j_k\}$. In contrast to $S(J)$, the order of the components of J is important.

4.1 Linear Optimization

The development of the simplex algorithm by George Dantzig (1947) marks the beginning of the age of modern optimization. This method makes it possible to analyze planning problems for large industrial and manufacturing systems in a systematic and efficient manner and to determine optimal solutions.

Dantzig's considerations appeared simultaneously with the development of the first digital computers, and the simplex method became one of the earliest important applications of this new and revolutionary technology.

Nowadays there exists sophisticated software for this kind of problem. The 'confidence' in it is occasionally so strong that the simplex method is even used if the problem is nonlinear.

From the variety of topics and variants we will only treat — relatively compactly — the *revised simplex algorithm*. We will not discuss the close and interesting relations to geometric questions either.

The Revised Simplex Method

Let $m, n \in \mathbb{N}$ with $m \leq n$, A a real (m, n)-matrix with the columns $a_1, \ldots, a_n$ and $\operatorname{rank}(A) = m$. Furthermore let $b \in \mathbb{R}^m$ and $c \in \mathbb{R}^n$.

Every *primal problem*

$$(P) \quad \begin{cases} c^T x \longrightarrow \min \\ Ax = b, \; x \geq 0 \end{cases}$$

has a corresponding *dual problem* (cf. section 2.4, p. 71)

$$(D) \quad \begin{cases} b^T u \longrightarrow \max \\ A^T u \leq c \end{cases} \quad .$$

For vectors x and u taken from the respective *feasible regions*

$$\mathcal{F}_P := \{x \in \mathbb{R}^n \mid Ax = b\,,\ x \geq 0\}$$

and

$$\mathcal{F}_D := \{u \in \mathbb{R}^m \mid A^T u \leq c\}$$

it holds that

$$c^T x \geq (u^T A)x = u^T(Ax) = u^T b = b^T u\,.$$

Hence, we get for the respective minimum $\min(P)$ of (P) and maximum $\max(D)$ of (D) the following *weak duality*:

$$\max(D) \leq \min(P) \tag{1}$$

Later on we will see that even $\max(D) = \min(P)$ holds if both sets $\mathcal{F}_P$ and $\mathcal{F}_D$ are *nonempty*, that is, there exist feasible points for both problems.

Let $N := \{1, \ldots, n\}$, $j_1, \ldots, j_m \in N$ and $J := (j_1, \ldots, j_m)$.

Definition

J is called a *basis* of (P) iff the matrix $A_J := (a_{j_1}, \ldots, a_{j_m})$ is invertible. The corresponding variables $x_{j_1}, \ldots, x_{j_m}$ are then called *basic variables*, the remaining variables are referred to as *nonbasic variables*.

Let J be a basis of (P) and $K = (k_1, \ldots, k_{n-m}) \in N^{n-m}$ with $S(K) \uplus S(J) = N$. Corresponding to K we define the matrix $A_K := (a_{k_1}, \ldots, a_{k_{n-m}})$. We split each vector $x \in \mathbb{R}^n$ into subvectors $x_J := (x_{j_1}, \ldots, x_{j_m})$ and $x_K := (x_{k_1}, \ldots, x_{k_{n-m}})$, where x_J and x_K refer to the basic and nonbasic variables, respectively. Then obviously

$$Ax = A_J x_J + A_K x_K$$

holds. Using this splitting for the equation $Ax = b$, the substitution $x_J = A_J^{-1}(b - A_K x_K)$ with $x_K \in \mathbb{R}^{n-m}$ gives a *parametrization* of the solution set $\{x \in \mathbb{R}^n \mid Ax = b\}$.

Corresponding to J there exists a unique *basic point* $\overline{x} = \overline{x}(J)$ to the linear system $Ax = b$ with $\overline{x}_K = 0$ and $A_J \overline{x}_J = b$.

Definition

A basis J of (P) is called *feasible* iff $\overline{x}(J) \geq 0$ holds, that is, all components of the corresponding basic point $\overline{x}(J)$ are nonnegative. Then $\overline{x}$ is called the *feasible basic point of* (P). If furthermore $\overline{x}_J > 0$ holds, $\overline{x}$ is called *nondegenerate*.

The objective function can be expressed by the *nonbasic variables*; substitution gives

$$\begin{aligned} c^T x = c_J^T x_J + c_K^T x_K &= c_J^T\big(A_J^{-1}(b - A_K x_K)\big) + c_K^T x_K \\ &= c_J^T \overline{x}_J + \underbrace{\left(c_K^T - c_J^T A_J^{-1} A_K\right)}_{=:\ \overline{c}^T} x_K\,. \end{aligned}$$

If J is feasible and $\overline{c} \geq 0$, then we have obviously found the minimum.

Algorithm

Let $J = (j_1, \dots j_m)$ be a feasible basis.

① Choose $K = (k_1, \dots, k_{n-m}) \in \mathbb{R}^{n-m}$ with $S(K) \uplus S(J) = N$.

Compute $\overline{b} := \overline{x}_J$ with $A_J \overline{x}_J = b$.

② $\overline{c}^T := c_K^T - c_J^T A_J^{-1} A_K$

If $\overline{c} \geq 0$: STOP. We have found the minimum (see above!).

③ Otherwise there exists an index $s = k_\sigma \in S(K)$ with $\overline{c}_\sigma < 0$. The index s enters the basis.

④ Compute the solution $\overline{a}_s = (\overline{a}_{1,s}, \dots, \overline{a}_{m,s})^T$ of $A_J \overline{a}_s = a_s$.

If $\overline{a}_s \leq 0$: STOP. The objective function is unbounded (see below!).

Otherwise determine a $\varrho \in \{1, \dots, m\}$ with $\displaystyle\min_{\overline{a}_{\mu,s}>0} \frac{\overline{b}_\mu}{\overline{a}_{\mu,s}} = \frac{\overline{b}_\varrho}{\overline{a}_{\varrho,s}}$;

$r := j_\varrho$. The index r leaves the basis:

$J' := (j_1, \dots, j_{\varrho-1}, s, j_{\varrho+1}, \dots, j_m)$

Update J: $J := J'$; go to ①.

Remark

For $\sigma \in \{1, \dots, n-m\}$ with $k_\sigma = s$ we consider $d \in \mathbb{R}^n$ with $d_J := -\overline{a}_s$ and $d_K := e_\sigma \in \mathbb{R}^{n-m}$.

Then $Ad = A_J d_J + A_K d_K = -a_s + a_{k_\sigma} = 0$ holds, and for $\tau \in \mathbb{R}_+$ we obtain $c^T(\overline{x} + \tau d) = c^T \overline{x} + \tau \underbrace{\left(c_J^T d_J + c_K^T d_K\right)}_{=\overline{c}_\sigma < 0}$:

$c_K^T = \overline{c}^T + c_J^T A_J^{-1} A_K$ holds by definition of $\overline{c}$; hence, $c_K^T d_K = \overline{c}^T e_\sigma + c_J^T A_J^{-1} a_s = \overline{c}_\sigma + c_J^T \overline{a}_s$ and therefore $c_J^T d_J + c_K^T d_K = \overline{c}_\sigma < 0$.

Hence, the term $c^T(\overline{x} + \tau d)$ is strongly antitone in τ.

From $A\overline{x} = b$ and $Ad = 0$ we get $A(\overline{x} + \tau d) = b$. We thus obtain:

$\overline{x} + \tau d \in \mathcal{F}_P \iff \overline{x}_J + \tau d_J = \overline{b} - \tau \overline{a}_s \geq 0$

If $\overline{a}_s \leq 0$, then all $\overline{x} + \tau d$ belong to $\mathcal{F}_P$, and because of $c^T(\overline{x} + \tau d) \longrightarrow -\infty$ $(\tau \to \infty)$ *the objective function* is *unbounded.*

Otherwise: $\overline{b} - \tau \overline{a}_s \geq 0 \iff \tau \leq \min\limits_{\overline{a}_{\mu,s}>0} \dfrac{\overline{b}_\mu}{\overline{a}_{\mu,s}}$.

Advantage of This Method

It saves much computing time (in comparison with the 'normal' simplex method) in the case of $m \ll n$. Additionally the data can be saved in a mass storage, if necessary.

We will now discuss how a solution to (P) can give a *solution to* (D):

Let J be a feasible basis of (P) with $c_K^T \geq c_J^T A_J^{-1} A_K$; therefore, (cf. p. 154) $\overline{x}(J)$ is a minimizer. Let $u \in \mathbb{R}^m$ be the solution to $A_J^T u = c_J$; then $A_K^T u = A_K^T A_J^{-T} c_J \leq c_K$ yields $A^T u \leq c$, that is, $u \in \mathcal{F}_D$. Furthermore

$$b^T u = u^T b = c_J^T A_J^{-1} b = c_J^T \overline{x}_J = \min(P)\,;$$

we conclude from (1) that $u = A_J^{-T} c_J$ gives a solution to (D).

Numerical Realization of the Method

Let J be a feasible basis; suppose that the feasible basis J' results from J via an exchange step of the kind described above. It holds that

$$A_J = A_{J'} T, \quad \text{and hence} \quad A_{J'}^{-1} = T A_J^{-1}, \tag{2}$$

where

$$T = \begin{pmatrix} 1 & & v_1 & & & \\ & \ddots & \vdots & & & \\ & & 1\ \vdots & & & \\ & & v_\varrho & & & \\ & & \vdots & 1 & & \\ & & \vdots & & \ddots & \\ & & v_m & & & 1 \end{pmatrix} \quad \text{with } v_\mu = -\frac{\overline{a}_{\mu,s}}{\overline{a}_{\varrho,s}} \quad (\mu \neq \varrho), \; v_\varrho = \frac{1}{\overline{a}_{\varrho,s}}\,.$$

Regular matrices which deviate from the identity matrix in only one column are called FROBENIUS *matrices.*

Proof of (2): With $w_\mu := a_{j_\mu}$ for $\mu = 1, \ldots, m$ we obtain

$A_J = (w_1, \ldots, w_m)$ and $A_{J'} = (w_1, \ldots, w_{\varrho-1}, a_s, w_{\varrho+1}, \ldots, w_m)$.

$Te_\mu = e_\mu$ for $\mu = 1, \ldots, m$ with $\mu \neq \varrho$ shows $A_J e_\mu = A_{J'} T e_\mu$. It remains to prove that $A_J e_\varrho = w_\varrho \overset{!}{=} A_{J'} T e_\varrho$:

$$\begin{aligned} A_{J'} T e_\varrho &= \frac{1}{\overline{a}_{\varrho,s}} \left[A_{J'}\left(-\overline{a}_s + (1+\overline{a}_{\varrho,s}) e_\varrho\right)\right] = \frac{1}{\overline{a}_{\varrho,s}} \left[A_{J'}(-\overline{a}_s) + (1+\overline{a}_{\varrho,s}) a_s\right] \\ &= \frac{1}{\overline{a}_{\varrho,s}} \left[(A_J - A_{J'})\overline{a}_s + \overline{a}_{\varrho,s} a_s\right] = \frac{1}{\overline{a}_{\varrho,s}} \left[\overline{a}_{\varrho,s}(w_\varrho - a_s) + \overline{a}_{\varrho,s} a_s\right] = w_\varrho \end{aligned}$$

□

Starting with $J_1 := J$, we carry out k exchange steps of this kind with FROBENIUS matrices $T_1, \ldots, T_k$ and get $A_{J_1} = A_{J_2} T_1$ with $T_1 := T$ and $J_2 := J'$ and hence

$$A_{J_1} = A_{J_{k+1}} T_k \cdots T_1 \quad \text{or} \quad T_k \cdots T_1 A_{J_1}^{-1} = A_{J_{k+1}}^{-1}.$$

One should *not* multiply the matrices T_κ explicitly, but only save the corresponding relevant columns. The procedure described above is disadvantageous as ill-conditioned intermediate matrices can lead to serious loss of accuracy (cf. the example in [St/Bu], p. 240 f). It is much better to revert to the original data: We do not need A_J^{-1} explicitly. Merely the following systems of linear equations need to be solved:

$$A_J \overline{b} = b, \quad A_J^T u = c_J, \quad A_J \overline{a}_s = a_s$$

With this we obtain

$$\overline{c}^T := c_K^T - u^T A_K.$$

We can solve these linear equations, for example, by using the following *QR-factorization* of A_J: $Q A_J = R$ with an orthogonal matrix Q and an upper triangular matrix R:

$$\begin{aligned} R\overline{b} &= Qb \\ R^T v &= c_J, \; u = Q^T v \\ R\overline{a}_s &= Q a_s \end{aligned}$$

Although the costs of the computation are slightly higher, this method is numerically more stable. We can utilize the obtained results to compute the QR-factorization of $A_{J'}$. If we modify the exchange step to

$$J' = (j_1, \ldots, j_{\varrho-1}, j_{\varrho+1}, \ldots, j_m, s),$$

then $Q A_{J'}$ has upper HESSENBERG form because of

$$Q A_{J'} = \left(\begin{array}{ccc|cccc} * & \cdots & * & * & \cdots & * & * \\ & \ddots & & \vdots & & \vdots & \vdots \\ & & * & * & & * & * \\ \hline & & & * & \cdots & * & * \\ & & & * & \ddots & * & * \\ & & & & \ddots & \ddots & \vdots \\ & & & & & * & * \end{array}\right)$$

and can be transformed to upper triangular form with $O(m^2)$ operations.

Example 1

$$\begin{array}{rcl} -3x_1 - x_2 - 3x_3 & \longrightarrow & \min \\ 2x_1 + x_2 + x_3 & \leq & 2 \\ x_1 + 2x_2 + 3x_3 & \leq & 5 \\ 2x_1 + 2x_2 + x_3 & \leq & 6 \end{array}$$

$$x_\nu \geq 0 \quad (1 \leq \nu \leq 3)$$

This problem is not in standard form (P). By introducing so-called *slack variables* $x_4, x_5, x_6 \geq 0$, we obtain:

$$\begin{array}{rcl} -3x_1 - x_2 - 3x_3 & \longrightarrow & \min \\ 2x_1 + x_2 + x_3 + x_4 & = & 2 \\ x_1 + 2x_2 + 3x_3 + x_5 & = & 5 \\ 2x_1 + 2x_2 + x_3 + x_6 & = & 6 \end{array}$$

$$x_\nu \geq 0 \quad (1 \leq \nu \leq 6)$$

$$A := \begin{pmatrix} 2\,1\,1\,1\,0\,0 \\ 1\,2\,3\,0\,1\,0 \\ 2\,2\,1\,0\,0\,1 \end{pmatrix}, \quad b := \begin{pmatrix} 2 \\ 5 \\ 6 \end{pmatrix}, \quad c := (-3, -1, -3, 0, 0, 0)^T$$

Now $(4, 5, 6)$ is a feasible basis.

1) $J_1 := (4, 5, 6)$, $K_1 := (1, 2, 3)$,
$A_{J_1} = I_3$, hence $Q_1 = R_1 = I_3$, $c_{J_1} = (0, 0, 0)^T$,
$\overline{b} = \overline{x}_{J_1} = b$, $u = 0$, $c^T \overline{x} = c_{J_1}^T \overline{b} = 0$,
$\overline{c}^T = c_{K_1}^T - u^T A_{K_1} = (-3, -1, -3)$; choose $\sigma = 1$, hence $s = k_1 = 1$,
$\overline{a}_1 = a_1 = (2, 1, 2)^T$, $\min\left\{\frac{2}{2}, \frac{5}{1}, \frac{6}{2}\right\} = 1$ yields $\varrho = 1$ and $r = j_1 = 4$.

2) $J_2 = (5, 6, 1)$, $K_2 := (2, 3, 4)$,

$$A_{J_2} = \begin{pmatrix} 0\,0\,2 \\ 1\,0\,1 \\ 0\,1\,2 \end{pmatrix}, \quad Q_2 := \begin{pmatrix} 0\,1\,0 \\ 0\,0\,1 \\ 1\,0\,0 \end{pmatrix}, \quad Q_2 A_{J_2} = \begin{pmatrix} 1\,0\,1 \\ 0\,1\,2 \\ 0\,0\,2 \end{pmatrix} =: R_2,$$

$$R_2 \overline{b} = Q_2 b = \begin{pmatrix} 5 \\ 6 \\ 2 \end{pmatrix}, \quad \overline{b} = \begin{pmatrix} 4 \\ 4 \\ 1 \end{pmatrix}, \quad c_{J_2}^T \overline{b} = -3,$$

$$R_2^T v = c_{J_2} = \begin{pmatrix} 0 \\ 0 \\ -3 \end{pmatrix}, \quad v = \begin{pmatrix} 0 \\ 0 \\ -3/2 \end{pmatrix}, \quad u = Q_2^T v = \begin{pmatrix} -3/2 \\ 0 \\ 0 \end{pmatrix},$$

$$\overline{c}^T = c_{K_2}^T - u^T A_{K_2} = (-1,-3,0) - (-3/2,0,0)\begin{pmatrix}1&1&1\\2&3&0\\2&1&0\end{pmatrix}$$
$$= (1/2,-3/2,3/2), \text{ hence } \sigma = 2 \text{ and } s = k_2 = 3.$$

$$R_2\,\overline{a}_3 = Q_2 a_3 = \begin{pmatrix}3\\1\\1\end{pmatrix}, \quad \overline{a}_3 = \begin{pmatrix}5/2\\0\\1/2\end{pmatrix}$$

$\min\left\{\frac{4}{5/2}, \frac{1}{1/2}\right\} = \frac{8}{5}$ yields $\varrho = 1$ and $r = j_1 = 5$.

3) $J_3 = (6,1,3)$, $K_3 := (2,4,5)$

$$A_{J_3} = \begin{pmatrix}0&2&1\\0&1&3\\1&2&1\end{pmatrix}, \quad Q_2 A_{J_3} = \begin{pmatrix}0&1&3\\1&2&1\\0&2&1\end{pmatrix},$$

$$\begin{pmatrix}0&1&0\\0&0&1\\1&0&0\end{pmatrix}\begin{pmatrix}0&1&3\\1&2&1\\0&2&1\end{pmatrix} = \begin{pmatrix}1&2&1\\0&2&1\\0&1&3\end{pmatrix}$$

$$\begin{pmatrix}1&0&0\\0&\frac{2}{\sqrt{5}}&\frac{1}{\sqrt{5}}\\0&-\frac{1}{\sqrt{5}}&\frac{2}{\sqrt{5}}\end{pmatrix}\begin{pmatrix}1&2&1\\0&2&1\\0&1&3\end{pmatrix} = \begin{pmatrix}1&2&1\\0&\sqrt{5}&\sqrt{5}\\0&0&\sqrt{5}\end{pmatrix} =: R_3, \quad Q_3 = \begin{pmatrix}0&0&1\\\frac{2}{\sqrt{5}}&\frac{1}{\sqrt{5}}&0\\-\frac{1}{\sqrt{5}}&\frac{2}{\sqrt{5}}&0\end{pmatrix}$$

$$R_3\,\overline{b} = Q_3 b = \begin{pmatrix}6\\9/\sqrt{5}\\8/\sqrt{5}\end{pmatrix}, \quad \overline{b} = \begin{pmatrix}4\\1/5\\8/5\end{pmatrix}, \quad c_{J_3}^T\,\overline{b} = -27/5$$

$$R_3^T v = c_{J_3} = \begin{pmatrix}0\\-3\\-3\end{pmatrix}, \quad v = \begin{pmatrix}0\\-3/\sqrt{5}\\0\end{pmatrix}, \quad u = Q_3^T v = \begin{pmatrix}-6/5\\-3/5\\0\end{pmatrix}$$

$$\overline{c}^T = c_{K_3}^T - u^T A_{K_3} = (-1,0,0) - (-6/5,-3/5,0)\begin{pmatrix}1&1&0\\2&0&1\\2&0&0\end{pmatrix}$$
$$= (7/5,6/5,3/5) \geq 0$$

Now the algorithm stops;
$\overline{x} = (1/5,0,8/5,0,0,4)^T$ gives a solution to (P) and $u = (-6/5,-3/5,0)^T$ a solution to (D); furthermore $\min(P) = \max(D) = -27/5$ holds. ◁

Calculation of a Feasible Basis

Assume that no feasible basis of the original problem

$$(P) \quad \begin{cases} c^T x \longrightarrow \min \\ Ax = b\,,\ x \geq 0 \end{cases}$$

is known. WLOG let $b \geq 0$; otherwise we multiply the μ-th row of A and b_μ by -1 for $\mu \in \{1, \ldots, m\}$ with $b_\mu < 0$.

Phase I: Calculation of a feasible basis

In phase I of the simplex algorithm we apply this method to an auxiliary problem $(\widehat{P})$ with a known initial feasible basis. A basis corresponding to a solution to $(\widehat{P})$ yields a feasible basis of (P).

$$(\widehat{P}) \quad \begin{cases} \Phi(x,w) := e^T w \longrightarrow \min \\ Ax + w = b \\ x \geq 0,\ w \geq 0 \end{cases} \qquad w = \begin{pmatrix} x_{n+1} \\ \vdots \\ x_{n+m} \end{pmatrix},\ e := \begin{pmatrix} 1 \\ \vdots \\ 1 \end{pmatrix}$$

$(0, b)^T$ is a feasible basic point to $(\widehat{P})$; hence $(\widehat{P})$ has a minimizer $\binom{\widehat{x}}{\widehat{w}}$ with $\min(\widehat{P}) \geq 0$. Furthermore: $\mathcal{F}_P \neq \emptyset \iff \min(\widehat{P}) = 0$

Let $\Phi(\widehat{x}, \widehat{w}) = 0$. Then $\widehat{x}$ is a feasible basic point to (P). If none of the *artificial variables* w_μ remains in the basis to the solution $\binom{\widehat{x}}{\widehat{w}}$, then we have found a feasible basis to (P). Otherwise we continue to iterate with this basis in *Phase II*. Artificial basic variables x_{j_ϱ} with $\overline{a}_{\varrho,s} \neq 0$ can be used as pivot elements. As $\overline{b}_j = 0$ holds for artificial basic variables x_j, we then get again a feasible basic point. Thus only those artificial variables x_{j_μ} with $\overline{a}_{\mu,s} = 0$ remain in the basis.

Example 2

$$\begin{array}{rrrrrcl} -x_1 & -x_2 & \longrightarrow \min & & & & \\ x_1 & & +2x_3 & +x_4 & & = & 1 \\ & x_2 & -x_3 & & +x_5 & = & 1 \\ x_1 & +x_2 & +x_3 & & & = & 2 \\ x \geq 0 & & & & & & \end{array}$$

Now we can proceed in a simpler way than in the general case ('worst case'):

Phase I: $w \longrightarrow \min$

$$\begin{array}{rrrrrrcl} x_1 & & +2x_3 & +x_4 & & & = & 1 \\ & x_2 & -x_3 & & +x_5 & & = & 1 \\ x_1 & +x_2 & +x_3 & & & +x_6 & = & 2 \qquad (w = x_6) \end{array}$$

$$x_\nu \geq 0 \quad (1 \leq \nu \leq 6)$$

$c := (0, 0, 0, 0, 0, 1)^T$

1) $J := (4, 5, 6),\ K := (1, 2, 3)$

$$A_J = I_3,\ \overline{b} = b = \begin{pmatrix} 1 \\ 1 \\ 2 \end{pmatrix},\ u = c_J = \begin{pmatrix} 0 \\ 0 \\ 1 \end{pmatrix}$$

Chapter 4

$$\overline{c}^T = (0,0,0) - (0,0,1)\begin{pmatrix} 1 & 0 & 2 \\ 0 & 1 & -1 \\ 1 & 1 & 1 \end{pmatrix} = (-1,-1,-1),\; s = k_1 = 1$$

$$\overline{a}_1 = a_1 = \begin{pmatrix} 1 \\ 0 \\ 1 \end{pmatrix},\; r = j_1 = 4$$

2) $J := (1,5,6),\; K := (2,3,4)$

$$A_J = \begin{pmatrix} 1\,0\,0 \\ 0\,1\,0 \\ 1\,0\,1 \end{pmatrix},\; \overline{b} = \begin{pmatrix} 1 \\ 1 \\ 1 \end{pmatrix},\; u = \begin{pmatrix} -1 \\ 0 \\ 1 \end{pmatrix}$$

$$\overline{c}^T = (0,0,0) - (-1,0,1)\begin{pmatrix} 0 & 2 & 1 \\ 1 & -1 & 0 \\ 1 & 1 & 0 \end{pmatrix} = (-1,1,1),\; s = k_1 = 2$$

$$\overline{a}_2 = \begin{pmatrix} 0 \\ 1 \\ 1 \end{pmatrix},\; r = j_2 = 5$$

3) $J := (1,2,6),\; K := (3,4,5)$

$$A_J = \begin{pmatrix} 1\,0\,0 \\ 0\,1\,0 \\ 1\,1\,1 \end{pmatrix},\; \overline{b} = \begin{pmatrix} 1 \\ 1 \\ 0 \end{pmatrix},\; u = \begin{pmatrix} -1 \\ -1 \\ 1 \end{pmatrix}$$

$$\overline{c}^T = (0,0,0) - (-1,-1,1)\begin{pmatrix} 2\,1\,0 \\ -1\,0\,1 \\ 1\,0\,0 \end{pmatrix} = (0,1,1) \geq 0$$

Phase II: $J := (1,2,6),\; K := (3,4,5)$

$$c := (-1,-1,0,0,0,0)^T,\; \overline{b} = \begin{pmatrix} 1 \\ 1 \\ 0 \end{pmatrix},\; u = \begin{pmatrix} -1 \\ -1 \\ 0 \end{pmatrix}$$

$$\overline{c}^T = (0,0,0) - (-1,-1,0)\begin{pmatrix} 2\,1\,0 \\ -1\,0\,1 \\ 1\,0\,0 \end{pmatrix} = (1,1,1) \geq 0$$

$$\min(P) = c_J^T\,\overline{b} = -2$$ ◁

The Active Set Method

The *idea* of the active set method[1], which is widely used in the field of linearly constrained optimization problems, consists of estimating the active set (referring to the Karush–Kuhn–Tucker conditions) at each iteration (cf. [Fle], p. 160ff).

[1] This method is also called the *working set algorithm*.

Problem Description

Starting from natural numbers n, m, vectors $c, a_\mu \in \mathbb{R}^n, b \in \mathbb{R}^m$ and a splitting of $\{1, \dots, m\}$ into disjoint subsets $\mathcal{E}$ (Equality) and $\mathcal{I}$ (Inequality), we consider for $x \in \mathbb{R}^n$:

$$\begin{aligned} f(x) &:= c^T x \longrightarrow \min \\ a_\mu^T x &= b_\mu \quad \text{for} \quad \mu \in \mathcal{E} \quad \text{and} \\ a_\mu^T x &\geq b_\mu \quad \text{for} \quad \mu \in \mathcal{I}. \end{aligned}$$

Assume that a feasible basis $J = (j_1, \dots, j_n)$ with $\mathcal{E} \subset S(J)$ and a vector $x \in \mathbb{R}^n$ with

$$a_\mu^T x = b_\mu \quad \text{for} \quad \mu \in S(J)$$

and

$$a_\mu^T x > b_\mu \quad \text{for} \quad \mu \in \mathcal{I} \setminus S(J)$$

exist. Hence, degeneracy shall be excluded for the moment. Then $S(J)$ contains exactly those constraints which are active at x. Each iteration step consists of a transition from one basic point to another which is geometrically illustrated by a move from one 'vertex' to a 'neighboring vertex' along a common 'edge'.

$A_J := \left(a_{j_1}, \dots, a_{j_n}\right)$, $A_J^{-T} =: (d_1, \dots, d_n)$ (will *not* be computed explicitly)

$$I_n = A_J^{-1} A_J = \begin{pmatrix} d_1^T \\ \vdots \\ d_n^T \end{pmatrix} \left(a_{j_1}, \dots, a_{j_n}\right) = \left(d_\sigma^T a_{j_\varrho}\right)$$

shows $\delta_{\varrho,\sigma} = d_\sigma^T a_{j_\varrho} = a_{j_\varrho}^T d_\sigma$.

$$u_J := A_J^{-1} c = \begin{pmatrix} d_1^T c \\ \vdots \\ d_n^T c \end{pmatrix} = \begin{pmatrix} u_{j_1} \\ \vdots \\ u_{j_n} \end{pmatrix}, \quad \text{hence} \quad c - A_J u_J = 0\,.$$

If $u_\mu \geq 0$ for all $\mu \in S(J) \cap \mathcal{I}$, then x gives a minimum (according to the Karush–Kuhn–Tucker conditions).

Otherwise there exists a $j_\sigma \in S(J) \cap \mathcal{I}$ with $u_{j_\sigma} < 0$.

As $f'(x) d_\sigma = c^T d_\sigma = u_{j_\sigma} < 0$, the vector d_σ is a *descent direction.* Let $u_{j_\sigma} = \min\left\{u_\mu \mid \mu \in S(J) \cap \mathcal{I}\right\}$ and $s := j_\sigma$. Consider $x' := x + \alpha d_\sigma$ for $\alpha > 0$.

For $\mu = j_\varrho \in S(J) \cap \mathcal{E}$ we have $\varrho \neq \sigma$ and therefore $a_\mu^T d_\sigma = 0$.

For $\mu \in S(J) \cap \mathcal{I}$ the inequality $a_\mu^T d_\sigma \geq 0$ holds.

For $\mu \in \mathcal{I} \setminus S(J)$ it needs to hold that $a_\mu^T x' = a_\mu^T x + \alpha\, a_\mu^T d_\sigma \geq b_\mu$.

Therefore $\alpha := \min M > 0$ for

$$M := \left\{ \frac{b_\mu - a_\mu^T x}{a_\mu^T d_\sigma} : \mu \in \mathcal{I} \setminus S(J) \ \wedge \ a_\mu^T d_\sigma < 0 \right\}$$

is the best possible step size; choose an index $r \in \mathcal{I} \setminus S(J)$ with $a_r^T d_\sigma < 0$ and

$$\alpha = \frac{b_r - a_r^T x}{a_r^T d_\sigma}$$

and define

$$J' := \big(j_1, \dots, j_{\sigma-1}, r, j_{\sigma+1}, \dots, j_n\big)\,.$$

Then for the objective function it holds that

$$f(x') = c^T x' = c^T x + \alpha c^T d_\sigma = c^T x + \alpha u_{j_\sigma} < c^T x.$$

If $M = \emptyset$, that is, $\alpha = \min \emptyset := \infty$, then the objective function is unbounded from below.

Example 3

$$\begin{aligned} 2x_1 + 4x_2 + 3x_3 &\longrightarrow \min \\ -x_1 + x_2 + x_3 &\geq 2 \\ 2x_1 + x_2 \quad &\geq 1 \\ x &\geq 0 \end{aligned}$$

Hence, we have $n = 3, m = 5$, $\mathcal{E} = \emptyset$, $\mathcal{I} = \{1, \dots, 5\}$, $b_1 = 2$, $b_2 = 1$, $b_3 = b_4 = b_5 = 0$, $c = (2, 4, 3)^T$ and

$$a_1 = \begin{pmatrix} -1 \\ 1 \\ 1 \end{pmatrix}, \ a_2 = \begin{pmatrix} 2 \\ 1 \\ 0 \end{pmatrix}, \ a_3 = e_1, \ a_4 = e_2, \ a_5 = e_3\,.$$

For $J := (1, 2, 4)$, $K := (3, 5)$ we obtain $A_J = \begin{pmatrix} -1 & 2 & 0 \\ 1 & 1 & 1 \\ 1 & 0 & 0 \end{pmatrix}$ and $b_J = \begin{pmatrix} 2 \\ 1 \\ 0 \end{pmatrix}$.

$a_\mu^T x = b_\mu$ for $\mu \in S(J)$ means $A_J^T x = b_J$, since this is equivalent to $a_{j_\nu}^T x = (A_J e_\nu)^T x = e_\nu^T A_J^T x = e_\nu^T b_J = b_{j_\nu}$ for $\nu = 1, \dots, n$. The two linear systems $A_J^T x = b_J$ and $A_J u_J = c$ can be solved as follows:

−1 1 1	2		−1 2 0	2	
2 1 0	1		1 1 1	4	
0 1 0	0		1 0 0	3	
−1 1 1	2	$x_1 = 1/2$	1 0 0	3	$u_1 = 3$
0 3 2	5	$x_3 = 5/2$	0 2 0	5	$u_2 = 5/2$
0 1 0	0	$x_2 = 0$	0 1 1	1	$u_4 = -3/2$

$x = \begin{pmatrix} 1/2 \\ 0 \\ 5/2 \end{pmatrix}$ and $u_J = \begin{pmatrix} 3 \\ 5/2 \\ -3/2 \end{pmatrix}$ yield $\sigma = 3$, hence $s = j_3 = 4$.

We need $d_3 = A_J^{-T} e_3$ and therefore solve $A_J^T d_3 = e_3$: $d_3 = (-1/2, 1, -3/2)^T$.

$a_3^T d_3 = e_1^T d_3 = -1/2$ and $a_5^T d_3 = e_3^T d_3 = -3/2$ yield

$$\alpha := \min\left\{\frac{b_3 - a_3^T x}{a_3^T d_3}, \frac{b_5 - a_5^T x}{a_5^T d_3}\right\} = \min\left\{\frac{0-1/2}{-1/2}, \frac{0-5/2}{-3/2}\right\} = 1.$$

This gives $r = 3$. Hence, we have

$$x' := x + \alpha d_3 = x + d_3 = \begin{pmatrix} 1/2 \\ 0 \\ 5/2 \end{pmatrix} + \begin{pmatrix} -1/2 \\ 1 \\ -3/2 \end{pmatrix} = \begin{pmatrix} 0 \\ 1 \\ 1 \end{pmatrix}.$$

$$J' := (1,2,3), \quad K' := (4,5), \quad A_{J'} = \begin{pmatrix} -1 & 2 & 1 \\ 1 & 1 & 0 \\ 1 & 0 & 0 \end{pmatrix}.$$

Solving the linear system $A_{J'} u_{J'} = c$ yields $u_{J'} = (3,1,3)^T \geq 0$:

−1 2 1	2	
1 1 0	4	
1 0 0	3	
1 0 0	3	$u_1 = 3$
0 2 1	5	$u_3 = 3$
0 1 0	1	$u_2 = 1$

Hence, x' is a minimal point with value $c^T x' = 7$. ◁

Calculation of a Feasible Basis

Let WLOG $\mathcal{E} = \emptyset$.

We will describe how to find an initial feasible basic point to the following problem:

$$\begin{cases} c^T x \longrightarrow \min \\ a_\mu^T x \geq b_\mu \quad (\mu \in \mathcal{I}) \end{cases} \quad \text{with } |\mathcal{I}| = m \geq n.$$

Phase I:

Let $x^{(k)} \in \mathbb{R}^n$ and $V_k := V\big(x^{(k)}\big) := \big\{\mu \in \mathcal{I} \mid a_\mu^T x^{(k)} < b_\mu\big\}$.

Hence, V_k contains the indices of the constraints *violated* at $x^{(k)}$.

We carry out one iteration step of the active set method for the following problem

$$(P_k) \quad \begin{cases} F_k(x) := \sum\limits_{\mu \in V_k} (b_\mu - a_\mu^T x) \longrightarrow \min \\ a_\mu^T x \leq b_\mu \quad (\mu \in V_k) \\ a_\mu^T x \geq b_\mu \quad (\mu \notin V_k). \end{cases}$$

Let a basis $J = (j_1, \ldots, j_n)$ with $S(J) \subset \mathcal{I} \setminus V_k$ and a corresponding basic point $x^{(k)}$ be given. The problem (P_k) will be updated after each iteration

Chapter 4

step. This will be repeated iteratively, until the current $x^{(k)}$ is feasible. If $u \geq 0$ occurs before that, then there exists no feasible point.

Example 4

$$\begin{aligned} &2x_1 + 5x_2 + 6x_3 \longrightarrow \min \\ &2x_1 \; + x_2 + 2x_3 \geq 3 \\ &\;\; x_1 + 2x_2 + 2x_3 \geq 1 \\ &\;\; x_1 + 3x_2 \; + x_3 \geq 3 \\ &x_1, x_2, x_3 \geq 0 \end{aligned}$$

Phase I:

1) $x^{(0)} := 0$, hence $V_0 = \{1, 2, 3\}$.

$$\begin{cases} \sum\limits_{\mu \in V_0} (b_\mu - a_\mu^T x) = 7 - (4x_1 + 6x_2 + 5x_3) \longrightarrow \min \\ a_\mu^T x \leq b_\mu \quad (\mu \in V_0) \\ a_\mu^T x \geq b_\mu \quad (\mu \notin V_0)\,, \quad \text{i.e., } x \geq 0 \end{cases}$$

$$J = (4, 5, 6)\,, \; A_J = I_3\,, \; c := \begin{pmatrix} -4 \\ -6 \\ -5 \end{pmatrix}$$

$c = A_J u_J = u_J$ yields $s = j_2 = 5$. With $d_2 = e_2 \in \mathbb{R}^3$ we get $\alpha = \min\{3/1,\; 1/2,\; 3/3\} = 1/2,\; r = 2$.

2) $x^{(1)} := x' = x^{(0)} + \alpha d_\sigma = \frac{1}{2} d_2 = \begin{pmatrix} 0 \\ 1/2 \\ 0 \end{pmatrix}$, hence $V_1 = \{1, 3\}$.

$$\begin{cases} \sum\limits_{\mu \in V_1} (b_\mu - a_\mu^T x) = 6 - (3x_1 + 4x_2 + 3x_3) \longrightarrow \min \\ a_\mu^T x \leq b_\mu \quad (\mu \in V_1) \\ a_\mu^T x \geq b_\mu \quad (\mu \notin V_1) \end{cases}$$

$$J = (2, 4, 6),\; A_J = \begin{pmatrix} 1\,1\,0 \\ 2\,0\,0 \\ 2\,0\,1 \end{pmatrix},\; A_J^{-T} = \begin{pmatrix} 0 & 1 & 0 \\ 1/2 & -1/2 & -1 \\ 0 & 0 & 1 \end{pmatrix},\; c := \begin{pmatrix} -3 \\ -4 \\ -3 \end{pmatrix}$$

$A_J u_J = c$ has the solution $u_J = (-2, -1, 1)^T$. Hence $\sigma = 1$ and $s = j_\sigma = 2,\; d_\sigma = \begin{pmatrix} 0 \\ 1/2 \\ 0 \end{pmatrix}$,

$$\alpha = \min\left\{\tfrac{3-1/2}{1/2},\; \tfrac{3-3/2}{3/2}\right\} = 1,\; r = 3.$$

3) $x^{(2)} := x^{(1)} + \alpha d_\sigma = \begin{pmatrix} 0 \\ 1/2 \\ 0 \end{pmatrix} + \begin{pmatrix} 0 \\ 1/2 \\ 0 \end{pmatrix} = \begin{pmatrix} 0 \\ 1 \\ 0 \end{pmatrix}$, $V_2 = \{1\}$, $J = (3, 4, 6)$

$$\begin{cases} \sum_{\mu \in V_2} (b_\mu - a_\mu^T x) = b_1 - a_1^T x = 3 - (2x_1 + x_2 + 2x_3) \longrightarrow \min \\ a_1^T x \le b_1 \\ a_\mu^T x \ge b_\mu \quad (2 \le \mu \le 6) \end{cases}$$

$$A_J = \begin{pmatrix} 1 & 1 & 0 \\ 3 & 0 & 0 \\ 1 & 0 & 1 \end{pmatrix}, \; A_J^{-T} = \begin{pmatrix} 0 & 1 & 0 \\ 1/3 & -1/3 & -1/3 \\ 0 & 0 & 1 \end{pmatrix}, \; c := \begin{pmatrix} -2 \\ -1 \\ -2 \end{pmatrix}$$

$A_J u_J = c$ has the solution $u_J = 1/3(-1,-5,-5)^T$. Hence $\sigma = 2$ and

$$s = j_\sigma = 4, \; d_\sigma = \begin{pmatrix} 1 \\ -1/3 \\ 0 \end{pmatrix}.$$

$$\alpha = \min\left\{ \frac{2}{5/3}, \frac{1}{1/3} \right\} = \frac{6}{5}, \; r = 1$$

$$x^{(3)} := x^{(2)} + \alpha d_\sigma = \begin{pmatrix} 0 \\ 1 \\ 0 \end{pmatrix} + \frac{6}{5} \begin{pmatrix} 1 \\ -1/3 \\ 0 \end{pmatrix} = \frac{1}{5} \begin{pmatrix} 6 \\ 3 \\ 0 \end{pmatrix}$$ is a *feasible* point,

as none of the constraints is violated, $J = (1,3,6)$.

Phase II:

$$J = (1,3,6), \; c := \begin{pmatrix} 2 \\ 5 \\ 6 \end{pmatrix}$$

$$A_J = \begin{pmatrix} 2 & 1 & 0 \\ 1 & 3 & 0 \\ 2 & 1 & 1 \end{pmatrix}, \; A_J u_J = c \text{ has the solution } u_J = \frac{1}{5} \begin{pmatrix} 1 \\ 8 \\ 20 \end{pmatrix} \ge 0. \text{ Hence}$$

$$x^{(3)} = \frac{1}{5} \begin{pmatrix} 6 \\ 3 \\ 0 \end{pmatrix}$$ is a minimizer with value $27/5$. ◁

4.2 Quadratic Optimization

Quadratic problems are linearly constrained optimization problems with a quadratic objective function. Quadratic optimization is often considered to be an essential field in its own right. More important, however, it forms the basis of several algorithms for general nonlinearly constrained problems.

Let the following optimization problem be given:

$$(QP) \quad \begin{cases} f(x) := \frac{1}{2} x^T C x + c^T x \longrightarrow \min \\ A^T x \le b \end{cases}$$

Suppose: $C \in \mathbb{R}^{n\times n}$ *symmetric positive semidefinite*, $c \in \mathbb{R}^n, b \in \mathbb{R}^m$,
$A \in \mathbb{R}^{n\times m}$ with $A = (a_1, \ldots, a_m)$.

We consider the *set of feasible points* $\mathcal{F} := \{x \in \mathbb{R}^n \mid A^T x \le b\}$.

Theorem 4.2.1 (BARANKIN–DORFMAN)[2]

Assume $\mathcal{F}$ to be nonempty and let f be bounded from below on $\mathcal{F}$. Then the function $f|_{\mathcal{F}}$ attains its minimum.

Proof[3]: For a positive ϱ we consider the set

$$\mathcal{F}_\varrho := \{x \in \mathcal{F} \mid \|x\|_2 \le \varrho\}.$$

Obviously this is a compact set. For sufficiently large ϱ it is nonempty.
$\varphi(\varrho) := \min\{f(x) \mid x \in \mathcal{F}_\varrho\} \downarrow \gamma := \inf\{f(x) \mid x \in \mathcal{F}\} \in \mathbb{R} \quad (\varrho \longrightarrow \infty)$

There exists an $x_\varrho \in \mathcal{F}_\varrho$ with $f(x_\varrho) = \varphi(\varrho)$ and
$\|x_\varrho\|_2 \le \|y\|_2$ for all $y \in \mathcal{F}_\varrho$ with $f(y) = \varphi(\varrho)$.

x_ϱ is a minimizer with minimal norm.

1) We will prove: There exists a $\varrho_0 > 0$ with $\|x_\varrho\|_2 < \varrho$ for all $\varrho > \varrho_0$.

Otherwise there exists an isotone, that is, monotone increasing, sequence (ϱ_k) of positive numbers such that

$$\|x_{\varrho_k}\|_2 = \varrho_k \longrightarrow \infty \text{ and } f(x_{\varrho_k}) \downarrow \gamma.$$

(i) For $k \in \mathbb{N}$ let $y_k := x_{\varrho_k}$ and $v_k := \frac{1}{\varrho_k} y_k$. Hence $\|v_k\|_2 = 1$ holds. WLOG let $v_k \longrightarrow v$ $(k \longrightarrow \infty)$ for a $v \in \mathbb{R}^n$. Obviously we have $\|v\|_2 = 1$.

$y_k \in \mathcal{F}_{\varrho_k} \subset \mathcal{F}$ implies $a_\mu^T y_k \le b_\mu$ for all $\mu \in \mathcal{I} := \{1, \ldots, m\}$,
hence $a_\mu^T v \le 0$ for all $\mu \in \mathcal{I}$, i.e., $A^T v \le 0$.

We consider

$$\mathcal{I}_0 := \{\mu \in \mathcal{I} \mid a_\mu^T v = 0\}.$$

For $\mu \in \mathcal{I} \setminus \mathcal{I}_0$ we have $a_\mu^T v < 0$, and hence $a_\mu^T v_k < a_\mu^T v/2$ and $\varrho_k a_\mu^T v/2 \le b_\mu$ for sufficiently large k, thus $a_\mu^T y_k = \varrho_k a_\mu^T v_k < \varrho_k a_\mu^T v/2 \le b_\mu$.

[2] Confer [Ba/Do]. Our proof is based on [Bl/Oe 1]. An alternative proof can be found in the book [Co/We].

[3] Without using the KARUSH–KUHN–TUCKER conditions and furthermore for a matrix C that is merely symmetric (*not necessarily positive semidefinite*).

Chapter 4

(ii) There exists a $\gamma' \in \mathbb{R}$ with $\gamma \le f(y_k) \le \gamma'$ for all $k \in \mathbb{N}$; hence $\gamma \le \varrho_k^2 \frac{1}{2} v_k^T C v_k + \varrho_k c^T v_k \le \gamma'$, thus, by $\varrho_k \frac{1}{2} v_k^T C v_k + c^T v_k \to 0$, $v^T C v = 0$.

$y_k + \lambda v \in \mathcal{F}$ for $\lambda \ge 0$ shows $\gamma \le f(y_k + \lambda v) = f(y_k) + \lambda (C y_k + c)^T v$ and so $(C y_k + c)^T v \ge 0$ by $\lambda \longrightarrow \infty$.

(iii) $a_\mu^T (y_k - \lambda v) = a_\mu^T y_k \le b_\mu$ for $\mu \in \mathcal{I}_0$ and $\lambda \ge 0$.

For $\mu \in \mathcal{I} \setminus \mathcal{I}_0$ and sufficiently large k we have shown $a_\mu^T y_k < b_\mu$. Hence, there exists a $\delta > 0$ with $a_\mu^T (y_k - \lambda v) \le b_\mu$ for $0 \le \lambda \le \delta$.

$\psi_k(\lambda) := \frac{1}{2} \|y_k - \lambda v\|_2^2 = \frac{1}{2}\lambda^2 - \lambda (y_k^T v) + \frac{1}{2}\|y_k\|_2^2$

$\psi_k'(0) = -y_k^T v = -\varrho_k v_k^T v$ is negative for sufficiently large k.

Thus there exists a $\lambda \in (0, \delta)$ with $y_k - \lambda v \in \mathcal{F}$ and $\|y_k - \lambda v\|_2 < \|y_k\|_2 = \varrho_k$. Because of

$$f(y_k - \lambda v) = f(y_k) - \lambda \underbrace{(C y_k + c)^T v}_{\ge 0} \le f(y_k)$$

we obtain a *contradiction* to the definition of $y_k = x_{\varrho_k}$.

2) We will show: $f(x_{\sigma_1}) = f(x_{\sigma_2})$ for $\varrho_0 < \sigma_1 < \sigma_2$.

From that it obviously follows that $f(x_\varrho) = \gamma$ for all $\varrho > \varrho_0$.

By construction $f(x_{\sigma_1}) \ge f(x_{\sigma_2})$ holds.

Let $f(x_{\sigma_1}) > f(x_{\sigma_2})$; then $\|x_{\sigma_1}\|_2 < \sigma_1 < \|x_{\sigma_2}\|_2 < \sigma_2$ holds:

The two outer inequalities hold by *1)*. For the inequality in the middle we consider: If $\|x_{\sigma_2}\|_2 \le \sigma_1$ holds, then x_{σ_2} is an element of $\mathcal{F}_{\sigma_1}$ with $f(x_{\sigma_2}) < f(x_{\sigma_1})$. This is a *contradiction* to the definition of x_{σ_1}.

$$\left.\begin{array}{r} \overline{\varrho} := \|x_{\sigma_2}\|_2 : \ \varrho_0 < \overline{\varrho} < \sigma_2 \Longrightarrow f(x_{\overline{\varrho}}) \ge f(x_{\sigma_2}) \\ x_{\sigma_2} \in \mathcal{F}_{\overline{\varrho}} \Longrightarrow f(x_{\overline{\varrho}}) \le f(x_{\sigma_2}) \end{array}\right\} \Longrightarrow f(x_{\overline{\varrho}}) = f(x_{\sigma_2})$$

This is a *contradiction* to the definition of x_{σ_2} (minimizer with minimal norm). □

Occasionally, we have already tacitly used the following formulation of the KKT conditions for the quadratic case:

Proposition 4.2.2 (Karush–Kuhn–Tucker)

A point $\overline{x} \in \mathcal{F}$ is a minimizer to problem (QP) iff there exist vectors $\overline{y}, \overline{u} \in \mathbb{R}_+^m$ with

$$C\overline{x} + c + A\overline{u} = 0,$$
$$A^T\overline{x} + \overline{y} = b \quad \textit{and}$$
$$\overline{u}^T\overline{y} = 0.$$

The constraint $A^Tx \le b$ is equivalent to $A^Tx + y = b$ with a *slack vector* $y = y(x) \in \mathbb{R}^m_+$.

Proof: Assuming the three equations, we conclude as follows (compare the proof of theorem 2.2.8): The objective function f is differentiable and convex, hence by the considerations on page 53

$$\begin{aligned} f(x) - f(\overline{x}) &\ge f'(\overline{x})(x - \overline{x}) \\ &= (C\overline{x} + c)^T(x - \overline{x}) = (A\overline{u})^T(\overline{x} - x) \\ &= \overline{u}^T A^T(\overline{x} - x) = \overline{u}^T(y - \overline{y}) = \overline{u}^T y \ge 0 \end{aligned}$$

for all $x \in \mathcal{F}$. For the other direction we firstly remark: The present problem (QP) with $\mathcal{E} := \emptyset$ and

$$g_\mu(x) := a_\mu^T x - b_\mu \text{ for } \mu \in \mathcal{I} := \{1, \ldots, m\}$$

has the form (P), which we have investigated in section 2.2. For a minimizer $\overline{x}$ to (P) it follows from theorem 2.2.5 that

$$\nabla f(\overline{x}) + \sum_{\mu=1}^{m} \overline{u}_\mu \nabla g_\mu(\overline{x}) = 0$$

and

$$\overline{u}_\mu \, g_\mu(\overline{x}) = 0 \text{ for } \mu \in \mathcal{I}$$

with a vector $\overline{u} \in \mathbb{R}^m_+$, as (MFCQ) or the Slater condition — in the case of linear constraints and $\mathcal{E} = \emptyset$ — are trivially fulfilled here.

With $\nabla g_\mu(\overline{x}) = a_\mu$ the first equation gives

$$C\overline{x} + c + A\overline{u} = 0\,.$$

With $\overline{y} := b - A^T\overline{x} \ge 0$ the second equation shows

$$\overline{u}^T\overline{y} = 0\,. \qquad \square$$

Now we will get to know a method which solves (QP) in a finite number of steps by constructing a solution to the Karush–Kuhn–Tucker conditions.

The Active Set Method

The solution strategy of this method — analogously to the active set method in the linear case of section 4.1 — is based on determining the optimum on varying submanifolds. In each iteration step we only use the constraints active at the current iteration point. All other constraints are ignored for the moment. These considerations date back to Zangwill (1967) (*manifold suboptimization*). An essential part of the following discussion also holds for the more general case of *convex* functions. Therefore we consider the following problem

$$(CP) \quad \begin{cases} f(x) \longrightarrow \min \\ A^T x \le b \end{cases}$$

with a *convex* function $f\colon \mathbb{R}^n \longrightarrow \mathbb{R}$, $b \in \mathbb{R}^m$, $A \in \mathbb{R}^{n \times m}$, $\mathcal{I} := \{1, \ldots, m\}$.

Let $\overline{x} \in \mathcal{F} := \big\{x \in \mathbb{R}^n \mid A^T x \le b\big\}$ and $\mathcal{A}(\overline{x}) := \big\{\mu \in \mathcal{I} \mid a_\mu^T \overline{x} = b_\mu\big\}$.

Then we obtain — using the same argument as on the preceding page:

$\overline{x}$ is a minimizer to (CP)

$\iff$ There exist $\overline{u}_\mu \ge 0$ for $\mu \in \mathcal{A}(\overline{x})$ with $\nabla f(\overline{x}) + \sum_{\mu \in \mathcal{A}(\overline{x})} \overline{u}_\mu a_\mu = 0$

$\iff$ $\overline{x}$ is a minimizer to $\begin{cases} f(x) \longrightarrow \min \\ a_\mu^T x \le b_\mu \text{ for all } \mu \in \mathcal{A}(\overline{x}). \end{cases}$

In general $\mathcal{A}(\overline{x})$ is unknown. We are going to describe an algorithm which determines $\mathcal{A}(\overline{x})$ iteratively and is based on the following *idea*: We solve a sequence of optimization problems

$$(P_k) \quad \begin{cases} f(x) \longrightarrow \min \\ a_\mu^T x = b_\mu \text{ for all } \mu \in J_k \,, \end{cases}$$

moving J_k to another index set successively, such that the minimal values of the objective function give an antitone sequence with $J_k = \mathcal{A}(\overline{x})$ after a finite number of steps.

Let $J \subset \mathcal{I}$ and y be a minimizer to

$$(P_J) \quad \begin{cases} f(x) \longrightarrow \min \\ a_\mu^T x = b_\mu \text{ for all } \mu \in J\,. \end{cases}$$

Then — with suitable $u_\mu \in \mathbb{R}$ for $\mu \in J$ — the KARUSH–KUHN–TUCKER conditions for (P_J) hold:

$$\nabla f(y) + \sum_{\mu \in J} u_\mu a_\mu = 0 \tag{3}$$

Definition

Let $y \in \mathcal{F}$ be a minimizer to (P_J). y is called *nondegenerate* iff the vectors $\big\{a_\mu \mid \mu \in \mathcal{A}(y)\big\}$ are linearly independent.

Remark

If y is nondegenerate, then the LAGRANGE *multipliers u_μ are uniquely determined.*

Lemma 4.2.3

Let $y \in \mathcal{F}$ be a minimizer to (P_J) and nondegenerate. In (3) let $u_s < 0$ for an index $s \in J$. If $\widetilde{y}$ is a minimizer to $(P_{\widetilde{J}})$ for $\widetilde{J} := \mathcal{A}(y) \setminus \{s\}$, then $f(\widetilde{y}) < f(y)$ and $a_s^T \widetilde{y} < b_s$ hold.

Proof: Because of (3) and $u_s < 0$ the point y is a minimizer to

$$\begin{cases} f(x) \longrightarrow \min \\ a_\mu^T x = b_\mu \quad \text{for all } \mu \in \widetilde{J} \\ a_s^T x \geq b_s \,. \end{cases} \tag{4}$$

For y the KARUSH–KUHN–TUCKER conditions to

$$\begin{cases} f(x) \longrightarrow \min \\ a_\mu^T x = b_\mu \quad \text{for all } \mu \in \widetilde{J} \\ a_s^T x \leq b_s \end{cases} \tag{5}$$

cannot be met as y is nondegenerate. Let z be a minimizer to (5). Then it follows:

a) $f(\widetilde{y}) < f(y)$: *Otherwise* we get $f(y) \leq f(\widetilde{y})$. As $(P_{\widetilde{J}})$ has less constraints than (5), $f(\widetilde{y}) \leq f(z)$ holds. Altogether, this gives $f(y) \leq f(z)$. Hence, y is a minimizer to (5) *contrary* to the above considerations.

b) $a_s^T \widetilde{y} < b_s$: *Otherwise* $a_s^T \widetilde{y} \geq b_s$ holds. Therefore $\widetilde{y}$ is feasible to (4) and thus also a minimizer to it. Hence, it follows $f(y) = f(\widetilde{y})$ *contrary* to *a)*.

□

Algorithm

0) Let $x^{(0)} \in \mathcal{F}$ and $J_0 := \mathcal{A}(x^{(0)})$.

If the points $x^{(0)}, \ldots, x^{(k)} \in \mathcal{F}$ and the corresponding index sets $J_0, \ldots, J_k$ are determined for a $k \in \mathbb{N}_0$, then we obtain $x^{(k+1)}$ and J_{k+1} as follows:

1) Determine a minimizer $y^{(k)}$ to

$$(P_k) \begin{cases} f(x) \longrightarrow \min \\ a_\mu^T x = b_\mu \quad \text{for all } \mu \in J_k \,. \end{cases}$$

2) Case 1: $y^{(k)} \in \mathcal{F}$

Solve $\sum_{\mu \in J_k} u_\mu \, a_\mu = -\nabla f(y^{(k)})$.

If $u_\mu \geq 0$ for all $\mu \in J_k$, then the point $y^{(k)}$ is a minimizer to (CP) and the algorithm stops.

Otherwise choose an index $s \in J_k$ *with* $u_s < 0$*, for example* $s \in J_k$ *minimal with* $u_s = \min\{u_\mu \mid \mu \in J_k\}$.
Set $J_{k+1} := \mathcal{A}(y^{(k)}) \setminus \{s\}$ and $x^{(k+1)} := y^{(k)}$ (deactivation step).

3) Case 2: $y^{(k)} \notin \mathcal{F}$
For $d := y^{(k)} - x^{(k)}$ *we determine a maximal* $\alpha > 0$ *with*

$$a_\mu^T(x^{(k)} + \alpha d) \le b_\mu \text{ for all } \mu \in \mathcal{I} \quad \text{(activation step)}$$

and set $x^{(k+1)} := x^{(k)} + \alpha d$ *and* $J_{k+1} := \mathcal{A}(x^{(k+1)})$.

Chapter 4

Example 5 $f(x) := x_1^2 - x_1x_2 + x_2^2 - 3x_1 \longrightarrow \min$ for $x = \begin{pmatrix} x_1 \\ x_2 \end{pmatrix} \in \mathbb{R}^2$

$$\begin{aligned} x_1 + x_2 &\le 2 \\ -x_1 \qquad &\le 0 \\ -x_2 &\le 0 \\ x_1 \qquad &\le 3/2 \end{aligned}$$

$$\nabla f(x) = \begin{pmatrix} 2x_1 - x_2 - 3 \\ -x_1 + 2x_2 \end{pmatrix}, \quad A = \begin{pmatrix} 1 & -1 & 0 & 1 \\ 1 & 0 & -1 & 0 \end{pmatrix}, \quad b = (2, 0, 0, 3/2)^T$$

0. 0) $x^{(0)} := 0 \in \mathcal{F}$, $J_0 = \mathcal{A}(x^{(0)}) = \{2, 3\}$; obviously this returns:

1) $y^{(0)} = 0$

2) $y^{(0)} \in \mathcal{F}$: $\quad u_2 \begin{pmatrix} -1 \\ 0 \end{pmatrix} + u_3 \begin{pmatrix} 0 \\ -1 \end{pmatrix} = \begin{pmatrix} 3 \\ 0 \end{pmatrix} \implies u_2 = -3,\ u_3 = 0$

For $s = 2$ we get $J_1 := \mathcal{A}(y^{(0)}) \setminus \{2\} = \{3\}$ and $x^{(1)} := y^{(0)} = 0$.

1. 1) $\left.\begin{array}{l} x_1^2 - x_1x_2 + x_2^2 - 3x_1 \longrightarrow \min \\ x_2 = 0 \end{array}\right\}$ leads to

$x_1^2 - 3x_1 \longrightarrow \min$ and thus $y^{(1)} = \begin{pmatrix} 3/2 \\ 0 \end{pmatrix}$, hence $\mathcal{A}(y^{(1)}) = \{3, 4\}$.

2) $y^{(1)} \in \mathcal{F}$: $u_3 \begin{pmatrix} 0 \\ -1 \end{pmatrix} = \begin{pmatrix} 0 \\ 3/2 \end{pmatrix}$ gives $u_3 = -3/2 < 0$. For $s = 3$ we get $J_2 := \mathcal{A}(y^{(1)}) \setminus \{3\} = \{4\}$ and $x^{(2)} := y^{(1)} = \begin{pmatrix} 3/2 \\ 0 \end{pmatrix}$.

2. 1) $\left.\begin{array}{l} x_1^2 - x_1x_2 + x_2^2 - 3x_1 \longrightarrow \min \\ x_1 = 3/2 \end{array}\right\}$ returns $y^{(2)} = \begin{pmatrix} 3/2 \\ 3/4 \end{pmatrix} \notin \mathcal{F}$.

3) (Case 2) $\quad d := y^{(2)} - x^{(2)} = \begin{pmatrix} 0 \\ 3/4 \end{pmatrix}$

$a_1^T d = 3/4$, $a_2^T d = 0$, $a_3^T d = -3/4$, $a_4^T d = 0$ give

$\alpha = \dfrac{2 - 3/2}{3/4} = 2/3$ and hence

$$x^{(3)} := x^{(2)} + \frac{2}{3} d = \begin{pmatrix} 3/2 \\ 1/2 \end{pmatrix},$$
$$J_3 := \mathcal{A}\big(x^{(3)}\big) = \{1, 4\}.$$

3. 1) $\left.\begin{array}{l} x_1^2 - x_1 x_2 + x_2^2 - 3x_1 \longrightarrow \min \\ x_1 + x_2 = 2 \\ x_1 = 3/2 \end{array}\right\} \Longrightarrow y^{(3)} = \begin{pmatrix} 3/2 \\ 1/2 \end{pmatrix}$

2) $y^{(3)} \in \mathcal{F}:\; u_1 \begin{pmatrix} 1 \\ 1 \end{pmatrix} + u_4 \begin{pmatrix} 1 \\ 0 \end{pmatrix} = -\begin{pmatrix} -1/2 \\ -1/2 \end{pmatrix} = \begin{pmatrix} 1/2 \\ 1/2 \end{pmatrix}$

$\left.\begin{array}{ll} u_1 + u_4 & = 1/2 \\ u_1 & = 1/2 \end{array}\right\}$ is solved by $u_1 = 1/2$, $u_4 = 0$.

Hence, $x^{(4)} := y^{(3)} = \begin{pmatrix} 3/2 \\ 1/2 \end{pmatrix}$ gives the optimum. ◁

Proposition 4.2.4[4]

Let $m \geq n$, $\operatorname{rank}(A) = n$ *and all the minimizers* $y^{(k)}$ *to the problems* (P_k) *which are feasible to* (CP) *be nondegenerate. Then the algorithm reaches an optimum of* (CP) *after a finite number of steps.*

Proof: Let $k \in \mathbb{N}_0$.

a) If $y^{(k)} \in \mathcal{F}$ with $u_\mu \geq 0$ for all $\mu \in J_k$, then $y^{(k)}$ meets the Karush–Kuhn–Tucker conditions for (CP). Hence, $y^{(k)}$ yields an optimum to (CP).

b) $x^{(k)}$ is a feasible point of (P_k). Hence, $f\big(y^{(k)}\big) \leq f\big(x^{(k)}\big)$ holds.

Furthermore we get $x^{(k+1)} = (1-\alpha)x^{(k)} + \alpha y^{(k)} \in \mathcal{F}_{P_k}$ with an $\alpha \in (0, 1]$. As f is convex, it follows that

$$f\big(y^{(k)}\big) \leq f\big(x^{(k+1)}\big) \leq (1-\alpha) f\big(x^{(k)}\big) + \alpha f\big(y^{(k)}\big) \leq f\big(x^{(k)}\big).$$

If $y^{(k)} \in \mathcal{F}$ with $\nabla f\big(y^{(k)}\big) + \sum_{\mu \in J_k} u_\mu a_\mu = 0$ and $u_s < 0$ for an index $s \in J_k$, then we first get $x^{(k+1)} := y^{(k)}$ and $J_{k+1} := \mathcal{A}\big(y^{(k)}\big) \setminus \{s\}$. By assumption $y^{(k)}$ is nondegenerate. Lemma 4.2.3 then gives that

$$f\big(y^{(k+1)}\big) < f\big(y^{(k)}\big) = f\big(x^{(k+1)}\big), \quad a_s^T y^{(k+1)} < b_s.$$

From that results, with a $\beta \in (0, 1]$,

$$\begin{aligned} f\big(x^{(k+2)}\big) &= f\big((1-\beta)x^{(k+1)} + \beta\, y^{(k+1)}\big) \\ &\leq (1-\beta) f\big(x^{(k+1)}\big) + \beta f\big(y^{(k+1)}\big) < f\big(x^{(k+1)}\big). \end{aligned}$$

[4] Confer [Bl/Oe 2], chapter 13.

c) Case 1 can only occur a finite number of times, since, if case 1 occurs for problem (P_k), we obtain $f\big(y^{(k)}\big) = f\big(x^{(k+1)}\big) > f\big(x^{(k+2)}\big) \ge f\big(y^{(k+1)}\big)$ by *b)* and thus $f\big(x^{(k+2)}\big) \ge f\big(x^{(k+\ell)}\big) \ge f\big(y^{(k+\ell)}\big)$ for all $\ell \ge 2$.

Therefore $\min(P_{k+\ell}) < \min(P_k)$ holds for all $\ell \in \mathbb{N}$. As $\mathcal{I}$ has only a finite number of subsets, case 1 can only appear a finite number of times.

d) Case 2 can at most occur m times in a row:
Let $y^{(k)} \notin \mathcal{F}$. Then $x^{(k+1)} = (1-\alpha)x^{(k)} + \alpha y^{(k)} \in \mathcal{F}_{P_k}$ with $\alpha \in (0,1)$ holds. There exists at least one $r \in \mathcal{I} \setminus J_k$ with $a_r^T y^{(k)} > b_r$ and $a_r^T x^{(k+1)} = b_r$. From that follows $J_{k+1} := \mathcal{A}(x^{(k+1)}) \supset J_k \cup \{r\}$, hence, $J_{k+1} \neq J_k$, where $J_k = \emptyset$ is possible. If case 2 occurs ℓ times in a row from the k-th to the $(k+\ell)$-th iteration, then $J_{k+\ell}$ contains at least ℓ elements. Therefore there exists an $\ell \le m$, such that $\{a_\mu \mid \mu \in J_{k+\ell}\}$ is a spanning set of the $\mathbb{R}^n$. From this follows $\mathcal{F}_{P_{k+\ell}} = \big\{x^{(k+\ell)}\big\}$ and $y^{(k+\ell)} = x^{(k+\ell)} \in \mathcal{F}$, that is, after at most m repetitions of case 2, we return to case 1. □

Chapter 4

The finiteness of the maximum number of steps does not mean that the algorithm is altogether finite, because it is possible that the subproblems (P_k) can only be solved iteratively.

From an applicational point of view *single exchange* is more advantageous than multiple exchange:

Algorithm (single exchange)

0) Let $x^{(0)} \in \mathcal{F}$, $J_0 := \mathcal{A}\big(x^{(0)}\big)$ and $\{a_\mu \mid \mu \in J_0\}$ be linearly independent.

If the points $x^{(0)}, \dots, x^{(k)} \in \mathcal{F}$ and the corresponding index sets $J_0, \dots, J_k$ are already determined for a $k \in \mathbb{N}_0$, we obtain $x^{(k+1)}$ and J_{k+1} as follows:

1) Determine a minimizer $y^{(k)}$ to

$$(P_k) \quad \begin{cases} f(x) \longrightarrow \min \\ a_\mu^T x = b_\mu \quad \text{for all } \mu \in J_k \,. \end{cases}$$

2) Case 1: $y^{(k)} \in \mathcal{F}$
Solve $\sum\limits_{\mu \in J_k} u_\mu \, a_\mu = -\nabla f(y^{(k)})$.

If $u_\mu \ge 0$ for all $\mu \in J_k$, then the point $y^{(k)}$ is a minimizer to (CP) and the algorithm stops.

Otherwise choose an index $s \in J_k$ with $u_s < 0$, for example $s \in J_k$ minimal with $u_s = \min\{u_\mu \mid \mu \in J_k\}$. Set $J_{k+1} := J_k \setminus \{s\}$ and $x^{(k+1)} := y^{(k)}$ (deactivation step).

3) *Case 2:* $y^{(k)} \notin \mathcal{F}$ (activation step)
For $d := y^{(k)} - x^{(k)}$ *we determine an index* $r \in \mathcal{I} \setminus J_k$ *with* $a_r^T d > 0$ *and*

$$\alpha := \min\left\{ \frac{b_\mu - a_\mu^T x^{(k)}}{a_\mu^T d} \;\middle|\; \mu \notin J_k\,,\, a_\mu^T d > 0 \right\} = \frac{b_r - a_r^T x^{(k)}}{a_r^T d} \qquad (0 \le \alpha < 1)$$

and set $x^{(k+1)} := x^{(k)} + \alpha d$ *and* $J_{k+1} := J_k \cup \{r\}$.

Remark

Let $\{a_\mu \mid \mu \in J_0\}$ *be linearly independent. Then the vectors* $\{a_\mu \mid \mu \in J_k\}$ *are also linearly independent for* $k \ge 1$. *Hence,* $\ell \le n$ *(for the* ℓ *according to d) in the proof of proposition 4.2.4) holds.*

Proof: by exercise 17.

Chapter 4

Example 6 (cf. Example 5)

0. (unaltered)

0) $x^{(0)} := 0 \in \mathcal{F}$, $J_0 = \mathcal{A}(x^{(0)}) = \{2, 3\}$; obviously this returns:

1) $y^{(0)} = 0$

2) $y^{(0)} \in \mathcal{F}$: $\quad u_2 \begin{pmatrix} -1 \\ 0 \end{pmatrix} + u_3 \begin{pmatrix} 0 \\ -1 \end{pmatrix} = \begin{pmatrix} 3 \\ 0 \end{pmatrix} \implies u_2 = -3,\; u_3 = 0$

For $s = 2$ we get
$J_1 := J_0 \setminus \{2\} = \{3\}$ and $x^{(1)} := y^{(0)} = 0$.

1. *1)* $\left.\begin{array}{l} x_1^2 - x_1 x_2 + x_2^2 - 3x_1 \longrightarrow \min \\ x_2 = 0 \end{array}\right\}$ again yields $y^{(1)} = \begin{pmatrix} 3/2 \\ 0 \end{pmatrix}$.

2) $y^{(1)} \in \mathcal{F}$: $u_3 \begin{pmatrix} 0 \\ -1 \end{pmatrix} = \begin{pmatrix} 0 \\ 3/2 \end{pmatrix}$ gives $u_3 = -3/2 < 0$. For $s = 3$ we get $J_2 := J_1 \setminus \{3\} = \emptyset$ and $x^{(2)} := y^{(1)} = \begin{pmatrix} 3/2 \\ 0 \end{pmatrix}$.

2. *1)* $x_1^2 - x_1 x_2 + x_2^2 - 3x_1 \longrightarrow \min$ (unconstrained minimum)
is solved by $y^{(2)} = \begin{pmatrix} 2 \\ 1 \end{pmatrix}$.

3) $y^{(2)} \notin \mathcal{F}$; $d := y^{(2)} - x^{(2)} = \begin{pmatrix} 1/2 \\ 1 \end{pmatrix}$

$a_1^T d = 3/2$, $a_2^T d = -1/2$, $a_3^T d = -1$, $a_4^T d = 1/2$

$\alpha := \min\left\{ \dfrac{2 - 3/2}{3/2}, \dfrac{3/2 - 3/2}{1/2} \right\} = 0$, $r = 4$

$x^{(3)} := x^{(2)} + \alpha d = x^{(2)} = \begin{pmatrix} 3/2 \\ 0 \end{pmatrix}$, $J_3 := J_2 \cup \{4\} = \{4\}$

3\. 1) $\left.\begin{array}{l} x_1^2 - x_1 x_2 + x_2^2 - 3x_1 \longrightarrow \min \\ x_1 = 3/2 \end{array}\right\} \Longrightarrow y^{(3)} = \begin{pmatrix} 3/2 \\ 3/4 \end{pmatrix}$

3) $y^{(3)} \notin \mathcal{F};\ d := y^{(3)} - x^{(3)} = \begin{pmatrix} 0 \\ 3/4 \end{pmatrix}$

$a_1^T d = 3/4,\ a_2^T d = 0,\ a_3^T d = -3/4,\ a_4^T d = 0$

$\alpha := \min\left\{\dfrac{2 - 3/2}{3/4}\right\} = 2/3,\ r = 1$

$x^{(4)} := \begin{pmatrix} 3/2 \\ 0 \end{pmatrix} + \dfrac{2}{3}\begin{pmatrix} 0 \\ 3/4 \end{pmatrix} = \begin{pmatrix} 3/2 \\ 1/2 \end{pmatrix},\ J_4 := J_3 \cup \{1\} = \{1, 4\}$

4\. 1) $\left.\begin{array}{l} x_1^2 - x_1 x_2 + x_2^2 - 3x_1 \longrightarrow \min \\ x_1 + x_2 = 2 \\ x_1 = 3/2 \end{array}\right\} \Longrightarrow y^{(4)} = \begin{pmatrix} 3/2 \\ 1/2 \end{pmatrix}$

2) $y^{(4)} \in \mathcal{F};\ u_1 \begin{pmatrix} 1 \\ 1 \end{pmatrix} + u_4 \begin{pmatrix} 1 \\ 0 \end{pmatrix} = -\begin{pmatrix} -1/2 \\ -1/2 \end{pmatrix} = \begin{pmatrix} 1/2 \\ 1/2 \end{pmatrix}$

is solved by $u_1 = 1/2,\ u_4 = 0$.

Hence, $x^{(5)} := y^{(4)} = \begin{pmatrix} 3/2 \\ 1/2 \end{pmatrix}$ gives the optimum! ◁

When executing the two algorithms just discussed, we had to solve a sequence of optimization problems subject to linear equality constraints. Therefore we will briefly treat this type of problem but only consider the quadratic case:

Minimization Subject to Linear Equality Constraints

We consider the following problem:

$$(QPE) \qquad \begin{cases} f(x) := \frac{1}{2} x^T C x + c^T x \longrightarrow \min \\ A^T x = b \end{cases}$$

Let $A = (a_1, \ldots, a_m) \in \mathbb{R}^{n \times m},\ m \le n,\ \operatorname{rank}(A) = m$,
$C \in \mathbb{R}^{n \times n}$ *symmetric positive semidefinite*, $c \in \mathbb{R}^n$ and $b \in \mathbb{R}^m$.

Elimination of Variables *(by means of QR-factorization):* The matrix A can be written in the factorized form $A = Q^T \widetilde{R}$ with an orthogonal (n, n)-matrix Q and an upper triangular matrix $\widetilde{R}$. Because of the assumption $\operatorname{rank}(A) = m$, the diagonal elements of $\widetilde{R}$ can be chosen to be positive. With that the factorization is unique, and $\widetilde{R}$ has the form

$$\widetilde{R} = \begin{pmatrix} R \\ 0 \end{pmatrix}$$

with an upper triangular matrix $R \in \mathbb{R}^{m \times m}$ and $0 \in \mathbb{R}^{(n-m) \times m}$. According to this, let Q be split up in

Chapter 4

$$Q = \begin{pmatrix} Q_1 \\ Q_2 \end{pmatrix}$$

with $Q_1 \in \mathbb{R}^{m \times n}$ and $Q_2 \in \mathbb{R}^{(n-m)\times n}$. $QA = \widetilde{R}$ then decomposes into $Q_1 A = R$ and $Q_2 A = 0$. Every $x \in \mathbb{R}^n$ can be represented uniquely as $x = Q^T \mathfrak{u}$ with a vector $\mathfrak{u} \in \mathbb{R}^n$. Decomposing $\mathfrak{u}$ in the form $\mathfrak{u} = \binom{u}{v}$ with $u \in \mathbb{R}^m$ and $v \in \mathbb{R}^{n-m}$, we get $x = Q^T \mathfrak{u} = Q_1^T u + Q_2^T v$. Thus

$$\begin{aligned} A^T x = b &\iff A^T Q_1^T u + A^T Q_2^T v = b \\ &\iff R^T u = b \iff u = R^{-T} b \end{aligned}$$

and, further,

$$x = Q_1^T R^{-T} b + Q_2^T v = x_0 + Q_2^T v \;\text{ with }\; x_0 := Q_1^T R^{-T} b = \bigl(R^{-1} Q_1\bigr)^T b.$$

For $Z := Q_2^T \in \mathbb{R}^{n \times (n-m)}$ the *substitution* $x = x_0 + Zv$ with $v \in \mathbb{R}^{n-m}$ leads to the minimization of the reduced function φ defined by

$$\varphi(v) := f(x_0 + Zv).$$

$\varphi'(v) = f'(x_0 + Zv)\,Z$ yields the *reduced gradient*

$$\nabla\varphi(v) = Z^T \nabla f(x_0 + Zv) = Z^T\bigl(C(x_0 + Zv) + c\bigr) = Z^T \nabla f(x_0) + Z^T C Z v$$

and the *reduced Hessian*

$$\nabla^2 \varphi(v) = Z^T C Z.$$

From that we obtain the 'TAYLOR series expansion'

$$\varphi(v) = f(x_0) + f'(x_0)\,Zv + \frac{1}{2} v^T Z^T C\, Zv.$$

Now the methods for the computation of 'unconstrained minima' can be applied to φ.

Example 7 $f(x) := x_1^2 + x_2^2 + x_3^2 \longrightarrow \min$

$$\begin{aligned} x_1 + 2x_2 - x_3 &= 4 \\ x_1 - x_2 + x_3 &= -2 \end{aligned}$$

Here we have the special case: $m = 2,\ n = 3,$

$$C = 2I_3,\ c = 0,\ \nabla f(x) = 2x,\ b = \begin{pmatrix} 4 \\ -2 \end{pmatrix} \text{ and } A = \begin{pmatrix} 1 & 1 \\ 2 & -1 \\ -1 & 1 \end{pmatrix}.$$

After a simple calculation[5] we obtain

[5] A widely used computation method for the QR-decomposition uses HOUSEHOLDER *matrices* $I - 2aa^T$ with $a^T a = 1$.

$$\begin{matrix} Q_1 \left\{ \begin{matrix} \\ \\ \end{matrix} \right. \\ Q_2 \left\{ \right. \end{matrix} \begin{pmatrix} 1/\sqrt{6} & 2/\sqrt{6} & -1/\sqrt{6} \\ 4/\sqrt{21} & -1/\sqrt{21} & 2/\sqrt{21} \\ \hline 1/\sqrt{14} & -2/\sqrt{14} & -3/\sqrt{14} \end{pmatrix} \begin{pmatrix} 1 & 1 \\ 2 & -1 \\ -1 & 1 \end{pmatrix} = \begin{pmatrix} \sqrt{6} & -\sqrt{6}/3 \\ 0 & \sqrt{21}/3 \\ \hline 0 & 0 \end{pmatrix}.$$

Hence, it follows that $R^{-1} = \begin{pmatrix} 1/\sqrt{6} & 1/\sqrt{21} \\ 0 & 3/\sqrt{21} \end{pmatrix}$ and $R^{-T}b = \begin{pmatrix} 4/\sqrt{6} \\ -2/\sqrt{21} \end{pmatrix}$.

$$x_0 := Q_1^T R^{-T} b = \frac{2}{7} \begin{pmatrix} 1 \\ 5 \\ -3 \end{pmatrix}, \quad x = x_0 + Zv = x_0 + Q_2^T v$$

$0 \stackrel{!}{=} \nabla\varphi(v) = Q_2 \nabla f(x_0) + Q_2 2\, I_3 Q_2^T v = Q_2 2x_0 + 2v \iff v = 0$, as in this case $Q_2 x_0 = 0$. Then $\overline{x} := x_0$ gives the optimum.

Calculation of the Lagrange *multipliers:*

With $\overline{g} := \nabla f(\overline{x})$ we get:

$$\begin{aligned} 0 &\stackrel{!}{=} \nabla f(\overline{x}) + \sum_{\mu=1}^{m} \overline{u}_\mu a_\mu = \overline{g} + A\overline{u} \iff A\overline{u} = -\overline{g} \\ &\iff \widetilde{R}\overline{u} = QA\overline{u} = -Q\overline{g} \implies R\overline{u} = -Q_1\overline{g} \\ &\iff \overline{u} = -R^{-1}Q_1\overline{g} \end{aligned}$$

for $\overline{u} = (\overline{u}_1, \ldots, \overline{u}_m) \in \mathbb{R}^m$. Then for our *Example 7* it follows that

$$\overline{g} := \nabla f(\overline{x}) = 2\overline{x} = \frac{4}{7} \begin{pmatrix} 1 \\ 5 \\ -3 \end{pmatrix}, \quad Q_1 \overline{g} = \frac{4}{7} \begin{pmatrix} 14/\sqrt{6} \\ -7/\sqrt{21} \end{pmatrix} = 4 \begin{pmatrix} 2/\sqrt{6} \\ -1/\sqrt{21} \end{pmatrix}$$

$$\overline{u} = -\begin{pmatrix} 1/\sqrt{6} & 1/\sqrt{21} \\ 0 & 3/\sqrt{21} \end{pmatrix} \cdot 4 \begin{pmatrix} 2/\sqrt{6} \\ -1/\sqrt{21} \end{pmatrix} = \frac{4}{7} \begin{pmatrix} -2 \\ 1 \end{pmatrix}. \qquad \triangleleft$$

The elimination method described above is often called the *zero space method* in the literature. It is advantageous if $m \approx n$ holds.

In the case of $m \ll n$ it is more useful to apply the so-called *range space method*. We start from the Karush–Kuhn–Tucker *conditions* as stated in proposition 4.2.2:

A point $\overline{x} \in \mathcal{F}$ is a minimizer to problem (QP) if and only if there exist vectors $\overline{y}, \overline{u} \in \mathbb{R}^m_+$ with

$$\begin{aligned} C\overline{x} + c + A\overline{u} &= 0, \\ A^T\overline{x} + \overline{y} &= b \text{ and} \\ \overline{u}^T\overline{y} &= 0. \end{aligned}$$

As we consider the *equality* constraint case $A^T x = b$, we have $\overline{y} = 0$, and therefore the system reduces to

Chapter 4

$$\begin{pmatrix} C & A \\ A^T & 0 \end{pmatrix} \begin{pmatrix} \overline{x} \\ \overline{u} \end{pmatrix} = \begin{pmatrix} -c \\ b \end{pmatrix}.$$

For the coefficient matrix of this linear system we will show:

1) If C is positive definite on $\operatorname{kernel}(A^T)$*, then*

$$\begin{pmatrix} C & A \\ A^T & 0 \end{pmatrix} \quad \textit{is invertible.}$$

2) If C is positive definite, then it holds that

$$\begin{pmatrix} C & A \\ A^T & 0 \end{pmatrix}^{-1} = \begin{pmatrix} U & X \\ X^T & W \end{pmatrix} \quad \textit{with} \quad \begin{cases} W = -(A^T C^{-1} A)^{-1} \\ X = -C^{-1} A W \\ U = C^{-1} - X A^T C^{-1}. \end{cases}$$

Proof: 1) From

$$\begin{pmatrix} C & A \\ A^T & 0 \end{pmatrix} \begin{pmatrix} x \\ y \end{pmatrix} = \begin{pmatrix} 0 \\ 0 \end{pmatrix}$$

for $x \in \mathbb{R}^n$ and $y \in \mathbb{R}^m$ it follows that $Cx + Ay = 0$ and $x \in \operatorname{kernel}(A^T)$. $0 = \langle x, Cx \rangle + \langle x, Ay \rangle = \langle x, Cx \rangle + \langle A^T x, y \rangle = \langle x, Cx \rangle$ shows $x = 0$. Hence, because of $\operatorname{rank}(A) = m$, $y = 0$ follows.

2) $\begin{pmatrix} U & X \\ X^T & W \end{pmatrix} \begin{pmatrix} C & A \\ A^T & 0 \end{pmatrix} = I_{n+m} = \begin{pmatrix} I_n & 0 \\ 0 & I_m \end{pmatrix}$ is equivalent to

$$\begin{cases} UC + XA^T = I_n, & UA = 0 \\ X^T C + WA^T = 0, & X^T A = I_m. \end{cases}$$

The third equation implies $X^T + WA^T C^{-1} = 0$ $(*)$ and thus $X^T A + WA^T C^{-1} A = 0$, hence, from the fourth equation $WA^T C^{-1} A = -I_m$ and $W = -\left(A^T C^{-1} A\right)^{-1}$. From $(*)$ follows $X = -C^{-1} A W^T = -C^{-1} A W$. The first equation shows $U = C^{-1} - X A^T C^{-1}$. □

For C positive definite we obtain the *solution to the* KARUSH–KUHN–TUCKER *conditions*

$$\begin{pmatrix} C & A \\ A^T & 0 \end{pmatrix} \begin{pmatrix} \overline{x} \\ \overline{u} \end{pmatrix} = \begin{pmatrix} -c \\ b \end{pmatrix}$$

from

$$\begin{pmatrix} \overline{x} \\ \overline{u} \end{pmatrix} = \begin{pmatrix} U & X \\ X^T & W \end{pmatrix} \begin{pmatrix} -c \\ b \end{pmatrix} = \begin{pmatrix} -Uc + Xb \\ -X^T c + Wb \end{pmatrix},$$

that is, $\overline{u} = -X^T c + Wb$ and $\overline{x} = -C^{-1}(A\overline{u} + c)$.

Continuation of Example 7

$$C^{-1} = \frac{1}{2} I_3 , \; c = 0 , \; b = \begin{pmatrix} 4 \\ -2 \end{pmatrix} , \; A = \begin{pmatrix} 1 & 1 \\ 2 & -1 \\ -1 & 1 \end{pmatrix}$$

$$W = -(A^T C^{-1} A)^{-1} = -2(A^T A)^{-1} = -2 \begin{pmatrix} 6 & -2 \\ -2 & 3 \end{pmatrix}^{-1} = -\frac{1}{7} \begin{pmatrix} 3 & 2 \\ 2 & 6 \end{pmatrix}$$

$$\overline{u} = Wb = -\frac{1}{7} \begin{pmatrix} 3 & 2 \\ 2 & 6 \end{pmatrix} \begin{pmatrix} 4 \\ -2 \end{pmatrix} = -\frac{1}{7} \begin{pmatrix} 8 \\ -4 \end{pmatrix} = \frac{4}{7} \begin{pmatrix} -2 \\ 1 \end{pmatrix}$$

$$\overline{x} = -\frac{1}{2} A \overline{u} = -\frac{1}{2} \begin{pmatrix} 1 & 1 \\ 2 & -1 \\ -1 & 1 \end{pmatrix} \frac{4}{7} \begin{pmatrix} -2 \\ 1 \end{pmatrix} = \frac{2}{7} \begin{pmatrix} 1 \\ 5 \\ -3 \end{pmatrix}$$

◁

These formulae for the solution of the KARUSH–KUHN–TUCKER conditions do not have the right form to *react flexibly to changes of A* within the active set method. This works better in the following way:

$QA = \begin{pmatrix} Q_1 \\ Q_2 \end{pmatrix} A = \begin{pmatrix} R \\ 0 \end{pmatrix}$ yields

$A = Q^T \begin{pmatrix} R \\ 0 \end{pmatrix} = (Q_1^T \; Q_2^T) \begin{pmatrix} R \\ 0 \end{pmatrix} = Q_1^T R$, hence, $A^T = R^T Q_1$.

$C\overline{x} = -(A\overline{u} + c) = -(Q_1^T R \overline{u} + c)$ then gives with $\overline{v} := R\overline{u}$:

$$\begin{aligned} \overline{v} = R\overline{u} &= R\big(-X^T c + Wb\big) \\ &= RW\big(A^T C^{-1} c + b\big) \quad \text{(definition of } X \text{ and symmetry of } W, C) \\ &= RW\big(R^T Q_1 C^{-1} c + b\big) \\ &= -R\big(R^T Q_1 C^{-1} Q_1^T R\big)^{-1} \big(R^T Q_1 C^{-1} c + b\big) \quad \text{(definition of } W) \\ &= -\big(Q_1 C^{-1} Q_1^T\big)^{-1} \big(Q_1 C^{-1} c + R^{-T} b\big). \end{aligned}$$

The CHOLESKY decomposition $C = LDL^T$ (L 'unit' lower triangular matrix, D diagonal matrix) and $S := Q_1 L^{-T}$ then yield

$Q_1 C^{-1} Q_1^T = Q_1 L^{-T} D^{-1} L^{-1} Q_1^T = S D^{-1} S^T$.

The Goldfarb–Idnani Method[6]

One of the main advantages of the following dual method is that it does not need a primal feasible starting vector. It is one of the best general-purpose methods for quadratic optimization. The method iteratively generates 'dually optimal', but possibly infeasible approximations with isotone values of the objective function.

[6] Confer [Go/Id].

Chapter 4

We consider the **problem**

$$(QP) \qquad \begin{cases} f(x) := \frac{1}{2}x^T C x + c^T x \longrightarrow \min \\ A^T x \leq b \end{cases} .$$

Let $A = (a_1, \ldots, a_m) \in \mathbb{R}^{n \times m}$, $c \in \mathbb{R}^n$, $b \in \mathbb{R}^m$,
$C \in \mathbb{R}^{n \times n}$ *symmetric positive definite.*

For $J \subset \mathcal{I} := \{1, \ldots, m\}$ we again consider the subproblem

$$(P_J) \qquad \begin{cases} f(y) \longrightarrow \min \\ A_J^T y \leq b_J . \end{cases}$$

Definition

For $x \in \mathbb{R}^n$

(x, J) is called an *S-pair to* (QP) iff $\begin{cases} \operatorname{rank}(A_J) = |J|, \\ A_J^T x = b_J, \\ x \text{ is a minimizer to } (P_J). \end{cases}$

Here *"S"* stands for *"solution"*.

The unconstrained problem $(P_\emptyset)$ with the minimizer $x = -C^{-1}c$ is a trivial *example*. $(x, \emptyset)$ is an S-pair to (QP). If (x, J) is an S-pair to (QP) with x feasible to (QP), then x is a minimizer to (QP).

Definition

Let $x \in \mathbb{R}^n$, $J \subset \mathcal{I}$, $p \in \mathcal{I} \setminus J$, $J' := J \cup \{p\}$. Then

(x, J, p) is called a *V-triple* iff $\begin{cases} A_J^T x = b_J,\ a_p^T x > b_p, \\ \{a_\mu \mid \mu \in J'\} \text{ linearly independent and} \\ \nabla f(x) + \sum_{\mu \in J} u_\mu a_\mu + \lambda a_p = 0 \\ \text{with suitable } u_\mu \geq 0 \text{ and } \lambda \geq 0 \end{cases}$

hold. Here *"V"* stands for *"violation"*.

Algorithm (model algorithm)

0) Let an S-pair (x, J) be given.[7]

1) If $A^T x \leq b$ holds, then x is a minimizer to (QP): STOP

2) Otherwise choose a $p \in \mathcal{I}$ with $a_p^T x > b_p$ (violated constraint):
$J' := J \cup \{p\}$
If $(P_{J'})$ does not have any feasible points, then (QP) does not have any feasible points either: STOP

[7] Choose, for example, $\left(-C^{-1}c, \emptyset \right)$.

3) *Otherwise* determine a new S-pair $(\overline{x}, \overline{J})$ with $p \in \overline{J} \subset J'$ and $f(\overline{x}) > f(x)$.
Set $x := \overline{x}$, $J := \overline{J}$ and go to *1)*.

The realization of the above algorithm naturally raises the questions of how — starting from a current S-pair (x, J) and a constraint violated by x — we can decide efficiently, whether the new problem has feasible vectors and how, if necessary, we can produce a new S-pair $(\overline{x}, \overline{J})$ with the properties mentioned above.

If we can ensure the realization, then the algorithm stops after a finite number of steps since there exist only a finite number of S-pairs, and after each step the value of the objective function increases strictly, hence, 'cycles' are excluded.

Implementation of step 3):

Let (x, J) be an S-pair and $p \in \mathcal{I}$ with $a_p^T x > b_p$.

According to this assumption $\big(x$ is a minimizer to $(P_J)\big)$ there exists a vector $u \geq 0$ with $\nabla f(x) + A_J u = 0$.

We define $\overline{\alpha} := a_p^T x - b_p$.

In the following, when determining a new S-pair, we will consider the two cases
(I) $\{a_\mu \mid \mu \in J'\}$ linearly independent and
(II) a_p linearly dependent on $\{a_\mu \mid \mu \in J\}$
separately.

For $\alpha \geq 0$ we consider the problem (P_α)
$$\begin{cases} f(y) \longrightarrow \min \\ A_J^T y = b_J \\ a_p^T y = a_p^T x - \alpha . \end{cases}$$

Remark

The problem (P_0) has the minimizer x with Lagrange multiplier $\binom{u}{0} \geq 0$, since $\nabla f(x) + A_{J'} \binom{u}{0} = 0$ holds.

Case I: $\{a_\mu \mid \mu \in J'\}$ linearly independent
Then (x, J, p) is a V-triple.

Let $x(\alpha)$ denote the minimizer to (P_α) and $\begin{pmatrix} u(\alpha) \\ \lambda(\alpha) \end{pmatrix}$ the corresponding Lagrange multiplier.

The Karush–Kuhn–Tucker conditions (cf. page 167) of (P_α) are:

$$C x(\alpha) + c + A_{J'} \begin{pmatrix} u(\alpha) \\ \lambda(\alpha) \end{pmatrix} = 0$$
$$A_{J'}^T x(\alpha) = \widetilde{b}_{J'} - \alpha e_{|J|+1} \text{ with } \widetilde{b}_{J'} := \begin{pmatrix} b_J \\ a_p^T x \end{pmatrix}$$

Chapter 4

This system of linear equations has a unique solution (cf. page 178), which is affinely linear with respect to α. Using the solution

$\begin{pmatrix} \widehat{x} \\ \widehat{u} \\ \widehat{\lambda} \end{pmatrix}$ for the right-hand side $\begin{pmatrix} 0 \\ -e_{|J|+1} \end{pmatrix}$, we obviously obtain

— with a suitable function $\lambda(\,\cdot\,) \geq 0$, which according to the above remarks is equal to zero in the beginning —

$$\begin{pmatrix} x(\alpha) \\ u(\alpha) \\ \lambda(\alpha) \end{pmatrix} = \begin{pmatrix} x + \alpha\,\widehat{x} \\ u + \alpha\,\widehat{u} \\ \lambda + \alpha\,\widehat{\lambda} \end{pmatrix}$$

and
$$C\,x + c + A_{J'}\begin{pmatrix} u \\ \lambda \end{pmatrix} = 0\,, \qquad C\,\widehat{x} + A_{J'}\begin{pmatrix} \widehat{u} \\ \widehat{\lambda} \end{pmatrix} = 0\,,$$
$$A_{J'}^T x = \widetilde{b}_{J'}\,, \qquad A_{J'}^T \widehat{x} = -e_{|J|+1}\,,$$

$$f(x(\alpha)) = f(x) + \alpha\, f'(x)\,\widehat{x} + \frac{1}{2}\,\alpha^2\,\widehat{x}^{\,T} C\,\widehat{x}\,,\ f'(x) = (C\,x + c)^T\,.$$

Remark

It holds that $f'(x)\,\widehat{x} = \lambda\,(\geq 0)$ *and* $\widehat{x}^{\,T} C\,\widehat{x} = \widehat{\lambda} > 0\,.$

Proof: $\widehat{x}$ is unequal to zero (because of $A_{J'}^T \widehat{x} = -e_{|J|+1}$), hence, we get

$$0 < \widehat{x}^{\,T} C\,\widehat{x} = -\widehat{x}^{\,T} A_{J'}\begin{pmatrix} \widehat{u} \\ \widehat{\lambda} \end{pmatrix} = e_{|J|+1}^T\begin{pmatrix} \widehat{u} \\ \widehat{\lambda} \end{pmatrix} = \widehat{\lambda},$$

$$0 \leq \lambda = (u^T, \lambda)\,e_{|J|+1} = -\begin{pmatrix} u \\ \lambda \end{pmatrix}^T A_{J'}^T \widehat{x} = f'(x)\,\widehat{x}\,. \qquad \Box$$

Conclusion:

The function $f \circ x$, that is, $(f \circ x)(\alpha) = f(x(\alpha)) = f(x) + \alpha\lambda + \frac{\alpha^2}{2}\,\widehat{\lambda}$ for $\alpha \geq 0$, *is strongly isotone.*

To guarantee $u(\alpha) = u + \alpha\,\widehat{u} \geq 0$, we define

$\widehat{\alpha} := \min\left\{ -\frac{u_\mu}{\widehat{u}_\mu} \mid \mu \in J,\ \widehat{u}_\mu < 0 \right\}$ (that is, $\widehat{\alpha} = \min\emptyset = \infty$, if $\widehat{u} \geq 0$).

For $0 \leq \alpha \leq \widehat{\alpha}$ we have $u(\alpha) \geq 0$ and $\lambda(\alpha) \geq 0$. Hence, $x(\alpha)$ is a minimizer to the problem

$$(P_\alpha^{\leq}) \quad \begin{cases} f(y) \longrightarrow \min \\ A_J^T y \leq b_J \\ a_p^T y \leq a_p^T x - \alpha\,. \end{cases}$$

Case Ia: $0 < \overline{\alpha} \le \widehat{\alpha}$

$\big(x(\overline{\alpha}),\ J \cup \{p\}\big)$ is an S-pair with $f(x(\overline{\alpha})) > f(x)$. Therefore, we set

$$\overline{x} := x(\overline{\alpha}),\ \overline{J} := J \cup \{p\}.$$

Case Ib: $0 \le \widehat{\alpha} < \overline{\alpha}$

There exists a $\mu \in J$ with $u_\mu(\widehat{\alpha}) = 0$ and $f(x(\widehat{\alpha})) \ge f(x)$. ('=' only for $\widehat{\alpha} = 0$)

Replace: x by $x(\widehat{\alpha})$
u by the vector arising from $u(\widehat{\alpha})$ when the component u_μ is removed
λ by $\lambda(\widehat{\alpha})$
J by $J \setminus \{\mu\}$
$\overline{\alpha}$ by $a_p^T x(\widehat{\alpha}) - b_p = \overline{\alpha} - \widehat{\alpha}$

Then (x, J, p) is a V-triple, that is, we are in the starting situation of case I, just with one constraint less. If we repeat the described procedure, after a finite number of steps — at the latest, if $J = \emptyset$ — case Ia will occur.

Case II: a_p depends linearly on $\{a_\mu \mid \mu \in J\}$, that is, $a_p = A_J v$ with a suitable $v \in \mathbb{R}^{|J|}$.

Case IIa: $v \le 0$

Then, respecting $A_J^T x = b_J$ and $a_p^T x > b_p$, the inequality system $A_J^T y \le b_J$ and $a_p^T y \le b_p$ is unsolvable, because otherwise we would get

$$\begin{aligned} A_J^T(y - x) &\le 0 \\ a_p^T(y - x) &< 0 \end{aligned}$$

and therefore a *contradiction*

$$0 > a_p^T(y - x) = v^T A_J^T(y - x) \ge 0.$$

Case IIb: $v \not\le 0$

As (x, J) is an S-pair, there exist $u_\ell \ge 0$ for $\ell \in J$ with $\nabla f(x) + \sum_{\ell \in J} u_\ell a_\ell = 0$. Let $\mu \in J$ with $v_\mu > 0$ and

$$\frac{u_\mu}{v_\mu} = \min\left\{ \frac{u_\ell}{v_\ell} \,\middle|\, \ell \in J,\ v_\ell > 0 \right\}.$$

From $a_p = A_J v = \sum_{\ell \ne \mu} v_\ell a_\ell + v_\mu a_\mu$ we get $a_\mu = \dfrac{1}{v_\mu} a_p - \sum_{\ell \ne \mu} \dfrac{v_\ell}{v_\mu} a_\ell$, hence

$$0 = \nabla f(x) + \sum_{\ell \neq \mu} u_\ell a_\ell + u_\mu a_\mu = \nabla f(x) + \sum_{\ell \neq \mu} \left(u_\ell - \frac{u_\mu v_\ell}{v_\mu} \right) a_\ell + \frac{u_\mu}{v_\mu} a_p .$$

All coefficients of this representation are — according to the choice of μ — nonnegative.

Replace J by $J \setminus \{\mu\}$. Now (x, J, p) is a V-triple, and we restart with step *3)*. The p-th constraint is still violated, but now we can continue with case I.

Example 8 $f(x) := x_1^2 - x_1 x_2 + x_2^2 - 3x_1 \longrightarrow \min$

$x_1 + x_2 \leq 2$

$x_1 \geq 0, \; x_2 \geq 0$

Here we have in particular $C = \begin{pmatrix} 2 & -1 \\ -1 & 2 \end{pmatrix}$, $c = \begin{pmatrix} -3 \\ 0 \end{pmatrix}$,

$\nabla f(x) = \begin{pmatrix} 2x_1 - x_2 - 3 \\ -x_1 + 2x_2 \end{pmatrix}$, $A = \begin{pmatrix} 1 & -1 & 0 \\ 1 & 0 & -1 \end{pmatrix}$, $b = (2, 0, 0)^T$.

0) $J := \emptyset$ and $x := -C^{-1}c = (2, 1)^T$ give a starting S-pair.

1) $x_1 + x_2 = 3 > 2$

2) Then we have $p = 1$, $\overline{\alpha} = a_1^T x - b_1 = 1$, $J' = \{1\}$.

3) For $\alpha \geq 0$ we solve the problem (P_α) $\begin{cases} f(y) \longrightarrow \min \\ y_1 + y_2 = 3 - \alpha . \end{cases}$

Now we have case I, since $\{a_\mu \mid \mu \in J'\} = \{a_1\}$ is linearly independent.

$C\widehat{x} + A_{J'} \begin{pmatrix} \widehat{u} \\ \widehat{\lambda} \end{pmatrix} = 0$ and $A_{J'}^T \widehat{x} = -e_{|J|+1} = -e_1 = -1$ result in this special case $\begin{pmatrix} 2 & -1 \\ -1 & 2 \end{pmatrix} \begin{pmatrix} \widehat{x}_1 \\ \widehat{x}_2 \end{pmatrix} + \widehat{\lambda} \begin{pmatrix} 1 \\ 1 \end{pmatrix} = 0$, $(1, \, 1) \begin{pmatrix} \widehat{x}_1 \\ \widehat{x}_2 \end{pmatrix} = -1$,

hence are solved by $\widehat{x}_1 = \widehat{x}_2 = -1/2$, $\widehat{\lambda} = 1/2$. We obtain

$x(\alpha) = \begin{pmatrix} 2 \\ 1 \end{pmatrix} + \alpha \begin{pmatrix} -1/2 \\ -1/2 \end{pmatrix} = \begin{pmatrix} 2 \\ 1 \end{pmatrix} - \frac{\alpha}{2} \begin{pmatrix} 1 \\ 1 \end{pmatrix}$ and $\widehat{\alpha} = \infty > \overline{\alpha} = 1$; therefore we have case I a. $x(1) = \begin{pmatrix} 3/2 \\ 1/2 \end{pmatrix}$ is feasible and so gives the solution.

Because of $f(x(\alpha)) = f(x) + \frac{\alpha^2}{4} = -3 + \frac{\alpha^2}{4}$, we obtain $f(x(1)) = \frac{-11}{4}$. ◁

For practice purposes, we have provided a *Matlab*® program for our readers. We refrained from the consideration of a number of technical details for the benefit of transparency. It is surprising how easily the Goldfarb–Idnani method can then be implemented:

```
function [x,u,J] = goldfarb(C,c,A,b)

% Experimental version of the Goldfarb-Idnani method
% for quadratic optimization
%
% Problem: min 1/2*x'*C*x + c'*x  s.t.  A'*x <= b

n = size(C,1);
x = C \ (-c);          % nabla(f) = 0  <==> C*x = -c
u = [];                % 0:0 - matrix
lambda = 0; J = [];
[alpha_, p] = max(A'*x - b);

E = 10^(-13);
while (alpha_ > E*norm(A(:,p)))
  ap = A(:,p); AJ = A(:,J); J_  = [J, p];

  if (rank(AJ) < rank([AJ, ap]))  % Case 1
    AJ_ = A(:,J_); Z = zeros(length(J_));
    M = [C, AJ_; AJ_', Z];
    rhs = [zeros(n,1); zeros(length(J),1); -1];
    xul = M \ rhs; x_ = xul(1:n);
    if (length(J) > 0)
      u_ = xul(n+1:end-1); else u_ = [];
    end;
    lambda_ = xul(end:end);

    I = find(u_ < 0);
    if (length(I) > 0)
      [alpha_hat, i] = min(-u(I) ./ u_(I))
    end;

    if (length(I) == 0) | (alpha_ <= alpha_hat)  % Case 1a
      x = x + alpha_*x_; u = u + alpha_*u_;
      lambda = lambda + alpha_*lambda_;
      u = [u; lambda];
      lambda = 0; J = J_;
      [alpha_, p] = max(A'*x - b);

    else  % Case 1b
      x = x + alpha_hat*x_; u = u + alpha_hat*u_;
      lambda = lambda + alpha_hat*lambda_;
      I = setdiff([1:length(J)],i);
      u = u(I); J = setdiff(J,J(i));
      alpha_ = alpha_ - alpha_hat;
    end;
```

```
  else  % Case 2
    v = AJ \ ap; I = find(v > 0);
    if (length(I) == 0) % Case 2a
      error('There exists no feasible solution.')
    else  % Case 2b
      [lambda, i] = min(u(I) ./ v(I));
      I = setdiff([1:length(J)],i);
      u = u(I) - lambda*v(I); J = setdiff(J,J(i));
    end;

  end;  % end if
end;  % end while
```


4.3 Projection Methods

We will now consider linearly constrained optimization problems with more general objective functions than linear or quadratic ones.

Let $x^{(0)}$ be a feasible, but not yet optimal point. Starting there, we try to move along the negative gradient of the objective function to get another feasible point with a smaller function value (in the case of minimization). If $x^{(0)}$ is an interior point of the feasible set, this is always possible. However, if $x^{(0)}$ is a boundary point, we may leave the feasible region by doing so.

The earliest proposals for *feasible direction methods* go back to Zoutendijk and Rosen (1960).

Let $n \in \mathbb{N}$, $m, p \in \mathbb{N}_0$, D an open subset of $\mathbb{R}^n$,
$f\colon D \longrightarrow \mathbb{R}$ a continuously differentiable function and
$a_\mu \in \mathbb{R}^n, b_\mu \in \mathbb{R}$ for $\mu \in \mathcal{I} \cup \mathcal{E}$ with
$\mathcal{I} := \{1, \ldots, m\}$, $\mathcal{E} := \{m+1, \ldots, m+p\}$.

Then we consider the problem

$$(PL) \quad \begin{cases} f(x) \longrightarrow \min \\ a_i^T x \le b_i \quad \text{for all } i \in \mathcal{I} \\ a_j^T x = b_j \quad \text{for all } j \in \mathcal{E} \end{cases} \tag{6}$$

with the feasible region

$$\mathcal{F} := \big\{x \in D \mid a_i^T x \le b_i \ (i \in \mathcal{I}),\ a_j^T x = b_j \ (j \in \mathcal{E})\big\}.$$

If we solely minimize with respect to linear *equality* constraints, that is, we have

$$(PLE) \quad \begin{cases} f(x) \longrightarrow \min \\ a_j^T x = b_j \quad \text{for all } j \in \mathcal{E}, \end{cases} \tag{7}$$

then, if we utilize the elimination process for the special case (QPE) on p. 175 ff (under the assumption: $a_{m+1}, \ldots, a_{m+p}$ are linearly independent), we can replace this problem by an unconstrained optimization problem.

For the general case we remember the already well-known concepts: For $x^{(0)} \in D$, a $d \in \mathbb{R}^n$ with $f'(x^{(0)})d < 0$ is called a *descent direction* at $x^{(0)}$. Then, for all $\tau > 0$ sufficiently small, it holds that $f(x^{(0)} + \tau d) < f(x^{(0)})$ (cf. p. 45). A vector $d \neq 0$ is called a *feasible direction* at $x^{(0)} \in \mathcal{F}$ if $x^{(0)} + \tau d \in \mathcal{F}$ holds for all $\tau > 0$ sufficiently small.

Using the index set $\mathcal{A}(x^{(0)}) := \{i \in \mathcal{I} \mid a_i^T x^{(0)} = b_i\}$ of the active inequality constraints at $x^{(0)}$, this obviously means:

d is a feasible direction at $x^{(0)} \in \mathcal{F}$ if and only if $a_i^T d \leq 0$ for all $i \in \mathcal{A}(x^{(0)})$ and $a_j^T d = 0$ for all $j \in \mathcal{E}$ hold.

A descent direction at $x^{(0)} \in \mathcal{F}$ which is at the same time a feasible direction is called a *feasible descent direction* at $x^{(0)}$.

We go back to the **active set method** (cf. the special case on p. 170 ff):

Algorithm (model algorithm)

0) Let $x^{(0)} \in \mathcal{F}$, $J_0 := \mathcal{A}(x^{(0)}) \cup \mathcal{E}$ and $(a_\mu \mid \mu \in J_0)$ be linearly independent.

If for a $k \in \mathbb{N}_0$ the points $x^{(0)}, \ldots, x^{(k)} \in \mathcal{F}$ and the corresponding index sets $J_0, \ldots, J_k$ with $(a_\mu \mid \mu \in J_k)$ linearly independent are already determined, then we obtain $x^{(k+1)}$ and J_{k+1} as follows:

1) Determine a *feasible descent direction* d_k at $x^{(k)}$ if one exists. *Otherwise* go to *4)*.

2) Choose an $r \notin J_k$ with $a_r^T d_k > 0$ and

$$\alpha_k := \begin{cases} \min\left\{ \dfrac{b_i - a_i^T x^{(k)}}{a_i^T d_k} \;\middle|\; i \notin J_k,\ a_i^T d_k > 0 \right\} = \dfrac{b_r - a_r^T x^{(k)}}{a_r^T d_k} \\ \infty\,, \quad \text{if } a_i^T d_k \leq 0 \text{ for all } i \notin J_k\,. \end{cases}$$

Determine $\lambda_k \in [0, \alpha_k]$ with
$f(x^{(k)} + \lambda_k d_k) = \min\{f(x^{(k)} + \lambda d_k) \mid 0 \leq \lambda \leq \alpha_k\}$.
That is, $\lambda_k = \infty$, if $\alpha_k = \infty$ and f is unbounded from below with respect to the direction d_k.
Otherwise: $x^{(k+1)} := x^{(k)} + \lambda_k d_k$

3) If $\lambda_k = \alpha_k < \infty$: $J_{k+1} := J_k \cup \{r\}$
Set $k := k + 1$ and go to *1)*.

4) Calculate the LAGRANGE *multipliers:* $\nabla f(x^{(k)}) + \sum_{i \in J_k} u_i a_i = 0$

5) If $u_i \geq 0$ for all $i \in J_k \cap \mathcal{I}$ holds, then $x^{(k)}$ is a KKT point: STOP

Otherwise there exists an $s \in J_k \cap \mathcal{I}$ with $u_s < 0$.

$J_{k+1} := J_k \setminus \{s\}$

$x^{(k+1)} := x^{(k)}$; set $k := k + 1$ and go to *1)*.

A feasible starting vector can be obtained — if necessary — via phase I of the simplex algorithm.

We will of course try to determine a descent direction d with $f'(x^{(0)})d$ *minimal.* However, this is only possible if d is bounded in some way. Therefore we introduce an additional *normalization constraint.*

In order to determine *locally optimal descent directions,* we consider the following method for $x^{(0)} \in \mathcal{F}$ and $g_0 := \nabla f(x^{(0)})$:

Zoutendijk's Method (1960)

$$\begin{cases} g_0^T d \longrightarrow \min \\ a_i^T d \leq 0 \text{ for } i \in \mathcal{A}(x^{(0)}) \\ a_j^T d = 0 \text{ for } j \in \mathcal{E} \\ -1 \leq d_\nu \leq 1 \quad (1 \leq \nu \leq n) \qquad \text{(unit ball with respect to the maximum norm)} \end{cases}$$

We can utilize a suitable version of the simplex algorithm to solve this problem.

If the minimum of $g_0^T d$ subject to these constraints is nonnegative, it follows that $\mathcal{C}_\ell(x^{(0)}) \cap \mathcal{C}_{dd}(x^{(0)}) = \emptyset$, that is, $x^{(0)}$ is a KKT point (cf. proposition 2.2.1).

Variants of this method are:

$$\begin{cases} g_0^T d \longrightarrow \min \\ a_i^T d \leq 0 \quad \text{for all } i \in \mathcal{A}(x^{(0)}) \\ a_j^T d = 0 \quad \text{for all } j \in \mathcal{E} \\ d^T d \leq 1 \quad \text{(unit ball with respect to the euclidean norm)} \end{cases} \tag{8}$$

$$\begin{cases} \frac{1}{2}(g_0 + d)^T (g_0 + d) \longrightarrow \min \\ a_i^T d \leq 0 \quad \text{for all } i \in \mathcal{A}(x^{(0)}) \\ a_j^T d = 0 \quad \text{for all } j \in \mathcal{E}. \end{cases} \tag{9}$$

Remark

a) Let d be a minimizer to (9). Then it holds that:

If $d = 0$, then d also yields a solution to (8).

If $d \neq 0$, then $d/\|d\|_2$ gives a solution to (8).

b) If d is a minimizer to (8), then there exists an $\alpha \geq 0$ such that αd yields a solution to (9).

Before we give the *proof*, observe: (8) and (9) are *convex* optimization problems fulfilling (SCQ). Hence, the KARUSH–KUHN–TUCKER conditions are necessary and sufficient for optimality. These *conditions* to (8) are:

$$\begin{cases} g_0 + \alpha d + A_1 u + A_2 v = 0 \\ \alpha \geq 0\,,\ u \geq 0 \\ A_1^T d \leq 0,\ A_2^T d = 0,\ d^T d \leq 1 \\ \alpha\,(d^T d - 1) = 0,\ u^T A_1^T d = 0 \end{cases} \tag{10}$$

with $A_1 := \big(a_i \mid i \in \mathcal{A}(x^{(0)})\big)$, $A_2 := \big(a_j \mid j \in \mathcal{E}\big)$ and
$A_J := \big(A_1, A_2\big)$ with $J := \mathcal{A}(x^{(0)}) \cup \mathcal{E}$.

The KARUSH–KUHN–TUCKER *conditions* to (9) are

$$\begin{cases} g_0 + d + A_1 u + A_2 v = 0 \\ u \geq 0,\ A_1^T d \leq 0,\ A_2^T d = 0 \\ u^T A_1^T d = 0\,. \end{cases} \tag{11}$$

Proof of the Remark: a) Suppose that $(\overline{d}, \overline{u}, \overline{v})$ meets (11).
If $\overline{d} = 0$, then $d = 0$, $\alpha = 0$, $u = \overline{u}$, $v = \overline{v}$ solve the equations in (10).
If $\overline{d} \neq 0$, then (10) is solved by $d := \overline{d}/\|\overline{d}\|_2$, $\alpha := \|\overline{d}\|_2$, $u := \overline{u}$, $v := \overline{v}$.
b) If $(\overline{d}, \overline{\alpha}, \overline{u}, \overline{v})$ solves (10), then $(\overline{\alpha}\,\overline{d}, \overline{u}, \overline{v})$ solves (11). □

Projected Gradient Method

ROSEN (1960) was the first to propose a projection of the gradient onto the boundary and to move along the projected gradient. Our discussion in this subsection will focus on the following equality constrained quadratic problem, which is a modification of problem (9):

$$\begin{cases} \frac{1}{2}(g_0 + d)^T (g_0 + d) \longrightarrow \min \\ A_J^T d = 0 \end{cases} \tag{12}$$

We **assume** that the column vectors of A_J are linearly independent.

Denote by L the subspace $\{d \in \mathbb{R}^n \mid A_J^T d = 0\}$, hence,

$$L = \text{kernel}\,\big(A_J^T\big)\,.$$

Chapter 4

The KARUSH–KUHN–TUCKER *conditions to (12)* are

$$\begin{aligned} 0 &= g_0 + d + A_J u \\ A_J^T d &= 0\,. \end{aligned}$$

$\langle A_J u, d\rangle = \langle u, A_J^T d\rangle = \langle u, 0\rangle = 0$ for any $d \in L$ shows $A_J u \in L^\perp$ (the *orthogonal complement* of L). $-g_0 = d + A_J u$ (with $d \in L$ and $A_J u \in L^\perp$) yields $-A_J^T g_0 = A_J^T d + A_J^T A_J u = A_J^T A_J u$ and thus $u = -(A_J^T A_J)^{-1} A_J^T g_0$; with

$$P_L := I_n - A_J (A_J^T A_J)^{-1} A_J^T\,,$$

the *orthogonal projector* onto the subspace L, we finally get $d = -P_L g_0$.

If $d = -P_L g_0 \neq 0$, then the vector d is a descent direction, because of $\langle g_0, d\rangle = -\langle g_0, P_L g_0\rangle = -\langle g_0, P_L^2 g_0\rangle = \langle d, d\rangle < 0$.

During the execution of the algorithm it is necessary to compute the projection matrices. Since the changes affect only *one* constraint at a time, it is important to have a *simple* way of obtaining the projection matrix from the preceding ones (cf. exercise 14).

Example 9 (cf. [Lu/Ye], p. 370 f)

$$\begin{aligned} &x_1^2 + x_2^2 + x_3^2 + x_4^2 - 2x_1 - 3x_4 \longrightarrow \min \\ &x_i \geq 0 \quad \text{for } 1 \leq i \leq 4 \\ &2x_1 + x_2 + x_3 + 4x_4 = 7 \\ &x_1 + x_2 + 2x_3 + x_4 = 6 \end{aligned}$$

$x^{(0)} := (2, 2, 1, 0)^T$ is feasible with $g_0 = (2, 4, 2, -3)^T$.

It holds that $\mathcal{A}(x^{(0)}) = \{4\}$, $J = \{4, 5, 6\}$ and $A_J^T = \begin{pmatrix} 0 & 0 & 0 & -1 \\ 2 & 1 & 1 & 4 \\ 1 & 1 & 2 & 1 \end{pmatrix}$.

As on p. 175 f we get the QR-factorization of A_J with

$$Q := \begin{pmatrix} Q_1 \\ \hline Q_2 \end{pmatrix} := \begin{pmatrix} 0 & 0 & 0 & -1 \\ 2/\sqrt{6} & 1/\sqrt{6} & 1/\sqrt{6} & 0 \\ -4/\sqrt{66} & 1/\sqrt{66} & 7/\sqrt{66} & 0 \\ \hline 1/\sqrt{11} & -3/\sqrt{11} & 1/\sqrt{11} & 0 \end{pmatrix}:$$

$$Q A_J = \begin{pmatrix} 1 & -4 & -1 \\ 0 & \sqrt{6} & 5/\sqrt{6} \\ 0 & 0 & 11/\sqrt{66} \\ \hline 0 & 0 & 0 \end{pmatrix} =: \widetilde{R} =: \begin{pmatrix} R \\ \hline 0 \end{pmatrix}, Q_1 A_J = R, Q_2 A_J = 0, A_J = Q_1^T R$$

From that we obtain $A_J^T A_J = R^T Q_1 Q_1^T R = R^T R$, $A_J \left(A_J^T A_J\right)^{-1} A_J^T = Q_1^T R \left(R^T R\right)^{-1} R^T Q_1 = Q_1^T Q_1$ and — with $I = Q^T Q = Q_1^T Q_1 + Q_2^T Q_2$ —

$$P_L = I - Q_1^T Q_1 = Q_2^T Q_2 = \frac{1}{11}\begin{pmatrix} 1 & -3 & 1 & 0 \\ -3 & 9 & -3 & 0 \\ 1 & -3 & 1 & 0 \\ 0 & 0 & 0 & 0 \end{pmatrix}, d = -P_L\, g_0 = \frac{8}{11}\begin{pmatrix} 1 \\ -3 \\ 1 \\ 0 \end{pmatrix}.$$

◁

What happens if $d = -P_L\, g_0 = 0$ *?*

Due to the KKT conditions, it holds that $0 = g_0 + \sum_{i \in J} u_i\, a_i$.

If $u_i \geq 0$ holds for all $i \in \mathcal{A}(x^{(0)})$, $x^{(0)}$ is a KKT point of (6) and hence a minimizer to the original problem.

Otherwise there exists an $s \in \mathcal{A}(x^{(0)})$ with $u_s < 0$.

With $\overline{J} := J \setminus \{s\}$, $\overline{A}_1 := \left(a_i \mid i \in \mathcal{A}(x^{(0)}) \setminus \{s\}\right)$, $A_{\overline{J}} := \left(\overline{A}_1, A_2\right)$ we consider $\overline{L} := \{d \in \mathbb{R}^n \mid A_{\overline{J}}^T d = 0\}$, $\overline{d} := -P_{\overline{L}}\, g_0$, $-g_0 = A_J u$ and $-g_0 = A_{\overline{J}}\, \overline{u} + \overline{d}$ and show $\overline{d} \neq 0$:

If $\overline{d} = 0$, we would have $A_J u = A_{\overline{J}}\, \overline{u}$, hence a_s as a linear combination of the remaining columns in *contradiction* to the assumption. □

Unfortunately, the projected gradient method may fail to converge to a KKT point or a minimizer because of the so-called *jamming* or *zigzagging* (cf. [Fle], section 11.3). The following example shows this phenomenon. We do not discuss strategies to avoid jamming.

Example 10 (Wolfe, 1972)

$f(x) := \frac{4}{3}\left(x_1^2 - x_1 x_2 + x_2^2\right)^{3/4} + x_3 \longrightarrow \min$

$x_1, x_2, x_3 \geq 0$

The objective function $f\colon \mathbb{R}^n \longrightarrow \mathbb{R}$ is convex and continuously differentiable (cf. exercise 19). The unique minimizer — with value 0 — is $(0,0,0)^T$.

We start from $x^{(0)} := (a, 0, c)^T$ where $0 < a \leq \frac{1}{2\sqrt{2}}$ and $c > \left(1 + \frac{1}{\sqrt{2}}\right)\sqrt{a}$ and are going to *prove:*

$$x^{(k)} = \begin{cases} (\alpha_k, 0, \gamma_k)^T, & \text{for } k \text{ even} \\ (0, \alpha_k, \gamma_k)^T, & \text{for } k \text{ odd} \end{cases}$$

$$\text{with } \alpha_k := \frac{a}{2^k}\,,\ \gamma_k := c - \frac{1}{2}\sqrt{a}\sum_{j=0}^{k-1}\left(\frac{1}{\sqrt{2}}\right)^j$$

Chapter 4

$d_k := -\nabla f(x^{(k)})$ are feasible descent directions.

$\alpha_k \longrightarrow 0,\ \gamma_k \longrightarrow c - \frac{\sqrt{a}}{2}\,\frac{1}{1-\frac{1}{\sqrt{2}}} = c - \sqrt{a}\,\left(1+\frac{1}{\sqrt{2}}\right) =: \gamma^* > 0$

$x^{(k)} \longrightarrow x^* = (0,0,\gamma^*)^T$ with $f(x^*) > 0$.

So we get a sequence of points converging to a point which has nothing to do with the solution to the problem!

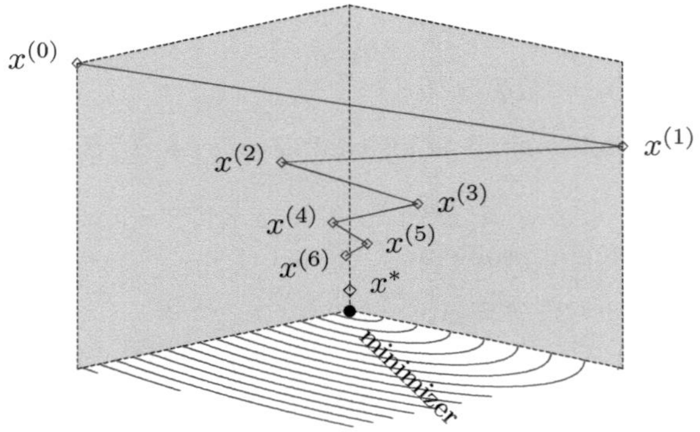

$$f'(x) = \left(\frac{2x_1 - x_2}{\sqrt[4]{x_1^2 - x_1x_2 + x_2^2}},\ \frac{2x_2 - x_1}{\sqrt[4]{x_1^2 - x_1x_2 + x_2^2}},\ 1\right) \quad \text{for } x \neq 0$$

$k = 0:\quad x^{(0)} = (a, 0, c)^T = (\alpha_0, 0, \gamma_0)^T,\ g_0 = \nabla f(x^{(0)}) = \left(2\sqrt{a}, -\sqrt{a}, 1\right)^T$

$\mathcal{A}(x^{(0)}) = \{2\};\ d_0 = -g_0$ is feasible, as $a_2^T d_0 = -\sqrt{a} \le 0$.

$x^{(0)} + \lambda d_0 = \left(a - 2\sqrt{a}\lambda, \lambda\sqrt{a}, c - \lambda\right)^T \in \mathcal{F} \iff 0 \le \lambda \le \sqrt{a}/2 < c$

So we get $\alpha_0 := \sqrt{a}/2$ (cf. *2)* on p. 187).

With $\varphi(\lambda) := f\left(x^{(0)} + \lambda d_0\right) = \frac{4}{3}a^{3/4}\left(7\lambda^2 - 5\sqrt{a}\lambda + a\right)^{3/4} + c - \lambda$ we are looking for a minimizer λ_0 of φ on $[0, \alpha_0]$:

$$\varphi'(\lambda) = a^{3/4}\frac{14\lambda - 5\sqrt{a}}{\sqrt[4]{7\lambda^2 - 5\sqrt{a}\lambda + a}} - 1,\quad \varphi''(\lambda) = a^{3/4}\frac{(14\lambda - 5\sqrt{a})^2 + 6a}{4(7\lambda^2 - 5\sqrt{a}\lambda + a)^{5/4}} > 0$$

This shows that φ' is strongly isotone, $\varphi'(\lambda) \le \varphi'(\alpha_0) = 2\sqrt{2}a - 1 \le 0$ and further that φ is antitone. Hence, $\lambda_0 = \alpha_0 = \sqrt{a}/2$ yields the minimum. With $x^{(1)} := x^{(0)} + \lambda_0 d_0 = \left(0, a/2, c - \sqrt{a}/2\right)^T$ instead of $x^{(0)}$ we get the next step in like manner. ◁

Reduced Gradient Method

The reduced gradient method was developed by Wolfe (1962) to solve nonlinear optimization problems with linear constraints. The method is closely related to the simplex method for linear problems because the variables are split into basic and nonbasic ones. Like gradient projection methods, it can be regarded as a steepest descent method generating feasible descent directions. The dimension of the problem is reduced, at each step, by representing all variables in terms of an independent subset of variables.

We consider the problem

$$(P) \quad \begin{cases} f(x) \longrightarrow \min \\ Ax = b, \ x \geq 0 \end{cases}$$

where $m, n \in \mathbb{N}$ with $m \leq n$, $A \in R^{m \times n}$, $b \in \mathbb{R}^m$ and $f\colon \mathbb{R}^n \longrightarrow \mathbb{R}$ continuously differentiable. So we have linear equality constraints for nonnegative variables.

We make the following **assumptions:**

1) $\mathcal{F} := \{x \in \mathbb{R}^n \mid Ax = b, \ x \geq 0\} \neq \emptyset$

2) Problem (P) is *nondegenerate* in the following sense: For each $x \in \mathcal{F}$ the set of restrictions active at x — including the equality constraints — is linearly independent.

The basic *idea* of the reduced gradient method is to consider, at each step, the problem only in terms of independent variables:

Let $N := \{1, \ldots, n\}$, $j_1, \ldots, j_m \in N$ and $J := (j_1, \ldots, j_m)$ be a feasible basis to A (cf. p. 153, simplex method). Hence, A_J is nonsingular. With $K = (k_1, \ldots, k_{n-m}) \in N^{n-m}$ such that $S(K) \uplus S(J) = N$ we have

$$A_J x_J + A_K x_K = Ax = b \iff x_J = \underbrace{A_J^{-1} b}_{=:\,\overline{b}} - \underbrace{A_J^{-1} A_K}_{=:\,\overline{A}_K} x_K \,.$$

x_J is the vector of basic or dependent variables, x_K the vector of nonbasic or independent variables. By assigning values to them, we get — via $x_J := \overline{b} - \overline{A}_K x_K$ — a unique solution to $Ax = b$.

With the reduced function $\widehat{f}$ given by

$$\widehat{f}(x_K) := f\big(x_J, x_K\big) = f\big(\overline{b} - \overline{A}_K x_K, x_K\big)$$

we consider the following **transformation** of (P):

$$(\widehat{P}) \quad \begin{cases} \widehat{f}(x_K) \longrightarrow \min \\ \overline{b} - \overline{A}_K x_K \geq 0 \\ x_K \geq 0 \end{cases}$$

The *reduced gradient of* f *is the gradient of* $\widehat{f}$*:*

$$x_j = \overline{b}_j - \sum_{\ell \in K} \overline{a}_{j\ell}\, x_\ell \qquad (j \in J)$$

$$\frac{\partial \widehat{f}}{\partial x_k} = \frac{\partial f}{\partial x_k} + \sum_{j \in J} \frac{\partial f}{\partial x_j} \cdot (-\overline{a}_{jk}) \qquad (k \in K)$$

$$\nabla \widehat{f}(x_K) = \nabla_{x_K} f(x) - \overline{A}_K^T \nabla_{x_J} f(x)$$

Chapter 4

Proposition 4.3.1

To each $x \in \mathcal{F}$ there exists a decomposition $S(K) \uplus S(J) = N$ with $x_J > 0$ and A_J regular. (Commonly, we do not have $x_K = 0$.)

Proof: Let $x \in \mathcal{F}$, WLOG $x_1 = \cdots = x_q = 0$ and $x_j > 0$ for $j = q+1, \ldots, n$ with a suitable $q \in \{0, \ldots, n\}$. Nondegeneracy means

$$\operatorname{rank}\left(\begin{array}{ccc|ccc} a_{11} & \ldots & a_{1q} & a_{1,q+1} & \ldots & a_{1n} \\ \vdots & & \vdots & \vdots & & \vdots \\ a_{m1} & \ldots & a_{mq} & a_{m,q+1} & \ldots & a_{mn} \\ \hline 1 & & & & & \\ & \ddots & & & 0 & \\ & & 1 & & & \end{array}\right) \begin{array}{l} \left.\vphantom{\begin{array}{c}a\\ \vdots\\ a\end{array}}\right\} m \\ \left.\vphantom{\begin{array}{c}1\\ \ddots\\ 1\end{array}}\right\} q \end{array} = m + q \quad (\leq n)\,.$$

This is equivalent to

$$\operatorname{rank}\begin{pmatrix} a_{1,q+1} & \ldots & a_{1,n} \\ \vdots & & \vdots \\ a_{m,q+1} & \ldots & a_{m,n} \end{pmatrix} = m$$

and further to the existence of $J \subset \{q+1, \ldots, n\}$ such that A_J is regular . □

If $x^{(0)}$ is feasible to (P) and $d \in$ kernel(A) with $d_i \geq 0$ if $x_i^{(0)} = 0$ (for $i \in N$), then $x^{(0)} + \varepsilon d \in \mathcal{F}$ for all $\varepsilon > 0$ sufficiently small. In other words: $d \in \mathbb{R}^n$ is a feasible direction if and only if $Ad = 0$ and $d_i \geq 0$ if $x_i^{(0)} = 0$ (for $i \in N$).

Proof: If $x_i^{(0)} > 0$, then we obtain $x_i^{(0)} + \varepsilon d_i > 0$ for all $\varepsilon > 0$ sufficiently small. If $x_i^{(0)} = 0$, then we have $d_i \geq 0$ and thereby $x_i^{(0)} + \varepsilon d_i \geq 0$ for all $\varepsilon > 0$. □

Let now $x^{(0)} \in \mathcal{F}$, $J \subset N$ with $x_J^{(0)} > 0$ and A_J regular be given. We determine a *feasible descent direction* d to (P) at $x^{(0)}$ in the following way: For $k \in K$, we define

$$d_k := \begin{cases} -\frac{\partial \widehat{f}}{\partial x_k}\big(x_K^{(0)}\big), & \textit{if } \frac{\partial \widehat{f}}{\partial x_k}\big(x_K^{(0)}\big) < 0 \textit{ or } x_k^{(0)} > 0 \\ 0, & \textit{if } \frac{\partial \widehat{f}}{\partial x_k}\big(x_K^{(0)}\big) \geq 0 \textit{ and } x_k^{(0)} = 0 \end{cases}$$

and set $d_J := -\overline{A}_K d_K$.

We are usually searching in the direction of the negative gradient. If we were able to make a 'small' move here from $x_K^{(0)}$ in the direction of the negative reduced gradient *without violating the constraints*, we would get a decrease of the value of f. Thus, we have to choose d_K in such a way that $\left\langle \nabla \widehat{f}(x_K^{(0)}), d_K \right\rangle < 0$ and that $d_i \geq 0$ if $x_i^{(0)} = 0$ (for $i \in K$).

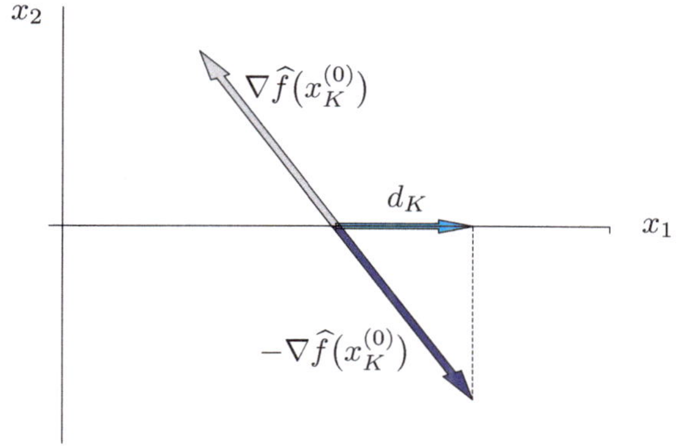

Proposition 4.3.2

1) *d_K is a feasible direction to $(\widehat{P})$ at $x_K^{(0)}$ and d is a feasible direction to (P) at $x^{(0)}$.*

2) *If $d_K \neq 0$, then d_K is a descent direction to $(\widehat{P})$ at $x_K^{(0)}$ and d is a descent direction to (P) at $x^{(0)}$.*

3) *$d_K = 0$ if and only if $x_K^{(0)}$ is a KKT point to $(\widehat{P})$ and further if and only if $x^{(0)}$ is a KKT point to (P).*

Proof:

1) We have $Ad = 0$ by definition of d_J. Since $\overline{b} - \overline{A}_K x_K^{(0)} = x_J^{(0)}$ is positive, it will remain positive for 'small' changes of $x_K^{(0)}$. Thus, we obtain $x_J^{(0)} + \varepsilon d_J > 0$ and (with the definition of d_K) $x_K^{(0)} + \varepsilon d_K \geq 0$ for sufficiently small $\varepsilon > 0$. Altogether we have for such ε: $x^{(0)} + \varepsilon d \in \mathcal{F}$ and $x_K^{(0)} + \varepsilon d_K$ is feasible to $(\widehat{P})$.

2) $$\begin{aligned} \left\langle d, \nabla f(x^{(0)}) \right\rangle &= \left\langle d_J, \nabla_{x_J} f(x^{(0)}) \right\rangle + \left\langle d_K, \nabla_{x_K} f(x^{(0)}) \right\rangle \\ &= \left\langle d_K, -\overline{A}_K^T \nabla_{x_J} f(x^{(0)}) + \nabla_{x_K} f(x^{(0)}) \right\rangle \\ &= \left\langle d_K, \nabla \widehat{f}(x_K^{(0)}) \right\rangle \overset{\checkmark}{\leq} -\left\langle d_K, d_K \right\rangle < 0 \end{aligned}$$

3) By the definition of d_K we have

$$d_K = 0 \iff \forall k \in K \begin{cases} x_k^{(0)} > 0 \Longrightarrow \frac{\partial \widehat{f}}{\partial x_k}(x_K^{(0)}) = 0 \\ x_k^{(0)} = 0 \Longrightarrow \frac{\partial \widehat{f}}{\partial x_k}(x_K^{(0)}) \geq 0 \,. \end{cases}$$

$x_K^{(0)}$ is a KKT point to $(\widehat{P})$ iff there exist vectors $u, v \geq 0$ such that:

$$x_K^{(0)} \geq 0, \quad \overline{b} - \overline{A}_K x_K^{(0)} \geq 0,$$

$$\nabla\widehat{f}(x_K^{(0)}) + \overline{A}_K^T u - v = 0,$$
$$\langle u, \overline{A}_K x_K^{(0)} - \overline{b}\rangle = 0 \text{ and } \langle v, x_K^{(0)}\rangle = 0.$$

Since $\overline{b} - \overline{A}_K x_K^{(0)} = x_J^{(0)} > 0$ and $u \geq 0$, it follows: $\left\langle u, \overline{A}_K x_K^{(0)} - \overline{b}\right\rangle = 0$ iff $u = 0$. In this case we have $v = \nabla\widehat{f}(x_K^{(0)})$. Thus, these KKT conditions reduce to

$$\nabla\widehat{f}(x_K^{(0)}) \geq 0 \text{ and } \left\langle \nabla\widehat{f}(x_K^{(0)}), x_K^{(0)}\right\rangle = 0.$$

Hence: $d_K = 0$ iff $x_K^{(0)}$ is a KKT point to $(\widehat{P})$.

$x^{(0)}$ is a KKT point to (P) iff there exist vectors $u \in \mathbb{R}^m$ and $v \in \mathbb{R}_+^n$ such that $\nabla f(x^{(0)}) + A^T u - v = 0$ and $\langle v, x^{(0)}\rangle = 0$.

This is equivalent to $\langle v_K, x_K^{(0)}\rangle + \langle v_J, x_J^{(0)}\rangle = 0$ and $\nabla_{x_K} f(x^{(0)}) + A_K^T u - v_K = 0$, $\nabla_{x_J} f(x^{(0)}) + A_J^T u - v_J = 0$.

Since $x_J^{(0)} > 0$ and $v_J \geq 0$, it follows: $\langle v_J, x_J^{(0)}\rangle = 0$ iff $v_J = 0$. Thus, these KKT conditions reduce to $\langle v_K, x_K^{(0)}\rangle = 0$ and $\nabla_{x_K} f(x^{(0)}) + A_K^T u - v_K = 0$, $\nabla_{x_J} f(x^{(0)}) + A_J^T u = 0$.

The last equation is uniquely solved by $u = -A_J^{-T}\nabla_{x_J} f(x^{(0)})$. With that the remaining KKT conditions are $\langle v_K, x_K^{(0)}\rangle = 0$ and $\nabla\widehat{f}(x_K^{(0)}) = \nabla_{x_K} f(x^{(0)}) - A_K^T A_J^{-T}\nabla_{x_J} f(x^{(0)}) = v_K$.

Hence: $x^{(0)}$ is a KKT point of (P) iff $x_K^{(0)}$ is a KKT point of $(\widehat{P})$. □

Now we can describe *one step of the reduced gradient method* as follows:

1) Determine d_K (cf. p. 194).

2) If $d_K = 0$, then STOP: The current point $x^{(0)}$ is a KKT point to (P).

3) $\alpha_1 := \max\{\alpha \geq 0 \mid x_J^{(0)} + \alpha d_J \geq 0\} > 0$ (since $x_J^{(0)} > 0$)

$\alpha_2 := \max\{\alpha \geq 0 \mid x_K^{(0)} + \alpha d_K \geq 0\} > 0$ $(x_k^{(0)} = 0 \Longrightarrow d_k \geq 0)$

$\widehat{\alpha} := \min\{\alpha_1, \alpha_2\}$

Search $\alpha_0 \in [0, \widehat{\alpha}]$ with $f(x^{(0)} + \alpha_0 d) = \min\{f(x^{(0)} + \alpha d) \mid 0 \leq \alpha \leq \widehat{\alpha}\}$

$x^{(1)} := x^{(0)} + \alpha_0 d$

4) If $\alpha_0 < \alpha_1$: J remains unchanged.

If $\alpha_0 = \alpha_1$: There exists an $s \in J$ with $x_s^{(0)} + \alpha_0 d_s = 0$.

The variable x_s leaves the basis.[8] One of the positive nonbasic variables x_r becomes a basic variable.

This is possible because of the assumed nondegeneracy.

[8] If there is more than one zero-component, each of them will be removed from the basis J via an exchange step.

Let us illustrate one step of the reduced gradient method with example 9 (cf. p. 190):

Example 11 We have $A = \begin{pmatrix} 2 & 1 & 1 & 4 \\ 1 & 1 & 2 & 1 \end{pmatrix}$, $b = \begin{pmatrix} 7 \\ 6 \end{pmatrix}$ and
$f(x) = x_1^2 + x_2^2 + x_3^2 + x_4^2 - 2x_1 - 3x_4$.
$x^{(0)} := (2,2,1,0)^T$ is feasible with $\nabla f(x^{(0)}) = (2,4,2,-3)^T$. To $J := (1, 2)$ (for example) we get: $K = (3, 4)$, $A_J = \begin{pmatrix} 2 & 1 \\ 1 & 1 \end{pmatrix}$, $A_K = \begin{pmatrix} 1 & 4 \\ 2 & 1 \end{pmatrix}$,

$$A_J^{-1} = \begin{pmatrix} 1 & -1 \\ -1 & 2 \end{pmatrix}, \ \overline{A}_K = A_J^{-1} A_K = \begin{pmatrix} -1 & 3 \\ 3 & -2 \end{pmatrix}$$

$$\nabla \widehat{f}(x_K^{(0)}) = \nabla_{x_K} f(x^{(0)}) - \overline{A}_K^T \nabla_{x_J} f(x^{(0)}) = \begin{pmatrix} 2 \\ -3 \end{pmatrix} - \begin{pmatrix} -1 & 3 \\ 3 & -2 \end{pmatrix} \begin{pmatrix} 2 \\ 4 \end{pmatrix} = \begin{pmatrix} -8 \\ -1 \end{pmatrix}$$

$$d_K = \begin{pmatrix} 8 \\ 1 \end{pmatrix}, \ \alpha_2 = \infty$$

$$d_J = -\overline{A}_K d_K = \begin{pmatrix} 5 \\ -22 \end{pmatrix}, \ \alpha_1 = \tfrac{1}{11}$$

Minimization of $f(x^{(0)} + \alpha d) = 5 - 65\alpha + 574\alpha^2$ yields $\alpha_0 = 65/1148$. Because of $\alpha_0 < \alpha_1$ the basis $J = (1,2)$ remains unchanged. $\triangleleft$

Remark

An overall disadvantage of the previous methods is that the convergence rate is at best as good as that of the gradient method. The so-called SQP methods, which are based on the same principles as NEWTON'S method, have proven more useful. In order to avoid redundancy in our discussion which is also important for chapter 5, we will for the moment only consider *general nonlinearly constrained optimization problems* in the following subsection. In the case of linearly constrained problems there will be some simplifications which will lead to feasible point methods.

Preview of SQP Methods

The abbreviation "SQP" stands for ***S**equential **Q**uadratic **P**rogramming* or ***S**uccessive **Q**uadratic **P**rogramming*. SQP methods date back to WILSON (1963) and belong to the most efficient methods for solving nonlinear optimization problems. We consider the following optimization problem

$$(P) \quad \begin{cases} f(x) \longrightarrow \min \\ c_i(x) \le 0\,, \ i \in \mathcal{I} := \{1, \ldots, m\} \\ c_i(x) = 0\,, \ i \in \mathcal{E} := \{m+1, \ldots, m+p\}. \end{cases} \qquad (13)$$

For that let $m, p \in \mathbb{N}_0$ (hence, $\mathcal{E} = \emptyset$ or $\mathcal{I} = \emptyset$ are permitted), and suppose that the functions $f, c_1, \ldots, c_{m+p}$ are twice continuously differentiable in an open subset D of $\mathbb{R}^n$. The vectors v from $\mathbb{R}^{m+p}$ are divided into the two components $v_{\mathcal{I}}$ and $v_{\mathcal{E}}$. Corresponding to that, suppose that the function c with the components $c_1, \ldots, c_{m+p}$ is written as $\binom{c_{\mathcal{I}}}{c_{\mathcal{E}}}$. With the notation introduced in chapter 2, it of course holds that $g = c_{\mathcal{I}}$ and $h = c_{\mathcal{E}}$.

As usual

$$L(x, \lambda) := f(x) + \lambda^T c(x) = f(x) + \langle \lambda, c(x) \rangle \qquad (\text{for } x \in D \text{ and } \lambda \in \mathbb{R}^{m+p})$$

denotes the corresponding *Lagrangian*. Let the following **optimality conditions** (cf. theorem 2.3.2) be fulfilled in an $x^* \in D$:

There exists a vector $\lambda^* \in \mathbb{R}^m$ with

1) $\nabla_x L(x^*, \lambda^*) = \nabla f(x^*) + \sum_{i=1}^{m+p} \lambda_i^* \nabla c_i(x^*) = 0$

$\lambda_i^* \geq 0, \; c_i(x^*) \leq 0$ and $\lambda_i^* c_i(x^*) = 0$ for all $i \in \mathcal{I}$

$c_i(x^*) = 0$ for all $i \in \mathcal{E}$.

2) For all $s \in \mathcal{C}_{\ell+}(x^*)$, $s \neq 0$, it holds that $s^T \nabla_{xx}^2 L(x^*, \lambda^*) s > 0$, where

$$\mathcal{C}_{\ell+}(x^*) := \left\{ s \in \mathbb{R}^n \;\middle|\; \begin{array}{l} c_i'(x^*) s \leq 0 \text{ for all } i \in \mathcal{A}(x^*) \setminus \mathcal{A}_+(x^*) \\ c_i'(x^*) s = 0 \text{ for all } i \in \mathcal{A}_+(x^*) \cup \mathcal{E} \end{array} \right\}$$

with $\mathcal{A}_+(x^*) := \{ i \in \mathcal{A}(x^*) \mid \lambda_i^* > 0 \}$.

Then x^* *yields a strict local minimum.*

The complementarity condition $\lambda_i^* c_i(x^*) = 0$ means $\lambda_i^* = 0$ or (in the nonexclusive sense) $c_i(x^*) = 0$. If exactly one of the λ_i^* and $c_i(x^*)$ is zero for all $i \in \mathcal{I}$, then *strict complementarity* is said to hold.

In addition we suppose:

3) $\lambda_i^* > 0$ for all $i \in \mathcal{A}(x^*)$ *(strict complementarity)*

$\left(\nabla c_i(x^*) \,\middle|\, i \in \mathcal{A}(x^*) \cup \mathcal{E} \right)$ linearly independent.

In this case we have $\mathcal{A}(x^*) = \mathcal{A}_+(x^*)$ and thus

$$\mathcal{C}_{\ell+}(x^*) = \left\{ s \in \mathbb{R}^n \;\middle|\; c_i'(x^*) s = 0 \text{ for all } i \in \mathcal{A}(x^*) \cup \mathcal{E} \right\}.$$

If we disregard the inequality constraints $\left(\lambda_i^* \geq 0,\, c_i(x^*) \leq 0 \text{ for all } i \in \mathcal{I}\right)$ for a moment, the KARUSH–KUHN–TUCKER conditions lead to the following system of nonlinear equations:

$$\Phi(x,\lambda) := \begin{pmatrix} \nabla_x L(x,\lambda) \\ \lambda_1 c_1(x) \\ \vdots \\ \lambda_m c_m(x) \\ c_{m+1}(x) \\ \vdots \\ c_{m+p}(x) \end{pmatrix} = 0$$

This is a nonlinear system of $n+m+p$ equations with $n+m+p$ variables.

We examine its *Jacobian*, respecting $\nabla_x L(x,\lambda) = \nabla f(x) + \sum_{i=1}^{m+p} \lambda_i \nabla c_i(x)$:

$$\Phi'(x,\lambda) = \left(\begin{array}{c|ccc|ccc} \nabla_{xx}^2 L(x,\lambda) & \nabla c_1(x) & \ldots & \nabla c_m(x) & \nabla c_{m+1}(x) & \ldots & \nabla c_{m+p}(x) \\ \hline \lambda_1 c_1'(x) & c_1(x) & \ldots & 0 & 0 & \ldots & 0 \\ \vdots & \vdots & \ddots & \vdots & \vdots & \ddots & \vdots \\ \lambda_m c_m'(x) & 0 & \ldots & c_m(x) & 0 & \ldots & 0 \\ \hline c_{m+1}'(x) & 0 & \ldots & 0 & 0 & \ldots & 0 \\ \vdots & \vdots & \ddots & \vdots & \vdots & \ddots & \vdots \\ c_{m+p}'(x) & 0 & \ldots & 0 & 0 & \ldots & 0 \end{array}\right)$$

Chapter 4

Under the assumptions made above, we can show that $\Phi'(x^*,\lambda^*)$ *is invertible*:

Proof: Let $s \in \mathbb{R}^n$, $\lambda \in \mathbb{R}^{m+p}$ with $\Phi'(x^*,\lambda^*)\binom{s}{\lambda} = 0$, that is:

$$\nabla_{xx}^2 L(x^*,\lambda^*)\, s + \sum_{i=1}^{m+p} \lambda_i \nabla c_i(x^*) = 0 \quad \text{and}$$
$$\lambda_i^* c_i'(x^*)\, s + \lambda_i c_i(x^*) = 0 \quad \text{for all } i \in \mathcal{I}, \quad c_i'(x^*)\, s = 0 \quad \text{for all } i \in \mathcal{E}.$$

If $s = 0$, then we get $\sum_{i=1}^{m+p} \lambda_i \nabla c_i(x^*) = 0$ and $\lambda_i c_i(x^*) = 0$ for all $i \in \mathcal{I}$. Hence, because of $\lambda_i = 0$ for all $i \in \mathcal{I} \setminus \mathcal{A}(x^*)$ and the linear independence of $\big(\nabla c_i(x^*) \mid i \in \mathcal{A}(x^*) \cup \mathcal{E}\big)$, $\lambda_1 = \cdots = \lambda_{m+p} = 0$ holds.

In the remaining case, $s \neq 0$, we have $\lambda_i^* > 0$ and $c_i(x^*) = 0$ for all $i \in \mathcal{A}(x^*)$ and thus $c_i'(x^*)s = 0$ for such i. Together with $c_i'(x^*)\, s = 0$ for all $i \in \mathcal{E}$ we obtain $s \in \mathcal{C}_{\ell+}(x^*)$.

By multiplying the first equation (from the left) by s^T we get:

$$s^T \nabla_{xx}^2 L(x^*,\lambda^*)\, s + \sum_{i=1}^{m+p} \lambda_i c_i'(x^*)\, s = 0$$

$c_i'(x^*)s = 0$ for $i \in \mathcal{A}(x^*) \cup \mathcal{E}$ and $\lambda_i = 0$ for $i \in \mathcal{I} \setminus \mathcal{A}(x^*)$ show $\sum_{i=1}^{m+p} \lambda_i c_i'(x^*)s = 0$, thus $s^T \nabla^2_{xx} L(x^*, \lambda^*)s = 0$ in *contradiction* to *2)*. □

Hence, we can apply NEWTON*'s method* to compute (x^*, λ^*) iteratively. In general we regard for a given iteration point $(x^{(k)}, \lambda^{(k)})$:

$$\Phi(x, \lambda) \approx \Phi(x^{(k)}, \lambda^{(k)}) + \Phi'(x^{(k)}, \lambda^{(k)}) \begin{pmatrix} x - x^{(k)} \\ \lambda - \lambda^{(k)} \end{pmatrix} = 0$$

In the foregoing special case this equation means:

$$\begin{aligned}
&\nabla_x L(x^{(k)}, \lambda^{(k)}) + \nabla^2_{xx} L(x^{(k)}, \lambda^{(k)})(x - x^{(k)}) + \sum_{i=1}^{m+p} (\lambda_i - \lambda_i^{(k)}) \nabla c_i(x^{(k)}) = 0 \\
&\lambda_i^{(k)} c_i(x^{(k)}) + \lambda_i^{(k)} c_i'(x^{(k)})(x - x^{(k)}) + (\lambda_i - \lambda_i^{(k)}) c_i(x^{(k)}) = 0 \ \text{ for all } i \in \mathcal{I} \\
&c_i(x^{(k)}) + c_i'(x^{(k)})(x - x^{(k)}) = 0 \ \text{ for all } i \in \mathcal{E}.
\end{aligned}$$

These conditions are equivalent to:

$$\begin{aligned}
&\nabla f(x^{(k)}) + \nabla^2_{xx} L(x^{(k)}, \lambda^{(k)})(x - x^{(k)}) + \sum_{i=1}^{m+p} \lambda_i \nabla c_i(x^{(k)}) = 0 \\
&\lambda_i \left[c_i(x^{(k)}) + c_i'(x^{(k)})(x - x^{(k)}) \right] - (\lambda_i - \lambda_i^{(k)}) c_i'(x^{(k)})(x - x^{(k)}) = 0 \ \ (i \in \mathcal{I}) \\
&c_i(x^{(k)}) + c_i'(x^{(k)})(x - x^{(k)}) = 0 \ \text{ for all } i \in \mathcal{E}.
\end{aligned}$$

Wilson's Lagrange–Newton Method (1963)

The following method consists of the successive minimization of second-order expansions of the Lagrangian, subject to first-order expansions of the constraints.

With the **notations**

$$\begin{aligned}
d &:= x - x^{(k)} \\
g_k &:= \nabla f(x^{(k)}), \quad A_k^T := c'(x^{(k)}) \\
B_k &:= \nabla^2 f(x^{(k)}) + \sum_{i=1}^{m+p} \lambda_i^{(k)} \nabla^2 c_i(x^{(k)}) = \nabla^2_{xx} L(x^{(k)}, \lambda^{(k)}) \\
c_i^{(k)} &:= c_i(x^{(k)}), \quad \ell_i^{(k)}(d) := c_i^{(k)} + c_i'(x^{(k)})\, d
\end{aligned}$$

the above equations — adding the condition $\lambda_i \geq 0$ and $\ell_i^{(k)}(d) \leq 0$ and neglecting the quadratic terms $(\lambda_i - \lambda_i^{(k)}) c_i'(x^{(k)})(x - x^{(k)})$ for all $i \in \mathcal{I}$ — can be written as follows:

$$0 = g_k + B_k d + A_k \lambda$$

Chapter 4

$$\lambda_i \geq 0\,,\ \ell_i^{(k)}(d) \leq 0\,,\ \lambda_i \ell_i^{(k)}(d) = 0 \quad \text{for all } i \in \mathcal{I}$$
$$\ell_i^{(k)}(d) = 0 \quad \text{for all } i \in \mathcal{E}\,.$$

These can be interpreted as the KARUSH–KUHN–TUCKER conditions of the following *quadratic problem*:

$$(QP)_k \qquad \begin{cases} \varphi_k(d) := \frac{1}{2} d^T B_k\, d + g_k^T\, d \longrightarrow \min \\ \ell_i^{(k)}(d) \leq 0 \quad \text{for all } i \in \mathcal{I} \\ \ell_i^{(k)}(d) = 0 \quad \text{for all } i \in \mathcal{E}\,. \end{cases}$$

For the rest of this section we are only interested in *linear constraints*, that is, with suitable $a_i \in \mathbb{R}^n$ and $b_i \in \mathbb{R}$ it holds that

$$c_i(x) = a_i^T x - b_i$$

for $x \in \mathbb{R}^n$ and $i = 1, \ldots, m+p$, and consider the problem (PL) (p. 186). Then evidently $B_k = \nabla^2 f\big(x^{(k)}\big)$ and $A_k = A := \big(a_1, \ldots, a_{m+p}\big)$. If additionally $x^{(k)} \in \mathcal{F}$, then problem $(QP)_k$ can be simplified as follows:

$$(QPL)_k \qquad \begin{cases} \varphi_k(d) := \frac{1}{2} d^T \nabla^2 f\big(x^{(k)}\big)\, d + g_k^T\, d \longrightarrow \min \\ \big(a_i^T x^{(k)} - b_i\big) + a_i^T\, d \leq 0 \quad \text{for all } i \in \mathcal{I} \\ a_i^T\, d = 0 \quad \text{for all } i \in \mathcal{E}\,. \end{cases}$$

The feasible region of $(QPL)_k$ is nonempty. If B_k is in addition *positive definite on* $\operatorname{kernel}(A_{\mathcal{E}}^T)$, then the objective function φ_k is bounded from below on the feasible region. The BARANKIN–DORFMAN theorem (cf. theorem 4.2.1) then gives the existence of a minimizer d to $(QPL)_k$. Together with the corresponding LAGRANGE multiplier λ we get:

$$0 = g_k + B_k\, d + A\lambda$$
$$\lambda_i \geq 0\,, \quad \big(a_i^T x^{(k)} - b_i\big) + a_i^T\, d \leq 0\,, \quad \lambda_i \big(a_i^T x^{(k)} - b_i + a_i^T d\big) = 0 \quad \text{for all } i \in \mathcal{I}$$
$$a_i^T\, d = 0 \quad \text{for all } i \in \mathcal{E}\,.$$

d is then a feasible direction to $\mathcal{F}$ in $x^{(k)}$. If $d = 0$, then $x^{(k)}$ is a KKT point to (PL). Otherwise d is a descent direction of f in $x^{(k)}$:

By multiplying $g_k = -B_k\, d - A\lambda$ (from the left) by d^T we get

$$\langle g_k, d\rangle = -\langle d, B_k\, d\rangle - \langle d, A\lambda\rangle = -\langle d, B_k d\rangle - \sum_{i=1}^{m+p} \lambda_i \langle a_i, d\rangle\,.$$

$\langle d, B_k\, d\rangle$ is positive because of our assumptions concerning B_k.
It furthermore holds that: $\langle a_i, d\rangle = 0$ for all $i \in \mathcal{E}$ and

$$\lambda_i \langle a_i, d\rangle = \lambda_i \big(b_i - a_i^T x^{(k)}\big) \geq 0 \quad \text{for all } i \in \mathcal{I}.$$

Hence $\langle g_k, d\rangle < 0\,.$

Chapter 4

Now a rudimentary algorithm can be specified as follows:

Algorithm (LAGRANGE–NEWTON SQP Method)

0) Let $x^{(0)} \in \mathcal{F}$ be given. Set $k := 0$.

1) Determine a solution d_k and a corresponding LAGRANGE multiplier $\lambda^{(k)}$ of $(QPL)_k$.

2) If $d_k = 0$: STOP ($x^{(k)}$ is a KKT point to (PL).)
Otherwise: $x^{(k+1)} := x^{(k)} + d_k$
Set $k := k+1$ and go to *1)*.

This method obviously yields a sequence $\left(x^{(k)}\right)$ of feasible points.

Similar to NEWTON's method, ***local*** *quadratic convergence* can be shown (cf. [Fle], theorem 12.3.1). One can try to extend the local convergence result to a global one:

1) Line search instead of fixed step size 1
The solution to $(QPL)_k$ produces a feasible descent direction d_k which is used to determine $\alpha_k \in [0,\,1]$ with

$$f\big(x^{(k)} + \alpha_k\, d_k\big) = \min\big\{f\big(x^{(k)} + \alpha\, d_k\big) \mid 0 \le \alpha \le 1\big\}.$$

With that, we define $x^{(k+1)} := x^{(k)} + \alpha_k\, d_k$. If $B_k := \nabla^2 f\big(x^{(k)}\big)$, we get the *damped* NEWTON *method*.

2) Quasi-NEWTON approximations to $\nabla^2 f\big(x^{(k)}\big)$
Quite a number of SQP methods use BFGS-like update formulae (cf. proposition 3.5.2'). With $q_k := \nabla_x L\big(x^{(k+1)}, \lambda^{(k+1)}\big) - \nabla_x L\big(x^{(k)}, \lambda^{(k)}\big)$ and $p_k := x^{(k+1)} - x^{(k)}$ it thereby holds

$$B_{k+1} := B_k + \frac{q_k\, q_k^T}{\langle p_k\,, q_k\rangle} - \frac{B_k\, p_k\, p_k^T\, B_k}{\langle p_k\,, B_k\, p_k\rangle}\,.$$

A very special choice is $B_k := I$. It leads to the projected gradient method from the beginning of this section.

Example 12 (cf. [MOT], p. 5–49 f)

$$\begin{aligned} f(x) &:= -x_1\, x_2\, x_3 \longrightarrow \min \\ -x_1 - 2x_2 - 2x_3 &\le 0 \\ x_1 + 2x_2 + 2x_3 &\le 72 \end{aligned}$$

This problem has the minimizer $x^* = (24, 12, 12)^T$ with the corresponding LAGRANGE multiplier $\lambda^* = (0, 144)^T$ and the active set $\mathcal{A}(x^{(*)}) = \{2\}$.

To solve it, we utilize the function fmincon from the *Matlab*® *Optimization Toolbox*. For the starting point $x^{(0)} := (10, 10, 10)^T$ we get:

k	$x_1^{(k)}$	$x_2^{(k)}$	$x_3^{(k)}$	$f(x^{(k)})$	α_k	$\mathcal{A}(x^{(k)})$
0	10.0000	10.0000	10.0000	−1000.00		$\emptyset$
1	33.4444	6.8889	6.8889	−1587.17	0.5	$\emptyset$
2	29.6582	10.5854	10.5854	−3323.25	1.0	{2}
3	29.3893	10.0580	11.2474	−3324.69	0.0625	{2}
4	29.2576	10.3573	11.0139	−3337.54	1.0	{2}
5	28.1660	11.2276	10.6894	−3380.38	1.0	{2}
6	26.1073	12.0935	10.8528	−3426.55	1.0	{2}
7	24.2705	12.4212	11.4436	−3449.87	1.0	{2}
8	23.9256	12.2145	11.8227	−3455.06	1.0	{2}
9	23.9829	12.0109	11.9976	−3556.00	1.0	{2}
10	23.9990	12.0001	12.0004	−3556.00	1.0	{2}
11	24.0000	12.0000	12.0000	−3556.00	1.0	{2}

The following two m-files illustrate the possible procedures:

```
f = @(x) -x(1)*x(2)*x(3);
A  = [-1 -2 -2; 1 2 2]; b = [0; 72]; x0 = [10,10,10];
Aeq = []; beq = []; lb = [];  ub = [];
options = ...
   optimset('Display','iter','LargeScale','off','OutputFcn',@outfun);
[x,fval] = fmincon(f,x0,A,b,Aeq,beq,lb,ub,[],options)
iteration
```

The output function outfun.m allows the logging of the course of the iteration and is activated by OutputFcn .

```
% cf. [MOT], p. 2-75
function stop = outfun(x,optimValues,state,varargin)
    stop = [];
    persistent history
    switch state
       case 'init'
          history = [];
       case 'iter'
          % Concatenate current point with history.
          % x must be a row vector.
          history = [history; x];
       case 'done'
          assignin('base','iteration',history);
       otherwise
    end
```

Exercises

1. Let the following linear program be given:

$$c^T x \longrightarrow \min$$
$$Ax = b, \ \ell \le x \le u$$

In the following we will use the notations from section 4.1 (cf. p. 152ff).

a) Show that this problem can be solved via a modification of the *revised simplex method*:

① Determine a feasible basic point $\overline{x}$ to a basis J with

$$\ell_j \le \overline{x}_j \le u_j \text{ for } j \in J \text{ and}$$
$$\overline{x}_j = \ell_j \text{ or } \overline{x}_j = u_j \text{ for } j \in K.$$

② Set $\overline{c}^T := c_K^T - y^T A_K$ with $y := A_J^{-T} c_J$. For all $s = k_\sigma \in K$ let $\overline{c}_\sigma \ge 0$ if $\overline{x}_s = \ell_s$ and $\overline{c}_\sigma \le 0$ if $\overline{x}_s = u_s$: STOP.

③ Otherwise there exists an index $s = k_\sigma \in K$ such that $\overline{x}_s = \ell_s$ and $\overline{c}_\sigma < 0$ (Case A) or $\overline{x}_s = u_s$ and $\overline{c}_\sigma > 0$ (Case B).

④ Define $d \in \mathbb{R}^n$ with $d_J := -\overline{a}_s$, $d_K := e_\sigma$ and $\overline{a}_s := A_J^{-1} a_s$.

A: Determine the largest possible $\tau \ge 0$ such that

$$\tau \le \frac{\overline{x}_{j_i} - u_{j_i}}{\overline{a}_{i,s}}, \quad \text{if } \overline{a}_{i,s} < 0,$$
$$\tau \le \frac{\overline{x}_{j_i} - \ell_{j_i}}{\overline{a}_{i,s}}, \quad \text{if } \overline{a}_{i,s} > 0.$$

Assume that the maximum is attained for $i = \varrho$. If $\tau \le u_s - \ell_s$: x_r with $r = j_\varrho$ leaves the basis, x_s enters. If $\tau > u_s - \ell_s$: Set $\tau := u_s - \ell_s$; J remains unchanged.

B: Determine the smallest possible $\tau \le 0$ such that

$$\tau \ge \frac{\overline{x}_{j_i} - \ell_{j_i}}{\overline{a}_{i,s}}, \quad \text{if } \overline{a}_{i,s} < 0,$$
$$\tau \ge \frac{\overline{x}_{j_i} - u_{j_i}}{\overline{a}_{i,s}}, \quad \text{if } \overline{a}_{i,s} > 0.$$

Assume that the minimum is reached for $i = \varrho$. If $\tau \ge \ell_s - u_s$: x_r with $r = j_\varrho$ leaves the basis, x_s enters. If $\tau < \ell_s - u_s$: Set $\tau := \ell_s - u_s$; J remains unchanged.

$\overline{x} := \overline{x} + \tau d$; update of J as described above; goto ②.

b) Derive the dual optimization problem from the primal problem given above and show how to obtain a solution to the dual problem with the help of the algorithm described in *a)*.

c) Solve the following linear optimization problem

$$\begin{array}{rcl} 2x_1 + x_2 + 3x_3 - 2x_4 + 10x_5 & \longrightarrow & \min \\ x_1 + x_3 - x_4 + 2x_5 & = & 5 \\ x_2 + 2x_3 + 2x_4 + x_5 & = & 9 \end{array}$$

$$0 \le x \le (7, 10, 1, 5, 3)^T.$$

Firstly, verify that $J := (1, 2)$, $K := (3, 4, 5)$ and $\overline{x}_K := (0, 0, 0)^T$ or $\overline{x}_K := (1, 0, 0)^T$ yield a feasible basic solution.

2. Let $m \in \mathbb{N}$, A an invertible (m, m)-matrix, $u, v \in \mathbb{R}^m$ and $\alpha := 1 + v^T A^{-1} u$. Show that $A + uv^T$ is invertible if and only if $\alpha \neq 0$, and that in this case the Sherman–Morrison–Woodbury *formula*

$$\left(A + uv^T\right)^{-1} = A^{-1} - \frac{1}{\alpha} A^{-1} u v^T A^{-1}$$

holds. Utilize this formula to determine — for $v_1, \ldots, v_m \in \mathbb{R}$ and $\varrho \in \{1, \ldots, m\}$ — the inverse of the matrix

$$T = \begin{pmatrix} 1 & & & v_1 & & & \\ & \ddots & & \vdots & & & \\ & & 1 & \vdots & & & \\ & & & v_\varrho & & & \\ & & & \vdots & 1 & & \\ & & & \vdots & & \ddots & \\ & & & v_m & & & 1 \end{pmatrix}$$

from section 4.1.

3. *Diet problems* which optimize the composition of feed in stock farming or the ration of components in industrial processes are a good example of the application of linear optimization.

Assume that the chef of the cafeteria of a well-known Californian university has only two types of food (to simplify the calculation), concentrated feed *Chaπ* and *fresh vegetables* from Salinas Valley. Important ingredients per unit, costs per unit and the students' daily demand are given in the following table:

	Carbohydrates	Proteins	Vitamins	Costs
Chaπ	20	15	5	10 dollars
Fresh Vegetables	20	3	10	7 dollars
Daily Demand	60	15	20	

The aim is to minimize the total costs while satisfying the daily demands of the students.

a) Visualize the problem and determine the solution with the help of the graph.

b) Formulate the corresponding problem in standard form by introducing three slack variables $x_3, x_4, x_5 \geq 0$ and start the (revised) simplex method with the basis $J := (1, 3, 4)$.

c) This simple example also illustrates the fact that the simplex method not only yields the optimal solution but also offers some insight as to how the solution and the optimal value respond to 'small' changes in the data of the problem (*sensitivity analysis*); see next exercise. Preparatorily: Determine A_J^{-1} for $J := (1, 2, 5)$ and the optimal solution to the dual problem.

4. We continue from the last exercise. In part *c)*

$$A_J^{-1} = \frac{1}{48}\begin{pmatrix} -0.6 & 4 & 0 \\ 3 & -4 & 0 \\ 27 & -20 & -48 \end{pmatrix} \quad \text{to} \quad A_J = \begin{pmatrix} 20 & 20 & 0 \\ 15 & 3 & 0 \\ 5 & 10 & -1 \end{pmatrix}$$

for $J := (1, 2, 5)$ and the optimizer

$$u = \frac{1}{48}(15, 12, 0)^T = (0.3125, 0.25, 0)^T$$

of the dual problem had been calculated preparatorily. The minimal value of the objective function (costs) was then 22.5 dollars. $x_J = \overline{b} = (0.5, 2.5, 7.5)^T$ showed $x_1 = 0.5$ and $x_2 = 2.5$.

With the help of the inverse A_J^{-1} or also $\widehat{A}_J^{-1}$ (cf. exercise 5) and the result of the last step of the simplex algorithm we can gain insight into the effects of 'small' changes in the data of the problem on the optimizer and the optimal value. We will illustrate this phenomenon with an example:

a) The chef of the cafeteria receives an offer of a new concentrated feed *Chaπplus* that costs 12 dollars per unit and contains 30 carbohydrates, 10 proteins and 10 vitamins (in suitable units of measurement). Would the use of this new feed be profitable?

b) How low would the price p (per unit) of the new feed have to be for the use of it to be profitable?
Note that although the optimal *value* changes continuously with p, the optimal *feeding plan* may depend discontinuously on p.

c) The shadow prices (optimizers of the dual problem) also give evidence as to how sensitively the optimal solution reacts to changes in the composition of the new feed. Show that an enrichment of the carbohydrates from 30 to 31 units is more profitable than an enrichment of the proteins by one unit from 10 to 11, while the concentration of vitamins is initially irrelevant.

d) Describe — in a general way — the influence of 'small' changes of b on a basic solution x_J and thus on the objective function: If J is the basis of the optimal solution, then u returns the *sensitivity* of the optimal costs with reference to 'small' changes of b.

e) It is a bit more complex to determine to what degree changes in the data influence the basic matrix. Calculate how much the previous price of 10 dollars for one unit of Chaπ may change without having to change the meal.

Remark: To do that, one can, for example, utilize the SHERMAN–MORRISON–WOODBURY formula for the inverse of the basic matrix to

$$\widehat{A}_J(\varepsilon) := \begin{pmatrix} 20 & 20 & 0 & 0 \\ 15 & 3 & 0 & 0 \\ 5 & 10 & -1 & 0 \\ -(10+\varepsilon) & -7 & 0 & 1 \end{pmatrix}$$

for the modified price of $10 + \varepsilon$.

5. For manual calculations it is sometimes in particular useful to set

$$\widehat{A} := \begin{pmatrix} A & 0 \\ -c^T & 1 \end{pmatrix} \in \mathbb{R}^{(m+1)\times(n+1)},\ \widehat{b} := \begin{pmatrix} b \\ 0 \end{pmatrix} \in \mathbb{R}^{m+1},\ \widehat{x} := \begin{pmatrix} x \\ z \end{pmatrix} \in \mathbb{R}^{n+1}$$

and with that rewrite the primal problem as follows:

$$z \longrightarrow \min$$
$$\widehat{A}\widehat{x} = \widehat{b},\ x \geq 0$$

To a basis J let

$$\widehat{A}_J := \begin{pmatrix} A_J & 0 \\ -c_J^T & 1 \end{pmatrix}$$

be formed accordingly. Verify that

$$\widehat{A}_J^{-1} = \begin{pmatrix} A_J^{-1} & 0 \\ u^T & 1 \end{pmatrix}$$

with $u := A_J^{-T} c_J$. The components of u are called *shadow prices.*

6. Implement the *active set method* as a *Matlab*® function beginning with the headline

```
function [x,u,J,iter] = LP(c,A,b,J0)
```

to solve the linear optimization problem

$$\begin{aligned} c^T x &\longrightarrow \min \\ A^T x &\geq b \end{aligned}$$

as described in section 4.1. Choose the matrices A, b, c and a feasible basis J_0 as input parameters. Output parameters are the optimal solution x, the corresponding LAGRANGE multipliers u, the optimal feasible basis J and the number of iterations `iter`. Modify this program for the case that J_0 is a nonfeasible basis, and write it as a *two-phase method.* Test these two versions with the examples 3 and 4 of chapter 4.

7. Consider the problem

$$\begin{aligned} f(x) &:= \tfrac{1}{2} x^T C x \longrightarrow \min \\ Ax &= b\,, \end{aligned}$$

where

$$C := \begin{pmatrix} 0 & -13 & -6 & -3 \\ -13 & 23 & -9 & 3 \\ -6 & -9 & -12 & 1 \\ -3 & 3 & 1 & -1 \end{pmatrix}, \quad A := \begin{pmatrix} 2 & 1 & 2 & 1 \\ 1 & 1 & 3 & -1 \end{pmatrix}, \quad b := (3,2)^T$$

and $x_0 := (1,1,0,0)^T$. Determine a matrix $Z \in \mathbb{R}^{4\times 2}$ the columns of which are a basis of the nullspace of A. Minimize the reduced objective function φ given by

$$\varphi(v) := f(x_0 + Zv)$$

and derive from its solution the solution to the original problem (cf. p. 175f).

8. Solve the quadratic optimization problem

$$\begin{aligned} \tfrac{1}{2}x_1^2 + \tfrac{1}{2}x_2^2 + 2x_1 + x_2 &\longrightarrow \min \\ A^T x &\leq b \end{aligned}$$

by means of the *active set method*, where

$$A := \begin{pmatrix} -1 & 0 & 1 & -1 & 1 & 0 \\ -1 & 1 & 1 & 1 & 0 & -1 \end{pmatrix} \text{ and } b := (0,2,5,2,5,1)^T.$$

Choose the feasible point $x^{(0)} := (5,0)^T$ as a starting vector.

In faded and hardly legible documents the following table to this exercise was found:

Chapter 4

k	$x^{(k)}$	J_k	$f(x^{(k)})$	$y^{(k)}$	u	α
0	$(5,0)$	$\{3,5\}$	22.5	$(5,0)\in\mathcal{F}$	$(0,0,-1,0,-6,0)$	
1	$(5,0)$	$\{3\}$	22.5	$(2,3)\notin\mathcal{F}$		$2/3$
2	$(3,2)$	$\{2,3\}$	14.5			
3	$(3,2)$	$\{2\}$	14.5			
4	$(0,2)$	$\{2,4\}$	4.0			
5	$(0,2)$	$\{4\}$	4.0			
6	$(-1,1)$	$\{1,4\}$	0.0			
7	$(-1,1)$	$\{1\}$	0.0			
8	$(-0.5,0.5)$	$\{1\}$	-0.25			

Confirm the values and complete the table.

Chapter 4

9. Show: For a symmetric, positive definite matrix $C\in\mathbb{R}^{n\times n}$ the *quadratic optimization problem*

$$\begin{aligned}&\tfrac{1}{2}x^TCx+c^Tx\longrightarrow\min\\&A^Tx\le b\end{aligned}$$

is equivalent to the *constrained least squares problem*

$$\begin{aligned}&\|Lx-d\|\longrightarrow\min\\&A^Tx\le b\end{aligned}$$

with suitable $L\in\mathbb{R}^{n\times n}$ and $d\in\mathbb{R}^n$.

10. Show that for a nonempty, convex and closed set $K\subset\mathbb{R}^n$:

a) To every $x\in\mathbb{R}^n$ there exists exactly one $z=:P_K(x)\in K$ such that $\|x-z\|_2\le\|x-y\|_2$ for all $y\in K$.

z is hence the element in K with minimal distance to x and is called the *orthogonal projection* of x onto K.

b) $z=P_K(x)$ is characterized by the fact that

$$\langle y-z,x-z\rangle\le 0\quad\text{for all } y\in K.$$

11. Let the following optimization problem be given:

$$(P)\quad\begin{cases}f(x)\longrightarrow\min\\A^Tx\le b,\end{cases}$$

with $f\colon\mathbb{R}^n\longrightarrow\mathbb{R}$ continuously differentiable, $A\in\mathbb{R}^{n\times m}$ and $b\in\mathbb{R}^m$. For $x\in\mathcal{F}:=\{y\in\mathbb{R}^n\mid A^Ty\le b\}$ the set $J:=\mathcal{A}(x)$ denotes the active set at x and $P_{H(x)}$ the projection on

$$H(x) := \{d \in \mathbb{R}^n \mid A_J^T d \le 0\}$$

in the sense of the preceding problem. Consider the following *algorithm*:

Choose $x_0 \in \mathcal{F}$, $k := 0$ and carry out the following loop:

Set $d_k := P_{H(x_k)}\big(-\nabla f(x_k)\big)$.

- If $d_k = 0$: STOP.
- Otherwise: Determine $\overline{\lambda}_k := \max\{\lambda > 0 \mid x_k + \lambda d_k \in \mathcal{F}\}$ and $x_{k+1} := x_k + \lambda_k d_k$ with $f(x_{k+1}) = \min\{f(x_k + \lambda d_k) \mid 0 \le \lambda \le \overline{\lambda}_k\}$.

Set $k := k+1$ and repeat the above calculation.

Show:

a) $d_k \neq 0$ is a feasible descent direction.

b) $\overline{x}$ is a KKT point of (P) if and only if $P_{H(\overline{x})}\big(-\nabla f(\overline{x})\big) = 0$ holds.

Hint to b): Utilize part *b)* of the preceding exercise and then apply Farkas' Theorem of the Alternative.

12. Solve the following problem

$$\begin{aligned} x_1^2 + 2x_2^2 &\longrightarrow \min \\ -x_1 + 4x_2 &\le 0 \\ -x_1 - 4x_2 &\le 0 \end{aligned}$$

by using the projection method from the preceding exercise. Choose $x^{(0)} := (4,1)^T$ as the starting point.

Hint: It holds that $x^{(k)} = \frac{1}{3^k}\big(4,(-1)^k\big)^T$.

13. Implement the *active set method* for quadratic optimization in *Matlab*® starting with the headline `function [x,u,J,iter] = QP(C,c,A,b,x0)` to solve the convex problem

$$\begin{aligned} &\tfrac{1}{2}x^T C x + c^T x \longrightarrow \min \\ &A^T x \le b \end{aligned}$$

as described in section 4.2. Choose the matrices C, c, A, b and a feasible point x_0 as input parameters. Output parameters are the optimizer x, the corresponding Lagrange multipliers u, the optimal feasible basis J and the number of iterations `iter`. Use the program to solve the problem in exercise 8. Choose the three points $(5,0)^T$, $(2,-1)^T$ and $(0,1)^T$ as starting vectors x_0.

14. An important subproblem of the implementation of optimization algorithms is the update of the QR-factorization of a matrix A if an extra column has to be attached to it or if one column of it is deleted. Come up with your own *Matlab*® procedures `QRins` bzw. `QRdel` for that and carry out the necessary rotations with the help of the *Matlab*® command givens . Compare your results with those of the standard functions qrinsert and qrdelete . Use this to solve the following problem:

a) Calculate the QR-factorization of the matrix

$$A = \begin{pmatrix} 1\;4\;2\;0\;4\;2\;1\;0 \\ 1\;1\;2\;1\;1\;2\;4\;1 \end{pmatrix}^T .$$

b) Based on Q and R we want to determine the QR-factorization of the matrix $\widetilde{A}$ which results from the addition of the following two columns

$$u = \big(0\;0\;2\;1\;3\;1\;1\;2\big)^T, \quad v = \big(1\;0\;3\;0\;4\;1\;2\;4\big)^T.$$

c) Finally we want to determine the QR-factorization of the matrix $\widehat{A}$ that results from the deletion of the second column of $\widetilde{A}$.

15. Use ROSEN's *projection method* and — with the starting value $x^{(0)} := (0,0)^T$ — solve the following optimization problem

$$\begin{aligned} \tfrac{1}{2}x_1^2 + \tfrac{1}{2}x_2^2 - 4x_1 - 6x_2 &\longrightarrow \min \\ x_1 + x_2 &\le 5 \\ 3x_1 + x_2 &\le 9 \\ x_1, x_2 &\ge 0 \end{aligned}$$

and visualize it.

16. Use the *reduced gradient method* to solve the following optimization problem with the starting values $x^{(0)} := (0,0,2)^T$ and $J_0 := \{3\}$:

$$\begin{aligned} x_1^2 - x_1 x_2 + x_2^2 - 3x_1 &\longrightarrow \min \\ x_1 + x_2 + x_3 &= 2 \\ x_1, x_2, x_3 &\ge 0 \end{aligned}$$

17. In the *active set method* (cf. p. 173f) assume that for the initial working set J_0 the vectors $\big(a_j \mid j \in J_0\big)$ are linearly independent. Show that a vector a_r which is added to this set of vectors is not linearly dependent on the other vectors, and hence prove inductively that the vectors $\big(a_j \mid j \in J_k\big)$ are linearly independent for $k = 1, 2, \ldots$.

18. Solve exercise 15

a) with the *active set method*,

b) with the help of the GOLDFARB–IDNANI *method.* For that discuss all of the possible solution paths.

19. Let α be a real number with $1/2 < \alpha < 1$ and $C \in \mathbb{R}^{n \times n}$ a symmetric positive definite matrix. Consider the function $f\colon \mathbb{R}^n \longrightarrow \mathbb{R}$ given by $f(x) := \langle x, Cx \rangle^{\alpha}$.

a) How does $f'(0)$ need to be defined in order for f to become continuously differentiable?

b) Prove that f is a convex function.

20. Solve the linearly constrained optimization problem (cf. [Col], p. 21 ff)

$$c^T x + x^T C x + d^T x^3 \longrightarrow \min$$
$$A^T x \le b, \ x \ge 0$$

where the vector x^3 arises from x raising each entry to the third power. Take the following test data:

$$C := \begin{pmatrix} 30 & -20 & -10 & 32 & -10 \\ -20 & 39 & -6 & -31 & 32 \\ -10 & -6 & 10 & -6 & -10 \\ 32 & -31 & -6 & 39 & -20 \\ -10 & 32 & -10 & -20 & 30 \end{pmatrix}, \quad c := \begin{pmatrix} -15 \\ -27 \\ -36 \\ -18 \\ -12 \end{pmatrix}, \quad d := \begin{pmatrix} 4 \\ 8 \\ 10 \\ 6 \\ 2 \end{pmatrix},$$

$$A := \begin{pmatrix} 16 & 0 & 3.5 & 0 & 0 & -2 & 1 & 1 & -1 & -1 \\ -2 & 2 & 0 & 2 & 9 & 0 & 1 & 2 & -2 & -1 \\ 0 & 0 & -2 & 0 & 2 & 4 & 1 & 3 & -3 & -1 \\ -1 & -4 & 0 & 4 & -1 & 0 & 1 & 2 & -4 & -1 \\ 0 & -2 & 0 & 1 & 2.8 & 0 & 1 & 1 & -5 & -1 \end{pmatrix},$$

$$b := \big(40,\ 2,\ 0.25,\ 4,\ 4,\ 1,\ 40,\ 60,\ -5,\ -1\big)^T.$$

To solve this, utilize — analogously to example 12 — the function fmincon of the *Matlab®* *Optimization Toolbox* and show the course of the iteration using the feasible starting point $x^{(0)} := (0,0,0,0,1)^T$.

Determine — for example by means of the symbolic power of *Maple®* — the *exact* minimizer of the optimization problem stated above.

5

Nonlinearly Constrained Optimization Problems

Chapter 5

We again assume $f, g_1, \ldots, g_m, h_1, \ldots, h_p$ to be continuously differentiable real-valued functions on $\mathbb{R}^n$ with $n \in \mathbb{N}$ and $m, p \in \mathbb{N}_0$, and consider the problem

$$(P) \quad \begin{cases} f(x) \longrightarrow \min \\ g_i(x) \le 0 & \text{for } i \in \mathcal{I} := \{1, \ldots, m\} \\ h_j(x) = 0 & \text{for } j \in \mathcal{E} := \{1, \ldots, p\} \end{cases}$$

or short

$$(P) \quad \begin{cases} f(x) \longrightarrow \min \\ x \in \mathcal{F} \end{cases}$$

with the *set of feasible points*

$$\mathcal{F} := \Big\{ x \in \mathbb{R}^n \mid g_i(x) \le 0 \text{ for } i \in \mathcal{I},\ h_j(x) = 0 \text{ for } j \in \mathcal{E} \Big\}.$$

The most difficult case is the general one in which both the objective function f and the constraint functions g_i and h_j are permitted to be nonlinear. We speak of *nonlinear optimization problems.*

In section 5.1 a constrained optimization problem is replaced by an unconstrained one. There are two different approaches to this: In *exterior penalty methods* a term is added to the objective function which 'penalizes' a violation of constraints. In *interior penalty methods* a barrier term prevents points from leaving the interior of the feasible region.

As a side product we get the 'starting point' for the development of so-called *primal–dual interior-point methods* which we will discuss in chapter 6 (linear case) and chapter 7 (semidefinite case).

W. Forst and D. Hoffmann, *Optimization—Theory and Practice*,
Springer Undergraduate Texts in Mathematics and Technology,
DOI 10.1007/978-0-387-78977-4_5,

By a *sequential quadratic programming method* — which we have already treated briefly in section 4.3 — a usually nonlinear optimization problem is converted into a sequence of quadratic optimization problems which are easier to solve. Many authors consider them to be the most important class of methods for solving nonlinear optimization problems without any particular structure. NEWTON's method is crucial for a lot of important algorithms.

5.1 Penalty Methods

Classic Penalty Methods (Exterior Penalty Methods)

We already got to know the special case of *quadratic* penalty functions in section 1.1 (cf. p. 16ff). This approach is based on an idea by COURANT (1943) who transformed a *constrained* minimization problem into an *unconstrained* one by 'inserting' the constraints as *penalty terms* into the objective function. At first, however, this idea was not followed up.

Suppose that the feasible region $\mathcal{F}$ is given in the form of $\mathcal{F} = C \cap S$ with *closed* sets C and S, where C is given by the 'simple' (for example, *convex* or linear) constraints and S by the 'difficult' ones. The problem

$$(P) \quad \begin{cases} f(x) \longrightarrow \min \\ x \in C \cap S \end{cases}$$

can then be transformed into the following *equivalent problem*

$$\begin{aligned} & f(x) + \delta_S(x) \longrightarrow \min \\ & x \in C \end{aligned}$$

with the help of the *indicator function*:

$$\delta_S(x) := \begin{cases} 0\,, & x \in S \\ \infty\,, & \text{else}\,. \end{cases}$$

In this section we will *assume* that *problem* (P) *is globally solvable.*

The minimization methods which we are familiar with require the objective function to be sufficiently *smooth* (and real-valued). Therefore, we will try to approximate the indicator function δ_S by (real-valued) functions which should be as smooth as possible. For example, if $\pi\colon \mathbb{R}^n \longrightarrow \mathbb{R}$ is a sufficiently smooth function with $\pi(x) = 0$ for all $x \in S$ and $\pi(x) > 0$ for all $x \notin S$, it holds that

$$\frac{1}{r}\,\pi(x) \longrightarrow \delta_S(x) \quad \text{for } 0 < r \longrightarrow 0\,.$$

Example

$S := \mathcal{F} = \{x \in \mathbb{R}^n \mid g_i(x) \le 0 \text{ for } i \in \mathcal{I},\ h_j(x) = 0 \text{ for } j \in \mathcal{E}\}$

$\pi(x) := \sum_{i=1}^{m} g_i^+(x)^\alpha + \sum_{j=1}^{p} |h_j(x)|^\alpha$

Here let $\alpha \ge 1$ (for example, $\alpha = 2$), $a^+ := \max\{0, a\}$ for a real number a and then obviously $g^+(x) := (g(x))^+$ for a real-valued map g and x from the domain of definition of g.

This is indicated in the following figures for
$n = 1$, $m = 2$, $p = 0$, $g_1(x) := -1 - x$, $g_2(x) := x - 1$; $\alpha = 2$ (left); $\alpha = 1$ (right).

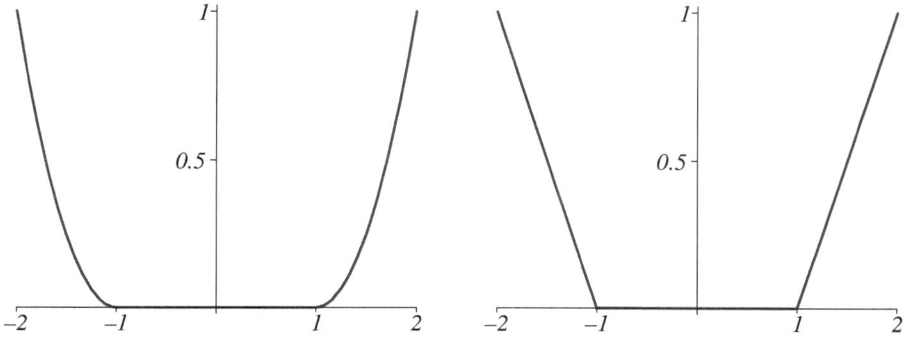

◁

Hence, with the *penalty function* Φ_r defined by

$$\Phi_r(x) := \Phi(x, r) := f(x) + \frac{1}{r}\,\pi(x)$$

for $r > 0$ and $x \in \mathbb{R}^n$ we get the following *subsidiary problems*

$$(P_r) \quad \begin{cases} \Phi_r(x) \longrightarrow \min_x \\ x \in C. \end{cases}$$

Here r is sometimes called the *penalty parameter.*

The violation of the constraints is 'penalized' by increasing 'costs' — when r is decreasing. Hence, the penalty function is the sum of the original objective function and a *penalty term* $\pi(x)/r$.

This approach is based on the intuitive expectation that the minimizer $\overline{x}(r)$ to (P_r) will converge to the minimizer $\overline{x}$ to (P) for $r \longrightarrow 0$.

Example 1 (cf. [Spe], p. 394ff)

$f(x) := x_1^2 + 4x_1x_2 + 5x_2^2 - 10x_1 - 20x_2 \longrightarrow \min$

$h_1(x) := x_1 + x_2 - 2 = 0$

Chapter 5

This problem can of course be solved very easily with *basic methods*: Substituting $x_2 = 2 - x_1$ in f gives $f(x) = 2x_1^2 - 2x_1 - 20$, hence, via

$x_1^2 - x_1 - 10 \longrightarrow \min$ for $x_1 \in \mathbb{R}$, yields $\overline{x}_1 = 1/2$, $\overline{x}_2 = 3/2$ and $f(\overline{x}) = -20.5$. By using the *KKT condition*

$$\nabla f(\overline{x}) + \overline{\mu}_1 \nabla h_1(\overline{x}) = 0$$

and $h_1(\overline{x}) = 0$, we also get $\overline{x}_1 = 1/2$, $\overline{x}_2 = 3/2$ and $\overline{\mu}_1 = 3$ very easily.

However, here we want to illustrate the *penalty method*:

$$\Phi_r(x) := f(x) + \frac{1}{r} h_1(x)^2$$

$$\nabla \Phi_r(x) = \nabla f(x) + \frac{2}{r} h_1(x) \nabla h_1(x) = \begin{pmatrix} 2x_1 + 4x_2 - 10 + \frac{2}{r}(x_1 + x_2 - 2) \\ 4x_1 + 10x_2 - 20 + \frac{2}{r}(x_1 + x_2 - 2) \end{pmatrix}$$

gives

$$\begin{pmatrix} 2 + \frac{2}{r} & 4 + \frac{2}{r} \\ 4 + \frac{2}{r} & 10 + \frac{2}{r} \end{pmatrix} \begin{pmatrix} \overline{x}_1(r) \\ \overline{x}_2(r) \end{pmatrix} = \begin{pmatrix} 10 + \frac{4}{r} \\ 20 + \frac{4}{r} \end{pmatrix}$$

for the zero $\overline{x}(r)$ of $\nabla \Phi_r$, hence,

$$\begin{cases} \overline{x}_1(r) = \frac{20 + \frac{4}{r}}{4 + \frac{8}{r}} = \frac{5r+1}{r+2} \\ \overline{x}_2(r) = \frac{\frac{12}{r}}{4 + \frac{8}{r}} = \frac{3}{r+2}. \end{cases}$$

Thus, we have $\overline{x}(r) \longrightarrow \begin{pmatrix} \frac{1}{2} \\ \frac{3}{2} \end{pmatrix} = \overline{x}$ for $r \longrightarrow 0$.

$\overline{x}_1(r) + \overline{x}_2(r) = \frac{5r+4}{r+2} = 2 + \frac{3r}{r+2} > 2$

Furthermore, it holds that $\frac{2}{r} h_1(\overline{x}(r)) = \frac{2}{r}\left(\frac{3r}{r+2}\right) \longrightarrow 3 = \overline{\mu}_1$.

The matrix $\nabla^2 f(\overline{x}) + \overline{\mu}_1 \underbrace{\nabla^2 h_1(\overline{x})}_{=0} = \begin{pmatrix} 2 & 4 \\ 4 & 10 \end{pmatrix}$ is positive definite.

The matrix $A_r := \nabla^2 \Phi_r(x) = \begin{pmatrix} 2 + \frac{2}{r} & 4 + \frac{2}{r} \\ 4 + \frac{2}{r} & 10 + \frac{2}{r} \end{pmatrix}$ has the *condition number*

$$\kappa = \frac{17r^2 + 10r + 2 + 2(3r+1)\sqrt{8r^2 + 4r + 1}}{r(r+2)} \sim \frac{2}{r},$$

that is, the Hessian is ill-conditioned for small r.

A short reminder: The condition number is — in this case — the quotient of the greatest and the smallest eigenvalue. The eigenvalues can be obtained from

$$\lambda^2 - \operatorname{trace}(A_r)\lambda + \det(A_r) = 0. \qquad \triangleleft$$

We derive a simple convergence property to the general penalty function approach.

Proposition 5.1.1

If there exist $(r_k) \in (0, \infty)^{\mathbb{N}}$ *with* $r_k \longrightarrow 0$ *and* $(x^{(k)}) \in C^{\mathbb{N}}$ *with*

$$\Phi_{r_k}(x^{(k)}) = \min\{\Phi_{r_k}(x) \mid x \in C\} \textit{ for each } k \in \mathbb{N},$$

then each accumulation point x^* *of the sequence* $(x^{(k)})$ *minimizes* (P).

Proof: Suppose that $\overline{x}$ is a (global) minimizer to (P) and x^* an accumulation point of the sequence $(x^{(k)})$. Since C is closed — according to the usual assumptions — x^* belongs to C. Let WLOG $x^{(k)} \longrightarrow x^*$. Taking into account $\pi(x) = 0$ for $x \in S$,

$$\Phi_{r_k}(x^{(k)}) \le \Phi_{r_k}(\overline{x}) = f(\overline{x})$$

holds by definition of $x^{(k)}$. From that it follows that

$$\limsup_{k\to\infty} \Phi_{r_k}(x^{(k)}) \le f(\overline{x}).$$

In addition $x^* \in C \cap S$; otherwise the sequence $\left(\Phi_{r_k}(x^{(k)})\right)$ would be unbounded. Finally,

$$\liminf_{k\to\infty} \Phi_{r_k}(x^{(k)}) \ge f(x^*) \ge f(\overline{x})$$

holds since

$$\Phi_{r_k}(x^{(k)}) = f(x^{(k)}) + \frac{1}{r_k}\pi(x^{(k)}) \ge f(x^{(k)}) \longrightarrow f(x^*).$$

Altogether, we get $\lim_{k\to\infty} \Phi_{r_k}(x^{(k)}) = f(x^*) = f(\overline{x})$; hence, x^* minimizes problem (P). □

Properties of Penalty Methods

1) Penalty methods substitute unconstrained optimization problems for constrained ones. The effective use of numerical methods for unconstrained problems requires the penalty function Φ_r to be *sufficiently differentiable.*

2) Unlimited reduction of the penalty parameter r increases ill-conditioning which is often inherent in the minimization process of the penalty function. Therefore there should exist an $\overline{r} > 0$ such that a local minimizer $\overline{x}$ to (P) minimizes locally each unconstrained problem (P_r) for $r < \overline{r}$ too. In this case the penalty function is called *exact.*

In general these two properties are not compatible. Now we are going to construct an exact penalty function to the optimization problem

$$(CP) \quad \begin{cases} f(x) \longrightarrow \min \\ g_i(x) \leq 0 \quad (i = 1, \ldots, m). \end{cases}$$

We have chosen the name (CP) because in applications of the following considerations the functions f and g_i will be *convex* in most cases.

Let the Lagrangian L to (CP) as always be defined by

$$L(x, \lambda) := f(x) + \sum_{i=1}^{m} \lambda_i g_i(x)$$

for $x \in \mathbb{R}^n$ and $\lambda \in \mathbb{R}^m_+$. We will use

$$\Phi_r(x) := f(x) + \frac{1}{r} \sum_{i=1}^{m} g_i^+(x)$$

as the penalty function for the moment.

Proposition 5.1.2 (Exactness of the Penalty Function)

Let $(\overline{x}, \overline{\lambda}) \in \mathbb{R}^n \times \in \mathbb{R}^m_+$ be a saddlepoint of L, that is,

$$L(\overline{x}, \lambda) \leq L(\overline{x}, \overline{\lambda}) \leq L(x, \overline{\lambda}) \quad \textit{for all } x \in \mathbb{R}^n \textit{ and } \lambda \in \mathbb{R}^m_+ .$$

For $r \in (0, \infty)$ it then holds that:

a) $\min\limits_{x \in \mathbb{R}^n} \Phi_r(x) = \Phi_r(\overline{x})$ *if* $r \, \|\overline{\lambda}\|_\infty \leq 1$.

b) If $r \, \|\overline{\lambda}\|_\infty < 1$ and $\Phi_r(x^*) = \min\limits_{x \in \mathbb{R}^n} \Phi_r(x)$, then x^* also minimizes (CP).

Φ_r is then called the 'exact penalty function'.

Proof: Following theorem 2.2.6, $\overline{x}$ yields a minimizer to (CP) and $L(\overline{x}, \overline{\lambda}) = f(\overline{x})$ holds.

a) It holds that $g_i^+(\overline{x}) = 0$ and $g_i(x) \leq g_i^+(x)$ for all $x \in \mathbb{R}^n$ and $i \in \mathcal{I}$. With $r \, \|\overline{\lambda}\|_\infty \leq 1$ it follows that

$$\begin{aligned} \Phi_r(\overline{x}) &= f(\overline{x}) = L(\overline{x}, \overline{\lambda}) \leq L(x, \overline{\lambda}) = f(x) + \sum_{i=1}^{m} \overline{\lambda}_i g_i(x) \\ &\leq f(x) + \sum_{i=1}^{m} \overline{\lambda}_i g_i^+(x) \leq f(x) + \sum_{i=1}^{m} \frac{1}{r} g_i^+(x) = \Phi_r(x). \end{aligned}$$

b) Let $r\|\overline{\lambda}\|_\infty < 1$ and $\Phi_r(x^*) = \min_{x\in\mathbb{R}^n} \Phi_r(x)$. If x^* is not a minimizer to (CP), it follows with *a)* that: $f(x^*) \le \Phi_r(x^*) = \min_{x\in\mathbb{R}^n} \Phi_r(x) = \Phi_r(\overline{x}) = f(\overline{x})$. Consequently, x^* is not feasible, that is, there exists an $i \in \mathcal{I}$ with $g_i(x^*) > 0$. The same chain of inequalities as in the proof of *a)* then yields the following for $x = x^*$:

$$\Phi_r(\overline{x}) \le f(x^*) + \sum_{i=1}^{m} \overline{\lambda}_i g_i^+(x^*) < f(x^*) + \sum_{i=1}^{m} \frac{1}{r} g_i^+(x^*) = \Phi_r(x^*),$$

hence, a *contradiction.* □

Remark

Since this Φ_r is not necessarily differentiable, it is not possible to apply methods for solving unrestricted nonlinear problems directly.

Example 2

$f(x) := x^2 - 10x \longrightarrow \min$
$x \le 1$

Here we are able to obtain the minimizer $\overline{x} = 1$ with $f(\overline{x}) = -9$ immediately (parabola).

Typical graphs of $\Phi(\,\cdot\,, r)$ — for different values of the penalty parameter r — are illustrated in the following figure:

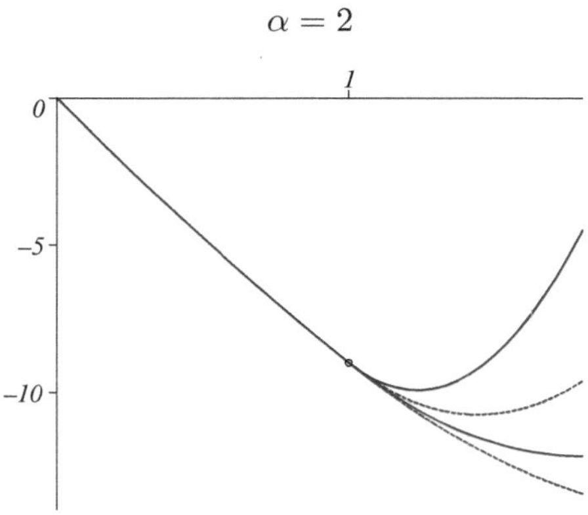

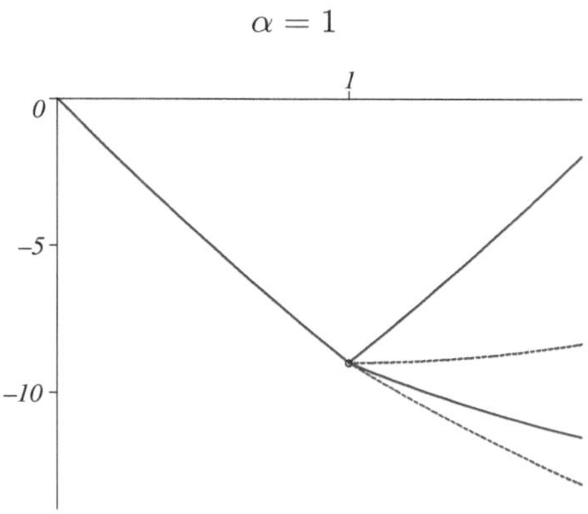

1) Let $g(x) := x - 1$ for $x \in \mathbb{R}$. For $r > 0$ consider at first

$$\Phi_r(x) := f(x) + \frac{1}{r} g^+(x)^2 = (x^2 - 10x) + \frac{1}{r}\big(\max(0, x-1)\big)^2.$$

Chapter 5

$$\begin{aligned}
\Phi_r' &= 2x - 10 + \frac{2}{r}\max(0, x-1) \\
&= \begin{cases} 2x - 10 & , \ x \le 1 \\ 2\left(\frac{1}{r}+1\right)x - 2\left(\frac{1}{r}+5\right) & , \ x > 1 \end{cases} \\
\overline{x}(r) &= \frac{\frac{1}{r}+5}{\frac{1}{r}+1} = 1 + \frac{4r}{1+r} \longrightarrow 1 = \overline{x} \\
\overline{\lambda}(r) &= \tfrac{2}{r}g(\overline{x}(r)) = \tfrac{8}{1+r} \longrightarrow 8 = \overline{\lambda}
\end{aligned}$$

2) For $L(x, \lambda) := f(x) + \lambda g(x)$ $(x \in \mathbb{R}, \lambda \in \mathbb{R}_+)$ the pair $(\overline{x}, \overline{\lambda}) := (1, 8)$ is a *saddlepoint* of L since

$$\begin{aligned}
L(x, \overline{\lambda}) &= x^2 - 10x + 8(x-1) = x^2 - 2x - 8 = (x-1)^2 - 9 \\
&\ge -9 = L(\overline{x}, \overline{\lambda}) = L(\overline{x}, \lambda)\,.
\end{aligned}$$

Hence, we can apply the observations from proposition 5.1.2 to

$$\Phi_r(x) := x^2 - 10x + \frac{1}{r}\max(0, x-1) = \begin{cases} x^2 - 10x & , \ x \le 1 \\ x^2 - 10x + \frac{1}{r}(x-1), & x > 1\,. \end{cases}$$

According to 5.1.2.a, we know

$$\Phi_r(1) = \min_{x \in \mathbb{R}} \Phi_r(x) \ \text{ if } r \le 1/8\,,$$

and by 5.1.2.b that the solutions to $\min_{x\in\mathbb{R}} \Phi_r(x)$ for $r < 1/8$ give the solution to (CP). ◁

Chapter 5

Barrier Methods (Interior Penalty Methods)

As with the exterior penalty methods, we consider the following minimization problem

$$(P) \quad \begin{cases} f(x) \longrightarrow \min \\ x \in C \cap S\,. \end{cases}$$

For that let C and S be *closed* subsets of $\mathbb{R}^n$; in addition let C be *convex*. When using barrier methods, we will consider the following kinds of *substitute problems* for $r > 0$:

$$\begin{aligned}
&\Phi_r(x) := \Phi(x, r) := f(x) + r\,B(x) \longrightarrow \min \\
&x \in C \cap \overset{\circ}{S}
\end{aligned}$$

Here, suppose that the *barrier function* $B\colon \overset{\circ}{S} \longrightarrow \mathbb{R}$ is continuous and $B(x) \longrightarrow \infty$ holds for $x \longrightarrow z$ for all $z \in \partial S$. Since the solutions to the substitute problems should always lie in $\overset{\circ}{S}$, we demand

$$\inf_{x \in C \cap \overset{\circ}{S}} f(x) = \min_{x \in C \cap S} f(x).$$

For example, $C \cap S = \overline{C \cap \overset{\circ}{S}}$ is sufficient for that.

Example

Let $S = \{x \in \mathbb{R}^n \mid g_i(x) \le 0 \quad (i = 1, \ldots, m)\}$ with continuous and *convex* functions $g_1, \ldots, g_m$ and $g_i(x_0) < 0$ $(i = 1, \ldots, m)$ for an $x_0 \in C$. Important barrier functions are (locally around x_0) the

inverse barrier function $B(x) = -\sum_{i=1}^{m} \frac{1}{g_i(x)}$ (CARROLL (1961)) and the

logarithmic barrier function $B(x) = -\sum_{i=1}^{m} \log(-g_i(x))$ (FRISCH (1955)).

With the same functions g_1 and g_2 as in the example on page 215 we get:

Inverse barrier function *Logarithmic barrier function*

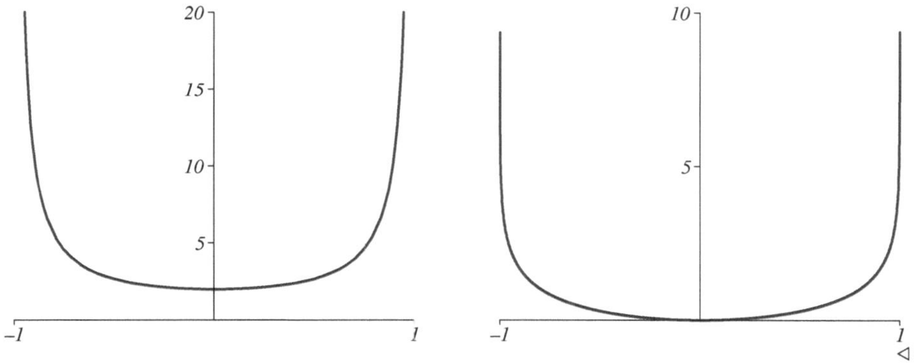

◁

The *barrier method* **algorithm** corresponds closely to the algorithms of exterior penalty methods:

To a given starting point $x^{(0)} \in \overset{\circ}{S}$ and a null sequence (r_k) with $r_k \downarrow 0$, we solve the following problems for $k \in \mathbb{N}$:

$$(P_k) \quad \begin{cases} \Phi_{r_k}(x) \longrightarrow \min \\ x \in C \cap \overset{\circ}{S} \end{cases}$$

with the starting value $x^{(k-1)}$. Suppose that we obtain the solution $x^{(k)} := x(r_k)$. Subject to suitable assumptions, we can also show that the accumulation points of the sequence $\left(x^{(k)}\right)$ yield solutions to (P).

Example 3

$f(x) = x \longrightarrow \min$

$-x + 1 \le 0$

Obviously this elementary problem has the solution $\overline{x} = 1$.

$\Phi_r(x) = x + \dfrac{r}{x-1}$

With $x(r) = 1 + \sqrt{r}$ we obtain an interior point of the feasible region from $\Phi_r'(x) = 1 - \dfrac{r}{(x-1)^2} = 0$.

For that it also holds that $\Phi_r''(x(r)) = \dfrac{2r}{(x(r)-1)^3} = \dfrac{2}{\sqrt{r}}$. ◁

The following example from [Fi/Co], p. 43ff illustrates more clearly how the barrier method works.

Example 4

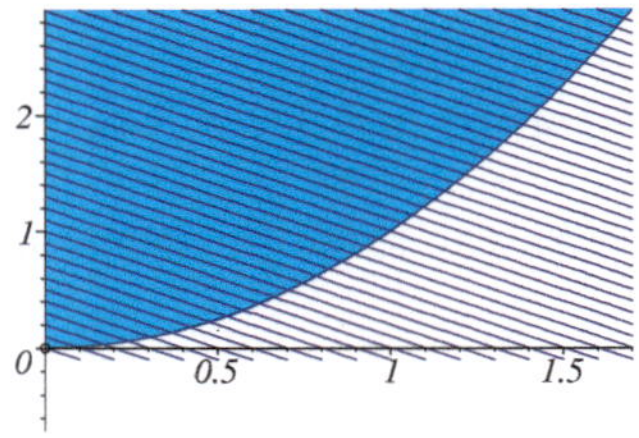

$f(x) := x_1 + x_2 \to \min$

$g_1(x) := x_1^2 - x_2 \le 0$

$g_2(x) := -x_1 \le 0$

$\Phi_r(x) := x_1 + x_2 - r\log(-x_1^2 + x_2) - r\log(x_1)$

$\nabla\Phi_r(x) = \begin{pmatrix}1\\1\end{pmatrix} + \dfrac{r}{-x_1^2+x_2}\begin{pmatrix}2x_1\\-1\end{pmatrix} + \dfrac{r}{x_1}\begin{pmatrix}-1\\0\end{pmatrix}$ shows:

$$\nabla\Phi_r(x) = 0 \iff \begin{cases} 1 + \dfrac{2rx_1}{-x_1^2+x_2} - \dfrac{r}{x_1} = 0 & (1)\\[2ex] 1 - \dfrac{r}{-x_1^2+x_2} = 0 & (2)\end{cases}$$

From (2) we obtain $-x_1^2 + x_2 = r$ or $x_2 = r + x_1^2$. Substitution into (1) gives the equation

$$1 + 2x_1 - \frac{r}{x_1} = 0\,;$$

from there we obtain $x_1(r) := \frac{1}{4}\big(-1 + \sqrt{1+8r}\big)$. The second solution to the quadratic equation is omitted since it is negative. The trajectory defined by

$$x(r) := \begin{pmatrix} x_1(r)\\ r + x_1(r)^2\end{pmatrix} \quad (r > 0)$$

is often called the *'central path'*. In addition, we note that

$$\lambda_1(r) := \frac{r}{-g_1(x(r))} = 1\,, \quad \lambda_2(r) := \frac{r}{-g_2(x(r))} = \frac{1}{2}\Big(1 + \sqrt{1+8r}\Big)$$

are approximations of the LAGRANGE multipliers since

$$\lim_{r\to 0} x(r) = \begin{pmatrix}0\\0\end{pmatrix} =: \overline{x}\,,\ \lim_{r\to 0}\lambda(r) = \begin{pmatrix}1\\1\end{pmatrix} =: \overline{\lambda}$$

yield a KKT pair $(\overline{x}, \overline{\lambda})$ — and hence, according to theorem 2.2.8, a solution — to our optimization problem. ◁

Chapter 5

In the more general case we will now consider the problem

$$(P) \quad \begin{cases} f(x) \longrightarrow \min \\ g_i(x) \le 0 \ \text{ for } \ i \in \mathcal{I} \end{cases}$$

and the corresponding logarithmic penalty function

$$\Phi_r(x) := f(x) - r \sum_{i=1}^{m} \log(-g_i(x)) \, .$$

We are looking for an optimal solution $x(r)$ to a given penalty parameter $r > 0$. For $x = x(r)$ the gradient of Φ_r is equal to 0, that is, it holds that

$$\nabla_x \Phi_r(x(r)) = \nabla f(x(r)) + \sum_{i=1}^{m} \frac{r}{-g_i(x(r))} \nabla g_i(x(r)) = 0. \qquad (3)$$

If we define an estimation of the Lagrange multipliers by

$$\lambda_i(r) := \frac{r}{-g_i(x(r))} \qquad \text{for } i \in \mathcal{I},$$

it is possible to write (3) as

$$\nabla f(x(r)) + \sum_{i=1}^{m} \lambda_i(r) \nabla g_i(x(r)) = 0 \, .$$

This condition is identical with the gradient condition $\nabla_x L(x, \lambda) = 0$ for the Lagrangian L of problem (P) defined by

$$L(x, \lambda) := f(x) + \sum_{i=1}^{m} \lambda_i g_i(x) \, .$$

Let us also look at the other KKT conditions:

$$g_i(x) \le 0 \quad \text{for } i \in \mathcal{I} \qquad (4)$$

$$\lambda_i \ge 0 \quad \text{for } i \in \mathcal{I} \qquad (5)$$

$$\lambda_i g_i(x) = 0 \quad \text{for } i \in \mathcal{I}. \qquad (6)$$

For $x = x(r)$ and $\lambda = \lambda(r)$ conditions (4) and (5) are evidently satisfied since $g_i(x(r)) < 0$ and $\lambda_i(r) > 0$ hold for all $i \in \mathcal{I}$. The complementary slackness condition (6), however, is only met 'approximately' because of

$$\lambda_i(r) \big(- g_i(x(r)) \big) = r \quad \text{for all } \ i \in \mathcal{I}$$

and r positive. If

$$\lim_{r \to 0} \big(x(r), \lambda(r) \big) = (\overline{x}, \overline{\lambda})$$

holds, $(\overline{x}, \overline{\lambda})$ is a KKT pair of (P). If we insert a *slack variable* into the constraints of (P), we get the equivalent problem:

$$(P') \quad \begin{cases} f(x) \longrightarrow \min \\ g_i(x) + s_i = 0 \text{ for } i \in \mathcal{I} \\ s_i \geq 0 \text{ for } i \in \mathcal{I}. \end{cases}$$

We obtain the following 'perturbed' KKT conditions of (P'):

$$\begin{aligned} \nabla f(x) + \sum_{i=1}^{m} \lambda_i \nabla g_i(x) &= 0 \\ g(x) + s &= 0 \\ \lambda_i s_i &= r \text{ for } i \in \mathcal{I} \\ \lambda,\, s &\geq 0. \end{aligned}$$

This is the starting point for the development of so-called *primal–dual* **interior-point methods** which we will discuss later on.

Chapter 5

5.2 SQP Methods

As already mentioned in section 4.3, 'SQP' stands for ***S**equential **Q**uadratic **P**rogramming*. SQP, however, does not denote a single algorithm but a general method. For the convenience of our readers, we repeat the essentials: Usually a nonlinear optimization problem is converted into a sequence of quadratic optimization problems which are easier to solve. SQP methods go back to R. B. Wilson (1963; PhD Thesis, Harvard University). Many authors consider them to be the most important class of methods for solving nonlinear optimization problems without any particular structure.

Consider the problem:

$$(P) \quad \begin{cases} f(x) \longrightarrow \min \\ c_i(x) \leq 0 \text{ for } i \in \mathcal{I} := \{1, \ldots, m\} \\ c_i(x) = 0 \text{ for } i \in \mathcal{E} := \{m+1, \ldots, m+p\}. \end{cases}$$

For that let $m, p \in \mathbb{N}_0$ (hence, $\mathcal{E} = \emptyset$ or $\mathcal{I} = \emptyset$ are permitted), and suppose that the functions $f, c_1, \ldots, c_{m+p}$ are twice continuously differentiable in an open subset D of the $\mathbb{R}^n$.

Hence, the set of feasible points is

$$\begin{aligned} \mathcal{F} &:= \big\{x \in D \mid c_i(x) \leq 0 \text{ for } i \in \mathcal{I},\ c_i(x) = 0 \text{ for } i \in \mathcal{E}\big\} \\ &= \big\{x \in D \mid c_{\mathcal{I}}(x) \leq 0,\ c_{\mathcal{E}}(x) = 0\big\}. \end{aligned}$$

For the Lagrangian L defined by

$$L(x, \lambda) := f(x) + \lambda^T c(x) \quad \text{for } x \in D \text{ and } \lambda \in \mathbb{R}^{m+p}$$

it holds that

$$\begin{aligned} \nabla_x L(x, \lambda) &= \nabla f(x) + \sum_{i=1}^{m+p} \lambda_i \nabla c_i(x) \\ \nabla_\lambda L(x, \lambda) &= c(x) \\ \nabla_{xx}^2 L(x, \lambda) &= \nabla^2 f(x) + \sum_{i=1}^{m+p} \lambda_i \nabla^2 c_i(x) \\ \nabla_{x\lambda}^2 L(x, \lambda) &= \big(c'(x)\big)^T. \end{aligned}$$

(x^*, λ^*) is a *KKT point* of (P) if and only if

$$\begin{aligned} &\nabla_x L(x^*, \lambda^*) = 0\,, \\ &\lambda_i^* c_i(x^*) = 0\,,\ \lambda_i^* \geq 0 \text{ and } c_i(x^*) \leq 0 \text{ for } i \in \mathcal{I} \text{ and} \\ &c_i(x^*) = 0 \text{ for } i \in \mathcal{E} \text{ hold.} \end{aligned}$$

The last two lines can be shortened to

$$\langle \lambda_{\mathcal{I}}^*, c_{\mathcal{I}}(x^*) \rangle = 0\,,\ \lambda_{\mathcal{I}}^* \geq 0\,,\ c_{\mathcal{I}}(x^*) \leq 0 \text{ and } c_{\mathcal{E}}(x^*) = 0\,.$$

To determine (x^*, λ^*) *iteratively,* we compute — starting from an 'approximation' $\big(x^{(k)}, \lambda^{(k)}\big)$ — a solution to the system

$$\begin{aligned} &0 = g_k + B_k(x - x^{(k)}) + A_k \lambda \\ &\lambda_i \geq 0,\quad \ell_i^{(k)}(x - x^{(k)}) \leq 0,\quad \lambda_i \ell_i^{(k)}(x - x^{(k)}) = 0 \text{ for } i \in \mathcal{I} \\ &\ell_i^{(k)}(x - x^{(k)}) = 0 \text{ for } i \in \mathcal{E} \end{aligned}$$

with

$$\begin{aligned} g_k &:= \nabla f\big(x^{(k)}\big) \\ A_k^T &:= c'\big(x^{(k)}\big) \\ B_k &:= \nabla^2 f\big(x^{(k)}\big) + \sum_{i=1}^{m+p} \lambda_i^{(k)} \nabla^2 c_i\big(x^{(k)}\big) \\ c_i^{(k)} &:= c_i\big(x^{(k)}\big);\ \ell_i^{(k)}(d) := c_i^{(k)} + c_i'\big(x^{(k)}\big) d \ \text{ for } d \in \mathbb{R}^n \text{ and } i \in \mathcal{I} \cup \mathcal{E} \end{aligned}$$

taking into account the 'linearizations' of $\nabla_x L(x, \lambda)$ and $c_i(x)$ by

$$\begin{aligned} &\nabla_x L\big(x^{(k)}, \lambda^{(k)}\big) + \nabla_{xx}^2 L\big(x^{(k)}, \lambda^{(k)}\big)(x - x^{(k)}) + \nabla_{x\lambda}^2 L\big(x^{(k)}, \lambda^{(k)}\big)(\lambda - \lambda^{(k)}) \\ &\quad = \big(g_k + A_k \lambda^{(k)}\big) + B_k(x - x^{(k)}) + A_k(\lambda - \lambda^{(k)}) \\ &\quad = g_k + B_k(x - x^{(k)}) + A_k \lambda \end{aligned}$$

and

$$c_i^{(k)} + c_i'\big(x^{(k)}\big)(x - x^{(k)}) = \ell_i^{(k)}(x - x^{(k)}).$$

If we set $d := x - x^{(k)}$, these are exactly the KKT conditions of the following *quadratic program:*

$$(QP)_k \quad \begin{cases} \varphi_k(d) := \frac{1}{2} d^T B_k d + g_k^T d \longrightarrow \min \\ \ell_i^{(k)}(d) \leq 0 \quad \text{for } i \in \mathcal{I} \\ \ell_i^{(k)}(d) = 0 \quad \text{for } i \in \mathcal{E}. \end{cases}$$

If B_k is positive semidefinite, $(QP)_k$ is a *convex quadratic* problem. Then the KKT conditions are also sufficient for a global minimizer.

These observations lead to the following **algorithm** (basic version):

0) Given: $x^{(0)}, \lambda^{(0)}$; $k := 0$

1) Determine the solution d *and the corresponding* LAGRANGE *multipliers* λ *to* $(QP)_k$.

2) If $d = 0$: STOP
Otherwise: $x^{(k+1)} := x^{(k)} + d$, $\lambda^{(k+1)} := \lambda$
With $k := k + 1$ *go to 1).*

We will of course stop if $\|d\|_2 < \varepsilon$ for a given $\varepsilon > 0$.

Similarly to NEWTON's *method* we can prove *locally* quadratic convergence. One speaks of the **Lagrange–Newton Method**.

Example 5

$$f(x) := -x_1 - x_2 \longrightarrow \min$$
$$c_1(x) := x_1^2 - x_2 \leq 0$$
$$c_2(x) := x_1^2 + x_2^2 - 1 \leq 0$$

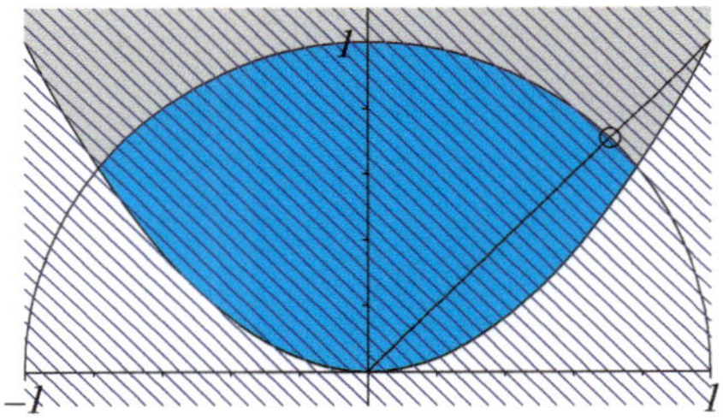

The figure shows the solution $x^* = \begin{pmatrix} 1/\sqrt{2} \\ 1/\sqrt{2} \end{pmatrix}$ of the minimization problem.

The corresponding LAGRANGE multiplier is $\lambda^* = \begin{pmatrix} 0 \\ 1/\sqrt{2} \end{pmatrix}$.

For $x^{(0)} := \begin{pmatrix} 1/2 \\ 1 \end{pmatrix}$ and $\lambda^{(0)} := \begin{pmatrix} 0 \\ 0 \end{pmatrix}$ it follows that

$$B_0 = 0, \; c_1^{(0)} = -3/4, \; c_2^{(0)} = 1/4, \; g_0 = \begin{pmatrix} -1 \\ -1 \end{pmatrix}.$$

Hence, in this case $(QP)_0$ is

$$\begin{cases} \varphi_0(d) = -d_1 - d_2 \longrightarrow \min \\ -3/4 + d_1 - d_2 \leq 0 \\ 1/4 + d_1 + 2d_2 \leq 0. \end{cases}$$

The solution $d = \begin{pmatrix} 5/12 \\ -1/3 \end{pmatrix}$, $\lambda^{(1)} = \begin{pmatrix} 1/3 \\ 2/3 \end{pmatrix}$ leads to

$x^{(1)} = x^{(0)} + d = \begin{pmatrix} 11/12 \\ 2/3 \end{pmatrix}$, $B_1 = \begin{pmatrix} 2 & 0 \\ 0 & 4/3 \end{pmatrix}$.

Hence, $(QP)_1$ now is

$$\begin{cases} \varphi_1(d) = d_1^2 + 2/3\, d_2^2 - d_1 - d_2 \longrightarrow \min \\ 25/144 + 11/6\, d_1 - d_2 \leq 0 \\ 41/144 + 11/6\, d_1 + 4/3\, d_2 \leq 0 \quad \text{(active!)}. \end{cases}$$

The solution $d = \begin{pmatrix} -0.1695 \\ 0.0196 \end{pmatrix}$, $\lambda^{(2)} = \begin{pmatrix} 0 \\ 0.7304 \end{pmatrix}$ yields $x^{(2)} = \begin{pmatrix} 0.7471 \\ 0.6863 \end{pmatrix}$. ◁

Possible problems are:

a) *The quadratic objective function is unbounded from below:*

Example 6

$$f(x) := x_1 + x_2 \longrightarrow \min$$
$$c_1(x) := x_1^2 - x_2 \leq 0$$

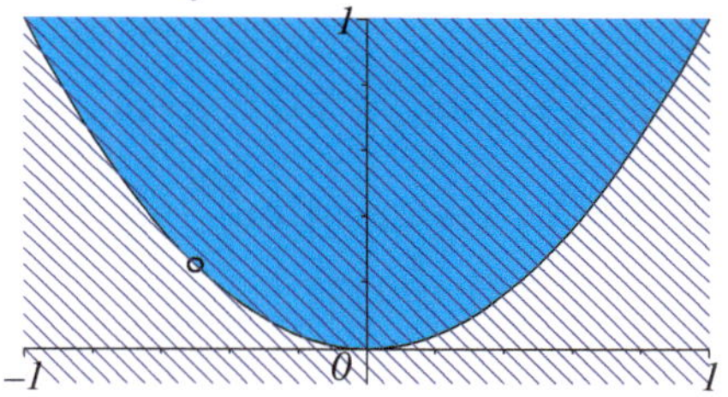

In the accompanying figure we observe that the objective function f attains its minimum at $x^* = (-1/2, 1/4)^T$.

$x^{(0)} := \begin{pmatrix} 0 \\ 0 \end{pmatrix}$, $\lambda^{(0)} := 0$ yield

$$\begin{cases} \varphi_0(d) = d_1 + d_2 \longrightarrow \min \\ -d_2 \leq 0. \end{cases}$$

◁

b) *The constraints of* $(QP)_0$ *are inconsistent:*

Example 7

$$c_1(x) := -x_1 \leq 0$$
$$c_2(x) := -x_2 \leq 0$$
$$c_3(x) := x_2 + (x_1 - 1)^3 \leq 0$$

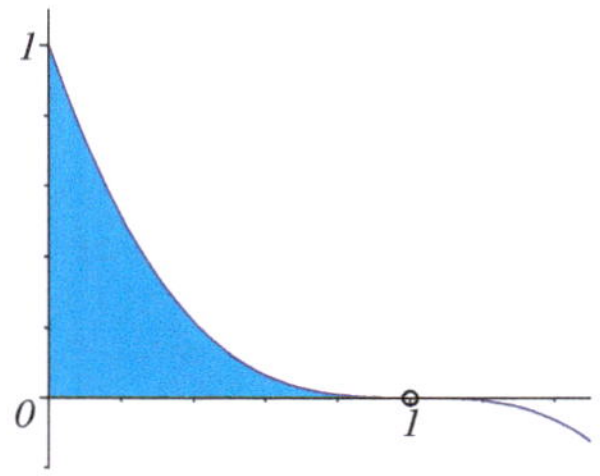

The point $\begin{pmatrix}1\\0\end{pmatrix}$ is feasible. For any f and $\lambda^{(0)}$ the point $x^{(0)} := \begin{pmatrix}-1\\0\end{pmatrix}$ leads to the obviously inconsistent system

$$\begin{cases} \ell_1^{(0)}(d) &= 1 - d_1 &\le 0 \\ \ell_2^{(0)}(d) &= -d_2 &\le 0 \\ \ell_3^{(0)}(d) &= -8 + 12d_1 + d_2 &\le 0. \end{cases}$$

◁

The **objectives** of the following considerations are to obtain *global convergence* and *avoid the second-order derivatives.* This will lead to *Quasi*-NEWTON *methods.*

If the point $x^{(k)}$ is still far away from x^*, we have to keep in mind that on the one hand f should be made as small as possible and on the other hand the constraints should be met as exactly as possible. In practice finding feasible points is more important than the minimization of f. We will try to strike a happy medium between these two objectives. We will see that the solution d_k to $(QP)_k$ is a descent direction of the following exact[1] penalty function:

$$\Phi_r(x) := f(x) + \sum_{i=1}^{m} \frac{1}{r_i}\, c_i^+(x) + \sum_{i=m+1}^{m+p} \frac{1}{r_i}\, |c_i(x)|$$

Here, $r_i > 0$ for $i \in \mathcal{I} \cup \mathcal{E}$ are of course given.

S.-P. HAN's **suggestion** (1976): Step size control via $\Phi_r(x)$

HAN **Algorithm**

0) Given: $x^{(0)}$ *and a positive definite* $B_0 \in \mathbb{R}^{n\times n}$; $k := 0$

1) Determine the solution d_k *and the corresponding* LAGRANGE *multiplier* $\lambda^{(k)}$ *to* $(QP)_k$.

2) Determine an $\alpha_k \in [0, 1]$ *with*

$$\Phi_r\big(x^{(k)} + \alpha_k d_k\big) = \min\Big\{\Phi_r\big(x^{(k)} + \alpha d_k\big) \mid 0 \le \alpha \le 1\Big\}$$

and set $p_k := \alpha_k d_k$, $x^{(k+1)} := x^{(k)} + p_k$.

3) Update B_k *to* B_{k+1} *such that the positive definiteness remains. With* $k := k + 1$ *go to 1).*

To 2): d_k is a *descent direction of the penalty function* Φ_r.

[1] We will not prove exactness here.

More precisely: *Let* $(d_k, \lambda^{(k)})$ *be a KKT pair of* $(QP)_k$ *with* $d_k \neq 0$ *and* $r_i |\lambda_i^{(k)}| \leq 1$ *for* $i \in \mathcal{E} \cup \mathcal{I}$. *Then for the directional derivative of* Φ_r *in* $x^{(k)}$ *in the direction of* d_k *it holds that*

$$\Phi_r(x^{(k)}; d_k) := \lim_{t \to 0+} \frac{\Phi_r(x^{(k)} + t\, d_k) - \Phi_r(x^{(k)})}{t} < 0\,.$$

Proof: At first we will consider the individual terms c_i^+ for $i \in \mathcal{I}$ and $|c_i(\ )|$ for $i \in \mathcal{E}$ and divide the set $\mathcal{I}$ into
$\mathcal{I}^+ := \{i \in \mathcal{I} \mid c_i^{(k)} > 0\}$, $\mathcal{I}^- := \{i \in \mathcal{I} \mid c_i^{(k)} < 0\}$, $\mathcal{I}^0 := \{i \in \mathcal{I} \mid c_i^{(k)} = 0\}$ and do the same for $\mathcal{E}$. For $i \in \mathcal{E}$, $t > 0$ sufficiently small and then passage to the limit $t \longrightarrow 0$, we have

$$\frac{|c_i(x^{(k)} + t\, d_k)| - |c_i^{(k)}|}{t} = \begin{cases} \dfrac{c_i(x^{(k)} + t\, d_k) - c_i^{(k)}}{t} \longrightarrow c_i'(x^{(k)})\, d_k\,, & i \in \mathcal{E}^+ \\ -\dfrac{c_i(x^{(k)} + t\, d_k) - c_i^{(k)}}{t} \longrightarrow -c_i'(x^{(k)})\, d_k\,, & i \in \mathcal{E}^- \\ \left|\dfrac{c_i(x^{(k)} + t\, d_k)}{t}\right| \longrightarrow |c_i'(x^{(k)}) d_k|\,, & i \in \mathcal{E}^0. \end{cases}$$

For $i \in \mathcal{I}$ we get analogously

$$\frac{c_i^+(x^{(k)} + t\, d_k) - c_i^{(k)+}}{t} = \begin{cases} \dfrac{c_i(x^{(k)} + t\, d_k) - c_i^{(k)}}{t} \longrightarrow c_i'(x^{(k)}) d_k\,, & i \in \mathcal{I}^+ \\ 0\,, & i \in \mathcal{I}^- \\ \left(\dfrac{c_i(x^{(k)} + t\, d_k)}{t}\right)^+ \longrightarrow (c_i'(x^{(k)}) d_k)^+\,, & i \in \mathcal{I}^0. \end{cases}$$

$(d_k, \lambda^{(k)})$ is a KKT pair of $(QP)_k$ which means the following:

$0 = g_k + B_k d_k + A_k \lambda^{(k)}$ and for $i \in \mathcal{I}$

$$\lambda_i^{(k)} \geq 0\,,\ \ell_i^{(k)}(d_k) = c_i^{(k)} + c_i'(x^{(k)})\, d_k \leq 0\,,\ \lambda_i^{(k)} \left[c_i^{(k)} + c_i'(x^{(k)}) d_k\right] = 0$$

as well as $c_i^{(k)} + c_i'(x^{(k)})\, d_k = 0$ for $i \in \mathcal{E}$.

These three lines together with the definition of A_k firstly yield

$g_k^T d_k = -d_k^T B_k d_k - \lambda^{(k)T} A_k^T d_k = -d_k^T B_k d_k - \sum_{i \in \mathcal{I} \cup \mathcal{E}} \lambda_i^{(k)} c_i'(x^{(k)})\, d_k$ and

$-\lambda_i^{(k)} c_i'(x^{(k)})\, d_k = \lambda_i^{(k)} c_i^{(k)}$ for $i \in \mathcal{I} \cup \mathcal{E}$.

For $i \in \mathcal{I}^0$ we have $c_i'(x^{(k)})\, d_k \leq 0$, hence, $(c_i'(x^{(k)})\, d_k)^+ = 0$.

For $i \in \mathcal{E}^0$ we have $c_i'(x^{(k)})\, d_k = 0$.

With that we get the following for the directional derivative of Φ_r:

Chapter 5

$$g_k^T d_k + \sum_{i\in\mathcal{E}^+} \frac{1}{r_i} c_i'(x^{(k)})\, d_k - \sum_{i\in\mathcal{E}^-} \frac{1}{r_i} c_i'(x^{(k)})\, d_k + \sum_{i\in\mathcal{I}^+} \frac{1}{r_i} c_i'(x^{(k)})\, d_k$$

$$= \underbrace{-d_k^T B_k d_k}_{<0} + \sum_{i\in\mathcal{E}^+} \underbrace{\Big(\lambda_i^{(k)} - \frac{1}{r_i}\Big) c_i^{(k)}}_{\leq 0} + \sum_{i\in\mathcal{E}^-} \underbrace{\Big(\lambda_i^{(k)} + \frac{1}{r_i}\Big) c_i^{(k)}}_{\leq 0}$$

$$+ \sum_{i\in\mathcal{I}^+} \Big(\lambda_i^{(k)} c_i^{(k)} + \frac{1}{r_i} \underbrace{c_i'(x^{(k)})\, d_k}_{\leq -c_i^{(k)}}\Big) + \sum_{i\in\mathcal{I}^-} \underbrace{\lambda_i^{(k)} c_i^{(k)}}_{\leq 0} < 0 \qquad \square$$

Addition: If $d_k = 0$, it holds that
$g_k + A_k \lambda^{(k)} = 0$,
$c_i^{(k)} \leq 0$, $\lambda_i^{(k)} \geq 0$, $\lambda_i^{(k)} c_i^{(k)} = 0$ for $i \in \mathcal{I}$ and
$c_i^{(k)} = 0$ for $i \in \mathcal{E}$.

Hence, $(x^{(k)}, \lambda^{(k)})$ *is a KKT pair of problem* (P).

Instead of *2)* we will often choose the more 'tolerant' search strategies (inexact line search instead of exact line search) which have already been discussed in section 3.2, for example, the following ARMIJO *step size rule:*

Let $\alpha \in (0, 1)$. Starting from $\lambda := 1$, we will halve λ until

$$\Phi_r\big(x^{(k)} + \lambda d_k\big) \leq \Phi_r\big(x^{(k)}\big) + \alpha\lambda \Phi_r\big(x^{(k)}; d_k\big)$$

holds.

To 3): With $q_k := \nabla L\big(x^{(k+1)}, \lambda^{(k)}\big) - \nabla L\big(x^{(k)}, \lambda^{(k)}\big)$

and the assumption $\langle p_k, q_k \rangle > 0$ we consider the BFGS updating formula

$$B_{k+1} := B_k + \frac{q_k q_k^T}{\langle p_k, q_k \rangle} - \frac{B_k p_k p_k^T B_k}{\langle p_k, B_k p_k \rangle},$$

which was discovered independently by BROYDEN, FLETCHER, GOLDFARB and SHANNO, using different methods (compare section 3.5).

Continuing from there, POWELL[2] proposed a number of *modifications* to make the algorithm more efficient:

To 1): The linear constraints in $(QP)_k$ can be inconsistent even if the nonlinear constraints in (P) are consistent. In this case we introduce a new variable ξ and replace the constraints $\ell_i^{(k)}(d) \leq 0$ for $i \in \mathcal{I}$ and $\ell_i^{(k)}(d) = 0$ for $i \in \mathcal{E}$ by:

[2] Confer [Pow 2].

Chapter 5

$$\left.\begin{array}{l} c_i^{(k)}\xi_i + c_i'(x^{(k)})d \le 0 \\ \text{with } \xi_i := \begin{cases} 1, & \text{if } c_i^{(k)} \le 0 \\ \xi, & \text{if } c_i^{(k)} > 0 \end{cases} \qquad (i \in \mathcal{I}) \\ c_i^{(k)}\xi + c_i'(x^{(k)})\,d = 0 \quad (i \in \mathcal{E}) \\ 0 \le \xi \le 1. \end{array}\right\} \tag{7}$$

For that we will *at first* choose ξ as the maximum and *then* determine d as the minimizer of the quadratic objective function restricted by (7). Hence, we only change the inequality constraints which are not met at the starting point. The factor ξ ensures that at least the pair $(\xi, d) = (0, 0)$ is feasible. If we thus get $\xi = 0$ *and* $d = 0$, the restrictions in (P) are considered to be inconsistent and the algorithm is terminated.

Example 8 (functions c_i from example 7)

$$\begin{array}{ll} c_1(x) := -x_1 \le 0 & \\ c_2(x) := -x_2 \le 0 & x^{(0)} := \begin{pmatrix} -1 \\ 0 \end{pmatrix} \\ c_3(x) := x_2 + (x_1 - 1)^3 \le 0 & \end{array}$$

$$c_1^{(0)} = 1\,,\ c_2^{(0)} = 0\,,\ c_3^{(0)} = -8$$

$$\left.\begin{array}{l} \xi - d_1 \le 0 \\ -d_2 \le 0 \\ -8 + 12d_1 + d_2 \le 0 \\ 0 \le \xi \le 1 \end{array}\right\} \Longrightarrow \xi = \tfrac{2}{3}\,;\ d = \begin{pmatrix} \frac{2}{3} \\ 0 \end{pmatrix} \text{ is feasible.}$$

◁

To 2): If we choose *suitable penalty parameters* r_i, it holds by theorem 14.3.1 in [Fle], p. 380, that x^* *yields a solution to* (P) *if and only if* x^* *is a minimal point of* Φ_r.

Powell suggests the following *modulation of the penalty parameters*:

$$k = 0: \quad \frac{1}{r_i} := |\lambda_i^{(0)}|$$

$$k \ge 1: \quad \frac{1}{r_i} := \max\left\{|\lambda_i^{(k)}|\,,\ \frac{1}{2}\left(\frac{1}{r_i} + |\lambda_i^{(k)}|\right)\right\}$$

Problem:

We are left with only one estimation $\Phi_{r^{(k)}}\big(x^{(k+1)}\big) < \Phi_{r^{(k)}}\big(x^{(k)}\big)$; however, it might happen that $\Phi_{r^{(k+1)}}\big(x^{(k+1)}\big) > \Phi_{r^{(k)}}\big(x^{(k)}\big)$ holds. Then the chain of inequalities does not fit anymore. So how can we obtain the convergence of $x^{(k)} \longrightarrow x^*$?

Furthermore:

- If $r^{(k)}$ is too small, the step size α_k can become extremely small since the Hessian becomes ill-conditioned. (In the ideal case we should have $\alpha_k \approx 1$.)
- If $r^{(k)}$ is too big, d_k might possibly not be a descent direction of $\Phi_{r^{(k)}}$ anymore.

We will of course hope for locally *superlinear* or *quadratic* convergence of the SQP method if we have chosen B_k suitably. However, due to the so-called MARATOS *Effect* (1978) this will not be the case:

Suppose that $r_i\,|\lambda_i^{(k)}| \leq 1 \quad (i = 1, \ldots, m+p)$ for all $k \geq 0$.
The following might happen:

- If all iteration points $x^{(k)}$ lie close to x^* and if we choose $x^{(k+1)} = x^{(k)} + d_k$, then $(x^{(k)})$ is *superlinearly* convergent.
- However, if we choose $\widetilde{x}^{(k+1)} = \widetilde{x}^{(k)} + \alpha_k d_k$ with

$$\Phi_r\big(\widetilde{x}^{(k+1)}\big) = \min_{0 \leq \alpha \leq 1} \Phi_r\big(\widetilde{x}^{(k)} + \alpha\, d_k\big),$$

  it can happen that $\big(\widetilde{x}^{(k)}\big)$ is *not* superlinearly convergent, since the line search of the penalty function prevents the optimal choice of $\alpha_k = 1$ because $\Phi_r\big(\widetilde{x}^{(k)} + d_k\big) > \Phi_r\big(\widetilde{x}^{(k)}\big)$.

Consider the following standard example by POWELL (cf. [Pow3]):

Example 9 (cf. exercise 8)
$f(x) := 2\big(x_1^2 + x_2^2 - 1\big) - x_1$
$h(x) := c_1(x) := x_1^2 + x_2^2 - 1 = 0$

The feasible region $\mathcal{F}$ is exactly the boundary of the unit circle in $\mathbb{R}^2$ in this case. For $x = (x_1, x_2)^T \in \mathcal{F}$ we have $f(x) = -x_1$. With that $x^* := (1,0)^T$ gives the minimum with value $f(x^*) = -1$. We obtain the LAGRANGE multiplier $\lambda^* = -3/2$.

With the positive definite matrices

$$B_k := \nabla_{xx}^2 L(x^*, \lambda^*) = \begin{pmatrix} 4 & 0 \\ 0 & 4 \end{pmatrix} + \lambda^* \begin{pmatrix} 2 & 0 \\ 0 & 2 \end{pmatrix} = \begin{pmatrix} 1 & 0 \\ 0 & 1 \end{pmatrix}$$

the quadratic subproblems are

$$\varphi_k(d) = g_k^T d + \frac{1}{2} d^T B_k d = f'\big(x^{(k)}\big) d + \frac{1}{2}\langle d, d\rangle \longrightarrow \min$$
$$\ell_1^{(k)}(d) = h\big(x^{(k)}\big) + h'\big(x^{(k)}\big) d = 0.$$

We will drop the index k in the following considerations; hence, we have to consider the following subproblem

$$\begin{aligned} f'(x)d + \frac{1}{2}\langle d, d\rangle &\longrightarrow \min \\ h(x) + \langle \nabla h(x), d\rangle &= 0. \end{aligned}$$

We obtain d via the corresponding KKT conditions

$$\begin{aligned} \nabla f(x) + d + \mu\nabla h(x) &= 0 \\ h(x) + \langle \nabla h(x), d\rangle &= 0. \end{aligned}$$

Due to the special (quadratic) structure, we get the following for h and f:

$$\begin{aligned} h(x+d) &= h(x) + h'(x)d + \langle d, d\rangle = \langle d, d\rangle \\ f(x+d) &= f(x) + f'(x)d + 2\langle d, d\rangle = f(x) - \langle d + \mu\nabla h(x), d\rangle + 2\langle d, d\rangle \\ &= f(x) - \mu\langle \nabla h(x), d\rangle + \langle d, d\rangle = f(x) + \mu h(x) + \langle d, d\rangle \end{aligned}$$

If x is now any *feasible* point,

$$\begin{aligned} f(x+d) &= f(x) + \langle d, d\rangle > f(x) \\ h(x+d) &= \langle d, d\rangle > 0 = h(x) \end{aligned}$$

hold for each $d \in \mathbb{R}^2 \setminus \{0\}$. For the penalty function Φ_r defined by

$$\Phi_r(x) := f(x) + \frac{1}{r}|h(x)| \qquad (x \in \mathbb{R}^2,\ r \in (0, \infty))$$

it hence holds that

$$\Phi_r(x+d) = f(x)+\langle d, d\rangle+\frac{1}{r}\langle d, d\rangle = f(x)+\Big(1+\frac{1}{r}\Big)\langle d, d\rangle > f(x) = \Phi_r(x).$$

Consequently, the full step size 1 will never be accepted, regardless of how close a feasible x is to x^*. This phenomenon can be explained by the discrepancy between the merit function Φ_r and the local quadratic model used to compute d.

Feasible region (black) and contours of the penalty function $\Phi_{\frac{1}{3}}$ (blue)

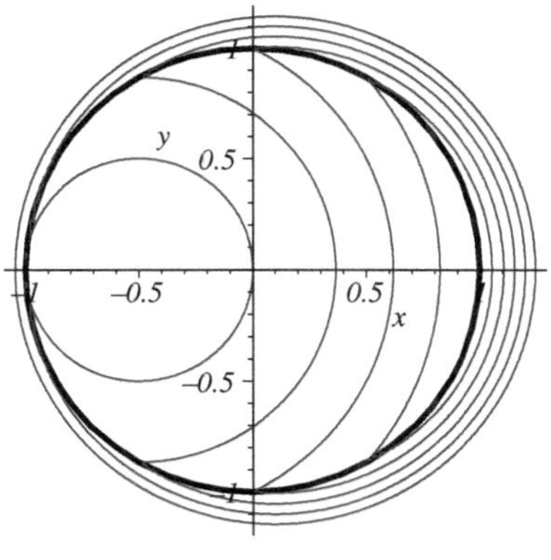

◁

Chapter 5

An Alternative Approach: **Fletcher's Sℓ_1QP Method** (cf. [Fle], p. 312ff)

Let $r_i|\lambda_i^*| < 1 \quad (1 \le i \le m+p)$. FLETCHER establishes a *connection between* *the trust-region method and the SQP method* in the following way. Instead of solving $(QP)_k$, he solves a trust-region problem in which an approximation of the exact ℓ_1-penalty function occurs:

$$\begin{aligned} &\varphi_k(d) + \sum_{i \in \mathcal{I}} \frac{1}{r_i} \big(\ell_i^{(k)}(d)\big)^+ + \sum_{i \in \mathcal{E}} \frac{1}{r_i} \big|\ell_i^{(k)}(d)\big| \longrightarrow \min \\ &\|d\|_\infty \le \Delta_k \end{aligned}$$

This problem is equivalent to a quadratic optimization problem with linear constraints:

$$(QP')_k \quad \begin{cases} \varphi_k(d) + \sum_{i=1}^{m+p} \frac{1}{r_i} \eta_i \longrightarrow \min \\ -\Delta_k \le d_\nu \le \Delta_k \quad (1 \le \nu \le n) \\ 0 \le \eta_i\,,\ \ell_i^{(k)}(d) \le \eta_i \quad \text{for } i \in \mathcal{I} \\ -\eta_i \le \ell_i^{(k)}(d) \le \eta_i \quad \text{for } i \in \mathcal{E}\,. \end{cases}$$

The feasible region of these problems is evidently always nonempty.

Consider $\Psi_r(d) := f\big(x^{(k)}\big) + \varphi_k(d) + \sum_{i \in \mathcal{I}} \frac{1}{r_i} \big(\ell_i^{(k)}(d)\big)^+ + \sum_{i \in \mathcal{E}} \frac{1}{r_i} \big|\ell_i^{(k)}(d)\big|$

and $\varrho_k := \dfrac{\Phi_r\big(x^{(k)} + d_k\big) - \Phi_r\big(x^{(k)}\big)}{\Psi_r(d_k) - \Psi_r(0)}\,.$

The numerator gives the actual reduction of the penalty function Φ_r, the denominator that of the modeling penalty function Ψ_r.

Model Algorithm

1) *Let* $x^{(k)}, \lambda^{(k)}$ *and* Δ_k *be given. Compute* $f\big(x^{(k)}\big)$, g_k, B_k, $c^{(k)}$ *and* A_k.
2) *The solution to* $(QP')_k$ *gives the KKT pair* $\big(d_k, \lambda^{(k+1)}\big)$.
3) *Compute* ϱ_k.
4) *If* $\varrho_k < \frac{1}{4}$: $\Delta_{k+1} := \Delta_k/2$ *and* $x^{(k+1)} := x^{(k)}$
   Otherwise: $x^{(k+1)} := x^{(k)} + d_k$
   If $\varrho_k > \frac{3}{4}$ *and* $\|d_k\|_\infty = \Delta_k$: $\Delta_{k+1} := 2\Delta_k$
   Otherwise: $\Delta_{k+1} := \Delta_k$
   With $k := k+1$ *go to 1).*

We will return to a *test problem* by Powell which we have already discussed in exercise 15 to chapter 2.

Example 10

We are looking for a triangle with the least possible area, containing two disjoint disks with radius 1. If $(0,0)$, $(x_1,0)$ and (x_2,x_3) denote the vertices of the triangle and (x_4,x_5) and (x_6,x_7) the centers of the disks, we have to solve the following problem:

$$f(x) := \frac{1}{2} x_1 x_3 \longrightarrow \min \quad \text{(area of the triangle)}.$$

WLOG let $x_1 \geq 0$ and $x_3 \geq 0$. Then we have the 'trivial' inequalities
$(x_4 - x_6)^2 + (x_5 - x_7)^2 \geq 4$ (disjoint disks), $x_5 \geq 1$ and $x_7 \geq 1$.

For the other four inequalities, we utilize the *point normal form* or *Hessian form* of a linear equation:

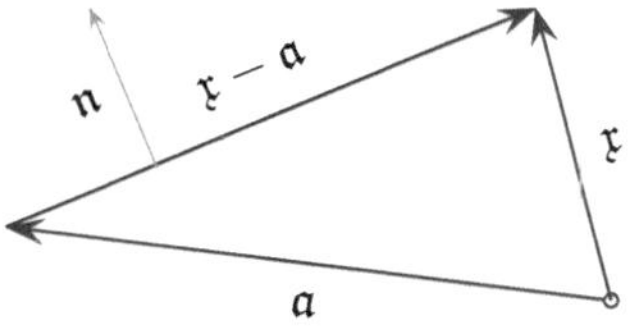

$$\langle \mathfrak{x} - \mathfrak{a}, \mathfrak{n} \rangle = 0$$

If the normal $\mathfrak{n}$ to the straight line is scaled with the norm $\| \;\|_2$, then

$$d := \langle \mathfrak{p} - \mathfrak{a}, \mathfrak{n} \rangle$$

returns the distance of a point $\mathfrak{p}$ to the straight line.

Here we have at first $\mathfrak{a} := (0,0)$ and $\mathfrak{x} := (x_2, x_3)$.
Then $\mathfrak{n} = (x_3, -x_2)/\sqrt{x_2^2 + x_3^2}$. For $\mathfrak{p} = (x, y)$ it thus follows that

$$1 \leq d = \langle \mathfrak{p}, \mathfrak{n} \rangle .$$

Insertion of the two points (x_4, x_5) and (x_6, x_7) gives

$$\frac{x_3 x_4 - x_2 x_5}{\sqrt{x_3^2 + x_2^2}} \geq 1 \quad \text{and} \quad \frac{x_3 x_6 - x_2 x_7}{\sqrt{x_3^2 + x_2^2}} \geq 1 .$$

Via the distance to the line through (x_2, x_3) and $(x_1, 0)$ with $\mathfrak{x} - \mathfrak{a} = (x_2 - x_1, x_3)$ and $\mathfrak{n} = (-x_3, x_1 - x_2)/\sqrt{(x_1 - x_2)^2 + x_3^2}$, we obtain accordingly

$$\frac{-x_3 x_4 + (x_2 - x_1) x_5 + x_1 x_3}{\sqrt{x_3^2 + (x_2 - x_1)^2}} \geq 1 \quad \text{and} \quad \frac{-x_3 x_6 + (x_2 - x_1) x_7 + x_1 x_3}{\sqrt{x_3^2 + (x_2 - x_1)^2}} \geq 1 .$$

The problem has, for example, the following minimal point:

$$x^* = \Big(4 + 2\sqrt{2},\ 2 + \sqrt{2},\ 2 + \sqrt{2},\ 1 + \sqrt{2},\ 1,\ 3 + \sqrt{2},\ 1\Big)^T$$

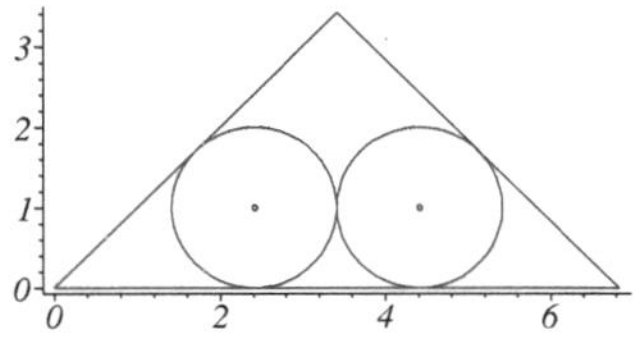

with $\mathcal{A}(x^*) = \{3, 4, 5, 6, 9\}$ and

$$\lambda^* = \Big(0,\ 0,\ \tfrac{1}{2} + \tfrac{1}{2\sqrt{2}},\ 2 + \sqrt{2},\ 2 + \sqrt{2},\ 2 + 2\sqrt{2},\ 0,\ 0,\ 2 + 2\sqrt{2}\Big)^T .$$

We obtain the following for the Hessian of the LAGRANGE function:

$$H^* = \begin{pmatrix} \frac{1+\sqrt{2}}{4} & \frac{1-\sqrt{2}}{4} & \frac{1-\sqrt{2}}{4} & 0 & 0 & -\frac{1}{2} & \frac{1}{2} \\ \frac{1-\sqrt{2}}{4} & -\frac{3-\sqrt{2}}{2} & 0 & \frac{1}{2} & \frac{1}{2} & \frac{1}{2} & -\frac{1}{2} \\ \frac{1-\sqrt{2}}{4} & 0 & \frac{1+\sqrt{2}}{2} & -\frac{1}{2} & -\frac{1}{2} & \frac{1}{2} & -\frac{1}{2} \\ 0 & \frac{1}{2} & -\frac{1}{2} & -\frac{2+\sqrt{2}}{2} & 0 & \frac{2+\sqrt{2}}{2} & 0 \\ 0 & \frac{1}{2} & -\frac{1}{2} & 0 & -\frac{2+\sqrt{2}}{2} & 0 & \frac{2+\sqrt{2}}{2} \\ -\frac{1}{2} & \frac{1}{2} & \frac{1}{2} & \frac{2+\sqrt{2}}{2} & 0 & -\frac{2+\sqrt{2}}{2} & 0 \\ \frac{1}{2} & -\frac{1}{2} & -\frac{1}{2} & 0 & \frac{2+\sqrt{2}}{2} & 0 & -\frac{2+\sqrt{2}}{2} \end{pmatrix}$$

A *Maple*® worksheet gives a good result for the POWELL starting points which are far away from a solution (with B_0 as the identity matrix) in only a few iteration steps. The attentive reader will of course have noticed that this is not the solution obtained above but the following alternative solution:

$$\left(2(1+\sqrt{2}),\ 0,\ 2(1+\sqrt{2}),\ 1,\ 1+\sqrt{2},\ 1+\sqrt{2},\ 1\right)^T$$

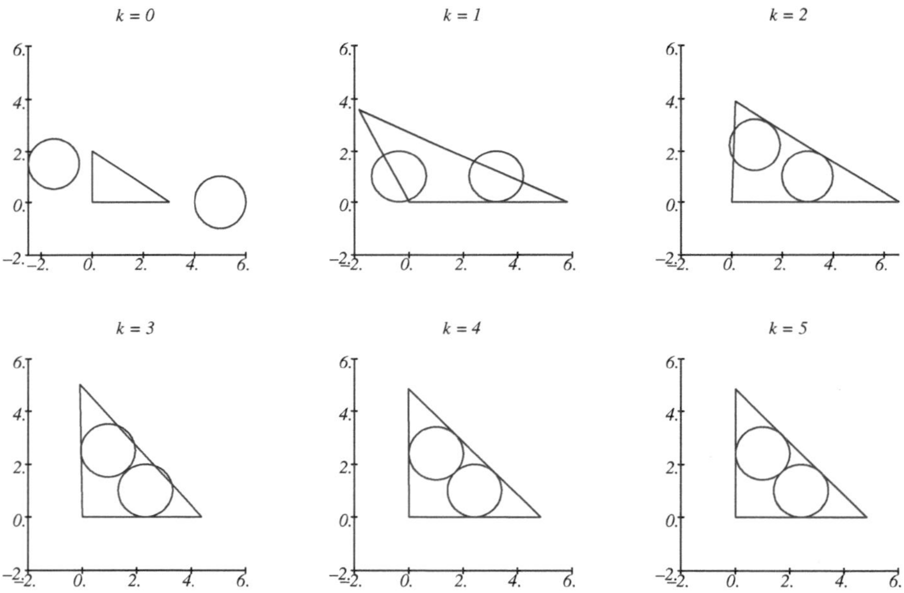

◁

Exercises

1. Let $\Phi_r(x) := \Phi(x, r) := f(x) + \frac{1}{r}\pi(x)$ be an outer penalty function for the problem

$$f(x) \longrightarrow \min$$
$$x \in S$$

and let furthermore a continuously decreasing sequence with $r_k \downarrow 0$ be given. We additionally assume that for each $k \in \mathbb{N}$ there exists a global minimizer x_k to $\Phi(\,\cdot\,, r_k)$, that is,

$$\Phi(x_k, r_k) = \min\left\{\Phi(x, r_k) \mid x \in S\right\}.$$

Show:

a) $\Phi(x_k, r_k) \le \Phi(x_{k+1}, r_{k+1})$

b) $\pi(x_k) \ge \pi(x_{k+1})$

c) $f(x_k) \le f(x_{k+1})$

d) $f(x) \ge \Phi(x_k, r_k) \ge f(x_k)$ for all $x \in S$, $k \in \mathbb{N}$.

2. Give a graphical illustration of the optimization problem

$$f(x) := -x_1 x_2 \longrightarrow \min$$
$$g_1(x) := x_1 + x_2^2 - 1 \le 0$$
$$g_2(x) := -x_1 - x_2 \le 0$$

and solve it by means of the *quadratic penalty method.* Compute its minimizer $x(r)$ and give estimates of the LAGRANGE multipliers.

3. Solve the following optimization problem by means of the *Barrier method*:

$$f(x) := -x_1 - 2x_2 \longrightarrow \min$$
$$g_1(x) := x_1 + x_2^2 - 1 \le 0$$
$$g_2(x) := -x_2 \le 0$$

Give estimates for the LAGRANGE multipliers and draw the central path.

4. Consider the following optimization problem:

$$\begin{cases} f(x) := \sum_{i=1}^n x_i^2 \longrightarrow \min \quad (x \in \mathbb{R}^n) \\ h(x) := \sum_{i=1}^n x_i - 1 = 0 \end{cases} \tag{8}$$

Determine the optimizer of (8) by means of the quadratic penalty function given by $\Phi_r(x) := f(x) + \frac{1}{2r}h(x)^2$ with the penalty parameter $r > 0$.

a) Prove that the corresponding optimization problems of the penalty method have a unique minimizer $x(r)$ for every $r > 0$.

b) Formulate a system of linear equations

$$B_r\,x(r) = b \quad \left(B_r \in \mathbb{R}^{n\times n},\ b \in \mathbb{R}^n\right)$$

determining $x(r)$ and solve it.

c) Calculate the condition number of the matrix B_r. Can you identify a potential 'weak point' of the penalty method by that?

d) Determine the minimizer x^* of problem (8).

e) Visualize for $n = 2$

- the contours of f and the feasible region,
- the contours of the objective function of the penalty method for the values $1/5$, $1/10$ and $1/15$ of the penalty parameter r. Plot x as a function of r and the point x^*.

What do you observe?

5. *Definite Quadratic Forms* (cf. [Deb])
Let $Q \in \mathbb{R}^{n\times n}$ be symmetric and $A \in \mathbb{R}^{m\times n}$ with $m \le n$. Prove the equivalence: Q is positive definite on kernel(A) iff there exists a number $\alpha \in \mathbb{R}$ such that $Q + \alpha A^T A$ is positive definite.

Hint: Show that the function $\varphi\colon K \longrightarrow \mathbb{R}$ defined by $\langle x, Qx\rangle / \langle Ax, Ax\rangle$ is bounded from below on $K := \big\{x \in \mathbb{R}^n \mid \langle x, x\rangle = 1 \text{ and } Ax \neq 0\big\}$.

6. Let the following optimization problem be given:

$$\begin{cases} f(x) \longrightarrow \min & \left(x \in \mathbb{R}^n\right) \\ h_j(x) = 0 & (j = 1, \ldots, p) \end{cases} \tag{9}$$

To that we define with $L(x, \mu) := f(x) + \mu^T h(x)$ the corresponding LAGRANGE function as well as with $L_A(x, \mu; r) := L(x, \mu) + \frac{1}{2r} h(x)^T h(x)$ the *augmented* LAGRANGE *function*.

a) Let (x^*, μ^*) satisfy the sufficient second-order optimality conditions to (9) (cf. theorem 2.3.2). Then there exists a number $r^* > 0$ such that for all $r \le r^*$ the point x^* is a strict local (unconstrained) minimizer of $L_A(\,\cdot\,, \mu^*; r)$.

b) If $h(\widehat{x}) = 0$ and $\widehat{x}$ is a stationary point of $L_A(\,\cdot\,, \widehat{\mu}; r)$ for some $\widehat{\mu} \in \mathbb{R}^p$, then $\widehat{x}$ is a KKT point to (9).

7. The preceding exercise suggests the following **algorithm**:

Let the starting points $x^{(0)}, \mu^{(0)}, r_0$ be given. Set $k = 0$.

- If $\nabla_x L(x^{(k)}, \mu^{(k)}) = 0$ and $h(x^{(k)}) = 0$: STOP.
- Calculate a minimizer $x^{(k+1)}$ of $L_A(\,\cdot\,, \mu^{(k)}; r_k) \longrightarrow \min$.
- $\mu^{(k+1)} := \mu^{(k)} + h(x^{(k+1)})/r_k$
  Choose r_{k+1} such that $r_k \le r_{k+1}$.
- Set $k = k+1$ and repeat the above calculation.

Use the algorithm to solve the following problem:

$$\begin{aligned} f(x) &:= x_1\, x_2^2 \longrightarrow \min \\ h(x) &:= x_1^2 + x_2^2 - 2 = 0 \end{aligned}$$

(cf. chapter 1, example 5). Choose $x^{(0)} := (-2, 2)^T$ and $\mu^{(0)} := 0$ as starting points and $r_k := 0.25$ for all $k \in \mathbb{N}_0$. Utilize, for example, the function fminunc from the *Matlab*® Optimization Toolbox.

8. *Maratos Effect*

We pick up on our discussion from example 9 and take a closer look at the phenomenon.

a) Show for $x \in \mathcal{F}$ that $d = (x_2^2, -x_1\, x_2)^T$ and

$$\Phi_r(x + t\, d) = \Phi_r(x) + \Big(\big(2 + \frac{1}{r}\big)\, t^2 - t\Big)\, \langle d, d\rangle$$

hold. Determine the Φ_r-optimal step size for HAN's algorithm.

b) Generalize the results of *a)* to the case $x \in \mathbb{R}^2$ with $\|x\| \ge 1$ and also verify the relation

$$\|x + t\, d\|^2 - 1 = (1 - t)\big(\|x\|^2 - 1\big) + t^2\, \langle d, d\rangle\ .$$

c) Do numerical experiments with the *full step* and *damped* NEWTON *method.* Choose, for example, $x^{(0)} := (\varepsilon + \cos\vartheta, \sin\vartheta)^T$ with $\varepsilon \in \{0, 0.01\}$, $\vartheta \in \{0.05, \frac{\pi}{2}\}$ as starting points and observe for $r = 1/2$ the behavior of the step sizes and the quotients

$$\frac{\big\|x^{(k+1)} - x^*\big\|}{\big\|x^{(k)} - x^*\big\|^2} \quad \text{and} \quad \frac{\big\|x^{(k+1)} - x^*\big\|}{\big\|x^{(k)} - x^*\big\|}\ .$$

9. Carry out two steps of the Sℓ_1QP method with example 5 (cf. p. 226) Choose the values $r_1 := r_2 := 1$ for the penalty parameters, and $x^{(0)} := (0.5, 1)^T$ as the starting point.

10. Solve POWELL's test example (cf. chapter 2, exercise 15) by means of the function fmincon from the *Matlab*® Optimization Toolbox. Apply the SQP method using the starting point $x^{(0)} := (3, 0, 2, -1.5, 1.5, 5, 0)^T$. Visualize the result of each iteration step. Define the objective function as an inline function and the constraints by an m-file.

Hints:

a) `func = @(x) 1/2*x(1)*x(3);`

b)
```
function [c,ceq] = con(x)
ceq = [];     % no equality constraints
c(1) = ... ; % c has 9 component functions!
```

c)
```
options = optimset('LargeScale','off','OutputFcn',@outfun);
x = fmincon(func,x0,[],[],[],[],[],[],@con,options);
iteration
```

d) The actual triangle and circles can be visualized by the commands:

```
N = size(iteration,1);
psi = linspace(0,2*pi,200);
for k = 1:N
   clf, hold on, axis equal
   xx = iteration(k,:);
   patch([xx(1),xx(2),0],[0,xx(3),0],'y');
   plot(xx(4)+cos(psi),xx(5)+sin(psi),'-');
   plot(xx(6)+cos(psi),xx(7)+sin(psi),'-');
   hold off, waitforbuttonpress
end
```

6

Interior-Point Methods for Linear Optimization

The development of the last 30 years has been greatly influenced by the aftermath of a "scientific earthquake" which was triggered in 1979 by the findings of the Russian mathematician KHACHIYAN (1952–2005) and in 1984 by those of the Indian-born mathematician KARMARKAR. The *New York Times*, which profiled KHACHIYAN's achievement in a November 1979 article entitled "Soviet Mathematician Is Obscure No More," called him "the mystery author of a new mathematical theorem that has rocked the world of computer analysis."

At first it *only* affected linear optimization and the up to that time unchallenged dominance of the simplex method. This method was seriously questioned for the first time ever when in 1972 KLEE and MINTY found examples in which the simplex algorithm ran through *all* vertices of the feasible region. This confirmed that the 'worst case complexity' depended *exponentially*

W. Forst and D. Hoffmann, *Optimization—Theory and Practice*,
Springer Undergraduate Texts in Mathematics and Technology,
DOI 10.1007/978-0-387-78977-4_6,

on the dimension of the problem. Afterwards people began searching for LP-algorithms with *polynomial* complexity.

Based on SHOR's *ellipsoid method,* it was KHACHIYAN who found the first algorithm of this kind. When we speak of the 'ellipsoid method' today, we usually refer to the 'Russian algorithm' by KHACHIYAN. In many applications, however, it turned out to be less efficient than the simplex method. In 1984 KARMARKAR achieved the breakthrough when he *announced* a *polynomial* algorithm which he claimed to be fifty times faster than the simplex method. This announcement was a bit of an exaggeration, but it stimulated very fruitful research activities.

GILL, MURRAY, SAUNDERS and WRIGHT proved the equivalence between KARMARKAR's method and the classical *logarithmic barrier methods,* in particular when applied to linear optimization. Logarithmic barrier methods are methods which — unlike the exterior penalty methods from section 5.1 — solve restricted problems by transforming a penalty or barrier term into a parameterized family of unbounded optimization problems the minimizers of which lie *in the interior of the feasible region.* First approaches to this method date back to FRISCH (1954). In the sixties FIACCO and MCCORMICK devised from that the so-called *interior-point methods.* Their book [Fi/Co] contains a detailed description of classical barrier methods and is regarded as the standard reference work. A disadvantage was that the Hessians of the barrier functions were ill-conditioned in the approximative minimizers. This is usually seen as the cause for large rounding errors. This flaw was probably the reason why people lost interest in these methods. Now, due to the reawakened interest, the special problem structure was studied again and it was shown that the rounding errors are less problematic if the implementation is thorough enough. Efficient interior-point methods have in the meantime been applied to larger classes of nonlinear optimization problems and are still topics of current research.

6.1 Linear Optimization II

We consider once more a linear problem in standard form, the *primal problem*

$$(P) \qquad \begin{cases} \langle c, x \rangle \longrightarrow \min \\ Ax = b,\ x \geq 0\,, \end{cases}$$

where A is a real (m, n)-matrix with $m, n \in \mathbb{N}$ and $m \leq n$, $b \in \mathbb{R}^m$, $c \in \mathbb{R}^n$ and $x \in \mathbb{R}^n$. Here, A, b and c are given, and the vector x is an unknown variable. In the following we assume that the *matrix A has full rank*, that is, $\operatorname{rank}(A) = m$.

Associated with any linear problem is another linear problem, called the *dual problem*, which consists of the same data objects arranged in a different way. The dual problem to (P) is

$$(D) \quad \begin{cases} \langle b, y \rangle \longrightarrow \max \\ A^T y \le c \, , \end{cases}$$

where $y \in \mathbb{R}^m$. In section 2.4 we obtained the dual problem with the help of the LAGRANGE function.

After introducing *slack variables* $s \in \mathbb{R}^n_+$, (D) can be written equivalently as

$$(D_e) \quad \begin{cases} \langle b, y \rangle \longrightarrow \max \\ A^T y + s = c \\ s \ge 0 \, , \end{cases}$$

where the index e stands for *extended* problem. We know (compare chapter 2, exercise 18) that the primal and dual problem are symmetric, i. e., the dual problem of (D) is again the primal problem. We show this once more:

The LAGRANGE function $\widetilde{L}$ to problem (D) is given by

$$\widetilde{L}(y, x) = -\langle b, y \rangle + \langle x, A^T y - c \rangle = \langle -c, x \rangle + \langle Ax - b, y \rangle \, ,$$

where in this case $x \in \mathbb{R}^n_+$ is the dual variable:

$$\inf_{y \in \mathbb{R}^m} \widetilde{L}(y, x) = \inf_{y \in \mathbb{R}^m} \{ -\langle c, x \rangle + \langle Ax - b, y \rangle \} = \begin{cases} -\langle c, x \rangle \, , & \text{if } Ax = b \\ -\infty \, , & \text{else.} \end{cases}$$

So we get the problem

$$\begin{cases} -\langle c, x \rangle \longrightarrow \max \\ Ax = b, \; x \ge 0 \end{cases}$$

which is equivalent to problem (P).

The Duality Theorem

Definition

The set of *feasible points* of (P) is defined by

$$\mathcal{F}_P := \{ x \in \mathbb{R}^n \mid Ax = b, \; x \ge 0 \} \, ,$$

and the set of *feasible points* of (D) is given by

$$\mathcal{F}_D := \{ y \in \mathbb{R}^m \mid A^T y \le c \}$$

or alternatively

$$\mathcal{F}_{D_e} := \big\{(y,s) \in \mathbb{R}^m \times \mathbb{R}^n \mid A^T y + s = c, s \geq 0\big\} .$$

Analogously we define the set of *strictly feasible points* of (P) and (D) by

$$\mathcal{F}_P^0 := \{x \in \mathbb{R}^n \mid Ax = b,\ x > 0\} ,$$

$$\mathcal{F}_D^0 := \big\{y \in \mathbb{R}^m \mid A^T y < c\big\}$$

and

$$\mathcal{F}_{D_e}^0 := \big\{(y,s) \in \mathbb{R}^m \times \mathbb{R}^n \mid A^T y + s = c,\ s > 0\big\} ,$$

respectively.

Any vector $x \in \mathcal{F}_P$ is called *primally feasible*, a vector $y \in \mathcal{F}_D$ or a pair $(y,s) \in \mathcal{F}_{D_e}$ is called *dually feasible*. The two problems (P) and (D) together are referred to as a *primal–dual system*.

A vector $y \in \mathcal{F}_D$ determines an $s \geq 0$ such that $A^T y + s = c$, and so we get a pair $(y,s) \in \mathcal{F}_{D_e}$ and vice versa. This holds analogously for the set of strictly feasible points.

For vectors $x \in \mathcal{F}_P$ and $y \in \mathcal{F}_D$ it holds that

$$\langle c, x\rangle \geq \big\langle A^T y, x\big\rangle = \langle y, Ax\rangle = \langle b, y\rangle .$$

Hence we have the following **weak duality**

$$\langle c, x\rangle \geq \langle b, y\rangle .$$

Chapter 6

Corollary

If $x^* \in \mathcal{F}_P$ and $y^* \in \mathcal{F}_D$ with $\langle c, x^*\rangle = \langle b, y^*\rangle$, then x^* and y^* are optimizers for their respective problems.

Proof: By weak duality we get $\langle c, x\rangle \geq \langle b, y^*\rangle = \langle c, x^*\rangle$ for any vector $x \in \mathcal{F}_P$ and $\langle b, y\rangle \leq \langle c, x^*\rangle = \langle b, y^*\rangle$ for any vector $y \in \mathcal{F}_D$, respectively. □

Hence, any y that is feasible to (D) provides a lower bound $\langle b, y\rangle$ for the values of $\langle c, x\rangle$ whenever x is feasible to (P). Conversely, any x that is feasible to (P) provides an upper bound $\langle c, x\rangle$ for the values of $\langle b, y\rangle$ whenever y is feasible to (D). The difference between the optimal value

$$p^* := v(P) := \inf(P) := \inf\{\langle c, x\rangle \mid x \in \mathcal{F}_P\}$$

of the primal problem and the optimal value

$$d^* := v(D) := \sup(D) := \sup\{\langle b, y\rangle \mid y \in \mathcal{F}_D\}$$

of the dual problem is called the *duality gap*. If the duality gap vanishes, i. e. $p^* = d^*$, we say that *strong duality* holds. For $s := c - A^T y$ the value $\langle s, x \rangle$ just measures the difference between the objective values $\langle c, x \rangle$ of the primal problem and $\langle b, y \rangle$ of the dual problem.

The following theorem gives information about an *optimality condition* for problems (P) and (D):

Theorem 6.1.1 *(Optimality condition)*

The following statements are equivalent:

a) The primal problem (P) *has a minimizer.*

b) The dual problem (D_e) *has a maximizer.*

c) The following optimality conditions have a solution:

$$\begin{aligned} Ax &= b, \\ A^T y + s &= c, \\ x_i s_i &= 0 \quad (i = 1, \ldots, n), \\ x, s &\geq 0 \end{aligned} \tag{1}$$

If a) to c) hold, then a minimizer x^* *to* (P) *and a maximizer* (y^*, s^*) *to* (D) *yield a solution* (x^*, y^*, s^*) *of c) and vice versa.*

Proof: *a)* $\Longleftrightarrow$ *c)*: Problem (P) is in particular a convex problem with linear constraints, so we know that a vector $x^* \in \mathbb{R}^n$ is a minimizer to (P) if and only if there exist multipliers $y^* \in \mathbb{R}^m$ and $s^* \in \mathbb{R}^n_+$ such that the triple (x^*, y^*, s^*) satisfies the KKT conditions of the primal problem (P). The LAGRANGE function L to problem (P) is given by $L(x, y, s) = \langle c, x \rangle + \langle y, b - Ax \rangle + \langle s, -x \rangle$. It is obvious that the KKT conditions of (P) are exactly the conditions (1):

$$\begin{aligned} \nabla_x L(x^*, y^*, s^*) &= c - s^* - A^T y^* = 0 \\ x^*, s^* &\geq 0 \\ Ax^* &= b \\ \langle x^*, s^* \rangle &= 0 \end{aligned}$$

b) $\Longleftrightarrow$ *c)*: We show this analogously: A vector $y^* \in \mathbb{R}^m$ is a maximizer to the dual problem (D) if and only if there exists a multiplier $x^* \in \mathbb{R}^n_+$ such that (x^*, y^*) satisfies the KKT conditions of the dual problem. The LAGRANGE function $\widetilde{L}$ to problem (D) is given by $\widetilde{L}(y, x) = \langle -b, y \rangle + \langle x, A^T y - c \rangle$. Hence we see that the KKT conditions of (D) are exactly the conditions (1):

$$\begin{aligned} \nabla_y \widetilde{L}(y^*, x^*) &= Ax^* - b = 0 \\ x^* &\geq 0 \\ A^T y^* &\leq c \\ \langle x^*, \underbrace{c - A^T y^*}_{=: s^*} \rangle &= 0 \end{aligned}$$

□

Chapter 6

We are now able to state the *Duality Theorem for Linear Programming:*

Theorem 6.1.2 (*Duality Theorem*)

For the primal problem (P) and the dual problem (D) exactly one of the following four cases holds:

a) Both problems have optimizers x^* and y^* respectively and the optimal values are the same, i. e., $\langle c, x^* \rangle = \langle b, y^* \rangle$ (normal case).

b) The primal problem has no feasible point, the dual problem has a feasible point and $d^* = \infty$.

c) The dual problem has no feasible point, the primal problem has a feasible point and $p^* = -\infty$.

d) Neither the primal nor the dual problem has feasible points.

Proof: We want to show this by using the Theorem of the Alternative (Farkas). To use it, we set

$$\widetilde{A} := \begin{pmatrix} A & 0 & 0 & 0 & 0 \\ 0 & A^T & -A^T & I & 0 \\ c^T & -b^T & b^T & 0 & 1 \end{pmatrix}, \quad \widetilde{x} := \begin{pmatrix} x \\ y_1 \\ y_2 \\ s \\ \omega \end{pmatrix}, \quad \widetilde{b} := \begin{pmatrix} b \\ c \\ 0 \end{pmatrix}.$$

Chapter 6

The normal case is given if and only if there exists a nonnegative solution $\widetilde{x}$ to

$$\widetilde{A}\widetilde{x} = \widetilde{b}. \tag{2}$$

If we set $y := y_1 - y_2$, we have $x \in \mathcal{F}_P$, $(y, s) \in \mathcal{F}_{D_e}$ and $\langle c, x \rangle - \langle b, y \rangle \leq 0$, thus by weak duality $\langle c, x \rangle = \langle b, y \rangle$. Then, with the corollary from page 244, we have that y is a maximizer to the dual problem and x a minimizer to the primal problem. If (2) has *no* nonnegative solution, then we know by the Theorem of the Alternative (Farkas) that there exists a vector $\widetilde{y} := (-y, x, \varrho)^T \in \mathbb{R}^{m+n+1}$ such that $\widetilde{A}^T\widetilde{y} \geq 0$ and $\widetilde{b}^T\widetilde{y} < 0$. This just means

$$\begin{aligned} -A^T y + \varrho c &\geq 0 \\ Ax - \varrho b &= 0 \\ x &\geq 0 \\ \varrho &\geq 0 \end{aligned}$$

and $\langle c, x \rangle < \langle b, y \rangle$. Suppose that $\varrho > 0$. Then we obtain $A^T(\frac{1}{\varrho}y) \leq c$ and $A(\frac{1}{\varrho}x) = b$ in the equation above. So we know that $\frac{1}{\varrho}x \in \mathcal{F}_P$ and

$\frac{1}{\varrho}y \in \mathcal{F}_D$. Weak duality yields $\langle c, x\rangle \geq \langle b, y\rangle$. This, however, is a contradiction to $\langle c, x\rangle < \langle b, y\rangle$. Hence we can conclude that $\varrho = 0$. By this we get $\langle c, x\rangle < \langle b, y\rangle$, $Ax = 0$, $x \geq 0$ and $A^T y \leq 0$. Now we consider two cases:

1. $\langle c, x\rangle < 0$: Then $\mathcal{F}_D = \emptyset$: If we assume that there exists a vector $\overline{y} \in \mathcal{F}_D$, then we have $A^T\overline{y} \leq c$. This yields $0 = \langle x, A^T\overline{y}\rangle \leq \langle c, x\rangle < 0$, and therefore we have a contradiction. If $\mathcal{F}_P = \emptyset$ as well, then we have case *d)*. Else, for $\mathcal{F}_P \neq \emptyset$, there exists an $\overline{x}$ with $A\overline{x} = b$. Hence we get $A(\overline{x} + \lambda x) = b$ for $\lambda \in \mathbb{R}_+$ and $\lim_{\lambda\to\infty} \langle c, \overline{x} + \lambda x\rangle = -\infty$. Thus we have case *c)*.

2. $\langle c, x\rangle \geq 0$: As $\langle c, x\rangle < \langle b, y\rangle$, this yields $\langle b, y\rangle > 0$. Then we have $\mathcal{F}_P = \emptyset$: Since, if we assume that there exists an $\overline{x} \in \mathcal{F}_P$, we have that $A\overline{x} = b$. By this we get $0 \geq \langle A^T y, \overline{x}\rangle = \langle y, A\overline{x}\rangle = \langle y, b\rangle > 0$ and we have a contradiction. If $\mathcal{F}_D = \emptyset$ as well, then we have case *d)*. Otherwise, there exists a $\overline{y} \in \mathcal{F}_D$, that is, $A^T\overline{y} \leq c$. Hence, we get $A^T(\overline{y} + \lambda y) = A^T\overline{y} + \lambda(\underbrace{A^T y}_{\leq 0}) \leq c$ for $\lambda \in \mathbb{R}_+$ and $\lim_{\lambda\to\infty} \langle b, \overline{y} + \lambda y\rangle = \infty$. This means that we have case *b)*.

As the four cases are mutually exclusive, exactly one of them occurs. □

Remark

In particular, by theorem 6.1.2 we know that *if both problems have feasible points $x \in \mathcal{F}_P$ and $y \in \mathcal{F}_D$, then both problems have optimizers x^* and y^* with $\langle c, x^*\rangle = \langle b, y^*\rangle$.*

Example 1

In the following we will give an example for each of the four cases of theorem 6.1.2:

a) Normal case: For $c := (3,4)^T$, $A := (1\ \ 2)$ and $b := 1$ we have $x = (0, \frac{1}{2})^T \in \mathcal{F}_P$ and $y = 2 \in \mathcal{F}_D$ with $\langle c, x\rangle = 2 = \langle b, y\rangle$.

b) For $c := (3,4)^T$, $A := (-1\ \ -2)$ and $b := 1$, the primal problem has no feasible point, and the dual has the optimal value $d^* = \infty$.

c) For $c := (-1,-1,-1)^T$, $A := \begin{pmatrix} 1 & 0 & -1 \\ 0 & 1 & 0 \end{pmatrix}$ and $b := (1,1)^T$, the dual problem has no feasible point and $p^* = -\infty$.

d) We consider A and c as in *c)*, set $b := (1,-1)^T$ and get: Both problems have no feasible points. ◁

From theorem 6.1.1 we can derive the *complementarity condition*:

Lemma 6.1.3 *(Complementary slackness optimality condition)*
If x is feasible to (P) and (y, s) is feasible to (D_e), then x is a minimizer for (P) and (y, s) is a maximizer for (D_e) if and only if

$$x_i \, s_i = 0 \quad \text{for } i = 1, \ldots, n.$$

Proof: If x is a minimizer of (P) and (y, s) is a maximizer of (D_e), we get $x_i \, s_i = 0$ by (1). If x is feasible to (P), (y, s) is feasible to (D_e) and $x_i \, s_i = 0$, then, by theorem 6.1.1, x is a minimizer of (P) and (y, s) is a maximizer of (D_e). □

To derive primal–dual interior-point methods later on, we rearrange the optimality condition (1) in the following way

$$F_0(x, y, s) := \begin{pmatrix} A^T y + s - c \\ A x - b \\ X s \end{pmatrix} = \begin{pmatrix} 0 \\ 0 \\ 0 \end{pmatrix}, \quad x \geq 0, \, s \geq 0 \, . \tag{3}$$

An uppercase letter denotes the diagonal matrix corresponding to the respective vectors, for example

$$X := \operatorname{Diag}(x) := \begin{pmatrix} x_1 & & & \\ & x_2 & & \\ & & \ddots & \\ & & & x_n \end{pmatrix}.$$

Hence we have $X e_\nu = x_\nu e_\nu$ for $\nu = 1, \ldots, n$ and $X e = x$ where $e := (1, \ldots, 1)^T$. In the following the norm $\| \;\|$ denotes the euclidean norm $\| \;\|_2$. We note that $X S e = X s = (x_1 s_1, \ldots, x_n s_n)^T =: x s$ (Hadamard product).

Chapter 6

The Interior-Point Condition

Definition

We say that the primal–dual system satisfies the *interior-point condition (IPC)* iff both problems (P) and (D) have strictly feasible points, that is, $\mathcal{F}_P^0 \neq \emptyset$ and $\mathcal{F}_D^0 \neq \emptyset$.

We are now going to introduce the *logarithmic barrier function* for the primal problem (P). This is the function Φ_μ defined by

$$\Phi_\mu(x) := \langle c, x \rangle - \mu \sum_{i=1}^{n} \log(x_i) \, ,$$

where $\mu > 0$ and x runs through all primally feasible vectors that are positive. The domain of Φ_μ is the set $\mathcal{F}_P^0$. μ is called the *barrier parameter*. The gradient of Φ_μ is given by

$$\nabla \Phi_\mu(x) = c - \mu X^{-1} e$$

and the Hessian by

$$\nabla^2 \Phi_\mu(x) = \mu X^{-2}.$$

Obviously, the Hessian is positive definite for any $x \in \mathcal{F}_P^0$. This means that Φ_μ is *strictly convex* on $\mathcal{F}_P^0$.

Analogously, the *logarithmic barrier function* for the dual problem (D) is given by

$$\widetilde{\Phi}_\mu(y) := \langle b, y \rangle + \mu \sum_{i=1}^n \log\left(c_i - \langle a_i, y \rangle\right),$$

where $c - A^T y > 0$, and $a_1, \dots, a_n$ denote the columns of the matrix A. The domain of $\widetilde{\Phi}_\mu$ is the set $\mathcal{F}_D^0$. The gradient of $\widetilde{\Phi}_\mu$ is given by

$$\nabla \widetilde{\Phi}_\mu(y) = b - \mu \sum_{i=1}^n \frac{a_i}{c_i - \langle a_i, y \rangle}$$

and the Hessian by

$$\nabla^2 \widetilde{\Phi}_\mu(y) = -\mu \sum_{i=1}^n \frac{a_i a_i^T}{(c_i - \langle a_i, y \rangle)^2}.$$

The Hessian is negative definite for any $y \in \mathcal{F}_D^0$: The assumption $\operatorname{rank}(A) = m$ gives $\langle a_i, z \rangle \neq 0$ for at least one i for any $z \neq 0$. Thus, for any $z \neq 0$ we have

$$\left\langle z, \nabla^2 \widetilde{\Phi}_\mu(y)\, z \right\rangle = -\mu \sum_{i=1}^n \frac{\langle a_i, z \rangle^2}{(c_i - \langle a_i, y \rangle)^2} < 0.$$

This means that $\widetilde{\Phi}_\mu$ is *strictly concave* on $\mathcal{F}_D^0$.

Theorem 6.1.4

Let $\mu > 0$. Then the following statements are equivalent:

a) Both $\mathcal{F}_P^0$ and $\mathcal{F}_D^0$ are not empty.

b) There exists a (unique) minimizer to Φ_μ on $\mathcal{F}_P^0$.

c) There exists a (unique) maximizer to $\widetilde{\Phi}_\mu$ on $\mathcal{F}_D^0$.

d) The system

$$F_\mu(x, y, s) := \begin{pmatrix} A^T y + s - c \\ Ax - b \\ Xs - \mu e \end{pmatrix} = \begin{pmatrix} 0 \\ 0 \\ 0 \end{pmatrix}, \quad x \geq 0,\, s \geq 0 \qquad (4)$$

has a (unique) solution.

If a) to d) hold, then the minimizer $x(\mu)$ to Φ_μ and the maximizer $y(\mu)$ to $\widetilde{\Phi}_\mu$ yield the solution $\big(x(\mu), y(\mu), s(\mu)\big)$ of (4), where $s(\mu) := c - A^T y(\mu)$.

Proof:

b) $\Longleftrightarrow$ *d)*: The definition of Φ_μ can be extended to the open set $\mathbb{R}^n_{++}$ and Φ_μ is differentiable there. On $\mathbb{R}^n_{++}$ we consider the system

$$(P_\mu) \quad \begin{cases} \Phi_\mu(x) \longrightarrow \min \\ Ax = b\,. \end{cases}$$

(P_μ) is a convex problem with linear constraints. So we know that a vector x is a (unique) minimizer to (P_μ) iff there exists a (unique) multiplier $y \in \mathbb{R}^m$ such that (x, y) satisfies the KKT conditions of problem (P_μ). The LAGRANGE function L to this problem is given by $L(x, y) = \Phi_\mu(x) + \langle y, b - Ax\rangle$. The KKT conditions to (P_μ) are

$$\begin{aligned} \nabla_x L(x, y) = c - \mu X^{-1}e - A^T y &= 0 \\ Ax = b. \end{aligned}$$

If we set $s := \mu X^{-1}e$ which is equivalent to $Xs = \mu e$, we get nothing else but (4). We have therefore shown that system (4) represents the optimality conditions of problem (P_μ).

c) $\Longleftrightarrow$ *d)*: $\widetilde{\Phi}_\mu$ has a (unique) maximizer $\widetilde{y}$ on $\mathcal{F}_D^0$ if and only if

$$0 = \nabla\widetilde{\Phi}_\mu(\widetilde{y}) = b - \mu\sum_{i=1}^{n} \frac{a_i}{c_i - \langle a_i, \widetilde{y}\rangle}\,. \tag{5}$$

For such a maximizer $\widetilde{y}$ we define $\widetilde{x}, \widetilde{s} \in \mathbb{R}^n$ to get a solution $(\widetilde{x}, \widetilde{y}, \widetilde{s})$ to system (4):

$$\widetilde{x}_i := \frac{\mu}{c_i - \langle a_i, \widetilde{y}\rangle} > 0, \quad \widetilde{s}_i := c_i - \langle a_i, \widetilde{y}\rangle > 0\,.$$

By this construction we have $A^T\widetilde{y} + \widetilde{s} - c = 0$ and in addition

$$A\widetilde{x} = \sum_{i=1}^{n} \widetilde{x}_i a_i = \mu\sum_{i=1}^{n} \frac{a_i}{c_i - \langle a_i, \widetilde{y}\rangle} \overset{(5)}{=} b\,.$$

Obviously, we have $\widetilde{x}_i\widetilde{s}_i = \mu$ for $i = 1, \ldots, n$. Hence we have obtained a solution to (4). For a solution (x, y, s) to (4) we have $A^T y + s - c = 0$. This means $s_i = c_i - \langle a_i, y\rangle$, thus $x_i = \mu s_i^{-1} = \mu(c_i - \langle a_i, y\rangle)^{-1}$ for $i = 1, \ldots, n$. Besides

$$0 = b - Ax = b - \sum_{i=1}^{n} x_i a_i = b - \mu\sum_{i=1}^{n} \frac{a_i}{c_i - \langle a_i, y\rangle} = \nabla\widetilde{\Phi}_\mu(y)$$

shows that y is the (unique) maximizer to $\widetilde{\Phi}_\mu$ on $\mathcal{F}_D^0$. Therefore we have shown that system (4) represents the optimality conditions for the problem

$$(D_\mu) \quad \begin{cases} \widetilde{\Phi}_\mu(y) \longrightarrow \max \\ A^T y + s = c, \; s > 0\,. \end{cases}$$

d) $\Longrightarrow$ *a)*: If system (4) has a solution (x, y, s), then we have $x \in \mathcal{F}_P^0$ and $y \in \mathcal{F}_D^0$.

a) $\Longrightarrow$ *b)*: Assuming *a)*, there exist vectors $x^0 \in \mathcal{F}_P^0$, $y^0 \in \mathcal{F}_D^0$ and hence $(y^0, s^0) \in \mathcal{F}_{D_e}^0$ for a suitable $s^0 > 0$. We define the level set $\mathcal{L}_K$ of Φ_μ by

$$\mathcal{L}_K := \left\{ x \in \mathcal{F}_P^0 \mid \Phi_\mu(x) \le K \right\}$$

with $K := \Phi_\mu(x^0)$. As we have $x^0 \in \mathcal{F}_P^0$, the set $\mathcal{L}_K$ is not empty. Since Φ_μ is continuous on its domain, it suffices to show that $\mathcal{L}_K$ is compact, because then Φ_μ has a minimizer and, since Φ_μ is strictly convex, this minimizer is unique. Obviously, $\mathcal{L}_K$ is closed in $\mathcal{F}_P^0$. Hence *it remains to show that* $\mathcal{L}_K$ *is bounded:* Let $x \in \mathcal{L}_K$. We have $\langle c, x \rangle - \langle b, y^0 \rangle = \langle x, s^0 \rangle$ and we thus get

$$\Phi_\mu(x) = \langle b, y^0 \rangle + \langle x, s^0 \rangle - \mu \sum_{i=1}^{n} \log(x_i)\,.$$

With $s := s^0/\mu$ the inequality $\Phi_\mu(x) \le K$ yields the boundedness of

$$h(x) := \langle x, s \rangle - \sum_{i=1}^{n} \log(x_i) = \sum_{i=1}^{n} \left(x_i s_i - \log(x_i) \right)$$

on $\mathcal{L}_K$. If $\mathcal{L}_K$ was unbounded, there would exist a sequence $\left(x^{(\kappa)}\right)$ in $\mathcal{L}_K$ with $\left\|x^{(\kappa)}\right\| \to \infty$ for $\kappa \to \infty$. By choosing a subsequence, we can obtain that for a $k \in \{1, \ldots, n\}$ WLOG exactly the sequences $\left(x_1^{(\kappa)}\right)$ to $\left(x_k^{(\kappa)}\right)$ will definitely diverge to ∞ or converge to 0, while the other sequences have a positive limit. In $\sum_{i=1}^{n} \left(x_i s_i - \log(x_i) \right)$ the first k summands will definitely diverge to ∞, while the others will converge and thus be bounded. Altogether — with $h\left(x^{(\kappa)}\right) \longrightarrow \infty$ for $\kappa \to \infty$ — we obtain a *contradiction.*

Graph of the function $x - \log(x)$

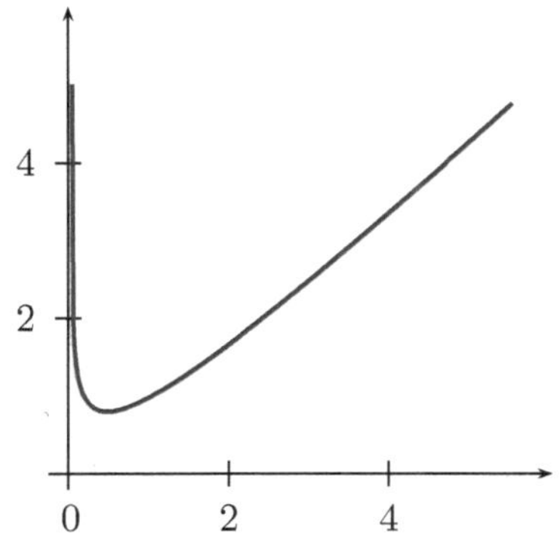

□

Chapter 6

Remark

As we have $x_i\, s_i = \mu > 0$ for $i = 1, \ldots, n$, we know that every solution to (4) satisfies $x > 0$ and $s > 0$. Thus, the second equation in (4) means nothing else but that x is primally feasible and the first that (y, s) is dually feasible. The third equation is called the *centering condition.*

The interior-point condition is not always fulfilled. We can easily see this with an example:

Example 2

For $A := (1\ 0\ 1)$, $b := 0$ and $c := (1, 0, 0)^T$ there exists no vector $x \in \mathcal{F}_P^0$ and no vector $y \in \mathcal{F}_D^0$: $Ax = b$ yields $x_1 = x_3 = 0$ and by $A^T y = (y, 0, y)^T$ we see that there exists no $y \in \mathbb{R}$ with $A^T y < c$. Hence the interior-point condition is not fulfilled.

In the following we will make the additional *assumption*:

> *Both problems (P) and (D) have strictly feasible points, that is,* $\mathcal{F}_P^0 \neq \emptyset$ *and* $\mathcal{F}_D^0 \neq \emptyset$.

6.2 The Central Path

We want to solve system (3) by means of NEWTON's method

$$\begin{pmatrix} x^{(k+1)} \\ y^{(k+1)} \\ s^{(k+1)} \end{pmatrix} = \begin{pmatrix} x^{(k)} \\ y^{(k)} \\ s^{(k)} \end{pmatrix} - DF_0\big(x^{(k)}, y^{(k)}, s^{(k)}\big)^{-1} \cdot F_0\big(x^{(k)}, y^{(k)}, s^{(k)}\big)\,,$$

where DF_0 is the Jacobian of F_0 given by

$$DF_0\,(x, y, s) = \begin{pmatrix} 0 & A^T & I \\ A & 0 & 0 \\ S & 0 & X \end{pmatrix}.$$

The nonlinear system (3) consists of $m+2n$ equations with $m+2n$ variables.

Any solution to (3) satisfies $x_i\, s_i = 0 \ \ (i = 1, \ldots, n)$ and therefore lies on the boundary of the *primal–dual feasible set*

$$\mathcal{F} := \big\{(x, y, s) \in \mathbb{R}_+^n \times \mathbb{R}^m \times \mathbb{R}_+^n \mid Ax = b,\ A^T y + s = c\big\}.$$

We define the set of *strictly primal–dual feasible points* as

$$\mathcal{F}^0 := \big\{(x, y, s) \in \mathbb{R}_{++}^n \times \mathbb{R}^m \times \mathbb{R}_{++}^n \mid Ax = b,\ A^T y + s = c\big\}.$$

Chapter 6

Theorem 6.2.1

The Jacobian

$$DF_0\,(x,y,s) \;=\; \begin{pmatrix} 0 & A^T & I \\ A & 0 & 0 \\ S & 0 & X \end{pmatrix}$$

is nonsingular if $x > 0$ *and* $s > 0$.

Proof: For $(u,v,w) \in \mathbb{R}^n \times \mathbb{R}^m \times \mathbb{R}^n$ with

$$\begin{pmatrix} 0 \\ 0 \\ 0 \end{pmatrix} = DF_0(x,y,s)\begin{pmatrix} u \\ v \\ w \end{pmatrix} = \begin{pmatrix} A^T v + w \\ Au \\ Su + Xw \end{pmatrix}$$

we have

$$\langle u, w\rangle = \big\langle u, -A^T v\big\rangle = -\langle Au, v\rangle = 0\,.$$

From $Su + Xw = 0$ we get $u = -S^{-1}Xw$ and, with the equality above, we obtain

$$0 = \langle u, w\rangle = \langle w, u\rangle = -\big\langle w, S^{-1}Xw\big\rangle\,.$$

Since $x > 0$ and $s > 0$, the matrix $S^{-1}X$ is positive definite and we can conclude that $w = 0$. From $0 = Su + Xw = Su$ we obtain $u = 0$ and from $0 = A^T v + w = A^T v$ we get $v = 0$ since $\operatorname{rank}(A) = m$. □

As mentioned above, the solution to system (3) lies on the boundary of the set $\mathcal{F}$ and thus it is *not* guaranteed that the Jacobian is invertible. We therefore consider the 'relaxed nonlinear system' (4). As we have $x_i s_i = \mu > 0$ for $i = 1,\dots,n$, we know that every solution to (4) satisfies $x > 0$ and $s > 0$. Now it is our aim to find a solution to system (4) by applying NEWTON's method. Theorem 6.1.4 gives that *the interior-point condition is independent of the barrier parameter* μ. Hence we can conclude that if a minimizer to Φ_μ or a maximizer to $\widetilde{\Phi}_\mu$ exists for some positive μ, then it exists for all positive μ. These solutions to system (4) are denoted as

$$\mathfrak{x}(\mu) = (x(\mu), y(\mu), s(\mu))\,.$$

Definition

The set

$$\mathcal{C} := \{\mathfrak{x}(\mu) \mid \mu > 0\}$$

of solutions to (4) is called the (primal–dual) *central path*.

The central path $\mathcal{C}$ is an 'arc' of strictly feasible points that is parameterized by a scalar $\mu > 0$, and each point $\mathfrak{x}(\mu)$ of $\mathcal{C}$ solves system (4). These conditions differ from the KKT conditions only in the term μ. Instead of the

Chapter 6

complementarity condition $x_i s_i = 0$ we demand that the products $x_i s_i$ have the same value for all indices i. By theorem 6.1.4 we know that the central path is well-defined. We also know that the equations (4) approximate (3) more and more closely as μ goes to zero. This lets us hope that $\big(\mathfrak{x}(\mu)\big)$ converges to an optimizer of (P) and (D) *if* $\big(\mathfrak{x}(\mu)\big)$ converges to any point for $\mu \downarrow 0$. In this case, the central path guides us to a minimizer or maximizer along a well-defined route. For $\mu > 0$ we have

$$n\,\mu = \sum_{i=1}^{n} \underbrace{x_i(\mu)\,s_i(\mu)}_{=\mu} = \langle x(\mu), s(\mu)\rangle = \langle c, x(\mu)\rangle - \langle b, y(\mu)\rangle .$$

This gives an idea of how far away $x(\mu)$ and $y(\mu)$ are from being optimal. For $\mu \downarrow 0$ we see that $\langle c, x(\mu)\rangle$ and $\langle b, y(\mu)\rangle$ strive against each other. With the *normalized duality gap*

$$\mu = \big(\langle c, x(\mu)\rangle - \langle b, y(\mu)\rangle\big)/n$$

we can see how close $\langle c, x(\mu)\rangle$ and $\langle b, y(\mu)\rangle$ are.

Example 3

We consider $A := (1\ 1\ 1)$, $b := 1$ and $c := (-1,-3,-4)^T$. Thus we get the primal problem

$$(P) \quad \begin{cases} -x_1 - 3x_2 - 4x_3 \longrightarrow \min \\ x_1 + x_2 + x_3 = 1,\ x \geq 0 \end{cases}$$

and the dual problem

$$(D_e) \quad \begin{cases} y & \longrightarrow \max \\ y + s_1 & = \quad -1 \\ y + s_2 & = \quad -3 \\ y + s_3 & = \quad -4 \\ s & \geq \quad 0\,. \end{cases}$$

Hence system (4) becomes

$$\begin{aligned} x_1 + x_2 + x_3 &= 1 \\ y + s_1 + 1 &= 0 \\ y + s_2 + 3 &= 0 \\ y + s_3 + 4 &= 0 \\ x_i\, s_i &= \mu \end{aligned}$$

where $x_i > 0$ and $s_i > 0$ for $i = 1, 2, 3$. This gives $s_1 = -1 - y$, $s_2 = -3 - y$, $s_3 = -4 - y$. $s > 0$ shows that $y < -4$. From $x_1 s_1 = x_2 s_2 = x_3 s_3 = \mu$ we obtain

$$\frac{1}{\mu} = \frac{x_1 + x_2 + x_3}{\mu} = \frac{1}{s_1} + \frac{1}{s_2} + \frac{1}{s_3} = -\left(\frac{1}{1+y} + \frac{1}{3+y} + \frac{1}{4+y}\right).$$

Multiplication with the common denominator gives a cubic equation from which $\widetilde{y}(\mu)$, with $\widetilde{y}(\mu) < -4$, can be obtained relatively easily. Then we can calculate $s_i(\mu)$ by $s_1(\mu) = -1 - \widetilde{y}(\mu)$, $s_2(\mu) = -3 - \widetilde{y}(\mu)$, $s_3(\mu) = -4 - \widetilde{y}(\mu)$, and with that evaluate $x_i(\mu)$ by $x_i(\mu) = \mu/s_i(\mu)$.

The figure below shows the central path. The triangular region is the feasible set of the primal problem. We see that the curve lies inside the feasible set and approaches the minimizer $(0,0,1)^T$ of problem (P).

Trajectory of the central path for (P)

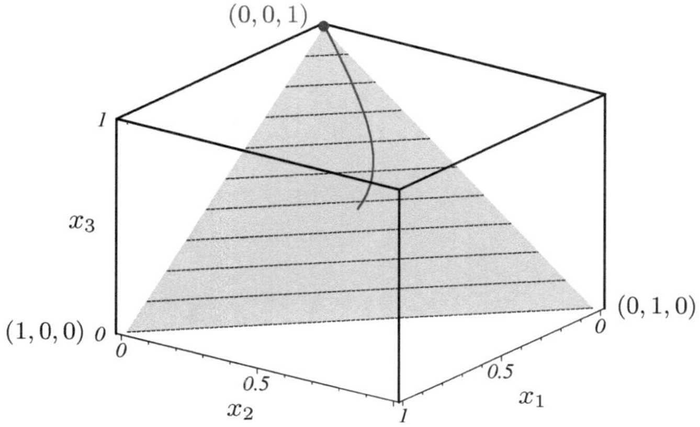

◁

If we change the objective function minimally by setting $c := (-1,-3,-3)^T$, then every point on the line connecting $(0,1,0)^T$ and $(0,0,1)^T$ is a minimizer. For $\mu \to 0$ the central path tends to the barycenter $(0,1/2,1/2)^T$ of this line. Here the starting point of the central path (i. e., $\mu \longrightarrow \infty$) is the barycenter $(1/3,1/3,1/3)^T$ of the feasible region.

Modified Objective Function

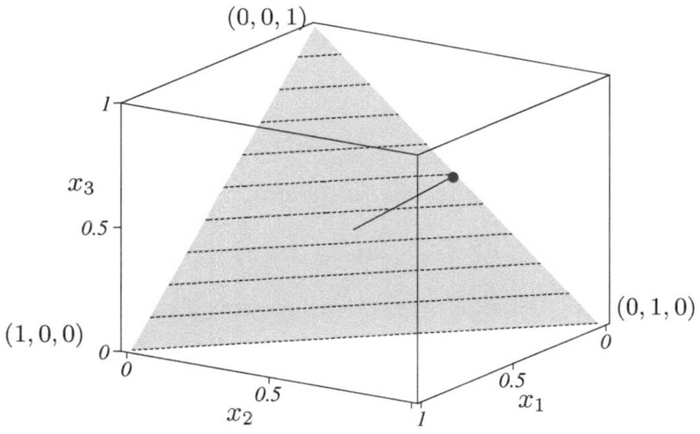

Chapter 6

Convergence of the Central Path

The central path is a 'curve' in the interior of the feasible region. It begins 'somewhere in its middle' — more precisely in its 'analytic center' — and ends 'somewhere in the middle of the optimal solutions' as $\mu \longrightarrow 0$. We would now like to examine this in more detail.

We therefore consider the sets of the respective optimizers

$$\mathcal{F}_P^{\text{opt}} := \{x \in \mathcal{F}_P \mid \langle c, x\rangle = p^*\} \text{ and } \mathcal{F}_D^{\text{opt}} := \{y \in \mathcal{F}_D \mid \langle b, y\rangle = d^*\}$$

and obtain as the main result:

Theorem 6.2.2

For $\mu > 0$ let $x(\mu)$ denote the unique minimizer of Φ_μ on $\mathcal{F}_P^0$ and $y(\mu)$ the unique maximizer of $\widetilde{\Phi}_\mu$ on $\mathcal{F}_D^0$. Then there exist the limits

$$x^* := \lim_{\mu \to 0+} x(\mu) \quad \textit{and} \quad y^* := \lim_{\mu \to 0+} y(\mu).$$

For them it holds that $x^* \in \mathcal{F}_P^{\text{opt}}$ and $y^* \in \mathcal{F}_D^{\text{opt}}$.

Before doing the *proof*, we will give some general remarks.

We had

$$\Phi_\mu(x) := \langle c, x\rangle - \mu \sum_{i=1}^n \log(x_i)$$

for $\mu > 0$ and $x \in \mathcal{F}_P^0$, hence

$$\Phi_\mu(x) = f(x) + \mu\, h(x)$$

with

$$f(x) := \langle c, x\rangle \quad \text{and} \quad h(x) := -\sum_{i=1}^n \log(x_i) \stackrel{\checkmark}{=} -\log\det(X).$$

Chapter 6

Lemma 6.2.3

For $0 < \mu < \lambda$ the following inequalities hold:

$$h\big(x(\lambda)\big) \le h\big(x(\mu)\big) \quad \textit{and} \quad f\big(x(\mu)\big) \le f\big(x(\lambda)\big).$$

The function $h \circ x$ is hence antitone, the function $f \circ x$ isotone.

Proof: $\Phi_\mu\big(x(\mu)\big) \le \Phi_\mu\big(x(\lambda)\big)$ and $\Phi_\lambda\big(x(\lambda)\big) \le \Phi_\lambda\big(x(\mu)\big)$ mean

$$f\big(x(\mu)\big) + \mu\, h\big(x(\mu)\big) \le f\big(x(\lambda)\big) + \mu\, h\big(x(\lambda)\big) \tag{6}$$

$$f\big(x(\lambda)\big) + \lambda\, h\big(x(\lambda)\big) \le f\big(x(\mu)\big) + \lambda\, h\big(x(\mu)\big).$$

Summation of the two equations gives

$$0 \geq (\lambda - \mu)\big(h(x(\lambda)) - h(x(\mu))\big), \text{ hence } h(x(\lambda)) \leq h(x(\mu)) \,.$$

Together with (6) this yields $f(x(\mu)) \leq f(x(\lambda))$. □

Lemma 6.2.4

For $\alpha \in \mathbb{R}$ the level sets

$$\{x \in \mathcal{F}_P \mid \langle c, x\rangle \leq \alpha\} \text{ and } \{y \in \mathcal{F}_D \mid \langle b, y\rangle \geq \alpha\}$$

are bounded.

Proof: For (P): *Otherwise* there exists a sequence $(x^{(k)})$ in $\mathcal{F}_P$ such that $\langle c, x^{(k)}\rangle \leq \alpha$ for $k \in \mathbb{N}$ and $0 < \|x^{(k)}\| \longrightarrow \infty$ $(k \longrightarrow \infty)$. We consider the sequence of the corresponding normalized vectors $p_k := x^{(k)}/\|x^{(k)}\|$. For them we have WLOG (otherwise choose a suitable subsequence!) $p_k \longrightarrow p$ for a vector $p \in \mathbb{R}^n$ with $\|p\| = 1$. We gain $p \geq 0, Ap = 0$ and $\langle c, p\rangle \leq 0$. With a $y^{(0)} \in \mathcal{F}_D^0$ the *contradiction* follows:

$$0 \overset{\checkmark}{<} \left\langle p, c - A^T y^{(0)} \right\rangle = \langle c, p\rangle - \left\langle Ap, y^{(0)} \right\rangle \leq 0$$

For (D): *Otherwise* there exists a sequence $(y^{(k)})$ in $\mathcal{F}_D$ such that $\langle b, y^{(k)}\rangle \geq \alpha$ for $k \in \mathbb{N}$ and $0 < \|y^{(k)}\| \longrightarrow \infty$ $(k \longrightarrow \infty)$. For the sequence of the corresponding normalized vectors $q_k := y^{(k)}/\|y^{(k)}\|$ we have WLOG $q_k \longrightarrow q$ for a $q \in \mathbb{R}^m$ with $\|q\| = 1$. Here we get $A^T q \leq 0$, $\langle b, q\rangle \geq 0$ and $A^T q \neq 0$, as $q \neq 0$ and $\operatorname{rank}(A) = m$. With an $x^{(0)} \in \mathcal{F}_P^0$ we obtain the *contradiction*

$$0 \overset{\checkmark}{>} \left\langle x^{(0)}, A^T q \right\rangle = \left\langle Ax^{(0)}, q \right\rangle = \langle b, q\rangle \geq 0. \qquad \square$$

Corollary

The sets of the optimal points $\mathcal{F}_P^{\text{opt}}$ and $\mathcal{F}_D^{\text{opt}}$ are nonempty, convex and compact.

Proof: By assumption (IPC) and the duality theorem the sets are nonempty. The convexity and closedness are immediately apparent. With that the compactness follows by the above lemma. □

We will now turn to the *proof of theorem 6.2.2* which will be divided into three parts:

(I) If (μ_k) is a null sequence in $\mathbb{R}_{++}$, the sequence $\big((x(\mu_k), y(\mu_k), s(\mu_k))\big)$ is bounded and each of its accumulation points $(\widetilde{x}, \widetilde{y}, \widetilde{s})$ gives a minimizer $\widetilde{x}$ of (P) and a maximizer $\widetilde{y}$ of (D).

Proof of (I): With an upper bound $\mu > 0$ for the sequence (μ_k) we have — by lemma 6.2.3 — $\langle c, x(\mu_k)\rangle \le \langle c, x(\mu)\rangle$ for $k \in \mathbb{N}$. In the same way it follows that $\langle b, y(\mu_k)\rangle \ge \langle b, y(\mu)\rangle$ for $k \in \mathbb{N}$. With lemma 6.2.4 we thus know that the sequences $\big(x(\mu_k)\big)$ and $\big(y(\mu_k)\big)$ are bounded and therefore the sequence $\big(s(\mu_k)\big)$ as well. From the central path conditions we thus get the following for each accumulation point $(\widetilde{x}, \widetilde{y}, \widetilde{s})$ by taking the limit of a corresponding subsequence:

$$\begin{aligned} A\widetilde{x} &= b, \quad \widetilde{x} \ge 0 \\ A^T\widetilde{y} + \widetilde{s} &= c, \quad \widetilde{s} \ge 0 \\ \langle \widetilde{x}, \widetilde{s}\rangle &= 0 \end{aligned}$$

With theorem 6.1.1 this gives the desired result. □

(II) With $\mathcal{B} := \big\{i \in \{1, \ldots, n\} \mid \exists x \in \mathcal{F}_P^{\mathrm{opt}}\ x_i > 0\big\}$ *it holds that:*

a) The barrier problem (B) $\begin{cases} \psi(x) := -\sum_{j\in\mathcal{B}} \log x_j \longrightarrow \min \\ x \in \mathcal{F}_P^{\mathrm{opt}},\ x_{\mathcal{B}} > 0 \end{cases}$

has a unique minimizer x^*.

b) $\lim_{\mu \to 0+} x(\mu) = x^*$

Proof of (II):

a) If $\mathcal{B} = \emptyset$, then $\mathcal{F}_P^{\mathrm{opt}} = \{0\}$ holds, and the assertions are trivially true.[1] Let therefore $\mathcal{B} \neq \emptyset$: For each $j \in \mathcal{B}$ there exists an $x^{(j)} \in \mathcal{F}_P^{\mathrm{opt}}$ with $x_j^{(j)} > 0$. Then

$$\widehat{x} := \frac{1}{|\mathcal{B}|} \sum_{j\in\mathcal{B}} x^{(j)}$$

is in $\mathcal{F}_P^{\mathrm{opt}}$ with $\widehat{x}_{\mathcal{B}} > 0$. The *existence* of a minimizer x^* of (B) follows once again from the assertion that level sets of (B) are compact. (Compare our considerations on p. 250f with $c := 0$, $\mu := 1$ and the index set $\mathcal{B}$ instead of $\{1, \ldots, n\}$.) The *uniqueness* follows directly from the strict convexity of the objective function ψ.

b) Let now (μ_k) be a null sequence in $\mathbb{R}_{++}$ and $\widetilde{x}$ an accumulation point of the corresponding sequence $\big(x(\mu_k)\big)$. By *(I)* it then holds that $\widetilde{x} \in \mathcal{F}_P^{\mathrm{opt}}$. WLOG $x(\mu_k) \longrightarrow \widetilde{x}$. We will show that $\widetilde{x}$ minimizes problem (B). $\widetilde{x} = x^*$ then follows from the already proven uniqueness of the minimizer which completes the proof.

i) $\widetilde{x}$ is feasible for (B): The KKT conditions to (P_μ) (cf. p. 250) yield the following relation for $\mu > 0$:

$$c - \mu X(\mu)^{-1} e - A^T y(\mu) = 0$$

[1] By convention we have $\sum_{\emptyset} \log x_j = 0$.

Chapter 6

For $x \in \mathcal{F}_P^{\text{opt}}$ it then holds that

$$0 = \langle y(\mu), A(x - x(\mu))\rangle = \big\langle A^T y(\mu), x - x(\mu)\big\rangle = \big\langle c - \mu X(\mu)^{-1}e, x - x(\mu)\big\rangle$$
$$= \langle c, x\rangle - \langle c, x(\mu)\rangle + \mu n - \mu \sum_{j\in\mathcal{B}} \tfrac{x_j}{x(\mu)_j}\,.$$

Since $\langle c, x\rangle \le \langle c, x(\mu)\rangle$, we then have $\sum_{j\in\mathcal{B}} \frac{x_j}{x(\mu)_j} \le n$ and thus $x(\mu)_j \ge \frac{1}{n} x_j$ for $j \in \mathcal{B}$. Setting $x := x^* \in \mathcal{F}_P^{\text{opt}}$ and $\mu := \mu_k$ gives the estimate $\widetilde{x}_j \ge \frac{1}{n} x_j^* > 0$ for $j \in \mathcal{B}$ as $k \longrightarrow \infty$.

ii) We will finally show that $\widetilde{x}$ minimizes problem (B):
For $\mu > 0$ and $x \in \mathcal{F}_P^{\text{opt}}$ it holds that (cf. *i)*)

$$\frac{1}{\mu}\langle c, x - x(\mu)\rangle \;=\; \big\langle X(\mu)^{-1}e, x - x(\mu)\big\rangle\,.$$

If we choose $x := x^*$ and $x := \widetilde{x}$, we get the following via subtraction and with the fact that $\langle c, x^*\rangle = \langle c, \widetilde{x}\rangle = p^*$

$$0 \;=\; \frac{1}{\mu}\langle c, x^* - \widetilde{x}\rangle \;=\; \big\langle X(\mu)^{-1}e, x^* - \widetilde{x}\big\rangle \;=\; \sum_{j\in\mathcal{B}} \frac{x_j^* - \widetilde{x}_j}{x(\mu)_j}\,.$$

With $\mu := \mu_k$ and $k \longrightarrow \infty$ we obtain

$$\sum_{j\in\mathcal{B}} \frac{x_j^*}{\widetilde{x}_j} \;=\; |\mathcal{B}| \;=:\, \ell\,.$$

The concavity of the logarithmic function then yields

$$\psi(\widetilde{x}) - \psi(x^*) \;=\; \textstyle\sum_{j\in\mathcal{B}} \log\Big(\frac{x_j^*}{\widetilde{x}_j}\Big) \;=\; \ell \sum_{j\in\mathcal{B}} \frac{1}{\ell}\log\Big(\frac{x_j^*}{\widetilde{x}_j}\Big)$$
$$\le\; \ell\, \log\Big(\textstyle\sum_{j\in\mathcal{B}} \frac{1}{\ell}\frac{x_j^*}{\widetilde{x}_j}\Big) \;=\; 0\,. \qquad \square$$

Analogously to *(II)* we get:

(III) With $\mathcal{N} := \big\{i \in \{1, \ldots, n\} \mid \exists y \in \mathcal{F}_D^{\text{opt}}\; c_i - \langle a_i, y\rangle > 0\big\}$ *we have:*

a) Problem $(\widetilde{B})$
$$\begin{cases} \widetilde{\psi}(y) := \sum_{j\in\mathcal{N}} \log\big(c_j - \langle a_i, y\rangle\big) \longrightarrow \max \\ y \in \mathcal{F}_D^{\text{opt}},\; (c - A^T y)_{\mathcal{N}} > 0 \end{cases}$$
has a unique maximizer y^*.

b) $\lim_{\mu\to 0+} y(\mu) \;=\; y^*$

Remark

The solution x^* to (B) is apparently the unique maximizer of

$$\max\Big\{\prod_{i\in\mathcal{B}} x_i \mid x \in \mathcal{F}_P^{\text{opt}}\Big\}\,.$$

In the same way the solution y^* to $\big(\widetilde{B}\big)$ is the unique maximizer of

$$\max\Big\{\prod_{i\in\mathcal{N}}\big(c_i-\langle a_i\,,y\rangle\big)\mid y\in\mathcal{F}_D^{\mathrm{opt}}\Big\}.$$

Following SONNEVEND (cf. [Son]) x^* is called the *analytic center* to $\mathcal{F}_P^{\mathrm{opt}}$ and y^* the *analytic center* to $\mathcal{F}_D^{\mathrm{opt}}$.

From (II) and (III) we derive an interesting **Corollary:**

For the index sets $\mathcal{B}$ and $\mathcal{N}$ it holds that $\mathcal{B}\cap\mathcal{N}=\emptyset$ as well as $\mathcal{B}\cup\mathcal{N}=\{1,\ldots,n\}$; we have in particular $x^*+s^*>0$ for $s^*:=c-A^Ty^*$.

Proof: The *optimality conditions* for the primal problem (P) and the dual problem (D) yield

$$\begin{aligned} Ax^* &= b, \quad x^*\geq 0\\ A^Ty^*+s^* &= c, \quad s^*\geq 0\\ \langle x^*\,,s^*\rangle &= 0. \end{aligned}\tag{7}$$

The complementarity condition $\langle x^*\,,s^*\rangle=0$ gives $\mathcal{B}\cap\mathcal{N}=\emptyset$. For $\mu>0$ the points of the central path meet the following condition:

$$\begin{aligned} Ax(\mu) &= b, \quad x(\mu)\geq 0\\ A^Ty(\mu)+s(\mu) &= c, \quad s(\mu)\geq 0\\ x(\mu)\cdot s(\mu) &= \mu\, e \end{aligned}\tag{8}$$

We have seen that

$$\lim_{\mu\to 0+}x(\mu)=x^*,\qquad \lim_{\mu\to 0+}y(\mu)=y^*.$$

With that we also have $\lim_{\mu\to 0+}s(\mu)=s^*$. The relation

$$\begin{aligned} \langle x(\mu)-x^*\,,s(\mu)-s^*\rangle &= -\big\langle x(\mu)-x^*\,,A^T\big(y(\mu)-y^*\big)\big\rangle\\ &= -\big\langle A\big(x(\mu)-x^*\big)\,,y(\mu)-y^*\big\rangle = 0 \end{aligned}$$

yields

$$\begin{aligned} \mu\cdot n &= \langle x(\mu)\,,s(\mu)\rangle = \langle x^*\,,s(\mu)\rangle+\langle x(\mu)\,,s^*\rangle\\ &= \sum_{i\in\mathcal{B}}x_i^*\,s(\mu)_i+\sum_{i\in\mathcal{N}}x(\mu)_i\,s_i^* \end{aligned}$$

and after dividing by $\mu=x(\mu)_i\,s(\mu)_i$:

$$n=\sum_{i\in\mathcal{B}}\frac{x_i^*}{x(\mu)_i}+\sum_{i\in\mathcal{N}}\frac{s_i^*}{s(\mu)_i}$$

Chapter 6

For $\mu \to 0+$ it follows that $n = |\mathcal{B}| + |\mathcal{N}|$ and thus with $\mathcal{B} \cap \mathcal{N} = \emptyset$ the assertion. □

From the considerations to theorem 6.2.2 we have thus gained the following proposition — albeit under the assumption that the Interior-Point Condition holds:

Goldman–Tucker Theorem

There exist optimizers x^* to (P) and y^* to (D) with $x^* + s^* > 0$ for $s^* := c - A^T y^*$.

In this context also see exercise 5.

6.3 Newton's Method for the Primal–Dual System

Our aim is to solve (4) for a fixed $\mu > 0$ by NEWTON's method.[2] A full step of the NEWTON iteration reads as follows:[3]

$$(x^+, y^+, s^+) := (x, y, s) - (\Delta x, \Delta y, \Delta s)$$

We will occasionally use the shortened notation

$$\mathfrak{x} := (x, y, s), \quad \mathfrak{x}^+ := (x^+, y^+, s^+) \text{ and } \Delta\mathfrak{x} := (\Delta x, \Delta y, \Delta s).$$

The above relation then reads

$$\mathfrak{x}^+ := \mathfrak{x} - \Delta\mathfrak{x}.$$

The NEWTON search direction $\Delta\mathfrak{x}$ is obtained by solving the system

$$DF_\mu(\mathfrak{x})\,\Delta\mathfrak{x} = F_\mu(\mathfrak{x}),$$

where we never calculate the inverse of the matrix $DF_\mu(\mathfrak{x})$ explicitly.

For $\mu > 0$ we define the *dual residual* r_d, the *primal residual* r_p and the *complementarity residual* r_c at $\mathfrak{x}$ as:

$$\begin{aligned} r_d &:= A^T y + s - c \\ r_p &:= Ax - b \\ r_c &:= Xs - \mu e \end{aligned}$$

Chapter 6

[2] In the following considerations we will drop the index k, that is, we write (x, y, s) instead of $\left(x^{(k)}, y^{(k)}, s^{(k)}\right)$ and (x^+, y^+, s^+) instead of $\left(x^{(k+1)}, y^{(k+1)}, s^{(k+1)}\right)$.

[3] We prefer this notation (with "−") which is slightly different from the widely used standard notation.

The NEWTON direction $\Delta\mathfrak{x}$ at $\mathfrak{x}$ is given as the solution to

$$DF_\mu(\mathfrak{x})\,\Delta\mathfrak{x} \;=\; \begin{pmatrix} r_d \\ r_p \\ r_c \end{pmatrix} \;=\; F_\mu(\mathfrak{x})$$

which we can restate as

$$\begin{pmatrix} 0 & A^T & I \\ A & 0 & 0 \\ S & 0 & X \end{pmatrix} \begin{pmatrix} \Delta x \\ \Delta y \\ \Delta s \end{pmatrix} = \begin{pmatrix} r_d \\ r_p \\ r_c \end{pmatrix}. \tag{9}$$

By theorem 6.2.1 the Jacobian is nonsingular here as we have $\mu > 0$ and therefore $x > 0$ and $s > 0$. We therefore know that the NEWTON step is well-defined. The system (9) is of size $2n + m$. It is possible to symmetrize $DF_\mu(x, y, s)$ by multiplying the last block row by S^{-1}:

$$\begin{pmatrix} 0 & A^T & I \\ A & 0 & 0 \\ I & 0 & S^{-1}X \end{pmatrix} \begin{pmatrix} \Delta x \\ \Delta y \\ \Delta s \end{pmatrix} = \begin{pmatrix} r_d \\ r_p \\ S^{-1}r_c \end{pmatrix}$$

In the following theorem we show an *alternative way of solving system (9):*

Theorem 6.3.1

The solution to system (9) is given by the following equations:

$$\begin{aligned} \Delta y &= (AXS^{-1}A^T)^{-1}(r_p - AS^{-1}(r_c - Xr_d)) \\ \Delta s &= r_d - A^T\Delta y \\ \Delta x &= S^{-1}(r_c - X\Delta s) \end{aligned}$$

Proof: (9) means

$$\begin{aligned} A^T\Delta y + \Delta s &= r_d \\ A\Delta x &= r_p \\ S\Delta x + X\Delta s &= r_c\,. \end{aligned}$$

Hence $\Delta s = r_d - A^T\Delta y$, $\Delta x = S^{-1}(r_c - X\Delta s) = x - \mu S^{-1}e - XS^{-1}\Delta s$. If we substitute Δx in the second equation above, we get $AS^{-1}(r_c - X\Delta s) = r_p$. Now we substitute Δs and obtain $AS^{-1}r_c - AS^{-1}Xr_d + AS^{-1}XA^T\Delta y = r_p$, hence $AS^{-1}XA^T\Delta y = r_p + AS^{-1}(-r_c + Xr_d)$ and so

$$\Delta y = (AXS^{-1}A^T)^{-1}(r_p + AS^{-1}(-r_c + Xr_d))\,. \qquad \square$$

With the positive definite matrix[4]

[4] For the positive definite matrix $M := AXS^{-1}A^T$ a CHOLESKY decomposition may be used. In the case of large sparse problems a preconditioned CG algorithm is suggested.

Chapter 6

$$M := AXS^{-1}A^T$$

and the right-hand side

$$\text{rhs} := r_p + AS^{-1}(-r_c + Xr_d)$$

the solution to

$$M\,\Delta y = \text{rhs}$$

firstly yields Δy. *Then we have*

$$\Delta s = r_d - A^T \Delta y \quad \text{and} \quad \Delta x = S^{-1}(r_c - X\Delta s).$$

In the *short-step method* (compare section 6.6) we firstly deal with full NEWTON steps and small updates of the parameter μ. Later on we derive a *long-step method* with damped NEWTON steps, that is, we introduce a step size $\alpha \in (0,1]$ and choose α such that $(x^+, y^+, s^+) := (x, y, s) - \alpha(\Delta x, \Delta y, \Delta s)$ satisfies $(x^+, s^+) > 0$. For $\alpha = 1$ we have the full NEWTON step.

6.4 A Primal–Dual Framework

For each $(x, y, s) \in \mathcal{F}^0$ we define the *duality measure* τ by

$$\tau := \tau_{(x\,s)} := \tau(x\,s) := \frac{1}{n}\sum_{i=1}^{n} x_i\, s_i = \frac{1}{n}\,\langle x\,, s\rangle$$

which gives the average value of the products $x_i s_i$. If we are *on* the central path, we obviously have $\mu = \tau$, but τ is also defined *off* the central path.

With the results we have gained we can now define a general framework for primal–dual algorithms. But first we introduce a *centering parameter* $\sigma \in [0,1]$. We now consider the system

$$\begin{pmatrix} 0 & A^T & I \\ A & 0 & 0 \\ S & 0 & X \end{pmatrix} \begin{pmatrix} \Delta x \\ \Delta y \\ \Delta s \end{pmatrix} = \begin{pmatrix} 0 \\ 0 \\ XSe - \sigma\tau e \end{pmatrix}. \tag{10}$$

The direction $\big(\Delta x, \Delta y, \Delta s\big)$ gives a NEWTON step toward a point

$$\big(x_{\sigma,\tau},\ y_{\sigma,\tau},\ s_{\sigma,\tau}\big) \in \mathcal{C}$$

with $\sigma\tau = \mu$.

General Interior-Point Algorithm

Let a starting vector $\left(x^{(0)}, y^{(0)}, s^{(0)}\right) \in \mathcal{F}^0$ *and an accuracy requirement* $\varepsilon > 0$ *be given.*

Initialize $(x, y, s) := (x^{(0)}, y^{(0)}, s^{(0)})$ *and* $\tau := \frac{1}{n}\langle x, s\rangle$.

while $\tau > \varepsilon$

Solve
$$\begin{pmatrix} 0 & A^T & I \\ A & 0 & 0 \\ S & 0 & X \end{pmatrix} \begin{pmatrix} \Delta x \\ \Delta y \\ \Delta s \end{pmatrix} = \begin{pmatrix} 0 \\ 0 \\ XSe - \sigma\tau e \end{pmatrix}$$

where $\sigma \in [0, 1]$.

Set $(x, y, s) := (x, y, s) - \alpha\left(\Delta x, \Delta y, \Delta s\right)$ *where* $\alpha \in (0, 1]$ *denotes a suitable step size which we choose such that* $(x, s) > 0$.

$\tau := \frac{1}{n}\langle x, s\rangle$

Depending on the choice of σ and α, we get different interior-point algorithms. We will discuss some of them later on. In principle, the term $\sigma\tau$ on the right-hand side of (10) plays the same role as the parameter μ in system (4). For $\sigma = 0$ we target directly at a point for which the optimality conditions (1) are fulfilled. This direction is also called the *affine-scaling direction.* At the other extreme, for $\sigma = 1$, the equations (10) define a NEWTON direction, which yields a step toward the point $(x_\tau, y_\tau, s_\tau) \in \mathcal{C}$ at which all the products $x_i s_i$ are identical to τ . Therefore this direction is called the *centering direction.* These directions usually make only little progress in reducing τ but move closer to the central path.

Chapter 6

Remark

To get a *suitable step size*, we choose an $\eta \in (0, 1)$ such that $\eta \approx 1$, for example $\eta = 0.99$, and set

$$\alpha := \min\left\{1, \, \eta \min_{\Delta x_\nu > 0} \frac{x_\nu}{\Delta x_\nu}, \, \eta \min_{\Delta s_\nu > 0} \frac{s_\nu}{\Delta s_\nu}\right\}.$$

So we have $x - \alpha\Delta x > 0$ and $s - \alpha\Delta s > 0$.

For the implementation it is useful to consider that

$$\min\left\{1, \, \eta \min_{\Delta x_\nu > 0} \frac{x_\nu}{\Delta x_\nu}, \eta \min_{\Delta s_\nu > 0} \frac{s_\nu}{\Delta s_\nu}\right\} = \frac{\eta}{\max\left\{\eta, \max\limits_\nu \frac{\Delta x_\nu}{x_\nu}, \max\limits_\nu \frac{\Delta s_\nu}{s_\nu}\right\}}.$$

In *Matlab*® the denominator of the right-hand side can be written in the form

$$\max\left([\eta;\ \Delta x./x;\ \Delta s./s]\right).$$

The next lemma shows that we can reduce τ, dependent on σ and α . We will use the notation

$$\begin{aligned}\big(x(\alpha), y(\alpha), s(\alpha)\big) &:= (x, y, s) - \alpha\big(\Delta x, \Delta y, \Delta s\big)\,,\\ \tau(\alpha) &:= \frac{1}{n}\,\langle x(\alpha), s(\alpha)\rangle\,.\end{aligned}$$

Lemma 6.4.1

The solution $(\Delta x, \Delta y, \Delta s)$ *to system (10) has the following properties:*

a) $\langle \Delta x, \Delta s\rangle = 0$

b) $\tau(\alpha) = \big(1 - \alpha(1-\sigma)\big)\tau$

Proof: a): We have

$$\begin{aligned} A\Delta x &= 0 \quad \text{and}\\ A^T\Delta y + \Delta s &= 0\,.\end{aligned}$$

Now we multiply the second equation with Δx and obtain

$$0 = \big\langle \Delta x, A^T\Delta y\big\rangle + \langle \Delta x, \Delta s\rangle = \langle A\Delta x, \Delta y\rangle + \langle \Delta x, \Delta s\rangle,$$

hence $\langle \Delta x, \Delta s\rangle = 0$.

b): The third row of (10) gives

$$S\Delta x + X\Delta s = XSe - \sigma\tau e\,.$$

Summation of the n components yields

$$\langle s, \Delta x\rangle + \langle x, \Delta s\rangle = \langle x, s\rangle - \sigma\langle x, s\rangle = (1-\sigma)\langle x, s\rangle\,.$$

With *a)* we therefore obtain

$$\begin{aligned} n\tau(\alpha) &= \langle x(\alpha), s(\alpha)\rangle\\ &= \langle x, s\rangle - \alpha\big(\langle s, \Delta x\rangle + \langle x, \Delta s\rangle\big)\\ &= \big(1 - \alpha(1-\sigma)\big)\langle x, s\rangle = n\big(1 - \alpha(1-\sigma)\big)\tau\,.\end{aligned}$$

□

The next theorem states that the algorithm has *polynomial complexity.*

Theorem 6.4.2

Suppose that $\big(x^{(k)}, y^{(k)}, s^{(k)}\big)$ *is a sequence generated by the algorithm,* $\varepsilon \in (0,1)$ *and that the parameters produced by the algorithm fulfill*

$$\tau_{k+1} \le \left(1 - \frac{\delta}{n^\omega}\right)\tau_k, \quad k = 0, 1, 2, \ldots \tag{11}$$

for some positive constants δ *and* ω. *If the initial vector* $\big(x^{(0)}, y^{(0)}, s^{(0)}\big)$ *satisfies*

$$\tau_0 \le \frac{1}{\varepsilon^\nu} \tag{12}$$

for a positive constant ν, *then there exists an index* $K \in \mathbb{N}$ *with*

$$K = O(n^\omega\,|\ln(\varepsilon)|)$$

and $\tau_k \le \varepsilon$ *for all* $k \ge K$.

Proof: We take logarithms on both sides in (11) and obtain

$$\ln \tau_{k+1} \le \ln\left(1 - \frac{\delta}{n^\omega}\right) + \ln \tau_k .$$

If we repeat this procedure and use (12) in the second inequality, we get

$$\ln \tau_k \le k \ln\left(1 - \frac{\delta}{n^\omega}\right) + \ln \tau_0 \le k \ln\left(1 - \frac{\delta}{n^\omega}\right) - \nu \ln \varepsilon.$$

We know that $t \ge \ln(1+t)$ for $t > -1$. This implies $\left(\text{for } t := -\frac{\delta}{n^\omega}\right)$ that

$$\ln \tau_k \le k\left(-\frac{\delta}{n^\omega}\right) - \nu \ln \varepsilon .$$

Hence, if we have

$$k\left(-\frac{\delta}{n^\omega}\right) - \nu \ln \varepsilon \le \ln \varepsilon ,$$

the convergence criterion $\tau_k \le \varepsilon$ is satisfied. This inequality holds for all k that satisfy

$$k \ge (1+\nu)\frac{n^\omega}{\delta} \left|\ln(\varepsilon)\right| .$$

With $K := \left\lceil (1+\nu)\frac{n^\omega}{\delta} \left|\ln(\varepsilon)\right| \right\rceil$ the proof is complete. □

Chapter 6

6.5 Neighborhoods of the Central Path

Path-following methods move along the central path in the direction of decreasing μ to the set of minimizers of (P) and maximizers of (D), respectively. These methods do not necessarily stay exactly *on* the central path but within a loose, well-defined neighborhood. Most of the literature does not examine the nature of these neighborhoods. We, however, will take a close look at them, especially at their geometric interpretation.

We want to measure the deviation of each $\mathfrak{x} = (x, y, s) \in \mathcal{F}^0$ from the central path: So far, we know that

$$\min_\mu \|F_\mu(\mathfrak{x})\| = \min_\mu \|XSe - \mu e\|$$

as $\mathfrak{x} \in \mathcal{F}^0$. The orthogonal projection from XSe onto $\{\mu e \mid \mu \in \mathbb{R}\}$ is given by $\frac{1}{n}\langle x, s\rangle e$. We thus get — with the duality measure τ —

$$\min_\mu \|F_\mu(\mathfrak{x})\| = \min_\mu \|XSe - \mu e\| = \left\|XSe - \frac{1}{n}\langle x, s\rangle e\right\| = \|XSe - \tau(xs)e\| .$$

Obviously, a triple $(x, y, s) \in \mathcal{F}^0$ lies on the central path if and only if $\|XSe - \tau(x\,s)e\| = 0$. Our aim is to control the approximation of the central path in such a way that the deviation converges to zero as $\langle x, s \rangle$ goes to zero. This leads to the following definition of the neighborhood $\mathcal{N}_2$ of the central path for $\beta \geq 0$:

$$\mathcal{N}_2(\beta) := \{(x, y, s) \in \mathcal{F}^0 \mid \|XSe - \tau(x\,s)e\| \leq \beta\tau(x\,s)\}$$

It is clear that

$$\mathcal{C} = \mathcal{N}_2(0) \subset \mathcal{N}_2(\beta_1) \subset \mathcal{N}_2(\beta_2) \subset \mathcal{F}^0 \quad \text{for} \quad 0 \leq \beta_1 \leq \beta_2 \leq 1\,.$$

The following picture shows the x-projection of $\mathcal{N}_2(\beta)$ corresponding to example 3 for the values $\beta = 1/2$ and $\beta = 1/4$:

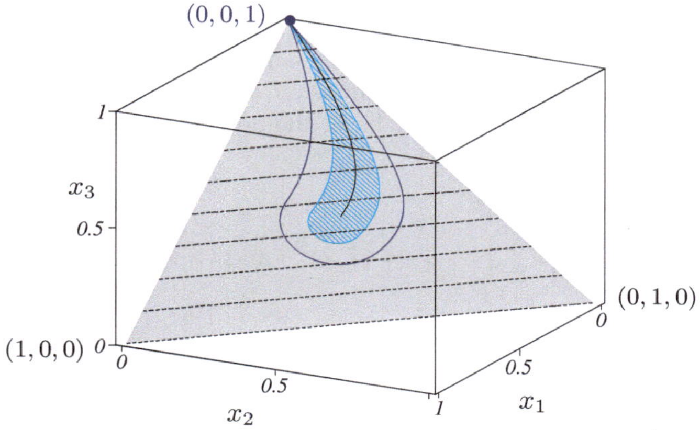

Another neighborhood which we will need later on is $\mathcal{N}_{-\infty}$ defined as

$$\mathcal{N}_{-\infty}(\gamma) := \{(x, y, s) \in \mathcal{F}^0 \mid x_\nu s_\nu \geq \gamma\tau_{(x\,s)} \quad \text{for} \quad \nu = 1, \ldots, n\}$$

for some $\gamma \in [0, 1]$. It is clear that

$$\mathcal{C} = \mathcal{N}_{-\infty}(1) \subset \mathcal{N}_{-\infty}(\gamma_2) \subset \mathcal{N}_{-\infty}(\gamma_1) \subset \mathcal{N}_{-\infty}(0) = \mathcal{F}^0$$

for $0 \leq \gamma_1 \leq \gamma_2 \leq 1$. We will illustrate this neighborhood with a picture (on the next page) for the same example and the values $\gamma = 0.3$ and $\gamma = 0.6$ as well.

Every one of the path-following algorithms which we will consider in the following restricts the iterates to one of the two neighborhoods of the central path $\mathcal{C}$ with $\beta \in [0, 1]$ and $\gamma \in [0, 1]$. The requirement in the $\mathcal{N}_{-\infty}$ neighborhood is not very strong:

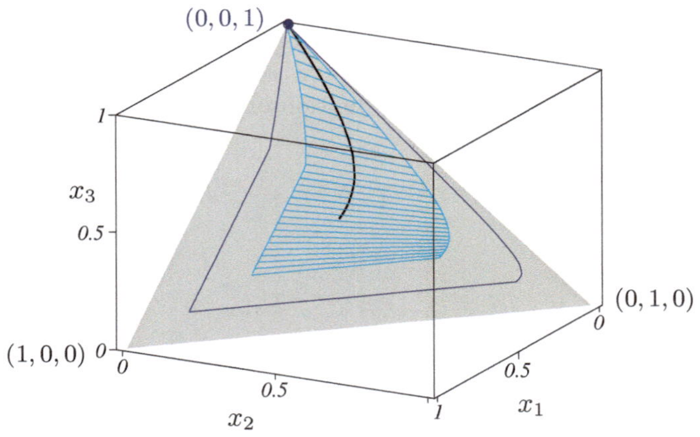

If we choose γ close to zero, we have almost the whole feasible set. By looking at $\mathcal{N}_2(\beta)$, we see that we cannot claim the same here for $\beta \in [0,1]$. In certain examples it does not matter how we choose β, there are always points in the strictly feasible region which do not lie in the neighborhood $\mathcal{N}_2(\beta)$:

Example 4

Let

$$A := \begin{pmatrix} 1 & 0 \\ 0 & 2 \end{pmatrix}, \quad b := (7,2)^T \text{ and } c := (1,1)^T.$$

Hence, we consider the primal problem

$$(P) \quad \begin{cases} x_1 + x_2 \longrightarrow \min \\ x_1 = 7 \\ 2x_2 = 2 \\ x \geq 0 \end{cases}$$

and the dual problem

$$(D_e) \quad \begin{cases} 7y_1 + 2y_2 \longrightarrow \max \\ y_1 + s_1 = 1 \\ 2y_2 + s_2 = 1 \\ s \geq 0. \end{cases}$$

Here we have $\mathcal{F}_P = \mathcal{F}_P^0 = \{(7,1)^T\}$. Hence we see that $x = (7,1)^T \in \mathcal{F}_P^0$ and $(y,s) = ((0,0)^T, (1,1)^T) \in \mathcal{F}_{D_e}^0$ with $(x,y,s) \in \mathcal{N}_{-\infty}(\gamma)$ for a suitable γ but $(x,y,s) \notin \mathcal{N}_2(\beta)$ for any $\beta \in [0,1]$: As we have $\tau = 1/2\,\langle x, s\rangle = 4$ we get with $\gamma := 1/4$ that $x_\nu s_\nu \geq \gamma\tau = 1$ for $\nu = 1, 2$. On the other hand, we have $\|XSe - \tau e\| = \sqrt{18} > \beta\tau$ for all $\beta \in [0,1]$. ◁

Since the condition $\|XSe - \tau(xs)e\| \leq \beta\tau(xs)$ in the definition of $\mathcal{N}_2(\beta)$ does not depend on (x,y,s), but only on xs, the vector of the products $x_\nu s_\nu$, we firstly consider the simpler set

$$\mathcal{N}_2'(\beta) := \left\{\omega \in \mathbb{R}^n_+ \mid \|\omega - \tau(\omega)e\| \leq \beta\tau(\omega)\right\}$$

with the mean value

$$\tau := \tau_\omega := \tau(\omega) := \frac{1}{n}\sum_{\nu=1}^{n}\omega_\nu = \frac{1}{n}\langle\omega, e\rangle = \frac{1}{n}e^T\omega \quad \text{for} \quad \omega \in \mathbb{R}^n_+ .$$

Here we ignore the fact that the constraints $Ax = b$ and $A^Ty + s = c$ are met. Likewise it is helpful to consider the simpler set

$$\mathcal{N}_{-\infty}'(\gamma) := \left\{\omega \in \mathbb{R}^n_+ \mid \omega_\nu \geq \gamma\tau_\omega \quad \text{for} \quad \nu = 1,\ldots,n\right\}$$

instead of $\mathcal{N}_{-\infty}(\gamma)$. Here the fact that the constraints are met is also ignored.

A Closer Look at These Neighborhoods

With

$$n^2\tau^2 = (e^T\omega)^2 = \langle\omega, ee^T\omega\rangle$$

we get for $\omega \in \mathcal{N}_2'(\beta)$

$$\|\omega - \tau e\|^2 = \langle\omega,\omega\rangle - 2\tau\langle\omega, e\rangle + \tau^2\langle e, e\rangle = \langle\omega,\omega\rangle - n\tau^2.$$

With that the inequality $\|\omega - \tau e\| \leq \beta\tau$ means

$$\langle\omega,\omega\rangle - (n+\beta^2)\tau^2 \leq 0 .$$

With the matrix

$$A := A_\beta := I - \left(\frac{1}{n} + \frac{\beta^2}{n^2}\right)ee^T$$

this can be written in the form

$$\langle\omega, A\omega\rangle \leq 0.$$

With $Ae = -\beta^2/n\, e$ we see that e is an eigenvector of A with eigenvalue $-\beta^2/n$. For any vector $u \in \mathbb{R}^n$ with $u \perp e$ we obtain $Au = u$, hence all vectors that are orthogonal to e are eigenvectors of A with eigenvalue 1. To $h_n := 1/\sqrt{n}\, e$ we choose $h_1,\ldots,h_{n-1}$ such that $h_1,\ldots,h_n$ is an orthonormal basis. We can therefore express any ω by

$$\omega = \sum_{\nu=1}^{n}\alpha_\nu h_\nu$$

Chapter 6

with $\alpha_\nu := \langle \omega, h_\nu \rangle$ and obtain with that

$$\tau(\omega) = \frac{1}{\sqrt{n}}\alpha_n \geq 0 \quad \text{and} \quad 0 \geq \langle \omega, A\omega \rangle = \alpha_1^2 + \cdots + \alpha_{n-1}^2 - \frac{\beta^2}{n}\alpha_n^2,$$

hence $\alpha_1^2 + \cdots + \alpha_{n-1}^2 \leq \frac{\beta^2}{n}\alpha_n^2$. With

$$\mathcal{C}_\beta := \left\{ \sum_{\nu=1}^{n} \alpha_\nu h_\nu \;\middle|\; \frac{\beta}{\sqrt{n}}\alpha_n \geq \sqrt{\alpha_1^2 + \cdots + \alpha_{n-1}^2};\ \alpha_1, \ldots, \alpha_n \in \mathbb{R} \right\}$$

we thus get $\mathcal{N}_2'(\beta) = \mathbb{R}_+^n \cap \mathcal{C}_\beta$. We will see that $\mathcal{C}_\beta$ is a Lorentz *cone*

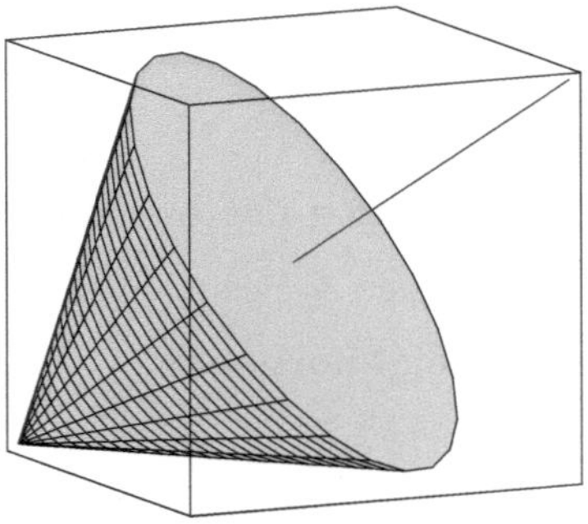

with apex 0 and symmetry axis $\{\lambda e \mid \lambda \in \mathbb{R}_+\}$, the bisecting line of the positive orthant, as well as a 'circular' cross section — orthogonal to the symmetry axis. For $\alpha_n = 1$ this cross section has radius $\beta/\sqrt{n}$ and lies in the hyperplane $H := \left\{x \in \mathbb{R}^n \mid \langle e, x \rangle = \sum_{\nu=1}^{n} x_\nu = \sqrt{n}\right\}$. If we now consider a circle K_r in H with center $M = h_n =: \overline{e}$ and radius $r \geq 0$, we have

$$K_r = \left\{\overline{e} + y \mid y \in \mathbb{R}^n, \langle e, y \rangle = 0,\ \|y\| \leq r\right\}.$$

We would like to find out when exactly $K_r \subset \mathbb{R}_+^n$ holds. Preparatorily, we prove the following

Lemma 6.5.1

For $y \in \mathbb{R}^n$ with $\langle e, y \rangle = 0$ and $\|y\| \leq 1$ it holds that $|y_k|^2 \leq 1 - \frac{1}{n}$ for $k = 1, \ldots, n$.

Proof: With the Cauchy–Schwarz inequality we firstly obtain

$$y_k^2 = (-y_k)^2 = \left(\sum_{\nu \neq k}^{n} 1 \cdot y_\nu\right)^2 \leq (n-1)\sum_{\nu \neq k} y_\nu^2 \leq (n-1)(1 - y_k^2).$$

From this it follows that $n y_k^2 \leq n - 1$, and thus the desired result. □

Now we are able to answer the question from above:

Chapter 6

Lemma 6.5.2

$K_r \subset \mathbb{R}^n_+$ holds if and only if the radius r fulfills the inequality $r \le \frac{1}{\sqrt{n-1}}$.

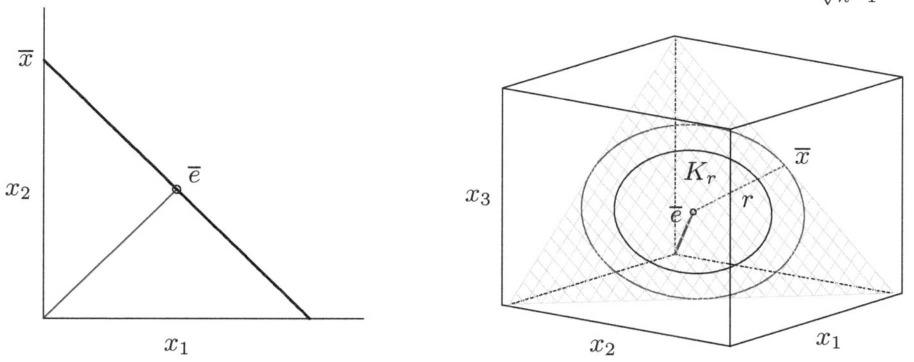

Proof: For $n = 2$ this can be directly deduced from the left picture. To prove the statement for arbitrary $n \ge 2$, we start from the assumption $K_r \subset \mathbb{R}^n_+$:

For
$$\overline{x} := \frac{1}{n-1} \sum_{\nu=2}^{n} \sqrt{n}\, e_\nu = \frac{\sqrt{n}}{n-1}(e - e_1)$$
it holds that $\langle e, \overline{x} - \overline{e} \rangle = 0$ and $\|\overline{x} - \overline{e}\| = \frac{1}{\sqrt{n-1}}$ since
$$\overline{x} - \overline{e} = \Big(\frac{\sqrt{n}}{n-1} - \frac{\sqrt{n}}{n}\Big)e - \frac{\sqrt{n}}{n-1} e_1 = \frac{\sqrt{n}}{n(n-1)}(e - n e_1)$$
and
$$\|\overline{x} - \overline{e}\|^2 = \frac{n}{n^2(n-1)^2}\big((n-1)^2 + (n-1)\big) = \frac{1}{n-1}\,.$$
Set
$$y := r\,\frac{\overline{x} - \overline{e}}{\|\overline{x} - \overline{e}\|} = \frac{r}{\sqrt{n(n-1)}}(e - n e_1)\,.$$
For that $\langle e, y \rangle = 0$ and $\|y\| = r$ hold, hence $x := \overline{e} + y \in K_r \subset \mathbb{R}^n_+$. In particular, we have
$$x_1 = \frac{1}{\sqrt{n}} - (n-1)\frac{r}{\sqrt{n(n-1)}} \ge 0\,,$$
from which we obtain $r \le \frac{1}{\sqrt{n-1}}$.

Let now $r \le \frac{1}{\sqrt{n-1}}$ and $x = \overline{e} + y$ with $\langle e, y \rangle = 0$ and $\|y\| \le r$, hence $x \in K_r$. With lemma 6.5.1 we have $\left|\frac{1}{r}\, y_k\right|^2 \le 1 - \frac{1}{n} = \frac{n-1}{n}$ for $k = 1, \ldots, n$.
$$|y_k| \le r\sqrt{\frac{n-1}{n}} \le \frac{1}{\sqrt{n}} \quad \text{for} \quad k = 1, \ldots, n$$
then yields $x = \overline{e} + y \in \mathbb{R}^n_+$. □

Chapter 6

Corollary

From the above we obtain the following condition for β *:*

$$\beta = \sqrt{n}\,r \le \sqrt{\frac{n}{n-1}}$$

Therefore $\mathcal{N}_2'(\beta) = \mathcal{C}_\beta$ *for* $\beta \in [0,1]$. *In addition* $\mathcal{N}_2'(\beta) \setminus \{0\} \subset \mathbb{R}^n_{++}$.

For the *polyhedral cone*

$$\mathcal{N}_{-\infty}'(\gamma) = \{\omega \in \mathbb{R}^n_+ \mid \omega_\nu \ge \gamma\,\tau_\omega \quad \text{for} \quad \nu = 1,\ldots,n\}$$

with $\gamma \in [0,1]$ we can show analogously

$$\begin{aligned}\mathcal{N}_{-\infty}'(\gamma) &= \{\omega \in \mathbb{R}^n_+ \mid \big(I - \gamma\, h_n h_n^T\big)\omega \ge 0\} \\ &= \big\{\textstyle\sum_{\nu=1}^n \alpha_\nu\, h_\nu \,\big|\, \alpha_n \ge 0,\ \sum_{\nu=1}^{n-1} \alpha_\nu\, h_\nu + (1-\gamma)\alpha_n\, h_n \ge 0\big\}.\end{aligned}$$

If $\beta \in [0,1]$, we can furthermore get the following with lemma 6.5.1:

$$\mathcal{N}_2'(\beta) \subset \mathcal{N}_{-\infty}'(\gamma) \quad \text{for} \quad \gamma \le 1 - \beta\,\sqrt{(n-1)/n}\,.$$

We will leave the *proof* as exercise 6 to the interested reader.

We will now look at an algorithm which is a special case of the general algorithm, the *short-step algorithm.* In this algorithm we choose $\alpha = 1$ and $\sigma = 1 - \frac{2}{5\sqrt{n}}$.

6.6 A Short-Step Path-Following Algorithm

In this section we consider a short-step path-following algorithm. This algorithm works with the 2-norm neighborhood. The duality measure τ is steadily reduced to zero. Each search direction is a NEWTON direction toward a point on the central path, a point for which the normalized duality gap μ is less or equal to the current duality measure τ. The short-step algorithm we discuss generates strictly feasible iterates $\big(x^{(k)}, y^{(k)}, s^{(k)}\big)$ that satisfy system (10). As the target value we use $\mu = \sigma\,\tau$, where σ is the centering parameter and τ the duality measure we introduced above. In the algorithm, we choose $\alpha = 1$ as the step length and for σ we choose a constant value $\sigma := 1 - 2/\big(5\sqrt{n}\big)$. We start at a point $\big(x^{(0)}, y^{(0)}, s^{(0)}\big) \in \mathcal{N}_2(\frac{1}{2})$ with $\tau_0 := \frac{1}{n}\big\langle x^{(0)}, s^{(0)}\big\rangle$. Then we take a NEWTON step, reduce τ and continue. The name *short-step algorithm* refers to algorithms which reduce τ in every step by a constant factor like $1 - 1/\sqrt{n}$ which only depends on the dimension of the problem.

Short-step algorithm

Let a starting vector $\left(x^{(0)}, y^{(0)}, s^{(0)}\right) \in \mathcal{N}_2(\frac{1}{2})$ *and an accuracy requirement* $\varepsilon > 0$ *be given.*

Initialize $(x, y, s) := \left(x^{(0)}, y^{(0)}, s^{(0)}\right)$ *and* $\tau := \frac{1}{n}\langle x, s\rangle$.

while $\tau > \varepsilon$

Determine the NEWTON *direction* $(\Delta x, \Delta y, \Delta s)$ *according to (10) and set*

$$x := x - \Delta x$$
$$y := y - \Delta y$$
$$s := s - \Delta s,$$

$\tau := \left(1 - \frac{2}{5\sqrt{n}}\right)\tau$.

If we apply the algorithm to example 3 (cf. page 254) with the starting points $x = (0.3, 0.45, 0.25)^T$, $s = (9, 7, 6)^T$ and $y = -10$, we get the following visualization (after 57 iteration steps):

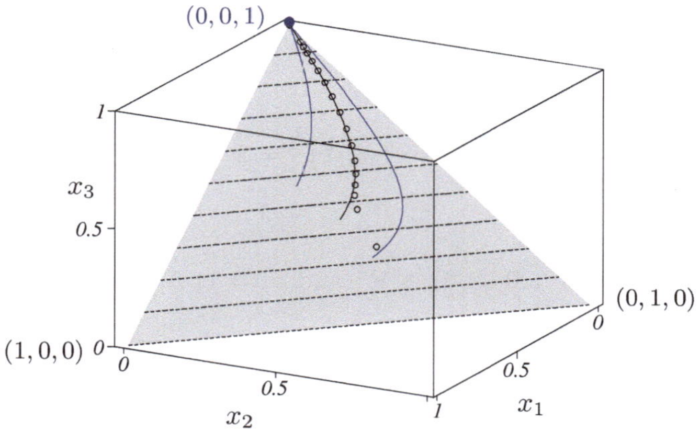

Lemma 6.6.1

Let $\left(x^{(0)}, y^{(0)}, s^{(0)}\right) \in \mathcal{N}_2(\frac{1}{2})$ *and*

$$\tau_0 \leq \frac{1}{\varepsilon^\nu}$$

for some positive constant ν *and a given* $\varepsilon \in (0,1)$. *Then there exists an index* $K \in \mathbb{N}$ *with* $K = O(\sqrt{n}\,|\ln(\varepsilon)|)$ *such that*

$$\tau_k \leq \varepsilon$$

for all $k \geq K$.

Proof: We set $\delta := \frac{2}{5}$ and $\omega := \frac{1}{2}$ and obtain the result with theorem 6.4.2 as we have — by our choice of σ and α — that $\tau_{k+1} = \sigma\tau_k = \left(1 - \frac{2}{5\sqrt{n}}\right)\tau_k$.
□

We want to show that all iterates $\left(x^{(k)}, y^{(k)}, s^{(k)}\right)$ of path-following algorithms stay in the corresponding neighborhoods, i.e., for $(x, y, s) \in \mathcal{N}_2(\beta)$ it also follows that $(x(\alpha), y(\alpha), s(\alpha)) \in \mathcal{N}_2(\beta)$ for all $\alpha \in [0,1]$ (cf. p. 265). In order to be able to state this, we will first provide some technical results:

Lemma 6.6.2

Let u and v be two vectors in $\mathbb{R}^n$ with $\langle u, v\rangle \geq 0$. Then

$$\|UVe\| \leq \sqrt{2^{-3}}\, \|u+v\|^2 . \tag{13}$$

Proof: We partition the index set $I := \{1, \ldots, n\}$ into $\mathcal{P} := \{i \in I \mid u_i v_i \geq 0\}$ and $\mathcal{M} := \{i \in I \mid u_i v_i < 0\}$. For $T \subset I$ we define the vector $z_T := \sum_{i \in T} u_i v_i e_i$. From $0 \leq \langle u, v\rangle = \|z_{\mathcal{P}}\|_1 - \|z_{\mathcal{M}}\|_1$ we get

$$\begin{aligned}
\|UVe\|^2 &= \sum_{i=1}^n (u_i v_i)^2 = \|z_{\mathcal{P}}\|^2 + \|z_{\mathcal{M}}\|^2 \\
&\overset{\checkmark}{\leq} \|z_{\mathcal{P}}\|_1^2 + \|z_{\mathcal{M}}\|_1^2 \leq 2\, \|z_{\mathcal{P}}\|_1^2 = 2\Big(\sum_{i \in \mathcal{P}} u_i v_i\Big)^2 \\
&\overset{\checkmark}{\leq} 2\Big(\sum_{i \in \mathcal{P}} \frac{1}{4}(u_i + v_i)^2\Big)^2 \\
&\leq 2^{-3}\Big(\sum_{i=1}^n (u_i + v_i)^2\Big)^2 = 2^{-3}\, \|u+v\|^4 .
\end{aligned}$$

□

Remark

For $n \geq 2$, $u := (1, -1, 0, \ldots, 0) \in \mathbb{R}^n$ and $v := (1, 1, 0, \ldots, 0) \in \mathbb{R}^n$ we get $\langle u, v\rangle = 0$ and

$$\|UVe\| = \sqrt{2^{-3}}\, \|u+v\|^2 ,$$

therefore *the inequality (13) is sharp* for $n \geq 2$.

Lemma 6.6.3

For $\beta \in [0,1]$, $\omega \in \mathcal{N}_2'(\beta)$ and $\tau := \tau_\omega$ we have $\min_\nu \omega_\nu \geq (1-\beta)\tau$.

Proof: For any index $\nu \in \{1, \ldots, n\}$ we get

$$\tau - \omega_\nu \leq |\omega_\nu - \tau| \leq \|\omega - \tau e\| \leq \beta\tau$$

and therefore $(1-\beta)\tau \leq \omega_\nu$. □

Corollary

For $\beta \in [0,1]$ we have $\mathcal{N}_2(\beta) \subset \mathcal{N}_{-\infty}(1-\beta)$.

Lemma 6.6.4

For $(x, y, s) \in \mathcal{N}_2(\beta)$ *we have*

$$\|\Delta X\,\Delta S e\| \;\le\; \frac{\beta^2 + n(1-\sigma)^2}{\sqrt{2^3}\,(1-\beta)}\,\tau\,.$$

Proof: We define the positive definite diagonal matrix $D := X^{\frac{1}{2}} S^{-\frac{1}{2}}$. If we multiply the third equation of (10) by $(XS)^{-\frac{1}{2}}$, we get

$$D^{-1}\Delta x + D\Delta s \;=\; (XS)^{-\frac{1}{2}}(XSe - \sigma\tau e)\,.$$

Now we set $u := D^{-1}\Delta x$ and $v := D\Delta s$ and thus — with the help of lemma 6.4.1 — we get $\langle u, v\rangle = 0$. Then we can apply lemma 6.6.2 and obtain

$$\begin{aligned}
\|\Delta X \Delta S e\| &= \big\|(D^{-1}\Delta X)(D\Delta S)e\big\| \\
&\le \sqrt{2^{-3}}\,\big\|D^{-1}\Delta x + D\Delta s\big\|^2 \\
&= \sqrt{2^{-3}}\,\Big\|(XS)^{-\frac{1}{2}}(XSe - \sigma\tau e)\Big\|^2 \\
&= \sqrt{2^{-3}}\sum_{\nu=1}^{n}\frac{(x_\nu s_\nu - \sigma\tau)^2}{x_\nu s_\nu} \;\le\; \sqrt{2^{-3}}\,\frac{\sum_{\nu=1}^{n}(x_\nu s_\nu - \sigma\tau)^2}{\min\limits_{\nu} x_\nu s_\nu} \\
&= \sqrt{2^{-3}}\,\frac{\|XSe - \sigma\tau e\|^2}{\min\limits_{\nu} x_\nu s_\nu}\,.
\end{aligned}$$

By assumption $(x, y, s) \in \mathcal{N}_2(\beta)$ and so we have

$$\begin{aligned}
\|XSe - \sigma\tau e\|^2 &= \|(XSe - \tau e) + (1-\sigma)\tau e\|^2 \\
&= \|XSe - \tau e\|^2 + 2(1-\sigma)\tau\,\langle e, XSe - \tau e\rangle + (1-\sigma)^2\tau^2\langle e, e\rangle \\
&= \|XSe - \tau e\|^2 + (1-\sigma)^2\tau^2 n \\
&\le \beta^2\tau^2 + (1-\sigma)^2\tau^2 n\,.
\end{aligned}$$

With lemma 6.6.3 we obtain the desired result. □

Lemma 6.6.5

For $(x, y, s) \in \mathcal{N}_2(\beta)$ *and* $\alpha \in [0, 1]$ *we have*

$$\begin{aligned}
\|X(\alpha)S(\alpha)e - \tau(\alpha)e\| &\le |1-\alpha|\,\|XSe - \tau e\| + \alpha^2\,\|\Delta X \Delta S e\| &\quad (14)\\
&\le |1-\alpha|\,\beta\tau + \alpha^2\left(\frac{\beta^2 + n(1-\sigma)^2}{\sqrt{8}\,(1-\beta)}\right)\tau. &\quad (15)
\end{aligned}$$

Proof: From lemma 6.4.1 we get

Chapter 6

$$\begin{aligned} X(\alpha)S(\alpha)e - \tau(\alpha)e &= (x-\alpha\,\Delta x)(s-\alpha\,\Delta s) - \big(1-\alpha(1-\sigma)\big)\tau e \\ &= xs - \alpha(s\,\Delta x + x\,\Delta s) + \alpha^2\Delta x\Delta s - \big(1-\alpha(1-\sigma)\big)\tau e \\ &\underset{(10)}{=} xs - \alpha(xs - \sigma\tau e) + \alpha^2\Delta x\Delta s - \big(1-\alpha+\alpha\sigma\big)\tau e \\ &= (1-\alpha)(XSe - \tau e) + \alpha^2\Delta X\Delta S e\,. \end{aligned}$$

This yields (14) and with lemma 6.6.4 we obtain (15). □

In the next theorem we will show that all new iterates stay in the neighborhood, even for a full step with $\alpha = 1$. We will also supply a relation between β and σ.

Theorem 6.6.6

Let the parameters $\beta \in (0,1)$ and $\sigma \in (0,1)$ be chosen to satisfy

$$\frac{\beta^2 + n(1-\sigma)^2}{\sqrt{8}(1-\beta)} \le \sigma\beta\,. \tag{16}$$

If $(x,y,s) \in \mathcal{N}_2(\beta)$, we have $\big(x(\alpha),y(\alpha),s(\alpha)\big) \in \mathcal{N}_2(\beta)$ for all $\alpha \in [0,1]$.

Proof: By substituting (16) into (15) we get

$$\begin{aligned} \|X(\alpha)S(\alpha)e - \tau(\alpha)e\| &\le (1-\alpha)\beta\tau + \alpha^2\sigma\beta\tau \\ &\le (1-\alpha+\sigma\alpha)\beta\tau \\ &= \beta\tau(\alpha) \quad \text{(by lemma 6.4.1.b)} \end{aligned} \tag{17}$$

for $\alpha \in [0,1]$. It remains to show that $x(\alpha) \in \mathcal{F}_P^0$, and $\big(y(\alpha),s(\alpha)\big) \in \mathcal{F}_{D_e}^0$. By (10) it holds that:

$$Ax(\alpha) = A(x-\alpha\,\Delta x) = Ax - \alpha A\Delta x = b \quad \text{and}$$

$$A^Ty(\alpha)+s(\alpha) = A^T(y-\alpha\,\Delta y)+s-\alpha\,\Delta s = A^Ty - \alpha(A^T\Delta y + \Delta s) + s = c\,.$$

We still need $x(\alpha) > 0$ and $s(\alpha) > 0$. We know that $(x,s) > 0$. With (17), lemma 6.6.3 and lemma 6.4.1 we get

$$x_\nu(\alpha)s_\nu(\alpha) \ge (1-\beta)\tau(\alpha) = (1-\beta)\big(1-\alpha(1-\sigma)\big)\tau > 0$$

for $\nu \in \{1,\dots,n\}$. It follows that we can neither have $x_\nu(\alpha) = 0$ nor $s_\nu(\alpha) = 0$ and therefore $x(\alpha) > 0$ and $s(\alpha) > 0$ by continuity, hence $(x(\alpha),y(\alpha),s(\alpha)) \in \mathcal{F}^0$. □

We can easily verify that the parameters we chose in the short-step method satisfy (16):

$$\frac{\left(\frac{1}{2}\right)^2 + n\left(\frac{2}{5\sqrt{n}}\right)^2}{\sqrt{8}(1-\frac{1}{2})} = \frac{41}{100\sqrt{2}} \le \frac{3}{10} \le \left(1-\frac{2}{5\sqrt{n}}\right)\frac{1}{2}$$

We illustrate inequality (16) with a little picture for the graphs of the functions f and g defined by $f(d) := \frac{\beta^2+n(1-\sigma)^2}{\sqrt{8}\,(1-\beta)}$ and $g(d) := \sigma\beta$ for $d \in [0,1]$ with $\sigma := 1 - \frac{d}{\sqrt{n}}$ and $\beta := \frac{1}{2}$ for $n = 1$:

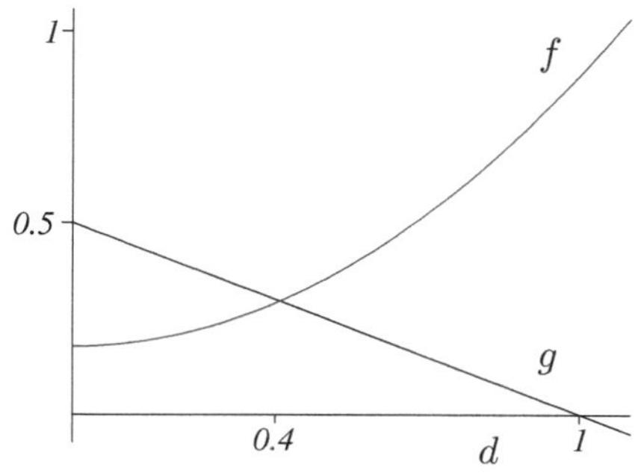

We see that $f(d) \le g(d)$ for all $d \in [0, 0.4]$.

6.7 The Mizuno–Todd–Ye Predictor-Corrector Method

In the short-step method we discussed above we chose a value slightly less than 1 for σ. By this choice we were able to ensure that the iterates did not approach the boundary of the feasible set too fast but at the same time τ could only slowly converge to zero. Therefore the algorithm made only slow progress toward the solution. Better results are obtained with the *predictor-corrector method* developed by MIZUNO, TODD and YE in 1993 (cf. [MTY]). It uses *two* nested 2-norm neighborhoods. In this method even-index iterates (i. e., $\left(x^{(k)}, y^{(k)}, s^{(k)}\right)$ with k even) are confined to the inner neighborhood, whereas the odd-index iterates are allowed to stay in the outer neighborhood but not beyond. Every second step of this method is a *predictor step* which starts in the inner neighborhood and moves to the boundary of a slightly larger neighborhood to compute a rough approximation. Between these predictor steps, the algorithm takes *corrector steps.* The corrector steps enable us to bring the iterates toward the central path and to get back into the inner neighborhood. Then we can start with the next predictor step. This algorithm is also a special case of the general algorithm. In the predictor step we choose $\sigma = 0$ to reduce τ. According to lemma 6.4.1, these steps reduce the value of τ by a factor of $(1-\alpha)$ where α is the step length. Corrector steps leave τ unchanged, we choose $\sigma = 1$ to improve centrality. By moving back into the inner neighborhood, the corrector step allows the algorithm to make better progress in the next predictor step.

The MIZUNO–TODD–YE predictor-corrector method is one of the most remarkable interior-point methods for linear and — more general — for quadratic optimization. It is based on a simple idea that is also used in the numerical solution of differential equations.

To describe the algorithm, suppose that two constants $\beta, \overline{\beta} \in [0,1]$ with $\beta < \overline{\beta}$ are given. We assume the inner neighborhood to be $\mathcal{N}_2(\beta)$ and the outer to be $\mathcal{N}_2(\overline{\beta})$.

Mizuno–Todd–Ye predictor-corrector algorithm

Let a starting vector $(x^{(0)}, y^{(0)}, s^{(0)}) \in \mathcal{N}_2(\beta)$ *and* $\varepsilon > 0$ *be given.*
Initialize $(x,y,s) := (x^{(0)}, y^{(0)}, s^{(0)})$ *and* $\tau := \frac{1}{n}\langle x^{(0)}, s^{(0)}\rangle$.
while $\tau > \varepsilon$

Predictor step

Solve (10) with $\sigma = 0$, *i. e., solve:*

$$\begin{pmatrix} 0 & A^T & I \\ A & 0 & 0 \\ S & 0 & X \end{pmatrix} \begin{pmatrix} \Delta x \\ \Delta y \\ \Delta s \end{pmatrix} = \begin{pmatrix} 0 \\ 0 \\ XSe \end{pmatrix} \tag{18}$$

Choose α *as the largest value in* $(0,1]$ *such that*

$$(x(\alpha), y(\alpha), s(\alpha)) \in \mathcal{N}_2(\overline{\beta})$$

and set $(x,y,s) := (x(\alpha), y(\alpha), s(\alpha))$, $\tau := (1-\alpha)\tau$.

Corrector step

Solve (10) with $\sigma = 1$, *i. e., solve*

$$\begin{pmatrix} 0 & A^T & I \\ A & 0 & 0 \\ S & 0 & X \end{pmatrix} \begin{pmatrix} \Delta x \\ \Delta y \\ \Delta s \end{pmatrix} = \begin{pmatrix} 0 \\ 0 \\ XSe - \tau e \end{pmatrix} \tag{19}$$

and set $(x,y,s) := (x,y,s) - (\Delta x, \Delta y, \Delta s)$.

We consider example 3 (cf. page 254) again to illustrate the algorithm. The picture shows the first iterates of the algorithm for the starting points $x = (0.32, 0.27, 0.41)^T$, $s = (11, 9, 8)^T$ and $y = -12$ with the neighborhoods $\mathcal{N}_2(1/4)$ and $\mathcal{N}_2(1/2)$:

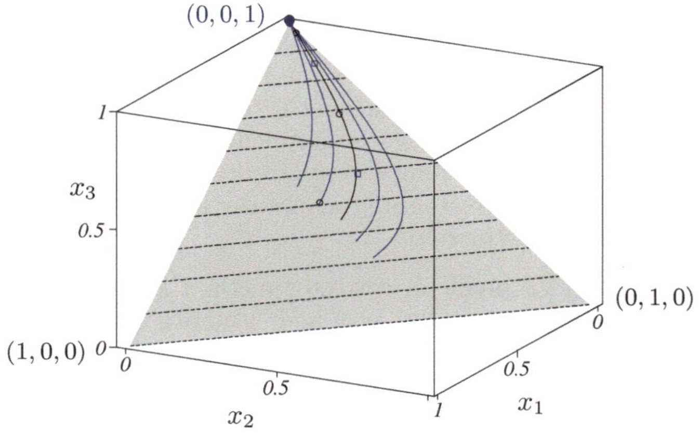

With $\beta := 1/4$ and $\overline{\beta} := 1/2$ we will have *a closer look at the predictor step:*

Lemma 6.7.1

Suppose that $(x, y, s) \in \mathcal{N}_2(1/4)$ and let $(\Delta x, \Delta y, \Delta s)$ be calculated from (10) with $\sigma = 0$ (predictor step). Then $(x(\alpha), y(\alpha), s(\alpha)) \in \mathcal{N}_2(1/2)$ holds for all $\alpha \in [0, \overline{\alpha}]$ with

$$\overline{\alpha} := \min\left(\frac{1}{2}, \left(\frac{\tau}{8\,\|\Delta X \Delta S e\|}\right)^{\frac{1}{2}}\right). \tag{20}$$

Hence, the predictor step has at least length $\overline{\alpha}$ and the new value of τ is at most $(1-\overline{\alpha})\tau$.

Proof: With (14) we get

$$\begin{aligned}
\|X(\alpha)S(\alpha)e - \tau(\alpha)e\| &\le |1-\alpha|\,\|XSe - \tau e\| + \alpha^2\,\|\Delta X \Delta S e\| \\
&\underset{(20)}{\le} |1-\alpha|\,\|XSe - \tau e\| + \frac{\tau}{8\,\|\Delta X \Delta S e\|}\,\|\Delta X \Delta S e\| \\
&\le \frac{1}{4}(1-\alpha)\tau + \frac{\tau}{8} \qquad \big(\text{as } (x,y,s) \in \mathcal{N}_2(1/4)\big) \\
&\le \frac{1}{4}(1-\alpha)\tau + \frac{1}{4}(1-\alpha)\tau \qquad \big(\text{since } \alpha \in [0, 1/2]\big) \\
&= \frac{1}{2}\tau(\alpha) \qquad \big(\text{by lemma 6.4.1.b with } \sigma = 0\big)
\end{aligned} \tag{21}$$

for $\alpha \in [0, \overline{\alpha}]$. So far we have shown that $(x(\alpha), y(\alpha), s(\alpha))$ satisfies the proximity condition for $\mathcal{N}_2(1/2)$. With (21), lemma 6.6.3 and lemma 6.4.1.b we get

$$x_\nu(\alpha)\, s_\nu(\alpha) \ge \frac{1}{2}\tau(\alpha) = \frac{1}{2}(1-\alpha)\tau > 0$$

for $\nu \in \{1, \ldots, n\}$. Using the same arguments as in the proof of theorem 6.6.6 we conclude $(x(\alpha), y(\alpha), s(\alpha)) \in \mathcal{F}^0$. □

From lemma 6.6.4 we get with $\beta = 1/4$ and $\sigma = 0$

$$\frac{\tau}{8\,\|\Delta X \Delta S e\|} \ge \frac{\tau}{8}\frac{\sqrt{8}\cdot 3/4}{1/16 + n} \cdot \frac{1}{\tau} = \frac{3\cdot\sqrt{2}}{1+16n} \ge \frac{3\cdot\sqrt{2}}{17n} \ge \frac{0.49^2}{n}.$$

This shows

$$\overline{\alpha} \ge \min\left(\frac{1}{2}, \frac{0.49}{\sqrt{n}}\right) = \frac{0.49}{\sqrt{n}}.$$

The following lemma gives a *description of the corrector steps* for the case of $\beta = \overline{\beta}^2$, hence especially for $\beta = 1/4$ and $\overline{\beta} = 1/2$. Any point of $\mathcal{N}_2(\overline{\beta})$ points toward the inner neighborhood $\mathcal{N}_2(\beta)$ without changing the value of τ:

Chapter 6

Lemma 6.7.2

Suppose that $(x, y, s) \in \mathcal{N}_2(\overline{\beta})$ *for* $\overline{\beta} \in \left(0, \frac{\sqrt{8}-1}{\sqrt{8}}\right]$ *and let* $(\Delta x, \Delta y, \Delta s)$ *be calculated from (10) with* $\sigma = 1$ *(corrector step). Then* $\big(x(1), y(1), s(1)\big)$ *belongs to* $\mathcal{N}_2(\overline{\beta}^2)$ *with* $\tau(1) = \tau$.[5]

Proof: $\tau(1) = \tau$ follows immediately from 6.4.1.b (with $\sigma = 1$). Then we substitute $\alpha = 1$ and $\sigma = 1$ into (15) and obtain

$$\|X(1)S(1)e - \tau(1)e\| \le \frac{\overline{\beta}^2}{\sqrt{8}(1-\overline{\beta})}\tau \overset{\checkmark}{\le} \overline{\beta}^2\tau = \overline{\beta}^2\tau(1). \tag{22}$$

So we know that $\big((x(1), y(1), s(1)\big)$ satisfies the proximity conditions for $\mathcal{N}_2(\overline{\beta}^2)$. As in the proof of theorem 6.6.6, we get that $\big(x(1), y(1), s(1)\big)$ belongs to $\mathcal{F}^0$. □

We know that the value of τ does not change in a corrector step, but because the predictor step reduces τ significantly, we can state the same kind of complexity result as for the short-step algorithm:

Lemma 6.7.3

Suppose that $\big(x^{(0)}, y^{(0)}, s^{(0)}\big) \in \mathcal{N}_2(\overline{\beta}^2)$. *If*

$$\tau_0 \le \frac{1}{\varepsilon^\nu}$$

holds for some positive constant ν, *then there exists an index* $K \in \mathbb{N}$ *with* $K = O(\sqrt{n}\,|\ln(\varepsilon)|)$ *such that*

$$\tau_k \le \varepsilon$$

for all $k \ge K$.

Proof: We have shown above that any corrector step leaves the value of τ unchanged. So we know that

$$\tau_{k+2} = \tau_{k+1} \quad \text{for} \quad k = 0, 2, 4, \ldots .$$

In the proof of lemma 6.6.4 we saw (with $\sigma = 0$ for a predictor step)

$$\|\Delta X \Delta S e\| \le \sqrt{2^{-3}} \left\|(XS)^{-\frac{1}{2}}(XSe)\right\|^2 \overset{\checkmark}{=} \sqrt{2^{-3}}\, n\tau .$$

With (20) we thus get for $n > 1$

$$\overline{\alpha} \ge \min\left(\frac{1}{2}, \frac{1}{\sqrt{n}\sqrt[4]{8}}\right) \overset{\checkmark}{=} \frac{1}{\sqrt{n}\sqrt[4]{8}}$$

[5] $\frac{\sqrt{8}-1}{\sqrt{8}} = 0.646\ldots$

Chapter 6

and therefore (compare lemma 6.7.1) we have

$$\tau_{k+2} = \tau_{k+1} \le \left(1 - \frac{1}{\sqrt{n}\sqrt[4]{8}}\right)\tau_k \quad \text{for} \quad k = 0, 2, 4, \dots .$$

For $n = 1$ we have $\overline{\alpha} \ge 1/2$ and so

$$\tau_{k+2} = \tau_{k+1} \le \frac{1}{2}\tau_k \quad \text{for} \quad k = 0, 2, 4, \dots .$$

If we set $\delta := 1/\sqrt[4]{8} = 0.5946\dots$ and $\omega := 1/2$ in the case of $n > 1$ and $\delta := 1/2$, ω arbitrary for $n = 1$, almost all the prerequisites for theorem 6.4.2 are given. The odd iterates are missing in the inequality above. This fact, however, does not affect the given proof. □

6.8 The Long-Step Path-Following Algorithm

Contrary to the short-step algorithm the long-step algorithm does not use the 2-norm neighborhoods $\mathcal{N}_2(\beta)$ but the neighborhoods $\mathcal{N}_{-\infty}(\gamma)$ defined by

$$\mathcal{N}_{-\infty}(\gamma) = \left\{(x, y, s) \in \mathcal{F}^0 \mid x_\nu s_\nu \ge \gamma\tau(xs) \text{ for all } \nu = 1, \dots, n\right\}$$

for some $\gamma \in [0, 1]$. This neighborhood is much more expansive compared to the $\mathcal{N}_2(\beta)$ neighborhood: The $\mathcal{N}_2(\beta)$ neighborhood contains only a small area of the strictly feasible set whereas with the $\mathcal{N}_{-\infty}(\gamma)$ neighborhood we almost get the entire strictly feasible set if γ is small. The long-step method can make fast progress because of the choice of the neighborhood for γ close to zero. We obtain the search direction by solving system (10) and choose the step length α as large as possible such that we still stay in the neighborhood $\mathcal{N}_{-\infty}(\gamma)$. Another difference with the short-step method is the choice of the centering parameter σ. In the short-step algorithm we choose σ slightly smaller than 1 whereas in the long-step algorithm we are a little more aggressive. We take a smaller value for σ which lies between an upper bound σ_{max} and a lower bound σ_{min}, independent of the dimension of the problem, to ensure a larger decrease in τ. The lower bound σ_{min} ensures that each search direction begins by moving away from the boundary of $\mathcal{N}_{-\infty}(\gamma)$ and closer to the central path. In the short-step method we have a constant value for the centering parameter and α is constantly 1, in the long-step method we choose a new σ in every iteration and have to perform a line search to choose α as large as possible such that we still stay in the neighborhood $\mathcal{N}_{-\infty}(\gamma)$.

Long-step algorithm

Let $\gamma \in (0,1)$, $\sigma_{\min}$ *and* $\sigma_{\max}$ *with* $0 < \sigma_{\min} \le \sigma_{\max} < 1$, *a starting vector* $\left(x^{(0)}, y^{(0)}, s^{(0)}\right) \in \mathcal{N}_{-\infty}(\gamma)$ *and an accuracy requirement* $\varepsilon > 0$ *be given.*

Initialize $(x,y,s) := \left(x^{(0)}, y^{(0)}, s^{(0)}\right)$ *and* $\tau := \frac{1}{n}\langle x, s\rangle$.

while $\tau > \varepsilon$

Choose $\sigma \in \left[\sigma_{\min}, \sigma_{\max}\right]$. *Determine the* NEWTON *step* $(\Delta x, \Delta y, \Delta s)$ *at* (x,y,s) *according to (10).*

Choose α *as the largest value in* $(0,1]$ *such that*

$$\left(x(\alpha), y(\alpha), s(\alpha)\right) \in \mathcal{N}_{-\infty}(\gamma)\,.$$

Set $(x,y,s) := \left(x(\alpha), y(\alpha), s(\alpha)\right)$ *and* $\tau := \left(1 - \alpha(1-\sigma)\right)\tau$.

Lemma 6.8.1

If $(x,y,s) \in \mathcal{N}_{-\infty}(\gamma)$, *then*

$$\|\Delta X \Delta S e\| \le 2^{-\frac{3}{2}}\left(1 + \frac{1}{\gamma}\right) n\tau\,.$$

Proof: As in the proof of lemma 6.6.4 we obtain the following by considering $\langle x, s\rangle = n\tau$, $\langle e, e\rangle = n$, $x_\nu s_\nu \ge \gamma\tau$ for $\nu = 1, \ldots, n$ and $\sigma \in (0,1)$:

$$\begin{aligned}
\|\Delta X \Delta S e\| &\le \sqrt{2^{-3}}\, \left\|(XS)^{-\frac{1}{2}}(XSe - \sigma\tau e)\right\|^2 \\
&= \sqrt{2^{-3}}\, \left\|(XS)^{\frac{1}{2}}e - \sigma\tau (XS)^{-\frac{1}{2}} e\right\|^2 \\
&\le \sqrt{2^{-3}}\left(\langle x, s\rangle - 2\sigma\tau n + \sigma^2\tau^2 \frac{n}{\gamma\tau}\right) \\
&= \sqrt{2^{-3}}\, n\tau\left(1 - 2\sigma + \frac{\sigma^2}{\gamma}\right) \\
&\le \sqrt{2^{-3}}\, n\tau\left(1 + \frac{1}{\gamma}\right)
\end{aligned}$$

□

Theorem 6.8.2

If $(x,y,s) \in \mathcal{N}_{-\infty}(\gamma)$, *then we have* $\left(x(\alpha), y(\alpha), s(\alpha)\right) \in \mathcal{N}_{-\infty}(\gamma)$ *for all* $\alpha \in \left(0, \overline{\alpha}\right]$ *with*

$$\overline{\alpha} := \sqrt{8}\gamma\sigma\frac{1-\gamma}{n(1+\gamma)}\,.$$

Proof: With lemma 6.8.1 it follows for $\nu = 1, \ldots, n$:

$$|\Delta x_\nu \, \Delta s_\nu| \;\le\; \|\Delta X \Delta s\| \;\le\; \sqrt{2^{-3}}\, n\tau\left(1+\frac{1}{\gamma}\right).$$

By using $S\Delta x + X\Delta s = XSe - \sigma\tau e$ (the third row of (10)) once more, we get with $x_\nu s_\nu \ge \gamma\tau$ that

$$\begin{aligned} x(\alpha)\, s(\alpha) &= xs - \alpha(s\Delta x + x\Delta s) + \alpha^2 \Delta x\,\Delta s \\ &\ge xs(1-\alpha) + \alpha\sigma\tau e - \alpha^2\,|\Delta x\,\Delta s| \\ &\ge \gamma\tau(1-\alpha)e + \alpha\sigma\tau e - \alpha^2\sqrt{2^{-3}}\, n\tau\left(1+\frac{1}{\gamma}\right)e. \end{aligned}$$

By lemma 6.4.1 we know that $\tau(\alpha) = \big(1-\alpha(1-\sigma)\big)\tau$. So we get

$$x(\alpha)\, s(\alpha) \ge \gamma\tau(\alpha)e \text{ if } \gamma\tau(1-\alpha)+\alpha\sigma\tau-\alpha^2\sqrt{2^{-3}}\,n\tau(1+\frac{1}{\gamma}) \ge \gamma(1-\alpha+\alpha\sigma)\tau,$$

and that is the case iff $\alpha \le \sqrt{8}\gamma\sigma\frac{1}{n}\frac{1-\gamma}{1+\gamma} = \overline{\alpha}$. Hence, for $\alpha \in \big(0, \overline{\alpha}\big]$ the triple $\big(x(\alpha), y(\alpha), s(\alpha)\big)$ satisfies the proximity condition. It remains to show that $\big(x(\alpha), y(\alpha), s(\alpha)\big) \in \mathcal{F}^0$ holds for these α: Analogously to the proof of theorem 6.6.6 we get $Ax(\alpha) = b$ and $A^T y(\alpha) + s(\alpha) = c$. As $\gamma \in (0,1)$, we know that $\gamma(1-\gamma) \le 1/4$ holds and so

$$\alpha \le \frac{\sqrt{8}}{n}\sigma\gamma\frac{1-\gamma}{1+\gamma} \le \frac{\sqrt{8}}{n}\frac{1}{4}\sigma\frac{1}{1+\gamma} < \frac{1}{\sqrt{2}n} < 1.$$

We have shown above that $x(\alpha)\, s(\alpha) \ge \gamma\big(1-\alpha(1-\sigma)\big)\tau e > 0$ holds for all $\alpha \in \big(0, \overline{\alpha}\big]$ and we know that $x(0) > 0$ and $s(0) > 0$. As in the proof of theorem 6.6.6 we conclude $\big(x(\alpha), y(\alpha), s(\alpha)\big) \in \mathcal{F}^0$. □

With this lemma we have obtained a lower bound $\overline{\alpha}$ for the step length α in the long-step algorithm.

Lemma 6.8.3

If the parameters γ, $\underline{\sigma} := \sigma_{\min}$ and $\overline{\sigma} := \sigma_{\max}$ are given, there exists a constant $\delta > 0$ independent of n such that

$$\tau_{k+1} \le \left(1 - \frac{\delta}{n}\right)\tau_k \quad \textit{for} \quad k = 1, 2, 3, \ldots .$$

Proof: By lemma 6.4.1 and as $\alpha \ge \overline{\alpha}$, we get

$$\tau_{k+1} = (1-\alpha(1-\sigma))\tau_k \le \left(1 - \frac{\sqrt{8}}{n}\gamma\frac{1-\gamma}{1+\gamma}\sigma(1-\sigma)\right)\tau_k.$$

The quadratic function $\sigma \longmapsto \sigma(1-\sigma)$ is strictly concave and therefore has its minimizer in the compact interval $[\underline{\sigma}, \overline{\sigma}] \subset (0,1)$ in one of the endpoints. So we obtain

$$\sigma(1-\sigma) \geq \min\left\{\underline{\sigma}(1-\underline{\sigma}), \overline{\sigma}(1-\overline{\sigma})\right\} =: \widehat{\sigma}$$

for all $\sigma \in [\underline{\sigma}, \overline{\sigma}]$. Now, $\delta := \sqrt{8}\gamma \frac{1-\gamma}{1+\gamma}\widehat{\sigma} > 0$ completes the proof. □

Theorem 6.8.4

For $\varepsilon, \gamma \in (0,1)$ suppose that the starting point $\left(x^{(0)}, y^{(0)}, s^{(0)}\right) \in \mathcal{N}_{-\infty}(\gamma)$ satisfies

$$\tau_0 \leq \frac{1}{\varepsilon^{\nu}}$$

for some positive constant ν. Then there exists an index K with $K = O(n \,|\ln \varepsilon|)$ such that

$$\tau_k \leq \varepsilon$$

for all $k \geq K$.

Proof: By theorem 6.4.2 and lemma 6.8.3 we get the result with $\omega := 1$. □

6.9 The Mehrotra Predictor-Corrector Method

In the algorithms we have discussed so far we always needed a strictly feasible starting point. For most problems such a starting point is difficult to find. For some problems a strictly feasible starting point does not even exist (compare example 2).

Chapter 6

Infeasible algorithms just require that $\left(x^{(0)}, s^{(0)}\right) > 0$ and neglect the feasibility. In this section we will look at an *infeasible interior-point* algorithm, the MEHROTRA *predictor-corrector method.* This algorithm generates a sequence of iterates $\left(x^{(k)}, y^{(k)}, s^{(k)}\right)$ permitted to be infeasible with $\left(x^{(k)}, s^{(k)}\right) > 0$. It departs from the general algorithm and also differs from the MIZUNO–TODD–YE algorithm. Unlike this method MEHROTRA's algorithm does not require an extra matrix factorization for the corrector step.

MEHROTRA obtained a breakthrough via a skillful synthesis of known ideas and path-following techniques. A lot of the currently used interior-point software for linear optimization (since 1992) is based on MEHROTRA's algorithm. The method is among the most effective for solving wide classes of linear problems. The improvement is obtained by including the effect of the second-order term. Unfortunately no rigorous convergence theory is known for MEHROTRA's algorithm. There is a gap between the practical behavior and the theoretical results, in favor of the practical behavior. Obviously, this gap is unsatisfactory.

MEHROTRA's approach uses three directions at each iteration, the predictor, the corrector and a centering direction, where the centering and the corrector step are accomplished together. As the method considers possibly infeasible points, we not only have to check that τ is small enough but also that the residuals $\|Ax-b\|$ and $\|A^Ty+s-c\|$ are small. One big advantage of this algorithm is the adaptive choice of σ. We can choose a new σ in every step instead of choosing a fixed value for it in the beginning and keeping it. In the *predictor step* we solve

$$\begin{pmatrix} 0 & A^T & I \\ A & 0 & 0 \\ S & 0 & X \end{pmatrix} \begin{pmatrix} \Delta x^{\mathrm{N}} \\ \Delta y^{\mathrm{N}} \\ \Delta s^{\mathrm{N}} \end{pmatrix} = \begin{pmatrix} A^T y + s - c \\ Ax - b \\ Xs \end{pmatrix} =: \begin{pmatrix} r_d \\ r_p \\ r_c \end{pmatrix}, \tag{23}$$

where $(\Delta x^{\mathrm{N}}, \Delta y^{\mathrm{N}}, \Delta s^{\mathrm{N}})$ denotes the *affine-scaling* or NEWTON *direction.* Then we calculate the largest step lengths α_p^{N}, $\alpha_d^{\mathrm{N}} \in (0,1]$ such that

$$x(\alpha_p^{\mathrm{N}}) := x - \alpha_p^{\mathrm{N}} \Delta x^{\mathrm{N}} \ge 0, \ s(\alpha_p^{\mathrm{N}}) := s - \alpha_d^{\mathrm{N}} \Delta s^{\mathrm{N}} \ge 0$$

by setting

$$\alpha_p^{\mathrm{N}} := \min\left\{1, \min_{\Delta x_i^{\mathrm{N}}>0} \frac{x_i}{\Delta x_i^{\mathrm{N}}}\right\}, \ \alpha_d^{\mathrm{N}} := \min\left\{1, \min_{\Delta s_i^{\mathrm{N}}>0} \frac{s_i}{\Delta s_i^{\mathrm{N}}}\right\}.$$

Thereby we get a *damped* NEWTON *step* for the optimality conditions (1). α_p^{N} and α_d^{N} give the maximum step lengths that can be taken without violating the nonnegativity condition. Now we compute $\tau^{\mathrm{N}} := \frac{1}{n}\langle x(\alpha_p^{\mathrm{N}}), s(\alpha_d^{\mathrm{N}})\rangle$ and with

$$\sigma := \left(\frac{\tau^{\mathrm{N}}}{\tau}\right)^3$$

obtain a heuristic choice of the centering parameter σ at each iteration. This choice has been shown to work well in practice. If τ^{N} is a lot smaller than τ, then we have a small σ and the duality gap is reduced significantly by the NEWTON step. The next step is to approximate the central path in a *corrector step.* We solve the equation

$$\begin{pmatrix} 0 & A^T & I \\ A & 0 & 0 \\ S & 0 & X \end{pmatrix} \begin{pmatrix} \Delta x \\ \Delta y \\ \Delta s \end{pmatrix} = \begin{pmatrix} A^T y + s - c \\ Ax - b \\ Xs - \sigma\tau e \end{pmatrix} + \begin{pmatrix} 0 \\ 0 \\ \Delta X^{\mathrm{N}} \Delta s^{\mathrm{N}} \end{pmatrix}. \tag{24}$$

We see that we use the same coefficient matrix in the corrector and the predictor step, only the right-hand sides differ slightly. Hence only *one* matrix factorization is necessary at each iteration. In this step we set — with $\eta \in (0,1)$ such that $\eta \approx 1$, for example $\eta = 0.99$ —

$$\alpha_p := \min\left\{1, \eta \min_{\Delta x_i>0} \frac{x_i}{\Delta x_i}\right\} \text{ and } \alpha_d := \min\left\{1, \eta \min_{\Delta s_i>0} \frac{s_i}{\Delta s_i}\right\}.$$

$\min_{\Delta x_i > 0} x_i/\Delta x_i$ and $\min_{\Delta s_i > 0} s_i/\Delta s_i$ give the maximum step length that can be taken without violating the nonnegativity condition $x \geq 0$ and $s \geq 0$. Here we choose values slightly less than this maximum.

In the context of general nonlinear equations, the idea of reusing the first derivative to accelerate the convergence of NEWTON's method is well-known, for example in [Ca/Ma] cited there as the CHEBYSHEV *method:* We can understand the corrector step as a modified CHEBYSHEV method: We want to find the zeros of $F_{\sigma\tau}$ given by

$$F_{\sigma\tau}(x, y, s) = \begin{pmatrix} A^T y + s - c \\ Ax - b \\ Xs - \sigma\tau e \end{pmatrix}.$$

It holds that

$$\begin{aligned} F_{\sigma\tau}(x - \Delta x, y - \Delta y, s - \Delta s) &= \begin{pmatrix} A^T(y - \Delta y) + (s - \Delta s) - c \\ A(x - \Delta x) - b \\ (X - \Delta X)(s - \Delta s) - \sigma\tau e \end{pmatrix} \\ &= F_{\sigma\tau}(x, y, s) - \begin{pmatrix} 0 & A^T & I \\ A & 0 & 0 \\ S & 0 & X \end{pmatrix} \begin{pmatrix} \Delta x \\ \Delta y \\ \Delta s \end{pmatrix} + \begin{pmatrix} 0 \\ 0 \\ \Delta X \Delta s \end{pmatrix} \\ &= F_{\sigma\tau}(x, y, s) - DF_{\sigma\tau}(x, y, s) \begin{pmatrix} \Delta x \\ \Delta y \\ \Delta s \end{pmatrix} + \tfrac{1}{2} D^2 F_{\sigma\tau}(x, y, s) \begin{pmatrix} \Delta x \\ \Delta y \\ \Delta s \end{pmatrix}^2. \end{aligned}$$

Hence we solve $\begin{pmatrix} 0 & A^T & I \\ A & 0 & 0 \\ S & 0 & X \end{pmatrix} \begin{pmatrix} \Delta x \\ \Delta y \\ \Delta s \end{pmatrix} = F_{\sigma\tau}(x, y, s) + \begin{pmatrix} 0 \\ 0 \\ \Delta X^{\mathrm{N}} \Delta s^{\mathrm{N}} \end{pmatrix}.$

Chapter 6

Mehrotra predictor-corrector algorithm

Let $(x^{(0)}, y^{(0)}, s^{(0)})$ *with* $(x^{(0)}, s^{(0)}) > 0$ *and an accuracy requirement* $\varepsilon > 0$ *be given. Initialize* $(x, y, s) := (x^{(0)}, y^{(0)}, s^{(0)})$ *and* $\tau := \frac{1}{n} \langle x, s \rangle$.

while $\max\left(\tau, \|Ax - b\|, \|A^T y + s - c\|\right) > \varepsilon$

Predictor step

Solve (23) for $(\Delta x^{\mathrm{N}}, \Delta y^{\mathrm{N}}, \Delta s^{\mathrm{N}})$.

Calculate the maximal possible step length α_p^{N}, α_d^{N} *and* τ^{N}.

Set $\sigma = \left(\frac{\tau^{\mathrm{N}}}{\tau_k}\right)^3$.

Corrector step

Solve (24) for $(\Delta x, \Delta y, \Delta s)$.

Calculate the primal and dual step length α_p *and* α_d.

Set $(x, y, s) := (x - \alpha_p \Delta x,\ y - \alpha_d \Delta y,\ s - \alpha_d \Delta s)$ *and* $\tau := \frac{1}{n} \langle x, s \rangle$.

In the following picture we can nicely see that the first of three iterates does not lie in the feasible set. Here we used the starting points $x = (0.3, 0.8, 0.7)^T$, $s = (11, 9, 8)^T$ and $y = -10$.

Iterates of the MEHROTRA *predictor-corrector algorithm*

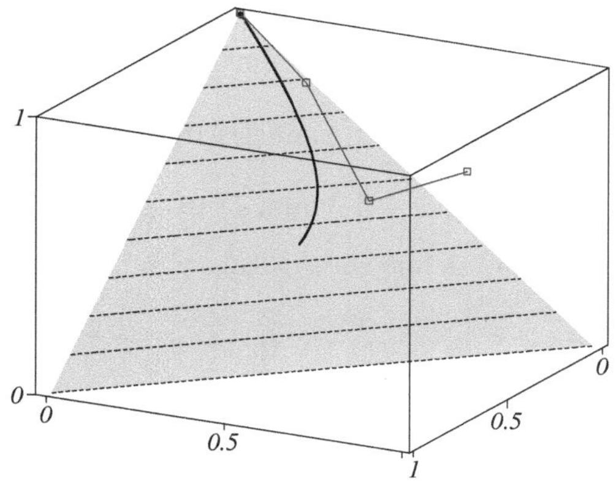

Starting Points

Problems (P) and (D) in canonical form can be embedded in a natural way into a slightly larger *self-dual homogeneous problem* which fulfills the interior-point condition (cf. [RTV], chapter 2). This approach has the following advantages:

- It is easy to find a starting point for the augmented problem. *Any* of the feasible interior points can be used to solve it.
- The solution of the self-dual problem gives us information about whether the original problems have solutions or not.
- In the affirmative case we are able to extract them from the solution of the augmented problem.

We, however, will not pursue this approach any further since our main interest lies with MEHROTRA's predictor-corrector method. Its main advantage over the methods discussed so far is that it does not necessarily need feasible starting points $\left(x^{(0)}, y^{(0)}, s^{(0)}\right)$ with $x^{(0)}, s^{(0)} \in \mathbb{R}^n_{++}$. The following heuristic developed by MEHROTRA has proved useful (cf. [Meh], p. 589, [No/Wr], p. 410):

We start from the least squares problems

$$\|x\| \longrightarrow \min_{x \in \mathbb{R}^n} \quad \text{such that} \quad Ax = b$$

and

$$\|s\| \longrightarrow \min_{y \in \mathbb{R}^m,\, s \in \mathbb{R}^n} \quad \text{such that} \quad A^T y + s = c\,.$$

The optimizers to these problems are given by

$$\widetilde{x} := A^T\big(AA^T\big)^{-1} b \quad \text{and} \quad \widetilde{y} := \big(AA^T\big)^{-1} Ac, \quad \widetilde{s} := c - A^T\widetilde{y}\,.$$

In these cases $\widetilde{x}, \widetilde{s} \in \mathbb{R}^n_{++}$ generally does not hold. By using

$$\delta_x := \max\big(-1.5\min(\widetilde{x}), 0\big) + 1/n\,,$$

$$\delta_s := \max\big(-1.5\min(\widetilde{s}), 0\big) + 1/n\,,$$

we adjust these vectors and define the starting point as follows

$$x^0 := \widetilde{x} + \delta_x e, \quad y^0 := \widetilde{y}, \quad s^0 := \widetilde{s} + \delta_s e\,.$$

For practice purposes, we will provide a *Matlab*® program for our readers. We refrained from the consideration of a number of technical details for the benefit of transparency. It is surprising how easily MEHROTRA's method can then be implemented:

```
function [x,y,s,f,iter] = PrimalDualLP(A,b,c)

% Experimental version of Mehrotra's primal-dual interior-point method
% for linear programming
%
% primal problem: min c'x  s.t.  Ax = b, x >= 0
% dual problem:   max b'y  s.t.  A'y + s = c,  s >= 0

Maxiter = 100; Tol = 1.e-8;
UpperB  = 1.e10*max([norm(A),norm(b),norm(c)]);
[m,n]   = size(A);

% starting point
e = ones(n,1); M = A*A'; R = chol(M);
x = A'*(R \ (R' \ b));  y = R \ (R' \ (A*c));  s = c - A'*y;
delta_x = max(-1.5*min(x),0)+1/n;  x = x+delta_x*e;
delta_s = max(-1.5*min(s),0)+1/n;  s = s+delta_s*e;

for iter = 0 : Maxiter


  f = c'*x;

  % residuals
  rp  = A*x-b;          % primal residual
```

```
rd  = A'*y+s-c;       % dual residual
rc  = x.*s;           % complementarity
tau = x'*s/n;         % duality measure

residue = norm(rc,1)/(1+abs(b'*y));
primalR = norm(rp)/(1+norm(x)); dualR = norm(rd)/(1+norm(y));
STR1 = 'iter %2i: f = %14.5e, residue = %10.2e';
STR2 = 'primalR = %10.2e, dualR = %10.2e\n';
fprintf([STR1 STR2], iter, f, residue, primalR, dualR);
if (norm(x)+norm(s) >= UpperB)
  error('Problem possibly infeasible!'); end
if (max([residue; primalR; dualR]) <= Tol) break; end

% coefficient matrix of the linear systems
M = A*diag(x./s)*A'; R = chol(M);  % M = R'R

% predictor step with maximal Newton step size
rhs = rp - A*((rc-x.*rd) ./ s);
dy  = R \ (R' \ rhs);  % dy = M \ rhs
ds  = rd-A'*dy; dx = (rc-x.*ds) ./ s;
alpha_p = 1/max([1; dx./x]); alpha_d = 1/max([1; ds./s]);
tau_N = ((x-alpha_p*dx)'*(s-alpha_d*ds))/n;

% corrector step: correct towards center path
sigma = (tau_N/tau)^3; % Mehrotra
rc  = rc - sigma*tau + dx.*ds;
rhs = rp - A*((rc-x.*rd) ./ s);
dy  = R \ (R' \ rhs);  % dy = M \ rhs
ds  = rd-A'*dy; dx = (rc-x.*ds) ./ s;
eta = max(0.99,1-tau);
alpha_p = eta/max([eta; dx./x]); alpha_d = eta/max([eta; ds./s]);
x = x-alpha_p*dx; y = y-alpha_d*dy; s = s-alpha_d*ds;

end % for-loop
```


6.10 A Short Look at Interior-Point Methods for Quadratic Optimization

So far we have dealt with linear problems, now we will have a look at quadratic problems. We consider the problem

$$(QP) \qquad \begin{cases} \frac{1}{2}\langle x, Cx\rangle + \langle c, x\rangle \longrightarrow \min \\ Ax = b, \ x \geq 0\,, \end{cases}$$

where C is a symmetric positive definite (n,n)-matrix, A an (m,n)-matrix with $m \leq n$, $c \in \mathbb{R}^n$, $b \in \mathbb{R}^m$, and $x \in \mathbb{R}^n$ is the unknown variable. In the following we again assume that the matrix A has full rank, that is, $\operatorname{rank}(A) = m$.

Theorem 6.10.1 (*Optimality condition*)

A given vector $x^* \in \mathbb{R}^n$ is a minimizer to problem (QP) iff there exist vectors $y^* \in \mathbb{R}^m$ and $s^* \in \mathbb{R}^n_+$ that solve

$$\begin{aligned} A^T y^* + s^* - Cx^* - c &= 0 \\ Ax^* - b &= 0 \\ x_i^* s_i^* &= 0 \qquad (i = 1, \ldots, n) \\ x^*,\, s^* &\geq 0. \end{aligned} \tag{25}$$

Proof: If x^* is a minimizer to problem (QP), there exist multipliers $y^* \in \mathbb{R}^m$ and $s^* \in \mathbb{R}^n_+$ such that the triple (x^*, y^*, s^*) satisfies the KKT conditions. The LAGRANGE function L to problem (QP) is given by

$$L(x, y, s) = \frac{1}{2} \langle x, Cx \rangle + \langle c, x \rangle + \langle y, b - Ax \rangle + \langle s, -x \rangle$$

for $x \in \mathbb{R}^n$, $s \in \mathbb{R}^n_+$ and $y \in \mathbb{R}^m$. Hence the KKT conditions are:

$$\begin{aligned} \nabla_x L(x^*, y^*, s^*) &= Cx^* + c - A^T y^* - s^* = 0 \\ Ax^* &= b \\ x_i^* s_i^* &= 0 \qquad (i = 1, \ldots, n) \\ x^*,\, s^* &\geq 0\,, \end{aligned}$$

and so we get (25). On the other hand, if we have (25), then (QP) attains its global minimum at x^* by theorem 2.2.8. We repeat the surprisingly simple argument: The objective function f defined by $f(x) := \frac{1}{2} \langle x, Cx \rangle + \langle c, x \rangle$ is differentiable and convex. For any feasible x we thus get

$$\begin{aligned} f(x) - f(x^*) &\geq \langle \nabla f(x^*), x - x^* \rangle \\ &= \langle Cx^* + c, x - x^* \rangle = \big\langle A^T y^* + s^*, x - x^* \big\rangle \\ &= \langle y^*, Ax - Ax^* \rangle + \langle s^*, x \rangle = \langle s^*, x \rangle \geq 0. \end{aligned}$$

Hence x^* is a (global) minimizer to (QP). □

For a given pair (y, s), the function $L(\,\cdot\,, y, s)$ is convex and differentiable. Therefore a necessary and sufficient condition for a minimizer x_0 to $L(\,\cdot\,, y, s)$ is that the gradient vanishes, that is,

$$0 = \nabla_x L(x_0, y, s) = Cx_0 + c - A^T y - s\,.$$

Hence,
$$\begin{aligned} \varphi(y, s) := \inf_{x \in \mathbb{R}^n} L(x, y, s) &= L(x_0, y, s) \\ &= \tfrac{1}{2} \langle x_0, Cx_0 \rangle + \langle c, x_0 \rangle + \langle y, b \rangle - \big\langle x_0, A^T y + s \big\rangle \\ &= \tfrac{1}{2} \langle x_0, Cx_0 \rangle + \langle c, x_0 \rangle + \langle y, b \rangle - \langle x_0, Cx_0 + c \rangle \\ &= \langle y, b \rangle - \tfrac{1}{2} \langle x_0, Cx_0 \rangle \\ &= \langle y, b \rangle - \tfrac{1}{2} \big\langle C^{-1}(A^T y + s - c), A^T y + s - c \big\rangle. \end{aligned}$$

Chapter 6

Thus, the corresponding *dual problem* can be written as follows:

$$(QD_e) \quad \begin{cases} \langle y, b\rangle - \frac{1}{2}\big\langle C^{-1}(A^Ty + s - c), A^Ty + s - c\big\rangle \longrightarrow \max \\ y \in \mathbb{R}^m, \; s \in \mathbb{R}^n_+ \end{cases}$$

As in section 6.1, we rearrange system (25) in the following way:

$$H_0(x, y, s) := \begin{pmatrix} A^Ty + s - Cx - c \\ Ax - b \\ Xs \end{pmatrix} = \begin{pmatrix} 0 \\ 0 \\ 0 \end{pmatrix}, \quad x, s \geq 0 \qquad (26)$$

Now we consider the logarithmic barrier function Ψ_μ for problem (QP) defined by

$$\Psi_\mu(x) := \frac{1}{2}\langle x, Cx\rangle + \langle c, x\rangle - \mu\sum_{i=1}^n \log(x_i) \qquad (x \in \mathcal{F}_P^0)$$

for $\mu > 0$. We will show in the same way as in the proof of theorem 6.1.4 that Ψ_μ has a (unique) minimizer iff the system

$$H_\mu(x, y, s) := \begin{pmatrix} A^Ty + s - Cx - c \\ Ax - b \\ Xs - \mu e \end{pmatrix} = \begin{pmatrix} 0 \\ 0 \\ 0 \end{pmatrix}, \quad x, s \geq 0 \qquad (27)$$

has a (unique) solution:

Theorem 6.10.2

There exists a minimizer to Ψ_μ on $\mathcal{F}_P^0$ iff system (27) has a solution. The minimizer to Ψ_μ and the solution of (27), if they exist, are unique.

Proof: The definition of Ψ_μ can be extended to the open set $\mathbb{R}^n_{++}$ and Ψ_μ is differentiable there. On $\mathbb{R}^n_{++}$ we consider the system

$$(QP_\mu) \quad \begin{cases} \Psi_\mu(x) \longrightarrow \min \\ Ax = b. \end{cases}$$

(QP_μ) is a convex problem with linear constraints. So we know that a vector $x \in \mathbb{R}^n_{++}$ is a (unique) minimizer to problem (QP_μ) iff there exists a (unique) multiplier $y \in \mathbb{R}^m$ such that the pair (x, y) satisfies the KKT conditions to problem (QP_μ). The LAGRANGE function L to this problem is given by $L(x, y) := \Psi_\mu(x) + \langle y, b - Ax\rangle$. The KKT conditions to (QP_μ) are

$$\begin{aligned} \nabla_x L(x, y) &= Cx + c - \mu X^{-1}e - A^Ty = 0 \\ Ax &= b. \end{aligned}$$

If we set $s := \mu X^{-1}e$ which is equivalent to $Xs = \mu e$, we get (27). So we have shown that system (27) represents the optimality conditions for problem (QP_μ). □

The *central path* is the set of solutions $\big\{(x(\mu), y(\mu), s(\mu)) \mid \mu > 0\big\}$ of (27).

Theorem 6.10.3

The Jacobian

$$DH_\mu(x, y, s) = \begin{pmatrix} -C & A^T & I \\ A & 0 & 0 \\ S & 0 & X \end{pmatrix}$$

is nonsingular if $x > 0$ *and* $s > 0$.

Proof: For $(u, v, w) \in \mathbb{R}^n \times \mathbb{R}^m \times \mathbb{R}^n$ with

$$\begin{pmatrix} 0 \\ 0 \\ 0 \end{pmatrix} = DH_\mu(x, y, s) \begin{pmatrix} u \\ v \\ w \end{pmatrix} = \begin{pmatrix} -Cu + A^T v + w \\ Au \\ Su + Xw \end{pmatrix}$$

we have $\langle u, w \rangle = \langle u, Cu - A^T v \rangle = \langle u, Cu \rangle$. On the other hand we have $\langle u, w \rangle = \langle u, -X^{-1} S u \rangle$. As C and $X^{-1}S$ are positive definite, we have for all $u \neq 0$ that $0 < \langle u, Cu \rangle = \langle u, -X^{-1} S u \rangle < 0$, hence we can conclude that $u = 0$. From $0 = Su + Xw = Xw$ and $x \neq 0$ we get $w = 0$ and from $0 = -Cu + A^T v + w = A^T v$ we get $v = 0$ since $\operatorname{rank}(A) = m$. □

Hence we obtain the NEWTON direction at a point $(x, y, s) = (x_{\sigma\tau}, y_{\sigma\tau}, s_{\sigma\tau})$ on the central path with $\mu = \sigma\tau$ by solving

$$\begin{pmatrix} -C & A^T & I \\ A & 0 & 0 \\ S & 0 & X \end{pmatrix} \begin{pmatrix} \Delta x \\ \Delta y \\ \Delta s \end{pmatrix} = \begin{pmatrix} r_q \\ r_p \\ r_c \end{pmatrix} \tag{28}$$

where $r_q := A^T y + s - Cx - c$, $r_p := Ax - b$ and $r_c := Xs - \sigma\tau e$.

The next iterate is given by

$$(x^+, y^+, s^+) := (x, y, s) - \alpha \left(\Delta x, \Delta y, \Delta s\right)$$

with a suitable $\alpha \in (0, 1]$ such that $(x^+, s^+) > 0$. We denote the *primal–dual feasible set* by

$$\mathcal{F}_q := \left\{(x, y, s) \in \mathbb{R}^n_+ \times \mathbb{R}^m \times \mathbb{R}^n_+ \mid Ax = b,\ A^T y - Cx + s = c\right\}$$

and the *strictly primal–dual feasible set* by

$$\mathcal{F}_q^0 := \left\{(x, y, s) \in \mathbb{R}^n_{++} \times \mathbb{R}^m \times \mathbb{R}^n_{++} \mid Ax = b,\ A^T y - Cx + s = c\right\}.$$

Now we can give a general framework for solving quadratic problems by using interior-point methods:

Chapter 6

General interior-point algorithm for quadratic problems

Let $\left(x^{(0)}, y^{(0)}, s^{(0)}\right) \in \mathcal{F}_q^0$ *and an accuracy requirement* $\varepsilon > 0$ *be given. Initialize* $(x, y, s) := \left(x^{(0)}, y^{(0)}, s^{(0)}\right)$ *and* $\tau := \frac{1}{n}\langle x, s\rangle$.

while $\tau > \varepsilon$

Solve (28) for $\sigma \in [0, 1]$.

Set $(x, y, s) := (x, y, s) - \alpha\left(\Delta x, \Delta y, \Delta s\right)$ *where* $\alpha \in (0, 1]$ *denotes a suitable step size such that* $(x, s) > 0$.

$\tau := \frac{1}{n}\langle x, s\rangle$

Exercises

1. Consider the *primal problem*

$$(P) \quad \begin{cases} \langle c, x \rangle \longrightarrow \min \\ Ax = b,\ x \geq 0 \end{cases}$$

and its *dual problem*

$$(D_e) \quad \begin{cases} \langle b, y \rangle \longrightarrow \max \\ A^T y + s = c \\ s \geq 0\,, \end{cases}$$

where $A := \begin{pmatrix} -1\,1\,1\,0 \\ 1\,1\,0\,1 \end{pmatrix}$, $\quad b := (2,4)^T \quad$ and $\quad c := (-1,-2,0,0)^T$.

a) Calculate the solution $\mathfrak{x}^* := (x^*, y^*, s^*)$ to the optimality conditions for (P) and (D) (cf. (1) in theorem 6.1.1) and determine $\mathcal{F}_P^{\text{opt}}$ and $\mathcal{F}_D^{\text{opt}}$. The nonnegativity constraints $x \geq 0$, $s \geq 0$ are essential in the optimality conditions. For otherwise — besides $\mathfrak{x}^*$ — we get solutions which are erroneous with respect to the linear optimization problems.

b) Show by means of the *implicit function theorem* that the *central path system* (cf. (4), theorem 6.1.4)

$$\begin{aligned} A^T y + s &= c \\ Ax &= b \\ Xs &= \mu e \\ x, s &\geq 0 \end{aligned}$$

has a unique solution $\mathfrak{x}(\mu) := (x(\mu), y(\mu), s(\mu))$ for $\mu \in \mathbb{R}$ in a suitable neighborhood of 0. Prove that $\mathfrak{x}^* = \lim_{\mu \to 0+} \mathfrak{x}(\mu)$.

Hint: When using *Maple®*, determine the functional matrix with the help of linalg[blockmatrix] .

c) Calculate the central path numerically and plot the projection $\big(x_1(\mu), x_2(\mu)\big)$.

d) Now choose $c := (-1,-1,0,0)^T$ and repeat *a) – c)*. Can you use the same approach as in *b)* again? If necessary, calculate $\lim_{\mu \to 0+} \mathfrak{x}(\mu)$ in a different way (cf. the remark on p. 259f).

2. Solve the linear optimization problem

$$(P) \quad \begin{cases} \langle c, x \rangle \longrightarrow \min \\ Ax = b,\ x \geq 0 \end{cases}$$

with the *projected gradient method* (cf. p. 189ff). Use the data from the preceding exercise for A, b, c and choose $x^{(0)} := (1,1,2,2)^T$ as the starting vector. Visualize this.

3. *Primal Affine-Scaling Method*

By modifying the projected gradient method, we can derive a *simple interior-point method.* Starting from $x^{(0)} \in \mathcal{F}_P^0$, we will describe one step of the iteration. By means of the *affine-scaling transformation* $x = X\overline{x}$ with $X := \operatorname{Diag}(x^{(0)})$ we get the following equivalent problem in $\overline{x}$-space:

$$\begin{cases} \langle \overline{c}, \overline{x} \rangle \longrightarrow \min \\ \overline{A}\,\overline{x} = b, \ \overline{x} \geq 0\,, \end{cases}$$

where $\overline{c} := Xc$ and $\overline{A} := AX$. Our current approximation in $\overline{x}$-space is apparently at the 'central point' $\overline{x}^{(0)} = X^{-1}x^{(0)} = e := (1,\dots,1)^T$. If we apply the projected gradient method to the transformed problem, we get $\overline{d} = -\overline{P}\,\overline{c}$ with

$$\overline{P} = I - \overline{A}^T\big(\overline{A}\,\overline{A}^T\big)^{-1}\overline{A} = I - XA^T\big(AX^2A^T\big)^{-1}AX\,.$$

Transformation back to x-space yields the *primal affine-scaling direction* $d = X\overline{d} = -X\overline{P}Xc$.

a) Show that $x^{(0)}$ is a KKT point if $d = 0$ and thus a *minimizer* of (P). Otherwise d is a descent direction. For $d \geq 0$ the objective function is unbounded from below. Determine the maximal $\alpha_{\max} > 0$ such that $x^{(0)} + \alpha_{\max} d \geq 0$, and set the step size $\alpha := \eta\alpha_{\max}$ with $\eta \in (0,1)$ (e.g., $\eta = 0.99$). In the update $x^{(1)} := x^{(0)} + \alpha d$ we thus get a strictly positive vector from $\mathcal{F}_P^0$ again, with which we can continue the iteration.

b) Implement the method sketched above in *Matlab*® and test the program with the data from exercise 1 and the starting vector $x^{(0)}$ from exercise 2.

4. *Primal* NEWTON *Barrier Method*

Assume that the interior-point condition holds, that is, $\mathcal{F}_P^0$ and $\mathcal{F}_D^0$ are nonempty. We apply the *logarithmic barrier method* to the primal problem (P):

$$\Phi_\mu(x) := \langle c, x\rangle - \mu \sum_{i=1}^n \log(x_i) \longrightarrow \min$$
$$Ax = b,\ x > 0\,,$$

where $\mu > 0$ is the *barrier parameter.*

a) Operate in the LAGRANGE–NEWTON framework (cf. p. 200ff) and derive the quadratic approximation to Φ_μ at x:

$$\tfrac{1}{2}\big\langle d, \mu X^{-2}d\big\rangle + \big\langle c - \mu X^{-1}e, d\big\rangle \longrightarrow \min$$
$$Ad = 0\,,$$

where X denotes the diagonal matrix $\operatorname{Diag}(x)$. Show that the NEWTON *direction*

Chapter 6

$$d = -\frac{1}{\mu} X \overline{P} X c + X \overline{P} e \quad \text{with} \quad \overline{P} = I - XA^T \big(AX^2A^T\big)^{-1} AX$$

is a minimizer of this problem.

The expression for d is the sum of a multiple of the *primal affine-scaling direction* and $X\overline{P}e$, which is called the *centering direction.* In part *c)* of this exercise we will see why.

b) Suppose $(y, s) \in \mathcal{F}_D^0$. Show that d can be written as

$$d = X\overline{P}\Big(e - \frac{1}{\mu} X s\Big).$$

If we additionally have $\Big\|e - \frac{1}{\mu} X s\Big\| < 1$, then $\big\|X^{-1}d\big\| < 1$ holds and the NEWTON iterate satisfies $x + d > 0$.

c) The problem

$$-\sum_{i=1}^{n} \log(x_i) \longrightarrow \min$$
$$Ax = b, \; x > 0$$

finds the *analytic center* of $\mathcal{F}_P$ (cf. the remark on p. 259f). Verify again that the NEWTON direction d of this problem yields the *centering direction* $X\overline{P}e$.

5. GOLDMAN–TUCKER *Theorem* (1956)

Let $\mathcal{F}_P$ and $\mathcal{F}_D$ be nonempty. Show that the following holds for

$$\mathcal{B} := \big\{i \in \{1, \ldots, n\} \mid \exists\, x \in \mathcal{F}_P^{\text{opt}} \;\; x_i > 0\big\}:$$

a) For every $k \notin \mathcal{B}$ there exists a $y \in \mathcal{F}_D^{\text{opt}}$ such that $c_k - \langle a_k, y\rangle > 0$.

b) There exist $x^* \in \mathcal{F}_P^{\text{opt}}$ and $y^* \in \mathcal{F}_D^{\text{opt}}$ with $x^* + s^* > 0$ for $s^* := c - A^T y^*$.

Hint to a): For $k \notin \mathcal{B}$ consider the linear optimization problem

$$\langle -e_k, x\rangle \longrightarrow \min$$
$$Ax = b$$
$$\langle c, x\rangle \le p^* := v(P), \; x \ge 0$$

and apply the Duality Theorem to it.
Hint to b): Firstly show with the help of *a)*

$$\forall\, k \in \{1, \ldots, n\} \; \exists\, x^{(k)} \in \mathcal{F}_P^{\text{opt}} \; \exists\, y^{(k)} \in \mathcal{F}_D^{\text{opt}} \;\; x_k^{(k)} + s_k^{(k)} > 0.$$

6. Show that:

a) $\mathcal{N}_2'(\beta_1) \subset \mathcal{N}_2'(\beta_2)$ for $0 \le \beta_1 < \beta_2$.

$\mathcal{N}_{-\infty}'(\gamma_2) \subset \mathcal{N}_{-\infty}'(\gamma_1)$ when $0 \le \gamma_1 < \gamma_2 \le 1$.

$\mathcal{N}_2'(\beta) \subset \mathcal{N}_{-\infty}'(\gamma)$ for $\gamma \le 1 - \beta$.

b) $\mathcal{N}_{-\infty}'(\gamma) = \{\omega \in \mathbb{R}_+^n \mid (I - \gamma\, h_n h_n^T)\,\omega \ge 0\}$

$= \{\sum_{\nu=1}^n \alpha_\nu h_\nu \mid \alpha_n \ge 0,\ \sum_{\nu=1}^{n-1} \alpha_\nu h_\nu + (1-\gamma)\alpha_n h_n \ge 0\}$

c) $\mathcal{N}_2'(\beta) \subset \mathcal{N}_{-\infty}'(\gamma)$ for $\gamma \le 1 - \beta\sqrt{(n-1)/n}$.

$\mathcal{N}_2'(\beta) = \mathbb{R}_+^n$ for $\beta \ge \sqrt{n(n-1)}$.

7. Implement the *short-step* or the *long-step path-following algorithm* as well as the MIZUNO–TODD–YE *predictor-corrector method.* Test them with the linear optimization problem

$$(P) \quad \begin{cases} \langle c, x\rangle \longrightarrow \min \\ Ax = b,\ x \ge 0\,, \end{cases}$$

where $A := \begin{pmatrix} 1\,1\,1\,0 \\ 2\,1\,0\,1 \end{pmatrix}$, $b := (3,2)^T$ and $c := (-1,-3,0,0)^T$.

Use $x^{(0)} := \big(\frac{1}{2}, \frac{1}{2}, 2, \frac{1}{2}\big)^T$, $y^{(0)} := \big(-2,-2\big)^T$ and $s^{(0)} := \big(5,1,2,2\big)^T$ as feasible starting points.

8. *Starting Points for* MEHROTRA's *algorithm* (cf. p. 287f)
Show that the solutions of the least squares problems

$$\|x\| \longrightarrow \min_{x \in \mathbb{R}^n} \quad \text{such that} \quad Ax = b \quad \text{and}$$

$$\|s\| \longrightarrow \min_{y \in \mathbb{R}^m,\, s \in \mathbb{R}^n} \quad \text{such that} \quad A^T y + s = c$$

are given by

$$\widetilde{x} := A^T\big(AA^T\big)^{-1} b \quad \text{and} \quad \widetilde{y} := \big(AA^T\big)^{-1} Ac, \quad \widetilde{s} := c - A^T \widetilde{y}\,.$$

9. Let the following optimization problem be given:

$$(QP) \quad \begin{cases} f(x) := \frac{1}{2} x^T C x + c^T x \longrightarrow \min \\ A^T x \le b \end{cases}$$

Suppose: $C \in \mathbb{R}^{n \times n}$ *symmetric positive semidefinite,* $c \in \mathbb{R}^n$, $b \in \mathbb{R}^m$, $A \in \mathbb{R}^{n \times m}$, C positive definite on $\text{kernel}(A^T)$.

a) Show that a point $x \in \mathcal{F} := \{v \in \mathbb{R}^n \mid A^T v \le b\}$ is a minimizer to (QP) iff there exist vectors $y, s \in \mathbb{R}^m_+$ with

$$\begin{aligned} Cx + c + Ay &= 0 \\ A^T x + s &= b \\ y_i s_i &= 0 \quad (i = 1, \ldots, n). \end{aligned}$$

b) Derive the *central path system* for $\mu > 0$:

$$F_\mu(x, y, s) := \begin{pmatrix} Cx + c + Ay \\ A^T x + s - b \\ Ys - \mu e \end{pmatrix} = \begin{pmatrix} 0 \\ 0 \\ 0 \end{pmatrix}, \quad y \ge 0,\ s \ge 0$$

c) Show that the Jacobian

$$DF_\mu(x, y, s) = \begin{pmatrix} C & A & 0 \\ A^T & 0 & I \\ 0 & S & Y \end{pmatrix}$$

is nonsingular if $y > 0$ and $s > 0$.

d) The NEWTON direction $(\Delta x, \Delta y, \Delta s)$ at (x, y, s) is given as the solution to

$$\begin{pmatrix} C & A & 0 \\ A^T & 0 & I \\ 0 & S & Y \end{pmatrix} \begin{pmatrix} \Delta x \\ \Delta y \\ \Delta s \end{pmatrix} = \begin{pmatrix} r_d \\ r_p \\ r_c \end{pmatrix}, \tag{29}$$

where the *dual residual* r_d, the *primal residual* r_p and the *complementarity residual* r_c are defined as

$$r_d := Cx + c + Ay, \quad r_p := A^T x + s - b, \quad r_c := Ys - \mu e.$$

Show that the solution to system (29) is given by the following equations

$$\begin{aligned} \Delta x &= \left(C + AYS^{-1}A^T\right)^{-1}\left(r_d - AS^{-1}(r_c - Yr_p)\right) \\ \Delta s &= r_p - A^T \Delta x \\ \Delta y &= S^{-1}\left(r_c - Y\Delta s\right). \end{aligned}$$

e) Translate MEHROTRA's method to the quadratic optimization problem (QP) and implement it in *Matlab*®. Use exercise 15 in chapter 4 to test your program.

7

Semidefinite Optimization

Semidefinite optimization (SDO) differs from linear optimization (LO) in that it deals with optimization problems over the cone of symmetric positive semidefinite matrices $\mathcal{S}_+^n$ instead of nonnegative vectors. In many cases the objective function is linear and SDO can be interpreted as an extension of LO. It is a branch of *convex optimization* and contains — besides LO — linearly constrained QP, *quadratically constrained* QP and — for example — *second-order cone programming* as special cases.

Problems with arguments from $\mathcal{S}_+^n$ are in particular of importance because many practically useful problems which are neither linear nor convex quadratic can be written as SDO problems. SDO, therefore, covers a lot of important applications in very different areas, for example, *system and control theory, eigenvalue optimization*

W. Forst and D. Hoffmann, *Optimization—Theory and Practice*,
Springer Undergraduate Texts in Mathematics and Technology,
DOI 10.1007/978-0-387-78977-4_7, © Springer Science+Business Media, LLC 2010

for affine matrix functions, *combinatorial optimization, graph theory, approximation theory* as well as *pattern recognition and separation by ellipsoids (cluster analyis).* This wide range of uses has quickly made SDO very popular — besides, of course, the fact that SDO problems can be solved efficiently via polynomially convergent interior-point methods, which had originally only been developed for LO.

Since the 1990s semidefinite problems (in optimization) have been intensively researched. First approaches and concepts, however, had already existed decades earlier, for example, in a paper by BELLMAN/FAN (1963). In 1994 NESTEROV/NEMIROVSKI provided a very general framework which could later on be simplified mainly due to the work and effort of ALIZADEH. The abundance of literature in this field can be overwhelming which makes it often difficult to separate the wheat from the chaff and find really new ideas.

Many concepts from linear optimization can be transferred to the more general case of semidefinite optimization. Note, however, that this cannot be done 'automatically' since the *duality theory* is weaker, and *strict complementarity* in the sense of the GOLDMAN–TUCKER theorem cannot generally be assumed. Therefore many considerations have to be done with great care.

The basic idea of many interior-point methods is to approximately follow the *central path* in the interior of the feasible set of an optimization problem which leads to an optimal point of the problem — mostly via variants of NEWTON's method. The technical difficulties that arise there and their extensive 'apparatus' will not be addressed here in detail. We content ourselves with a short illustration of the problems in comparison with chapter 6. We limit our attention in this chapter to some basic facts and only provide a first introduction to a very interesting field with a focus on the parallels to linear optimization.

7.1 Background and Motivation

Basics and Notations

An obvious difference between linear optimization and semidefinite optimization[1] is the feasible set. While in linear optimization the feasible set is a subset of $\mathbb{R}^n$, the feasible set of a semidefinite problem is a subset of the set of the real symmetric $n \times n$ matrices. Therefore we have to look at real symmetric — especially semidefinite — matrices:

Note that a quadratic real symmetric matrix A is *positive definite* if and only if $x^T A x > 0$ for all $x \in \mathbb{R}^n \setminus \{0\}$, and *positive semidefinite* if and only if $x^T A x \geq 0$ for all $x \in \mathbb{R}^n$.

[1] Semidefinite optimization is also referred to as *semidefinite programming* (SDP), which is a historically entrenched term.

We define:

$$
\begin{aligned}
S^n &:= \{A \in \mathbb{R}^{n\times n} \mid A \text{ symmetric}\}\\
S^n_+ &:= \{A \in S^n \mid A \text{ positive semidefinite}\}\\
S^n_{++} &:= \{A \in S^n \mid A \text{ positive definite}\}\\
A \succeq B &:\Longleftrightarrow A - B \in S^n_+\\
A \succ B &:\Longleftrightarrow A - B \in S^n_{++}
\end{aligned}
$$

The relation $\succeq$ is often referred to as the LÖWNER *partial order.*

Definition

- The standard *inner product* $\langle\ ,\ \rangle_{S^n} : S^n \times S^n \to \mathbb{R}$ is given by

$$\langle A, B\rangle_{S^n} := \operatorname{tr}(A^T B) = \operatorname{tr}(AB) = \sum_{i=1}^n \sum_{j=1}^n a_{ij} b_{ij}\,.$$

- This inner product yields the FROBENIUS *norm* $\|\ \|_F$ defined by

$$\|A\|_F := \sqrt{\operatorname{tr}(A^T A)} = \sqrt{\sum_{i=1}^n \sum_{j=1}^n a_{ij}^2}\,.$$

$\langle A, B\rangle := \operatorname{tr}(A^T B)$ can also be defined for nonsymmetric matrices. Recall that the trace of a matrix is the sum of its eigenvalues. The eigenvalues of A^2 are the squares of the eigenvalues of A counted according to their multiplicity. Therefore it holds for any $A \in S^n$:

$$\|A\|_F = \sqrt{\operatorname{tr}(A^2)} = \sqrt{\sum_{i=1}^n \lambda_i(A)^2}$$

where here and below $\lambda_i(A)$ denotes the i-th largest eigenvalue of A.

S^n is a vector space with dimension $\widetilde{n} := n(n+1)/2$. For example, we can identify S^n and $\mathbb{R}^{\widetilde{n}}$ using the map *svec* defined by

$$\operatorname{svec}(A) := \left(a_{11}, \sqrt{2}\,a_{12}, \ldots, \sqrt{2}\,a_{1n}, a_{22}, \sqrt{2}\,a_{23}, \ldots, a_{nn}\right)^T.$$

The factor $\sqrt{2}$ in front of all nondiagonal elements in the definition of *svec* is motivated by the fact that the standard inner products of S^n and $\mathbb{R}^{\widetilde{n}}$ are compatible, i.e., for all A, $B \in S^n$ it holds that

$$\langle A, B\rangle_{S^n} = \langle \operatorname{svec}(A), \operatorname{svec}(B)\rangle_{\mathbb{R}^{\widetilde{n}}}\,.$$

The operator *svec* is invertible. Positive semidefiniteness in S^n is equivalent to the property that all eigenvalues are nonnegative. Correspondingly, a symmetric matrix is positive definite if and only if all eigenvalues are positive.

Remark

S^n_+ is a closed convex cone. Its interior is S^n_{++}.

Proof: It is obvious that S^n_+ is a closed convex cone. It remains to show that its interior is S^n_{++} (cf. exercise 4). □

Theorem 7.1.1

S^n_+ is self-dual. In other words: A symmetric matrix A is positive semidefinite if and only if $\langle A, B \rangle_{S^n} \geq 0$ for all $B \succeq 0$.

We provide two lemmas to prove self-duality:

Lemma 7.1.2

For any $A \in S^n_+$ the positive semidefinite root denoted by $A^{1/2}$ exists and is unique (cf. exercise 2).

Lemma 7.1.3

For $A \in \mathbb{R}^{m \times n}$ and $B \in \mathbb{R}^{n \times m}$ it holds:

$$\operatorname{tr}(AB) = \operatorname{tr}(BA)$$

Proof: $(AB)_{jj} = \sum_{i=1}^{n} a_{ji} b_{ij}$ implies $\operatorname{tr}(AB) = \sum_{j=1}^{m} \sum_{i=1}^{n} a_{ji} b_{ij}$ and accordingly $\operatorname{tr}(BA) = \sum_{i=1}^{n} \sum_{j=1}^{m} b_{ij} a_{ji}$. □

Proof of 7.1.1: i) Let $A, B \in S^n_+$. Due to lemma 7.1.2 the roots $A^{1/2}$ and $B^{1/2}$ exist and are positive semidefinite, too. Thus

$$\begin{aligned}\langle A, B \rangle_{S^n} &= \operatorname{tr}(AB) = \operatorname{tr}\left(A^{1/2} A^{1/2} B^{1/2} B^{1/2}\right) \underset{(7.1.3)}{=} \operatorname{tr}\left(B^{1/2} A^{1/2} A^{1/2} B^{1/2}\right) \\ &= \operatorname{tr}\left((A^{1/2} B^{1/2})^T A^{1/2} B^{1/2}\right) = \left\| A^{1/2} B^{1/2} \right\|_F^2 \geq 0 .\end{aligned}$$

ii) Let $x \in \mathbb{R}^n$. The matrix $x x^T \in \mathbb{R}^{n \times n}$ is positive semidefinite because it holds for all $v \in \mathbb{R}^n$ that

$$v^T x x^T v = (x^T v)^T x^T v = \left\| x^T v \right\|^2 \geq 0.$$

By lemma 7.1.3 we get

$$x^T A x = \operatorname{tr}(x^T A x) = \operatorname{tr}(A x x^T) = \langle A, x x^T \rangle_{S^n} \geq 0,$$

i.e., A is positive semidefinite. □

Boundary of the cone S^2_+

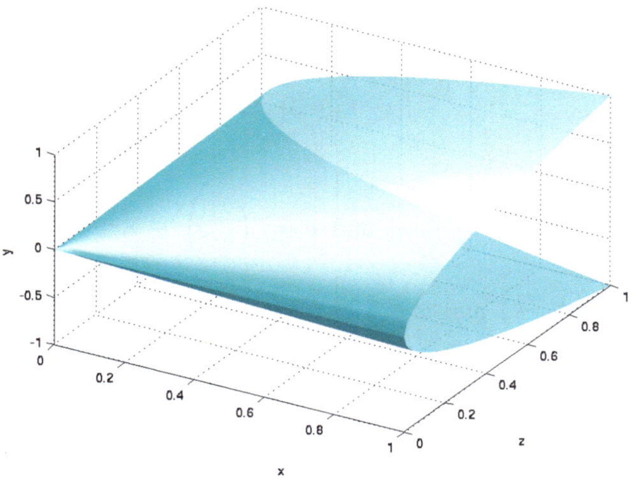

To create the figure, we used the following condition: A matrix $X = \begin{pmatrix} x & y \\ y & z \end{pmatrix}$ belongs to S^2_+ iff $x\,z \geq y^2$ and $x,\ z \geq 0$.

To keep it simple, we denote the *standard inner product on* S^n in the following by $\langle\ ,\ \rangle$ and mostly do not differentiate between the standard inner product for symmetric matrices and the standard inner product for vectors in our notation.

The following technical lemma will be instrumental in proofs:

Lemma 7.1.4 *For* $A, B \in S^n_+$ *it holds:*

$$\lambda_1(A)\,\lambda_n(B) \leq \lambda_1(A)\,\mathrm{tr}(B) \leq \langle A, B\rangle \leq \lambda_n(A)\,\mathrm{tr}(B) \leq n\,\lambda_n(A)\,\lambda_n(B)$$

Proof: Let Q be an orthogonal matrix which diagonalizes A, i.e., $A = QDQ^T$, where D is a diagonal matrix formed by the eigenvalues of A. It holds:

$$\begin{aligned} \langle A, B\rangle &= \mathrm{tr}(AB) = \mathrm{tr}(QDQ^TB) = \mathrm{tr}(DQ^TBQ) \quad \geq \lambda_1(A)\,\mathrm{tr}(Q^TBQ) \\ &= \lambda_1(A)\,\mathrm{tr}(B) \geq \lambda_1(A)\,\lambda_n(B) \end{aligned}$$

In the same manner we get the remaining inequalities. □

The following remark will be useful for applications of semidefinite optimization.

Remark

A block diagonal matrix is symmetric positive (semi-) definite iff all its blocks are symmetric positive (semi-) definite, i.e., for $X_\kappa \in \mathbb{R}^{n_\kappa \times n_\kappa}$, $1 \le \kappa \le k$,

$$X := \begin{pmatrix} X_1 & 0 & \dots & 0 \\ 0 & X_2 & \dots & 0 \\ \vdots & \vdots & \ddots & \vdots \\ 0 & 0 & 0 & X_k \end{pmatrix} \succeq 0 \ (\succ 0) \iff X_1, \dots, X_k \succeq 0 \ (\succ 0).$$

Proof: Let $\alpha := n_1 + \cdots + n_k$ and $v =: \begin{pmatrix} v_1 \\ \vdots \\ v_k \end{pmatrix} \in \mathbb{R}^\alpha$ with $v_\kappa \in \mathbb{R}^{n_\kappa}$ for $1 \le \kappa \le k$.

(*i*) Let $\kappa_0 \in \{1, \dots, k\}$ and $w \in \mathbb{R}^{n_{\kappa_0}}$. With $v_{\kappa_0} := w$ and $v_\kappa := 0$ for $\kappa \ne \kappa_0$ it holds that $\langle w, X_{\kappa_0} w \rangle = \langle v, Xv \rangle$. Thus if X is positive semidefinite, then $\langle w, X_{\kappa_0} w \rangle \ge 0$.
For $X \succ 0$, we get $\langle w, X_{\kappa_0} w \rangle = 0$ iff $v = 0 \in \mathbb{R}^\alpha$, especially $w = 0 \in \mathbb{R}^{\kappa_0}$.

(*ii*) By $\langle v, Xv \rangle = \sum_{\kappa=1}^k \langle v_\kappa, X_\kappa v_\kappa \rangle$ we get $X \succeq 0$, if $X_1, \dots, X_k \succeq 0$. Since $\langle v_\kappa, X_\kappa v_\kappa \rangle \ge 0$, it holds that $\langle v, Xv \rangle = \sum_{\kappa=1}^k \langle v_\kappa, X_\kappa v_\kappa \rangle = 0$ if and only if all summands are zero. If in addition all X_κ are positive definite, $v_\kappa = 0$ must hold for $1 \le \kappa \le k$, i.e., $v = 0$ and therefore $X \succ 0$. □

Primal Problem and Dual Problem

Let $b \in \mathbb{R}^m$, $A^{(j)} \in S^n$ $(1 \le j \le m)$ and $C \in S^n$. We consider the problem

$$(SDP') \quad \begin{cases} \langle C, X \rangle \longrightarrow \min \\ \langle A^{(j)}, X \rangle = b_j \quad (1 \le j \le m) \\ X \succeq 0. \end{cases}$$

The constraint $X \succeq 0$ is convex, but nonlinear.
Let the linear operator $\mathcal{A} : S^n \longrightarrow \mathbb{R}^m$ be defined by

$$\mathcal{A}(X) := \begin{pmatrix} \langle A^{(1)}, X \rangle \\ \vdots \\ \langle A^{(m)}, X \rangle \end{pmatrix} = \sum_{j=1}^m \big\langle A^{(j)}, X \big\rangle e_j, \tag{1}$$

where $e_j \in \mathbb{R}^m$ is the j-th unit vector. Then we get the final form of our semidefinite problem:

$$(SDP) \quad \begin{cases} \langle C, X \rangle \longrightarrow \min \\ \mathcal{A}(X) = b \\ X \succeq 0 \end{cases}$$

We denote the set of *feasible points* of (SDP) by

$$\mathcal{F}_P := \big\{X \in \mathcal{S}^n : \mathcal{A}(X) = b \text{ and } X \succeq 0\big\}.$$

Remark

There is no loss of generality by assuming symmetry of the matrix C: If C is not symmetric, we can replace C by $1/2(C + C^T)$, since $\langle C, X \rangle = \langle C^T, X \rangle$. The same holds for the matrices $A^{(j)}$, $1 \le j \le m$.

The unique *adjoint operator* $\mathcal{A}^* \colon \mathbb{R}^m \longrightarrow \mathcal{S}^n$ to $\mathcal{A}$ is given by the property

$$\big\langle y, \mathcal{A}(X) \big\rangle = \big\langle \mathcal{A}^*(y), X \big\rangle \text{ for all } X \in \mathcal{S}^n \text{ and } y \in \mathbb{R}^m.$$

Remark

The adjoint operator $\mathcal{A}^*$ has the form

$$\mathcal{A}^*(y) = \sum_{j=1}^m y_j A^{(j)}.$$

Proof: By (1) it holds:

$$\big\langle \mathcal{A}^*(y), X \big\rangle = \langle y, \mathcal{A}(X) \rangle = \left\langle y, \sum_{j=1}^m \left\langle A^{(j)}, X \right\rangle e_j \right\rangle = \sum_{j=1}^m \left\langle A^{(j)}, X \right\rangle \langle y, e_j \rangle$$
$$= \sum_{j=1}^m \left\langle A^{(j)}, X \right\rangle y_j = \left\langle \sum_{j=1}^m y_j A^{(j)}, X \right\rangle$$
□

For convenience we write $\mathcal{A}X$ for $\mathcal{A}(X)$ and $\mathcal{A}^*y$ for $\mathcal{A}^*(y)$.

Let us now consider the (LAGRANGE) *dual problem* belonging to (SDP):

The LAGRANGE function L of (SDP) is given by

$$L(X, y) := \langle C, X \rangle + \langle y, b - \mathcal{A}X \rangle$$

where $y \in \mathbb{R}^m$ is a multiplier. Let φ be the dual function, that is,

$$\varphi(y) := \inf_{X \succeq 0} L(X, y) = \inf_{X \succeq 0} \big(\langle y, b \rangle + \langle C - \mathcal{A}^*y, X \rangle\big).$$

If $C - \mathcal{A}^*y \succeq 0$, then we get by theorem 7.1.1 $\langle C - \mathcal{A}^*y, X \rangle \ge 0$ for any $X \succeq 0$. Thus $\varphi(y) = \langle b, y \rangle$. Otherwise, also by theorem 7.1.1, there exists an $X \succeq 0$ with $\langle C - \mathcal{A}^*y, X \rangle < 0$. For any $k \in \mathbb{N}$ it holds that $kX \succeq 0$ and so

$$\langle C - \mathcal{A}^*y, kX \rangle = k \langle C - \mathcal{A}^*y, X \rangle \longrightarrow -\infty \text{ for } k \longrightarrow \infty.$$

Thus $\varphi(y) = -\infty$. Altogether, we have

$$\varphi(y) = \begin{cases} \langle y, b \rangle, & \text{if } C - \mathcal{A}^*y \succeq 0 \\ -\infty \;, & \text{otherwise.} \end{cases}$$

For the effective domain of φ we get

$$\mathcal{F}_D = \{ y \in \mathbb{R}^m \mid C - \mathcal{A}^*y \succeq 0 \}$$

and with that the LAGRANGE dual problem

$$(DSDP) \quad \begin{cases} \langle b, y \rangle \longrightarrow \max \\ C - \mathcal{A}^*y \succeq 0. \end{cases} \tag{2}$$

The condition $C - \mathcal{A}^*y \succeq 0$ can be rewritten as $\mathcal{A}^*y + S = C$ for some $S \in \mathcal{S}^n_+$ called the *slack variable.*

If $b \in \mathcal{R}(\mathcal{A})$, then we have:

Lemma 7.1.5 *The dual problem (DSDP) itself is a semidefinite problem.*

Proof: Let $X \in \mathcal{S}^n$ be fixed such that $\mathcal{A}X = b$. With that the following holds for $y \in \mathbb{R}^m$ and $S := C - \mathcal{A}^*y$:

$$\langle b, y \rangle = \langle \mathcal{A}X, y \rangle = \langle X, \mathcal{A}^*y \rangle = \langle X, C - S \rangle$$

The dual problem can then be written in the form

$$\begin{cases} \langle X, C - S \rangle \longrightarrow \max \\ S \succeq 0, \, C - S \in \mathcal{R}(\mathcal{A}^*) \end{cases} \quad \text{or} \quad \begin{cases} \langle X, S \rangle \longrightarrow \min \\ C - S \in \mathcal{R}(\mathcal{A}^*), \, S \succeq 0. \end{cases}$$

Let now — in the nontrivial case $\mathcal{R}(\mathcal{A}^*) \neq \mathcal{S}^n$ — $F^{(1)}, \ldots, F^{(k)}$ be a basis of $\mathcal{N}(\mathcal{A})$. Then we have $C - S \in \mathcal{R}(\mathcal{A}^*) = \mathcal{N}(\mathcal{A})^\perp$ if and only if $\langle C - S, F^{(\kappa)} \rangle = 0$ or $\langle F^{(\kappa)}, S \rangle = \langle C, F^{(\kappa)} \rangle =: f_\kappa$ holds for all $\kappa = 1, \ldots, k$.

Thus we get the dual problem in the desired form

$$\begin{cases} \langle X, S \rangle \longrightarrow \min \\ \langle F^{(\kappa)}, S \rangle = f_\kappa \text{ for } \kappa = 1, \ldots, k \\ S \succeq 0. \end{cases}$$ □

If the matrices $A^{(1)}, \ldots, A^{(m)}$ are linearly independent, then the assumption $b \in \mathcal{R}(\mathcal{A})$ is given, as $\mathcal{R}(\mathcal{A}) = \mathbb{R}^m$.

Proof: For $y \in \mathbb{R}^m \setminus \{0\}$ it holds that $\mathcal{A}^*y = \sum_{j=1}^m y_j A^{(j)} \neq 0$, that is, $\mathcal{N}(\mathcal{A}^*) = \{0\}$, hence $\mathcal{R}(\mathcal{A}) = \mathcal{N}(\mathcal{A}^*)^\perp = \mathbb{R}^m$. □

Demanding the linear independence of the matrices $A^{(1)}, \ldots, A^{(m)}$ — which we will do later on anyway — is not an essential restriction: If these matrices are linearly dependent, we can — in the nontrivial case — choose WLOG $\{A^{(1)}, \ldots, A^{(\ell)}\}$ as the basis of $\langle A^{(1)}, \ldots, A^{(m)} \rangle$ for a suitable $\ell \in \{1, \ldots, m-1\}$. If $\langle b, y \rangle = 0$ holds for all $y \in \mathcal{N}(\mathcal{A}^*)$, then the two problems can be rewritten equivalently as

$$(SDP') \quad \begin{cases} \langle C, X \rangle \longrightarrow \min \\ \left\langle A^{(j)}, X \right\rangle = b_j \quad (1 \le j \le \ell) \\ X \succeq 0 \end{cases}$$

and

$$(DSDP) \quad \begin{cases} \langle b, y \rangle \longrightarrow \max \\ C - \sum_{j=1}^{\ell} y_j A^{(j)} \succeq 0 \quad (y \in \mathbb{R}^\ell). \end{cases}$$

If, however, $\langle b, \widetilde{y} \rangle \neq 0$ for some $\widetilde{y} \in \mathcal{N}(\mathcal{A}^*)$, hence WLOG $\langle b, \widetilde{y} \rangle > 0$, then the dual problem is unbounded if it has feasible points.

Remark

The dual problem of $(DSDP)$ formulated in equality form with a slack variable is the primal problem (SDP).

Proof: We proceed in the same way as above when we constructed $(DSDP)$. If we write $(DSDP)$ as a minimization problem, we get the LAGRANGE function L of $(DSDP)$ by

$$L(y, X, S) := \langle -b, y \rangle + \langle X, \mathcal{A}^* y + S - C \rangle,$$

where X is a multiplier. WLOG we can assume that X is symmetric (compare the first remark on page 305). For the dual function denoted by φ we get

$$\begin{aligned} \varphi(X) &:= \inf_{S \succeq 0,\, y \in \mathbb{R}^m} L(y, X, S) \\ &= -\langle C, X \rangle + \inf_{y \in \mathbb{R}^m} \langle \mathcal{A}X - b, y \rangle + \inf_{S \succeq 0} \langle X, S \rangle. \end{aligned}$$

For $\mathcal{A}X - b \neq 0$ we have $\inf_{y \in \mathbb{R}^m} \langle \mathcal{A}X - b, y \rangle = -\infty$. Otherwise this infimum is zero. If X is *not* positive semidefinite, then we have $\inf_{S \succeq 0} \langle X, S \rangle = -\infty$. We obtain this in the same way as on page 305f. If $X \in \mathcal{S}_+^n$ holds, then this term is zero. This yields

$$\varphi(X) = \begin{cases} -\langle C, X \rangle, & \mathcal{A}X = b \text{ and } X \succeq 0 \\ -\infty, & \text{otherwise.} \end{cases}$$

Hence, the effective domain of φ is the set $\{X \in \mathcal{S}_+^n : \mathcal{A}X = b\}$. To sum up, the dual problem to $(DSDP)$ is

$$
\begin{aligned}
-\langle C, X\rangle &\longrightarrow \max \\
\mathcal{A}X &= b,\ X \succeq 0 .
\end{aligned}
$$

If we exchange max and min, we get exactly the primal problem. □

Weak duality can be confirmed easily:

$$\langle C, X\rangle - \langle b, y\rangle = \langle C, X\rangle - \langle \mathcal{A}X, y\rangle = \langle C - \mathcal{A}^*y, X\rangle \geq 0, \tag{3}$$

where $X \in \mathcal{F}_P$ and $y \in \mathcal{F}_D$.

$\langle C - \mathcal{A}^*y, X\rangle = \langle S, X\rangle$ is called the *duality gap* for feasible X and y. We denote

$$
\begin{aligned}
p^* &:= \inf\{\langle C, X\rangle \mid X \in \mathcal{F}_P\} \\
d^* &:= \sup\{\langle b, y\rangle \mid y \in \mathcal{F}_D\}.
\end{aligned}
$$

If the *duality gap* $p^* - d^*$ vanishes, that is, $p^* = d^*$, we say that *strong duality* holds.

A matrix $X \in \mathcal{F}_P$ with $\langle C, X\rangle = p^*$ is called a *minimizer* for (SDP). Correspondingly, a vector $y \in \mathcal{F}_D$ with $\langle b, y\rangle = d^*$ is called a *maximizer* for $(DSDP)$.

Lemma 7.1.6

If there are feasible X^* and y^* with $\langle C, X^*\rangle = \langle b, y^*\rangle$, i.e., the duality gap for X^* and y^* is zero, then X^* and y^* are optimizers for the primal and dual problem, respectively.

Proof: By weak duality we get for each $X \in \mathcal{F}_P$ that $\langle C, X\rangle \geq \langle b, y^*\rangle = \langle C, X^*\rangle$. For any $y \in \mathcal{F}_D$ it holds that $\langle b, y\rangle \leq \langle C, X^*\rangle = \langle b, y^*\rangle$, respectively. □

Later on we will see that the other implication — in contrast to linear optimization — is not always true. In addition, duality results are weaker for semidefinite optimization than for linear optimization.

Definition

A matrix X is called *strictly (primal) feasible* iff $X \in \mathcal{F}_P$ and $X \succ 0$.
A vector y is called *strictly (dual) feasible* iff $y \in \mathcal{F}_D$ and $C - \mathcal{A}^*y \succ 0$.
The condition to have strictly feasible points for both the primal and the dual problem is called the SLATER *condition.*

The following Duality Theorem will be given without proof. For a proof an extension of the Theorem of the Alternative (FARKAS) is needed which can be proved by separation theorems and then applied to cones which are not polyhedral — unlike in linear optimization:

Theorem 7.1.7 (*Duality Theorem*)

Assume that both the primal and the dual semidefinite problem have feasible points.

a) If the dual problem has a strictly feasible point y, then a minimizer to the primal problem exists and

$$\min\{\langle C, X\rangle : X \in \mathcal{F}_P\} = \sup\{\langle b, y\rangle : y \in \mathcal{F}_D\}.$$

b) If the primal problem has a strictly feasible point X, then a maximizer to the dual problem exists and

$$\inf\{\langle C, X\rangle : X \in \mathcal{F}_P\} = \max\{\langle b, y\rangle : y \in \mathcal{F}_D\}.$$

c) If both problems have strictly feasible points, then both have optimizers, whose values coincide, that is,

$$\min\{\langle C, X\rangle : X \in \mathcal{F}_P\} = \max\{\langle b, y\rangle : y \in \mathcal{F}_D\}.$$

Instead of pursuing these theoretical aspects extensively any further, we will now firstly have a look at some special cases, examples and supplementary considerations.

7.2 Selected Special Cases

Semidefinite optimization is not only important due to its variety of practical applications[2], it also includes many other branches of optimization, e.g., linear optimization, quadratic optimization with both linear and quadratic constraints. *Second-order cone problems*, which contain quasi-convex nonlinear problems, can also be interpreted as semidefinite problems. Semidefinite optimization itself belongs to *cone optimization*, which is part of convex optimization.

In the following, several connections will be pointed out. Below $\operatorname{Diag}(x)$ denotes the diagonal matrix whose diagonal elements are the entries of a vector $x \in \mathbb{R}^n$ and $\operatorname{diag}(X)$ denotes the column vector consisting of the diagonal elements of a matrix $X \in \mathbb{R}^{n\times n}$.

Chapter 7

Linear Optimization and Duality Complications

In the following considerations we use the fact that

$$\langle \operatorname{Diag}(a), \operatorname{Diag}(b)\rangle_{\mathcal{S}^n} = \langle a, b\rangle_{\mathbb{R}^n}$$

[2] Confer [Be/Ne].

holds for all $a,\, b \in \mathbb{R}^n$.

Linear optimization is a special case of semidefinite optimization. If we have a linear problem

$$(LP) \quad \begin{cases} \langle c, x \rangle \longrightarrow \min \\ Ax = b,\, x \geq 0 \end{cases}$$

with $A \in \mathbb{R}^{m \times n}$, $b \in \mathbb{R}^m$ and $c \in \mathbb{R}^n$, we can transfer it to a problem in (SDP) form:

If we set $C := \operatorname{Diag}(c)$, $X := \operatorname{Diag}(x)$ and $A^{(j)} := \operatorname{Diag}\left(a^{(j)}\right)$, where ${a^{(j)}}^T$ is the j-th row of A for $1 \leq j \leq m$, we get

$$\langle c, x \rangle = \langle C, X \rangle\,,\; Ax = \mathcal{A}X.$$

Finally, $x \geq 0$ means $X \succeq 0$.

On the other hand, an (SDP) can be formulated as an (LP) — to a certain degree. To this end the *vec*-operator, which identifies $\mathbb{R}^{n \times n}$ with $\mathbb{R}^{n^2}$, can be used:

$$\operatorname{vec}\colon \mathbb{R}^{n \times n} \longrightarrow \mathbb{R}^{n^2}$$
$$\operatorname{vec}(A) := \begin{pmatrix} a_1 \\ \vdots \\ a_n \end{pmatrix}$$

where $a_1, \ldots, a_n$ are the columns of A. By this definition we get for $1 \leq k \leq m$

$$\langle \operatorname{vec}(C), \operatorname{vec}(X) \rangle = \sum_{j=1}^n \sum_{i=1}^n c_{ij}\, x_{ij} = \langle C, X \rangle$$
$$\left\langle A^{(k)}, X \right\rangle = \left\langle \operatorname{vec}(A^{(k)}), \operatorname{vec}(X) \right\rangle.$$

In general, however, the eponymous constraint $X \succeq 0$ cannot simply be translated.

If the matrices C and $A^{(k)}$ are *diagonal matrices*, the objective function and the equality constraint of (SDP) correspond to the ones in (LP):

With $c := \operatorname{diag}(C)$, $x := \operatorname{diag}(X)$, $a^{(j)} := \operatorname{diag}(A^{(j)})$ for $1 \leq j \leq m$ we have

$$\langle C, X \rangle = \langle c, x \rangle$$
$$\left\langle A^{(j)}, X \right\rangle = \left\langle a^{(j)}, x \right\rangle.$$

Notwithstanding these similarities, one must be careful about promptly transferring the results gained for linear optimization to semidefinite optimization, since there are, in particular, differences in duality properties to be aware of. Such differences will be shown in the following three examples. The crucial point there is the existence of strictly feasible points.

Example 1 (cf. [Todd 2]) $\mathcal{F}_P = \emptyset$, $\max(D)$ exists

We consider the following problem in standard form

$$\left\langle \begin{pmatrix} 0 & 0 \\ 0 & 0 \end{pmatrix}, X \right\rangle \longrightarrow \min$$

$$\left\langle \begin{pmatrix} 1 & 0 \\ 0 & 0 \end{pmatrix}, X \right\rangle = 0, \left\langle \begin{pmatrix} 0 & 1 \\ 1 & 0 \end{pmatrix}, X \right\rangle = 2, \; X \succeq 0.$$

In other words the constraints are $x_{11} = 0$ and $x_{21} = x_{12} = 1$ for $X \in \mathcal{S}^2$, but such a matrix is not positive semidefinite. So the feasible set is empty and the infimum is infinite. The corresponding dual problem is

$$2\,y_2 \longrightarrow \max$$

$$y_1 \begin{pmatrix} 1 & 0 \\ 0 & 0 \end{pmatrix} + y_2 \begin{pmatrix} 0 & 1 \\ 1 & 0 \end{pmatrix} \preceq \begin{pmatrix} 0 & 0 \\ 0 & 0 \end{pmatrix}.$$

We get

$$S = \begin{pmatrix} -y_1 & -y_2 \\ -y_2 & 0 \end{pmatrix} \succeq 0.$$

Thus it must hold: $y_2 = 0$ and $y_1 \le 0$. Therefore the supremum is zero and a maximizer is $y = (0, 0)^T$. ◁

Example 2 (cf. [Hel])

$\max(D)$ exists, $\min(P)$ does not exist, though $\inf(P)$ is finite

We consider the problem

$$x_{11} \longrightarrow \min$$

$$\begin{pmatrix} x_{11} & 1 \\ 1 & x_{22} \end{pmatrix} \succeq 0$$

and formulate it in 'standard' form: We have $x_{11} = \langle C, X \rangle$ for

$$C := \begin{pmatrix} 1 & 0 \\ 0 & 0 \end{pmatrix}.$$

With

$$A^{(1)} := \begin{pmatrix} 0 & 1 \\ 1 & 0 \end{pmatrix}$$

it holds that

$$\left\langle A^{(1)}, X \right\rangle = x_{21} + x_{12} = 2x_{12}\,.$$

Finally, if we set $b = 2$, we receive the standard form of the primal problem. With

$$C - y_1 A^{(1)} = \begin{pmatrix} 1 & -y_1 \\ -y_1 & 0 \end{pmatrix}$$

we get the dual problem

Chapter 7

$$2y_1 \longrightarrow \max$$
$$\begin{pmatrix} 1 & -y_1 \\ -y_1 & 0 \end{pmatrix} \succeq 0.$$

The dual problem's optimum is zero, and for a maximizer y it holds that $y_1 = 0$. The primal problem has a strictly feasible point (for example $x_{11} = x_{22} = 2$). Because of the semidefiniteness constraint the principal determinants must be nonnegative: $x_{11} \geq 0$ and $x_{11}x_{22} - 1 \geq 0$. (Here, it can be seen that the constraint $X \succeq 0$ is not just a simple extension of the linear constraint $x \geq 0$.) Since $x_{11} \geq 1/x_{22}$, the infimum is zero, but it is not attained. ◁

This example shows that unlike in the case of linear optimization a maximizer of the dual problem might exist but the infimum of the primal problem will not be attained if no strictly dual feasible point exists.

Example 3 (cf. [Hel]) max(D) and min(P) exist with $p^* \neq d^*$

As in linear optimization a zero duality gap at feasible points X and y implies that they are optimizers (see lemma 7.1.6) but *the following example illustrates that in contrast to linear optimization optimality does no longer imply a zero duality gap in general. The condition of strictly feasible points cannot be omitted.* The considered primal problem is

$$x_{12} \longrightarrow \min$$
$$\begin{pmatrix} 0 & x_{12} & 0 \\ x_{12} & x_{22} & 0 \\ 0 & 0 & 1+x_{12} \end{pmatrix} \succeq 0.$$

The standard form of the objective function is given by

$$C := \begin{pmatrix} 0 & 1/2 & 0 \\ 1/2 & 0 & 0 \\ 0 & 0 & 0 \end{pmatrix}.$$

The constraints for the entries of the matrix X are ensured by the definition of the following matrices $A^{(1)}, \ldots, A^{(4)}$:

$$A^{(1)} := \begin{pmatrix} 0 & -1/2 & 0 \\ -1/2 & 0 & 0 \\ 0 & 0 & 1 \end{pmatrix} \text{ and } b_1 := 1 \text{ yields } x_{33} = 1 + x_{12},$$

$$A^{(2)} := \begin{pmatrix} 1 & 0 & 0 \\ 0 & 0 & 0 \\ 0 & 0 & 0 \end{pmatrix} \text{ and } b_2 := 0 \text{ gives } x_{11} = 0,$$

$$A^{(3)} := \begin{pmatrix} 0 & 0 & 1 \\ 0 & 0 & 0 \\ 1 & 0 & 0 \end{pmatrix} \text{ and } b_3 := 0 \text{ shows } x_{13} = 0,$$

$$A^{(4)} := \begin{pmatrix} 0 & 0 & 0 \\ 0 & 0 & 1 \\ 0 & 1 & 0 \end{pmatrix} \quad \text{and } b_4 := 0 \text{ yields } x_{23} = 0.$$

As the dual problem we get

$$y_1 \longrightarrow \max$$

$$S = \begin{pmatrix} -y_2 & (1+y_1)/2 & -y_3 \\ (1+y_1)/2 & 0 & -y_4 \\ -y_3 & -y_4 & -y_1 \end{pmatrix} \succeq 0$$

with $S := C - y_1 A^{(1)} - y_2 A^{(2)} - y_3 A^{(3)} - y_4 A^{(4)}$. A necessary condition for a matrix X to be primally feasible is $x_{12} = 0$. Thus a *strictly* primal feasible point does not exist. The infimum is attained iff $x_{12} = 0$ and $x_{22} \geq 0$. If a vector y is dually feasible, $y_1 = -1$ must hold. By SARRUS' rule we get $\det(S) = y_2 y_4^2$. Thus there exists no strictly feasible y: Otherwise, $\det(S) > 0$ would hold. Since $y_2 < 0$, we would get $y_4^2 < 0$. The dual's supremum is -1 and a maximizer is found with $y_2 = y_3 = -1$, for example. The duality gap is always -1 although optimizers do exist. ◁

Second-Order Cone Programming

Definition

The *Second-Order Cone (Ice Cream Cone* or LORENTZ *Cone)* — see the figure on page 270 — is defined[3] by

$$\mathcal{L}^n := \left\{(u,t) \in \mathbb{R}^{n-1} \times \mathbb{R} \,:\, \|u\|_2 \leq t\right\}. \tag{4}$$

To show that the second-order cone can be embedded in the cone of semidefinite matrices, we use the SCHUR *Complement*:

Definition

If

$$M = \begin{pmatrix} A & B \\ B^T & C \end{pmatrix},$$

where A is symmetric positive definite and C is symmetric, then the matrix

$$C - B^T A^{-1} B$$

is called the SCHUR *Complement of A in M*.

[3] SeDuMi 's definition differs slightly from ours:
$$\mathcal{L}^n := \left\{(t,u) \in \mathbb{R} \times \mathbb{R}^{n-1} \,:\, t \geq \|u\|_2\right\}$$

Chapter 7

Theorem 7.2.1

In the situation above the following statements are equivalent:

a) M is symmetric positive (semi-) definite.
b) $C - B^T A^{-1} B$ is symmetric positive (semi-) definite.

Proof: Since $A \succ 0$, the matrix A is invertible. Setting $D := -A^{-1}B$ gives

$$\begin{pmatrix} I & 0 \\ D^T & I \end{pmatrix}\begin{pmatrix} A & B \\ B^T & C \end{pmatrix}\begin{pmatrix} I & D \\ 0 & I \end{pmatrix} = \begin{pmatrix} A & 0 \\ 0 & C - B^T A^{-1} B \end{pmatrix} =: N\,.$$

The matrix $\begin{pmatrix} I & D \\ 0 & I \end{pmatrix}$ is regular, thus M is (symmetric) positive (semi-) definite iff N is (symmetric) positive (semi-) definite. N is (symmetric) positive (semi-) definite iff its diagonal blocks are (symmetric) positive (semi-) definite by the remark on page 304. This is the case iff the SCHUR Complement $C - B^T A^{-1} B$ is positive (semi-) definite. □

With $(u,t) \in \mathbb{R}^{n-1} \times \mathbb{R}_+$ we look at

$$E_{t,u} := \begin{pmatrix} t I_{n-1} & u \\ u^T & t \end{pmatrix}.$$

Due to $t \geq 0$, $(u,t) \in \mathcal{L}^n$ is equivalent to $t^2 - u^T u \geq 0$. For $t > 0$ this is equivalent to $t - u^T \frac{I_{n-1}}{t} u \geq 0$, and we get by theorem 7.2.1:

$$(u,t) \in \mathcal{L}^n \iff E_{t,u} \succeq 0 \tag{5}$$

If $t = 0$, it holds that $u = 0$ and $E_{0,0} \succeq 0$. Therefore, the second-order cone can be embedded in the cone of symmetric positive semidefinite matrices.

For $c \in \mathbb{R}^n$, $A^{(j)} \in \mathbb{R}^{n_j \times n}$, $a^{(j)} \in \mathbb{R}^{n_j}$, $b^{(j)} \in \mathbb{R}^n$ and $d^{(j)} \in \mathbb{R}$ $(1 \leq j \leq m)$ a *second-order cone problem* has the form

$$(SOCP) \quad \begin{cases} \langle c, x \rangle \longrightarrow \min \\ \|A^{(j)} x + a^{(j)}\|_2 \leq \langle b^{(j)}, x \rangle + d^{(j)}. \end{cases}$$

The constraint is called the *second-order cone constraint.* With (5) the second-order cone contraint is equivalent to

$$\begin{pmatrix} \left(\langle b^{(j)}, x \rangle + d^{(j)}\right) I_{n_j} & A^{(j)} x + a^{(j)} \\ \left(A^{(j)} x + a^{(j)}\right)^T & \langle b^{(j)}, x \rangle + d^{(j)} \end{pmatrix} \succeq 0 \quad \text{for } 1 \leq j \leq m,$$

that is, we can reformulate it by a positive semidefiniteness constraint:

Let $a_i^{(j)} \in \mathbb{R}^{n_j}$ be the i-th column of $A^{(j)}$ for $1 \leq i \leq n$. With

$$A_i^{(j)} := \begin{pmatrix} -b_i^{(j)} I_{n_j} & -a_i^{(j)} \\ -(a_i^{(j)})^T & -b_i^{(j)} \end{pmatrix} \quad \text{and} \quad C^{(j)} := \begin{pmatrix} d^{(j)} I_{n_j} & a^{(j)} \\ (a^{(j)})^T & d^{(j)} \end{pmatrix}$$

we get the equivalent problem

$$\begin{cases} \langle c, x \rangle \longrightarrow \min \\ C^{(j)} - \sum\limits_{i=1}^{n} x_i A_i^{(j)} \succeq 0 \text{ for } 1 \le j \le m\,. \end{cases}$$

The m constraints can be combined in *one* condition $C - \sum_{i=1}^n x_i A_i \succeq 0$, where C and A_i are the block matrices formed by $C^{(j)}$ and $A_i^{(j)}$ for $1 \le i \le n$ and $1 \le j \le m$, correspondingly (compare the remark on page 304). This yields a dual problem in standard form, which is itself a semidefinite problem by lemma 7.1.5:

$$\begin{cases} \langle -c, x \rangle \longrightarrow \max \\ C - \sum\limits_{i=1}^{n} x_i A_i \succeq 0 \end{cases}$$

Thus $(SOCP)$ is a special case of (SDP).[4]

Lemma 7.2.2

a) Second-order cone optimization contains linear optimization.

b) Second-order cone optimization includes quadratic optimization with both quadratic and linear inequality constraints.

Proof: a) Setting $A^{(j)} := 0$ and $a^{(j)} := 0$ for $1 \le j \le m$ yields a general linear problem (in dual form).

b) A quadratic problem with quadratic inequality constraints is

$$(QP_I) \quad \begin{cases} \langle x, Cx \rangle + 2\langle c, x \rangle \longrightarrow \min \\ \langle x, A_j x \rangle + 2\langle c_j, x \rangle + d_j \le 0 \quad (1 \le j \le m) \end{cases} \tag{6}$$

where C, $A_j \in \mathcal{S}^n_{++}$, c, $c_j \in \mathbb{R}^n$ and $d_j \in \mathbb{R}$ for $1 \le j \le m$. Since C and A_j are symmetric positive definite, there exist $C^{1/2}$, $C^{-1/2}$, $A_j^{1/2}$ and $A_j^{-1/2}$. We can write problem (QP_I) as

$$\begin{cases} \left\|C^{1/2}x + C^{-1/2}c\right\|^2 - \langle c, C^{-1}c \rangle \longrightarrow \min \\ \left\|A_j^{1/2}x + A_j^{-1/2}c_j\right\|^2 - \langle c_j, A_j^{-1}c_j \rangle + d_j \le 0 \quad (1 \le j \le m). \end{cases} \tag{7}$$

[4] Solving an $(SOCP)$ via (SDP), however, is *not* expedient: Interior-point methods which solve $(SOCP)$ directly have a much better worst-case complexity than (SDP) interior-point methods applied to the semidefinite formulation of $(SOCP)$; compare the literature in [LVBL].

We linearize the objective function by minimizing a nonnegative value t such that $t^2 \geq \|C^{1/2}x + C^{-1/2}c\|^2$. Up to a square and a constant, problem (7) equals the second-order cone problem

$$\begin{cases} t \longrightarrow \min_{(t,x)} \\ \|C^{1/2}x + C^{-1/2}c\| \leq t \\ \left\|A_j^{1/2}x + A_j^{-1/2}c_j\right\| \leq \sqrt{\langle c_j, A_j^{-1}c_j\rangle - d_j} \qquad (1 \leq j \leq m). \end{cases} \tag{8}$$

If q^* is the optimal value of the subsidiary problem, $p^* := q^{*2} - \langle c, C^{-1}c\rangle$ is the optimal value of (6). □

Other important special cases are *semidefinite relaxation of quadratic optimization with equality constraints, quasi-convex nonlinear optimization* and *max-cut problems.* We will, however, not go into any more detail here and refer the interested reader to the special literature on this topic.

7.3 The $\mathcal{S}$-Procedure

Minimal Enclosing Ellipsoid of Ellipsoids

Given a finite set of ellipsoids $\mathcal{E}_1, \ldots, \mathcal{E}_m$, we consider the problem of finding the ellipsoid $\mathcal{E}$ of minimal volume which contains $\mathcal{E}_1, \ldots, \mathcal{E}_m$. This topic is important in statistics and cluster theory, for example. It can be formulated as a semidefinite program with a *nonlinear* objective function. Let $A \in S^n_{++}$ and $c \in \mathbb{R}^n$. An *ellipsoid* $\mathcal{E}$ with center c (and full dimension) is defined by[5]

$$\mathcal{E} := \mathcal{E}(A, c) := \{x \in \mathbb{R}^n \mid \langle x - c, A(x - c)\rangle \leq 1\}$$

or alternatively

$$\mathcal{E} = \{x \in \mathbb{R}^n \mid \langle x, Ax\rangle - 2\langle c, Ax\rangle + \langle c, Ac\rangle - 1 \leq 0\}.$$

We consider the condition $\mathcal{E}_j \subset \mathcal{E}$ for $1 \leq j \leq m$. To this end, we use the so-called *$\mathcal{S}$-Procedure:*

[5] The definition here differs minimally from the one in section 3.1, A instead of the matrix A^{-1}.

Lemma 7.3.1 (*$\mathcal{S}$-Procedure*)

Let $A_1, A_2 \in S^n_{++}$, $c_1, c_2 \in \mathbb{R}^n$ and $d_1, d_2 \in \mathbb{R}$. If there exists an $\widetilde{x} \in \mathbb{R}^n$ with $\langle \widetilde{x}, A_1 \widetilde{x} \rangle + 2 \langle c_1, \widetilde{x} \rangle + d_1 < 0$, then the following statements are equivalent:

a) The following implication holds:

$$\langle x, A_1 x \rangle + 2 \langle c_1, x \rangle + d_1 \leq 0 \implies \langle x, A_2 x \rangle + 2 \langle c_2, x \rangle + d_2 \leq 0$$

b) There exists a $\lambda \geq 0$ such that

$$\begin{pmatrix} A_2 & c_2 \\ c_2^T & d_2 \end{pmatrix} \preceq \lambda \begin{pmatrix} A_1 & c_1 \\ c_1^T & d_1 \end{pmatrix}.$$

Proof: The implication from $b)$ to $a)$ is 'trivial': For $x \in \mathbb{R}^n$ and $j = 1, 2$ we have

$$(x^T, 1) \begin{pmatrix} A_j & c_j \\ c_j^T & d_j \end{pmatrix} \begin{pmatrix} x \\ 1 \end{pmatrix} = \langle x, A_j x \rangle + 2 \langle c_j, x \rangle + d_j .$$

Note that the existence of $\widetilde{x}$ is not necessary for this implication. □

We postpone the proof of the other direction for the moment and firstly have a look at the important application of the $\mathcal{S}$-Procedure to the ellipsoid problem.

The existence of a point $\widetilde{x}$ in the lemma corresponds to the requirement that the interior of the corresponding ellipsoid is nonempty. With

$$A_1 := A_j, \; c_1 := -A_j c_j, \; d_1 := \langle c_j, A_j c_j \rangle - 1,$$

$$A_2 := A, \; c_2 := -Ac, \; d_2 := \langle c, Ac \rangle - 1$$

and $(\mathcal{E}_j)^\circ \neq \emptyset$ item $a)$ in the lemma means $\mathcal{E}_j \subset \mathcal{E}$. Then it follows directly that $\mathcal{E}_j \subset \mathcal{E}$ holds if and only if there exists a $\lambda_j \geq 0$ such that

$$\begin{pmatrix} A & -Ac \\ (-Ac)^T & \langle c, Ac \rangle - 1 \end{pmatrix} - \lambda_j \begin{pmatrix} A_j & -A_j c_j \\ (-A_j c_j)^T & \langle c_j, A_j c_j \rangle - 1 \end{pmatrix} \preceq 0 . \qquad (9)$$

In this section we use an alternative formulation of the theorem about the SCHUR complement:

Let E, G be symmetric matrices and $G \succ 0$. Then the matrix M given by

$$M := \begin{pmatrix} E & F \\ F^T & G \end{pmatrix}$$

is positive semidefinite if and only if $E - FG^{-1}F^T$ is positive semidefinite.

Chapter 7

We use this to 'linearize' (9):

Lemma 7.3.2

With $b := -Ac$ and $b_j := -A_j c_j$ for $1 \le j \le m$ condition (9) is equivalent to

$$\begin{pmatrix} A & b & 0 \\ b^T & -1 & b^T \\ 0 & b & -A \end{pmatrix} - \lambda_j \begin{pmatrix} A_j & b_j & 0 \\ b_j^T & \langle c_j, A c_j \rangle - 1 & 0 \\ 0 & 0 & 0 \end{pmatrix} \preceq 0. \qquad (10)$$

Proof: For clarification we denote the (n,n)-matrix of zeros by $0_{n\times n}$ and the n-dimensional vector of zeros by 0_n here. We formulate (10) as

$$M := \begin{pmatrix} \lambda_j A_j - A & \lambda_j b_j - b & 0_{n\times n} \\ \lambda_j b_j^T - b^T & \lambda_j \langle c_j, A_j c_j \rangle - \lambda_j + 1 & -b^T \\ 0_{n\times n} & -b & A \end{pmatrix} \succeq 0. \qquad (11)$$

With

$$E := \begin{pmatrix} \lambda_j A_j - A & \lambda_j b_j - b \\ \lambda_j b_j^T - b^T & \lambda_j \langle c_j, A_j c_j \rangle - \lambda_j + 1 \end{pmatrix} \quad \text{and} \quad F := \begin{pmatrix} 0_{n\times n} \\ -b^T \end{pmatrix}$$

we write the matrix M in block form

$$M = \begin{pmatrix} E & F \\ F^T & A \end{pmatrix}.$$

It holds that

$$F A^{-1} F^T = \begin{pmatrix} 0_{n\times n} & 0_n \\ 0_n^T & \langle b, A^{-1} b \rangle \end{pmatrix}.$$

Thus, we have

$$E - F A^{-1} F^T = \lambda_j \begin{pmatrix} A_j & b_j \\ b_j^T & \langle c_j, A_j c_j \rangle - 1 \end{pmatrix} - \begin{pmatrix} A & b \\ b^T & \langle b, A^{-1} b \rangle - 1 \end{pmatrix}.$$

By definition of b it holds that $\langle c, Ac \rangle - 1 = \langle b, A^{-1} b \rangle - 1$. If we use the alternative formulation of the SCHUR complement, we see that condition (11) is equivalent to (9). □

From section 3.1 we know:

$$\operatorname{vol}(\mathcal{E}) = \omega_n \sqrt{\det A^{-1}},$$

where ω_n is the volume of the n-dimensional unit ball. Thus the volume $\operatorname{vol}(\mathcal{E})$ of the ellipsoid $\mathcal{E}$ is proportional to $\sqrt{\det A^{-1}}$, and minimizing $\operatorname{vol}(\mathcal{E})$ means minimizing $\det A^{-1}$ for $A \in S^n_{++}$. Since the logarithm log is strictly isotone, we consider the objective function given by $\log\det A^{-1} = -\log\det A$, which is a *convex* function (see section 7.4).

The semidefinite optimization problem (with *nonlinear* objective function) used for finding the minimal enclosing ellipsoid of ellipsoids is given by

$$\left\{\begin{array}{l} -\log\det A \longrightarrow \min\limits_{(\lambda,c,A)} \\ \begin{pmatrix} A & b & 0 \\ b^T & -1 & b^T \\ 0 & b & -A \end{pmatrix} - \lambda_j \begin{pmatrix} A_j & b_j & 0 \\ b_j^T & \langle c_j\,, A c_j\rangle - 1 & 0 \\ 0 & 0 & 0 \end{pmatrix} \preceq 0 \ \text{ for } \ 1 \le j \le m \\ A \succ 0,\ \lambda = (\lambda_1, \ldots, \lambda_m)^T \in \mathbb{R}_+^m . \end{array}\right. \tag{12}$$

In the same way as in the considerations in section 7.2 this problem can be transferred to a semidefinite problem with constraints in standard form. Semidefinite programs with a *nonlinear* objective function and linear constraints can also be solved by primal–dual interior-point methods.

We limit ourselves to *linear* semidefinite programs. For continuative studies we refer to the work of YAMASHITA, YABE and HARADA, who consider general nonlinear semidefinite optimization (cf. [YYH]), and the work of TOH and VANDENBERGHE, BOYD and WU, who concentrate their considerations on the special case of determinant maximization problems with linear matrix inequality constaints (cf. [Toh] and [VBW]).

In the following we will give some auxiliary considerations as a preparation for the proof of the nontrivial direction of the central lemma 7.3.1 ($\mathcal{S}$-Procedure), which is also an important tool in other branches of mathematics (e.g., control theory). The proof, however, is only intended for readers with a special interest in mathematics. In the literature, one can find a number of proofs which employ unsuitable means and are therefore difficult to understand. The following clear and elementary proof was given by MARKUS SIGG on the basis of [Ro/Wo].

Lemma 7.3.3

$$\sum_{x\,\in\,\{-1,1\}^n} x x^T = 2^n I$$

Proof: On the left the *diagonal entries* of all 2^n summands are $x_i^2 = 1$. For the *off-diagonal entries* the number of entries $x_i x_j = 1$ obtained by the products $1 \cdot 1$ and $(-1)\cdot(-1)$ equals the number of entries $x_i x_j = -1$ obtained by the products $(-1)\cdot 1$ and $1 \cdot (-1)$. □

Lemma 7.3.4

For any $A \in \mathbb{R}^{n\times n}$ *it holds that* $\quad 2^n \operatorname{tr} A = \sum\limits_{x\,\in\,\{-1,1\}^n} \langle x\,, Ax\rangle\,.$

Proof: Lemma 7.3.3 combined with the linearity of the trace and 7.1.3 yields

$$2^n \operatorname{tr} A = 2^n \operatorname{tr}(I\,A) = \sum_{x\,\in\,\{-1,1\}^n} \operatorname{tr}(x x^T A) = \sum_{x\,\in\,\{-1,1\}^n} \langle x\,, Ax\rangle\,.$$

□

Chapter 7

Lemma 7.3.5

Let $P,\, Q \in \mathcal{S}^n$ with $\operatorname{tr} Q \le 0 < \operatorname{tr} P$. Then there exists a vector $y \in \mathbb{R}^n$ with $\langle y, Qy\rangle \le 0 < \langle y, Py\rangle$.

Proof: Let $U \in \mathbb{R}^{n\times n}$ be an orthogonal matrix diagonalizing Q, i.e., $Q = U^T D U$ where $D := \operatorname{Diag}\big((\lambda_1,\dots,\lambda_n)^T\big)$ with the eigenvalues $\lambda_1,\dots,\lambda_n$ of Q. By lemma 7.3.4 we get

$$0 < 2^n \operatorname{tr} P = 2^n \operatorname{tr}(UPU^T) = \sum_{x\in\{-1,1\}^n} \langle x, UPU^T x\rangle = \sum_{x\in\{-1,1\}^n} \big\langle U^T x, P(U^T x)\big\rangle .$$

Hence there exists an $x \in \{-1,1\}^n$ with $\langle y, Py\rangle > 0$ for $y := U^T x$ and

$$\langle y, Qy\rangle = \big\langle U^T x, U^T D U U^T x\big\rangle \overset{\checkmark}{=} \langle x, Dx\rangle \overset{\checkmark}{=} \operatorname{tr} D = \operatorname{tr} Q \le 0 . \qquad \square$$

Lemma 7.3.6

Let $A, B \in \mathcal{S}^n$ with:

a) There exists an $\widetilde{x} \in \mathbb{R}^n$ with $\langle \widetilde{x}, A\widetilde{x}\rangle < 0$.

b) For any $\varepsilon > 0$ there exists a $\lambda \ge 0$ with $B \preceq \lambda A + \varepsilon I$.

Then $B \preceq \lambda A$ for a suitable $\lambda \ge 0$.

Proof: By *b)* there exists a $\lambda_k \ge 0$ with $B \preceq \lambda_k A + \frac{1}{k} I$ for any $k \in \mathbb{N}$. In particular we have $\big\langle \widetilde{x}, (B - \frac{1}{k} I)\widetilde{x}\big\rangle \le \lambda_k \langle \widetilde{x}, A\widetilde{x}\rangle$. As the left-hand side converges for $k \to \infty$, there exists a convergent subsequence (λ_{k_ν}) of (λ_k) by *a)*. With the corresponding limit $\lambda \ge 0$ we have for any $x \in \mathbb{R}^n$

$$\langle x, Bx\rangle = \lim_{\nu\to\infty} \Big\langle x, \Big(B - \frac{1}{k_\nu} I\Big) x\Big\rangle \le \lim_{\nu\to\infty} \lambda_{k_\nu} \langle x, Ax\rangle = \lambda \langle x, Ax\rangle . \qquad \square$$

Chapter 7

Now we prove lemma 7.3.1 in the *homogeneous case*, i. e.:

Proposition 7.3.7

Let $A, B \in \mathcal{S}^n$. If there exists an $\widetilde{x} \in \mathbb{R}^n$ with $\langle \widetilde{x}, A\widetilde{x}\rangle < 0$, then the following statements are equivalent:

a) For any $x \in \mathbb{R}^n$ with $\langle x, Ax\rangle \le 0$ it holds that $\langle x, Bx\rangle \le 0$.

b) There exists a $\lambda \ge 0$ such that $B \preceq \lambda A$.

Proof: The implication from *b)* to *a)* is 'trivial' (cf. p. 317). For the other implication — *a)* $\Longrightarrow$ *b)* — we consider the optimization problem

$$(D) \qquad \begin{cases} \mu \longrightarrow \max \\ B \preceq \lambda A - \mu I \qquad (\lambda,\, \mu \in \mathbb{R}) \\ \lambda \ge 0 . \end{cases}$$

If we prove $d^* = \sup\{\mu \in \mathbb{R} \mid \exists \lambda \in \mathbb{R}_+ \;\; B \preceq \lambda A - \mu I\} \geq 0$, *then* we obtain *b*) by lemma 7.3.6. For $\lambda, \mu \in \mathbb{R}$ we define

$$y := \begin{pmatrix} \lambda \\ \mu \end{pmatrix}, \quad b := \begin{pmatrix} 0 \\ 1 \end{pmatrix}, \quad C := \begin{pmatrix} 0 & 0 \\ 0 & -B \end{pmatrix}$$

and

$$\mathcal{A}^* y := y_1 A_1 + y_1 A_2 := \lambda \begin{pmatrix} -1 & 0 \\ 0 & -A \end{pmatrix} + \mu \begin{pmatrix} 0 & 0 \\ 0 & I \end{pmatrix}$$

with $C, A_1, A_2 \in \mathcal{S}^{1+n}$. Problem (D) is a (dual) semidefinite problem in standard form

$$(D) \quad \begin{cases} \langle b, y \rangle \longrightarrow \max \\ C - \mathcal{A}^* y \succeq 0 \end{cases} \qquad (y \in \mathbb{R}^2).$$

The corresponding primal problem is given by

$$(P) \quad \begin{cases} \langle C, X \rangle \longrightarrow \min \\ \mathcal{A}X = b, \; X \succeq 0 \end{cases} \qquad (X \in \mathcal{S}^{1+n}).$$

Any $X \in \mathcal{S}^{1+n}$ can be written as $X = \begin{pmatrix} \xi & x^T \\ x & X_1 \end{pmatrix}$ with $\xi \in \mathbb{R}$, $x \in \mathbb{R}^n$ and $X_1 \in \mathcal{S}^n$. With that we get

$$(P) \quad \begin{cases} -\langle B, X_1 \rangle \longrightarrow \min \\ \xi + \langle A, X_1 \rangle = 0 \\ \operatorname{tr} X_1 = 1 \\ X \succeq 0 \end{cases} \qquad (\xi \in \mathbb{R}, \; x \in \mathbb{R}^n, \; X_1 \in \mathcal{S}^n).$$

The dual problem (D) has a strictly feasible point: For $\mu < -\|A - B\|$ and $\lambda := 1$ we have $B \prec \lambda A - \mu I$ since it holds for all $x \in \mathbb{R}^2 \setminus \{0\}$ that

$$\langle x, (\lambda A - B)x \rangle = \langle x, (A - B)x \rangle \overset{\checkmark}{\geq} -\|A - B\| \, \|x\|^2 > \mu \, \|x\|^2 = \langle x, \mu I x \rangle .$$

The primal problem (P) has a feasible point, for example:

$$x := 0, \quad X_1 := \frac{1}{\|\widetilde{x}\|^2} \widetilde{x}\widetilde{x}^T \succeq 0, \quad \xi := -\langle A, X_1 \rangle = -\frac{1}{\|\widetilde{x}\|^2} \langle \widetilde{x}, A\widetilde{x} \rangle \geq 0 .$$

Hence, by theorem 7.1.7 (Duality Theorem, part *a*)) problem (P) has a minimizer

$$X^* = \begin{pmatrix} \xi^* & (x^*)^T \\ x^* & X_1^* \end{pmatrix} \quad \text{with} \quad \langle C, X^* \rangle = d^* .$$

Thus we have $-\langle B, X_1^* \rangle = d^*$. $X^* \succeq 0$ yields $\xi^* \geq 0$ and $X_1^* \succeq 0$. We define $W := (X_1^*)^{1/2}$, $Q := WAW$ and $P := WBW$ and get

$$\begin{aligned} \operatorname{tr} Q &= \operatorname{tr}(WAW) = \operatorname{tr}(AW^2) = \langle A, X_1^* \rangle = -\xi^* \leq 0 \\ \operatorname{tr} P &= \operatorname{tr}(WBW) = \langle B, X_1^* \rangle = -d^* . \end{aligned}$$

Assume $d^* < 0$, i.e., $\operatorname{tr} P > 0$. By lemma 7.3.5 there would exist a vector $y \in \mathbb{R}^n$ with

$$\langle Wy, A(Wy) \rangle = \langle y, Qy \rangle \leq 0 < \langle y, Py \rangle = \langle Wy, B(Wy) \rangle ,$$

which is a contradiction to *a)* with $x := Wy$. Hence $d^* \geq 0$ holds which completes the proof. □

We define $\mathbb{E} := \{x \in \mathbb{R}^{n+1} : x_{n+1} = 1\}$ and reformulate lemma 7.3.1:

Let $A', B' \in \mathcal{S}^n_{++}$, $a, b \in \mathbb{R}^n$ *and* $\alpha, \beta \in \mathbb{R}$. *If there exists an* $\widetilde{x} \in \mathbb{E}$ *with* $\langle \widetilde{x}, A\widetilde{x} \rangle < 0$ *for*

$$A := \begin{pmatrix} A' & a \\ a^T & \alpha \end{pmatrix} \quad \textit{and} \quad B := \begin{pmatrix} B' & b \\ b^T & \beta \end{pmatrix},$$

then the following statements are equivalent:

a) For any $x \in \mathbb{E}$ *with* $\langle x, Ax \rangle \leq 0$ *it holds that* $\langle x, Bx \rangle \leq 0$.

b) There exists a $\lambda \geq 0$ *with* $B \preceq \lambda A$.

We define $\mathbb{E}' := \{x \in \mathbb{R}^{n+1} : x_{n+1} \neq 0\}$.

Proof of 7.3.1: Let *a)* hold. For all $x \in \mathbb{E}'$ with $\langle x, Ax \rangle \leq 0$ we have $\langle x, Bx \rangle \leq 0$ by homogeneity. Any $x \in \mathbb{R}^{n+1} \setminus \mathbb{E}'$ can be written as $x = \big((x')^T, 0\big)^T$ with $x' \in \mathbb{R}^n$. From $\langle x, Ax \rangle \leq 0$ we get $\langle x', A'x' \rangle = \langle x, Ax \rangle \leq 0$. As $A' \succ 0$, it follows that $x' = 0$ which yields $x = 0$ and so $\langle x, Bx \rangle = 0$. Hence, $\langle x, Bx \rangle \leq 0$ holds for all $x \in \mathbb{R}^{n+1}$ with $\langle x, Ax \rangle \leq 0$. Now, we can apply the $\mathcal{S}$-Lemma in the homogeneous version (7.3.7) which directly yields lemma 7.3.1. □

Note, we did not use the requirement that B' is positive definite.

Remark

A matrix $A \in \mathcal{S}^{n+1}$ *is positive semidefinite if and only if* $\langle x, Ax \rangle \geq 0$ *holds for all* $x \in \mathbb{E}$.

Proof: If $\langle x, Ax \rangle \geq 0$ holds for all $x \in \mathbb{E}$, then we get $\langle x, Ax \rangle \geq 0$ for all $x \in \mathbb{E}'$ by homogeneity. We write $x \in \mathbb{R}^{n+1} \setminus \mathbb{E}'$ as $x = \big((x')^T, 0\big)^T$ with an $x' \in \mathbb{R}^n$. For $k \in \mathbb{N}$ and $x_k := \big((x')^T, 1/k\big)^T \in \mathbb{E}'$ it follows that $0 \leq \lim_{k \to \infty} \langle x_k, Ax_k \rangle = \langle x, Ax \rangle$. The other implication is trivial. □

This remark shows that it is no restriction to require the existence of a suitable $\widetilde{x} \in \mathbb{E}$ in the inhomogeneous $\mathcal{S}$-Lemma.

7.4 The Function $\log \circ \det$

In this section we consider the function $\varphi\colon \mathcal{S}^n_{++} \longrightarrow \mathbb{R}$ defined by

$$\varphi(X) := -\log\det(X) \quad \text{for} \quad X \in \mathcal{S}^n_{++}.$$

Lemma 7.4.1

a) For $X \in \mathcal{S}^n_{++}$ and $H \in \mathcal{S}^n$, where $\|H\|$ is sufficiently small, it holds that

$$\varphi(X+H) = \varphi(X) + \langle -X^{-1}, H\rangle + o(\|H\|).$$

Thus, φ is differentiable, and the derivative at a point $X \in \mathcal{S}^n_{++}$ can be identified with $-X^{-1}$ in the following sense:

$$\varphi'(X)H = \langle -X^{-1}, H\rangle \quad \textit{for all } H \in \mathcal{S}^n.$$

b) φ is strictly convex.

Here, $\|\ \|$ denotes the operator norm belonging to an arbitrarily chosen norm on $\mathbb{R}^n$.

The following proof is in fact based on a simple idea. To do the proof mathematically clean, however, requires great care, which tends to be 'overlooked' in a number of versions found in the literature.

Proof: a) Let $X \in \mathcal{S}^n_{++}$ and $H \in \mathcal{S}^n$ with $\|H\|$ sufficiently small such that $X + H \in \mathcal{S}^n_{++}$. It holds:

$$\begin{aligned}\varphi(X+H) - \varphi(X) &= -\big(\log\det(X+H) - \log\det(X)\big)\\ &= -\big(\log\det(X+H) + \log\det(X^{-1})\big)\\ &= -\log\det(X^{-1}(X+H))\\ &= -\log\det(I + X^{-1}H)\end{aligned}$$

λ is an eigenvalue of $X^{-1}H$ iff $\lambda+1$ is an eigenvalue of $I + X^{-1}H$ (counted according to their multiplicity). We abbreviate $\lambda_i(X^{-1}H)$ to λ_i and get

$$-\log\det(I + X^{-1}H) = -\log\prod_{i=1}^n(1+\lambda_i) = -\sum_{i=1}^n \log(1+\lambda_i).$$

By the differentiability of the logarithm at $x = 1$ we have $\log(1+\lambda) = \log(1) + \lambda + o(\lambda) = \lambda + o(\lambda)$, and with that we get

$$-\sum_{i=1}^n \log(1+\lambda_i) = -\sum_{i=1}^n \lambda_i + \sum_{i=1}^n o(\lambda_i) = -\operatorname{tr}(X^{-1}H) + \sum_{i=1}^n o(\lambda_i).$$

It remains to show that $\sum_{i=1}^n o(\lambda_i) = o(\|H\|)$: Note that $|\lambda_i| \le \|X^{-1}H\| \le \|X^{-1}\|\,\|H\|$ holds for any eigenvalue λ_i. We write $o(\lambda_i) = |\lambda_i| r_i(\lambda_i)$, where $r_i \colon \mathbb{R} \longrightarrow \mathbb{R}$ is a function with $r_i(\lambda) \longrightarrow 0$ for $\lambda \longrightarrow 0$. We get

$$\left|\sum_{i=1}^n |\lambda_i| r_i(\lambda_i)\right| \le \lambda_{\max} \sum_{i=1}^n |r_i(\lambda_i)| \le \|X^{-1}\|\,\|H\| \sum_{i=1}^n |r_i(\lambda_i)|\,.$$

Due to $|\lambda_i| \le \|X^{-1}\|\,\|H\| \to 0$ for $i = 1, \ldots, n$, we have $|\sum_{i=1}^n |r_i(\lambda_i)| \to 0$ if $\|H\| \longrightarrow 0$. Altogether we get

$$\sum_{i=1}^n o(\lambda_i) = o(\|H\|)\,.$$

b) For $X, Y \in S^n_{++}$ it holds that

$$\begin{aligned} \varphi(Y) - \varphi(X) &= -\log\det(Y) + \log\det(X) = -\log\det(X^{-1}Y) \\ &= -\log\prod_{i=1}^n \lambda_i = -\sum_{i=1}^n \log(\lambda_i) = -n\Big(\frac{1}{n}\sum_{i=1}^n \log(\lambda_i)\Big), \end{aligned} \tag{13}$$

where $\lambda_1, \ldots, \lambda_n$ are the eigenvalues of $X^{-1}Y$ (counted according to their multiplicity). Since the logarithm function is concave, we have

$$\frac{1}{n}\sum_{i=1}^n \log\lambda_i \le \log\left(\frac{1}{n}\sum_{i=1}^n \lambda_i\right). \tag{14}$$

The well-known fact $\log(x) \le x - 1$ for any $x \in \mathbb{R}_{++}$ yields

$$\log\left(\frac{1}{n}\sum_{i=1}^n \lambda_i\right) \le \frac{1}{n}\sum_{i=1}^n \lambda_i - 1\,. \tag{15}$$

From (13) and (15) we get

$$\begin{aligned} \varphi(Y) - \varphi(X) &= -n\Big(\frac{1}{n}\sum_{i=1}^n \log(\lambda_i)\Big) \ge -\sum_{i=1}^n \lambda_i + n \\ &= \operatorname{tr}(-X^{-1}Y + I) = \operatorname{tr}(-X^{-1}(Y - X)) \\ &= \varphi'(X)(Y - X)\,. \end{aligned}$$

Thus, φ is convex. It remains to show that φ is *strictly* convex:

Let $X, Y \in S^n_{++}$ with $\varphi(Y) - \varphi(X) = \varphi'(X)(Y - X)$. Then equality holds in all the inequalities above. Due to (14), we get $\lambda_1 = \cdots = \lambda_n$ as the logarithm is strictly concave. $\log(x) = x - 1$ holds iff $x = 1$. Hence, (15) yields $\frac{1}{n}\sum_{i=1}^n \lambda_i = 1$ and thus $\sum_{i=1}^n \lambda_i = n$. Finally, $\lambda_1 = \cdots = \lambda_n = 1$. Since all eigenvalues of $X^{-1}Y$ are one (and $X^{-1}Y$ is diagonalizable), $X^{-1}Y = I$ holds, i.e., $X = Y$. Therefore, the function φ is *strictly* convex. □

Chapter 7

7.5 Path-Following Methods

We assume that

‖ *the matrices* $A^{(1)},\ldots, A^{(m)}$ *are linearly independent.*

Lemma 7.5.1

The adjoint operator $\mathcal{A}^*$ *is injective. In particular a vector* y *which is a pre-image of a given matrix is unique.*

Proof: The matrices $A^{(j)}$ are linearly independent and we have $\mathcal{A}^* y = \sum_{j=1}^m y_j A^{(j)}$ for all $y \in \mathbb{R}^m$. Thus the operator $\mathcal{A}^*$ is injective. □

Primal–Dual System

Similar to the situation of linear programming we can formulate *a necessary and sufficient condition for optimality* under our assumptions:

Lemma 7.5.2

X and y are optimizers for (SDP) and (DSDP) if and only if there exists a matrix $S \succeq 0$ such that

$$\begin{aligned} \mathcal{A}X &= b,\ X \succeq 0, \quad i.e.,\ X \in \mathcal{F}_P \\ \mathcal{A}^*y + S &= C,\ S \succeq 0,\ i.e.,\ y \in \mathcal{F}_D \\ \langle S, X \rangle &= 0. \end{aligned}$$

The last equality is called the *complementary slackness condition*, sometimes also referred to as the *centering condition*.

Proof: In the proof of weak duality we have seen: $\langle C, X \rangle - \langle b, y \rangle = \langle C - \mathcal{A}^*y, X \rangle = \langle S, X \rangle$ for $X \in \mathcal{F}_P$, $y \in \mathcal{F}_D$ and $S := C - \mathcal{A}^*y$. If $\langle S, X \rangle = 0$, we have a zero duality gap and so by lemma 7.1.6 X and y are optimizers. Let X and y be optimizers. Then $\langle C, X \rangle = \langle b, y \rangle$ holds by the Duality Theorem, that is, $\langle S, X \rangle = 0$. □

The complementary slackness condition can be reformulated:

Lemma 7.5.3

For $X, S \succeq 0$ it holds that:

$$\langle S, X \rangle = 0 \iff XS = 0$$

Proof: *i*) From $XS = 0$ it follows that $0 = \operatorname{tr}(XS) = \langle X, S \rangle = \langle S, X \rangle$.
ii) Let $\langle X, S \rangle = \langle S, X \rangle = 0$. The proof of theorem 7.1.1 has shown $\langle X, S \rangle =$

Chapter 7

$\|X^{1/2}S^{1/2}\|_F^2$, thus $X^{1/2}S^{1/2} = 0$ and so $XS = X^{1/2}\left(X^{1/2}S^{1/2}\right)S^{1/2} = 0$.
□

We change our system while 'relaxing' the third condition. With $\mu > 0$ we can formulate a (preliminary) primal–dual system for semidefinite optimization:

$$\begin{aligned} \mathcal{A}X &= b \\ \mathcal{A}^*y + S &= C \\ XS &= \mu I \\ X \succeq 0,\ S &\succeq 0 \end{aligned}$$

$X \succeq 0$ and $S \succeq 0$ can be replaced by $X \succ 0$ and $S \succ 0$: By $XS = \mu I$ it holds that the matrices X and S are regular, thus positive definite.

The final form of our *primal–dual system* is:

$$\begin{aligned} \mathcal{A}X &= b \\ \mathcal{A}^*y + S &= C \\ XS &= \mu I \\ X \succ 0,\ S &\succ 0 \end{aligned}$$

Barrier Functions

A second way to obtain the primal–dual system and the corresponding central path is to look at barrier functions which result either from the primal or from the dual problem.

First, we regard the *logarithmic barrier function* Φ_μ for the primal problem (SDP) defined by

$$\Phi_\mu(X) := \langle C, X\rangle - \mu \log\det(X),$$

where $\mu > 0$ and $X \in \mathcal{F}_P^0 := \left\{X \in S_{++}^n \mid \mathcal{A}X = b\right\}$.

Analogously, the *logarithmic barrier function* for the dual problem $(DSDP)$ is given by

$$\widetilde{\Phi}_\mu(y) := \langle b, y\rangle + \mu \log\det\left(C - \mathcal{A}^*y\right),$$

where $y \in \mathcal{F}_D^0 := \left\{y \in \mathbb{R}^m \mid C - \mathcal{A}^*y \in S_{++}^n\right\}$. The following theorem yields the connection between the primal–dual system and the corresponding *barrier problems:*

Theorem 7.5.4

Let $\mu > 0$. Then the following statements are equivalent:

a) $\mathcal{F}_P^0$ and $\mathcal{F}_D^0$ are nonempty (SLATER *condition*).

b) There exists a (unique) minimizer to Φ_μ on $\mathcal{F}_P^0$.

c) There exists a (unique) maximizer to $\widetilde{\Phi}_\mu$ on $\mathcal{F}_D^0$.

d) The primal–dual system

$$F_\mu(x, y, s) := \begin{pmatrix} \mathcal{A}^* y + S - C \\ \mathcal{A}X - b \\ XS - \mu I \end{pmatrix} = \begin{pmatrix} 0 \\ 0 \\ 0 \end{pmatrix} \quad (X, S \in \mathcal{S}_{++}^n) \tag{16}$$

has a (unique) solution.

If a) to d) hold, then the minimizer $X(\mu)$ to Φ_μ and the maximizer $y(\mu)$ to $\widetilde{\Phi}_\mu$ yield the solution $\big(X(\mu), y(\mu), S(\mu)\big)$ of (16), where $S(\mu) := C - \mathcal{A}^* y(\mu)$.

The proof runs parallel to that of theorem 6.1.4:

Proof: $b) \iff d)$: The definition of Φ_μ can be extended to the open set $\mathcal{S}_{++}^n$ and Φ_μ is differentiable there (cf. lemma 7.4.1). On $\mathcal{S}_{++}^n$ we consider the problem

$$(P_\mu) \quad \begin{cases} \Phi_\mu(X) \longrightarrow \min \\ \mathcal{A}X = b. \end{cases}$$

(P_μ) is a *convex* problem (cf. lemma 7.4.1) with (affinely) *linear* constraints. Therefore the KKT conditions are necessary and sufficient for a minimizer to (P_μ). The LAGRANGE function L is given by $L(X, y) := \Phi_\mu(X) + \langle y, b - \mathcal{A}X \rangle$ for $X \in \mathcal{S}_{++}^n$ and $y \in \mathbb{R}^m$. The KKT conditions to (P_μ) are

$$\begin{aligned} \nabla_X L(X, y) &= C - \mu X^{-1} - \mathcal{A}^* y = 0 \\ \mathcal{A}X &= b. \end{aligned}$$

If we set $S := \mu X^{-1} \succ 0$ which is equivalent to $XS = \mu I$, we get nothing else but (16). Thus, an X is a KKT point to (P_μ) iff (X, y, S) solves (16). The triple (X, y, S) is unique: X is unique, as (P_μ) is *strictly* convex. Since $XS = \mu I$ must hold, S is unique and then by lemma 7.5.1 the vector y is unique. The proof of $c) \iff d)$ can be done in like manner (cf. the proof of theorem 6.1.4) and will be left to the interested reader as an exercise.
$d) \Longrightarrow a)$: If system (16) has a solution (X, y, S), then we have $X \in \mathcal{F}_P^0$ and $y \in \mathcal{F}_D^0$.

$a) \Longrightarrow b)$: Let X^* and (y^*, S^*) be strictly feasible to (SDP) and $(DSDP)$, respectively. For $X \in \mathcal{F}_P$ it holds that

$$\langle S^*, X \rangle = \langle C, X \rangle - \langle b, y^* \rangle.$$

So $\langle C, X \rangle - \mu \log\det(X) = \Phi_\mu(X)$ can be replaced by $\langle S^*, X \rangle - \mu \log\det(X)$ in problem (P_μ). We may add the constraint

Chapter 7

$$\langle S^*, X \rangle - \mu \log\det(X) \le \langle S^*, X^* \rangle - \mu \log\det(X^*) =: \alpha$$

and obtain the subsidiary problem

$$(P_\mu^*) \quad \begin{cases} \langle S^*, X \rangle - \mu \log\det(X) \longrightarrow \min \\ \mathcal{A}X = b\,,\ X \succ 0 \\ \langle S^*, X \rangle - \mu \log\det(X) \le \alpha\,. \end{cases}$$

We aim to show that the feasible set of (P_μ^*) — a level set of the objective function denoted by $\mathcal{F}_\mu^*$ — is compact. $\mathcal{F}_\mu^*$ is nonempty since X^* is feasible. With $\sigma := \lambda_1(S^*)$ it holds that $\sigma > 0$. For $X \in \mathcal{F}_\mu^*$ we get

$$\sum_{i=1}^{n} \big(\sigma \lambda_i(X) - \mu \log \lambda_i(X)\big) = \sigma \operatorname{tr} X - \mu \log\det(X)$$

because of $\det X = \prod_{i=1}^n \lambda_i(X)$. By lemma 7.1.4 we have

$$0 < \sigma \operatorname{tr} X \le \langle S^*, X \rangle\,.$$

Due to $X \in \mathcal{F}_\mu^*$, this yields

$$\sum_{i=1}^{n} \big(\sigma \lambda_i(X) - \mu \log \lambda_i(X)\big) \le \alpha\,. \tag{17}$$

We define $f(\tau) := \sigma\tau - \mu \log\tau$ for $\tau \in \mathbb{R}_{++}$.

$f(\tau) = \sigma\tau - \mu \log\tau$ for $\sigma = 1$, $\mu = 0.2$

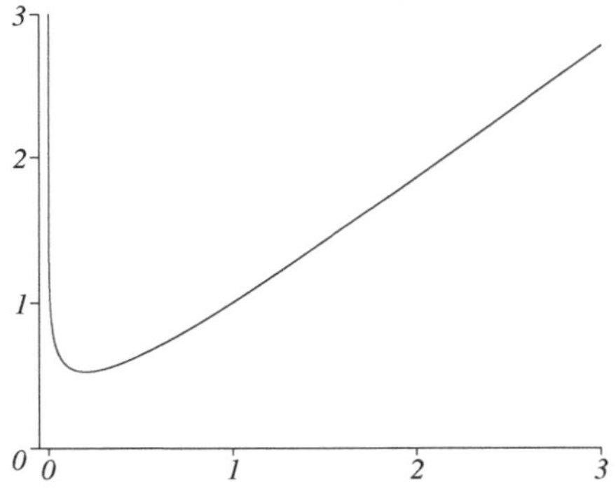

The function f is strictly convex and has a unique minimizer $\tau^* := \mu/\sigma$. It holds that $f(\tau) \longrightarrow \infty$ if $\tau \to 0$ or $\tau \to \infty$. With $v^* := f(\tau^*)$, we have $\alpha \ge nv^*$. Since $\alpha - (n-1)v^* \ge v^*$ holds, there exist $\tau_1, \tau_2 \in \mathbb{R}_{++}$ such that $f(\tau) > \alpha - (n-1)v^*$ for $\tau \in (0, \tau_1)$ and $\tau \in (\tau_2, \infty)$. $\sum_{i=1}^{n} f(\lambda_i(X)) \le \alpha$ then shows $\lambda_i(X) \in [\tau_1, \tau_2]$ for $i = 1, \ldots, n$. It follows

$$\|X\|_F^2 = \sum_{i=1}^{n} \lambda_i(X)^2 \le n\tau_2^2 .$$

So the set $\mathcal{F}_\mu^*$ is bounded. The subset of S^n defined by $\mathcal{A}X = b$ is closed. Since the function $\log \circ \det$ is continuous on S_{++}^n and $\lambda_i(X) \ge \tau_1$ for $i = 1, \dots, n$, the set defined by the condition $\langle S^*, X \rangle - \mu \log\det(X) \le \alpha$ is closed, too. Therefore, the feasible set is compact. As the objective function given by $\langle S^*, X \rangle - \mu \log\det(X)$ is continuous on $\mathcal{F}_\mu^*$, it has a minimizer. □

Central Path

From now on we assume that

> *the* SLATER *condition holds, that is, both the primal and the dual problem have strictly feasible points, and the matrices $A^{(1)}, \dots, A^{(m)}$ are linearly independent.*

Following theorem 7.5.4 there exists then a unique solution $\big(X(\mu), y(\mu), S(\mu)\big)$ to (16) for all $\mu > 0$.

Definition

The set
$$\big\{\big(X(\mu), y(\mu), S(\mu)\big) : \mu > 0\big\}$$
of solutions to the primal–dual system is called the *(primal–dual) central path* of the semidefinite problems (SDP) and $(DSDP)$.

The central path is well-defined, if for *some* $\mu > 0$ equation (16) has a solution.

Remark

For any $\mu > 0$ the matrices $X(\mu)$ and $S(\mu)$ lie in the interior of S_+^n.

We use the following idea to solve the semidefinite program: We reduce $\mu > 0$ and hope that a sequence of corresponding points on the central path will lead to a pair of solutions to problems (SDP) and $(DSDP)$. In order to show that a suitable *subsequence* converges to a pair of optimizers, we need:

Lemma 7.5.5

For $X_1, X_2 \in S^n$ with $\mathcal{A}X_1 = \mathcal{A}X_2$ and $S_1, S_2 \in C - \mathcal{R}(\mathcal{A}^*)$ it holds that

$$\langle X_1 - X_2, S_1 - S_2 \rangle = 0 .$$

Proof: $X_1 - X_2 \in \mathcal{N}(\mathcal{A})$ and $S_1 - S_2 \in \mathcal{R}(\mathcal{A}^*) \subset \mathcal{N}(\mathcal{A})^\perp$. □

Theorem 7.5.6

For any sequence (μ_k) in $\mathbb{R}_{++}$ with $\mu_k \longrightarrow 0$ for $k \to \infty$ a suitable subsequence of the corresponding points

$$(X_k, y_k, S_k) := (X(\mu_k), y(\mu_k), S(\mu_k))$$

of the central path converges to a pair of optimizers of (SDP) and $(DSDP)$.

Proof: WLOG we can assume that $\mu_k \downarrow 0$. Let X_0 be strictly feasible to (SDP) and (y_0, S_0) strictly feasible to $(DSDP)$. By lemma 7.5.5 it holds that

$$0 = \langle X_k - X_0, S_k - S_0 \rangle = \langle X_k, S_k \rangle - \langle X_k, S_0 \rangle - \langle X_0, S_k \rangle + \langle X_0, S_0 \rangle .$$

With $\langle X_k, S_k \rangle = \operatorname{tr}(X_k S_k) = \operatorname{tr}(\mu_k I) = n\mu_k$ we get

$$\begin{aligned} n\mu_1 + \langle X_0, S_0 \rangle \geq n\mu_k + \langle X_0, S_0 \rangle &= \langle X_k, S_0 \rangle + \langle X_0, S_k \rangle \\ &\underset{(7.1.4)}{\geq} \lambda_n(X_k)\lambda_1(S_0) + \lambda_n(S_k)\lambda_1(X_0). \end{aligned}$$

Therefore the eigenvalues of the matrices X_k and S_k are uniformly bounded from above by some constant M. It follows that

$$\|X_k\|_F^2 = \sum_{i=1}^n \lambda_i(X_k)^2 \leq nM^2 \quad \text{holds for any } k \in \mathbb{N}.$$

Accordingly the sequence (S_k) is bounded and with it the sequence (y_k). Thus the sequence (X_k, y_k, S_k) is bounded. Therefore a suitable subsequence $(X_{k_\ell}, y_{k_\ell}, S_{k_\ell})$ converges to a triple $(\widetilde{X}, \widetilde{y}, \widetilde{S}) \in \mathcal{S}_+^n \times \mathbb{R}^m \times \mathcal{S}_+^n$. By the continuity of the inner product it holds that

$$\left\langle \widetilde{X}, \widetilde{S} \right\rangle = \lim_{\ell\to\infty} \langle X_{k_\ell}, S_{k_\ell} \rangle = \lim_{\ell\to\infty} \mu_{k_\ell} \cdot n = 0 .$$

So $(\widetilde{X}, \widetilde{S})$ fulfills the complementary slackness condition. By continuity of $\mathcal{A}$ and $\mathcal{A}^*$ the matrix $\widetilde{X}$ is feasible for (SDP) and the pair $(\widetilde{y}, \widetilde{S})$ is feasible for $(DSDP)$. By lemma 7.5.2 the points $\widetilde{X}$ and $\widetilde{y}$ are optimizers to (SDP) and $(DSDP)$, respectively. □

Chapter 7

It was shown in [HKR] that, unlike in linear optimization, the central path in semidefinite optimization does *not* converge to the analytic center of the optimal set in general. The authors analyze the limiting behavior of the central path to explain this phenomenon.

7.6 Applying Newton's Method

We have seen in the foregoing section that the primal–dual system is solvable and the solution is unique for any $\mu > 0$. If then the corresponding triples

(X_μ, y_μ, S_μ) converge for $\mu \in \mathbb{R}_{++}$ and $\mu \longrightarrow 0$, X_μ and Y_μ converge to a pair of optimizers of (SDP) and $(DSDP)$. Theoretically, we just need to solve a sequence of primal–dual systems, that is, follow the central path until we reach a limit. Practically, however, we face several difficulties when solving

$$\begin{aligned} \mathcal{A}^* y + S &= C \\ \mathcal{A}X &= b \\ XS &= \mu I \\ X \succ 0,\ S &\succ 0\,. \end{aligned} \tag{18}$$

The first two block equations are linear, while the third one is nonlinear. Hence, a NEWTON step seems to be a natural idea for an iteration algorithm. Using path-following methods we solve system (18) approximately and then reduce μ. Starting with $X \succ 0$ and $S \succ 0$, we aim to get *primal and dual directions* $\varDelta X$ and $\varDelta S$, respectively, that satisfy $X - \varDelta X,\ S - \varDelta S \succeq 0$ as well as the *linearized system*

$$\begin{aligned} \mathcal{A}^* \varDelta y + \varDelta S &= \mathcal{A}^* y + S - C \\ \mathcal{A}\varDelta X &= \mathcal{A}X - b \\ X\varDelta S + \varDelta X\, S &= X\,S - \mu I\,. \end{aligned} \tag{19}$$

A crucial observation is that the above system might have no *symmetric* solution $\varDelta X$. This is a serious problem! The second condition gives m equations and the first condition yields $\widetilde{n} = n(n+1)/2$ equations since $\mathcal{A}^* \varDelta y$ and thus $\varDelta S$ are symmetric. The third block equation contains n^2 equations as the product XS is in general not symmetric even if X and S are symmetric. We, however, have only $m + 2\widetilde{n} = m + n(n+1)$ unknowns and so system (19) is overdetermined while we require $\varDelta X$ to be symmetric. Therefore NEWTON's method cannot be applied directly. There are many ways to solve this problem, which has caused a great deal of research.[6]

Another difficulty arises as in practice the equation $XS = \mu I$ will hold only approximately. In theory the three relations

$$XS = \mu I, \quad SX = \mu I, \quad XS + SX = 2\mu I$$

are equivalent but linearizations of these three equations will lead to different search directions if $XS \approx \mu I$.

There are two natural ways — used by the first SDO algorithms — to handle the problem. A first possibility is to drop the symmetry condition for $\varDelta X$. Then the system can be solved. If $\widetilde{\varDelta X}$ is a solution to the relaxed system of equations, we take the symmetric part $\varDelta X$ of $\widetilde{\varDelta X}$, i.e.,

$$\varDelta X := \frac{\widetilde{\varDelta X} + \widetilde{\varDelta X}^T}{2}\,.$$

[6] A comparison of the best known search directions was done by [Todd 1].

Chapter 7

This search direction is called *HKM-direction.*[7] A second approach proposed by ALIZADEH, HAEBERLY and OVERTON is to start with the equivalent formulation of the relaxed complementary slackness condition $XS + SX = 2\mu I$ where the left-hand side is now symmetric. After linearizing, the so-called AHO-direction is the solution to the equation (in addition to the feasibility equations)[8]

$$\Delta X S + S \Delta X + X \Delta S + \Delta S X = (X S + S X) - 2\mu I .$$

In the development of their analysis, the search directions have been grouped into families, e.g., the MONTEIRO–ZHANG-*family (MZ-family)* or the MONTEIRO–TSUCHIYA-*family (MT-family)*. The AHO- and HKM-directions belong to these families. Due to good theoretical and practical results lots of considerations concentrate on these families of search directions. The 'famous' *NT-direction* (cf. NESTEROV/TODD (1997)), which is one of the directions with the best theoretical results, belongs to the MZ-family, too.

In the concluding section of this chapter we restrict ourselves to describing only the general principles underlying many algorithms.

7.7 How to Solve SDO Problems?

We rewrite the 'relaxed' complementary slackness condition $XS = \mu I$ in the symmetric form (cf. exercise 9) as

$$X \circ S := \frac{1}{2}\big(XS + SX\big) = \mu I .$$

It is easy to see that the binary operation $\circ$ is *commutative* and *distributive* with respect to addition $+$, but *not* associative.

In the NEWTON approach the equations

$$\begin{aligned} \mathcal{A}^*(y - \Delta y) + (S - \Delta S) &= C \\ \mathcal{A}(X - \Delta X) &= b \\ (X - \Delta X) \circ (S - \Delta S) &= \mu I \end{aligned}$$

lead — after linearization — to

[7] Also referred to as *HRVW/KSH/M-direction*, since [HRVW] (joint paper by HELMBERG, RENDL, VANDERBEI, WOLKOWICZ (1996)) and [KSH] (joint paper by KOJIMA, SHINDOH, HARA (1997)) proposed it independently, and it was rediscovered by [Mon] (MONTEIRO 1997).

[8] For convenience we often drop the feasibility equations for search directions in the following.

Chapter 7

$$\begin{aligned} \mathcal{A}^*\Delta y + \Delta S &= \mathcal{A}^* y + S - C =: R_d \\ \mathcal{A}\Delta X &= \mathcal{A}X - b =: r_p \\ X \circ \Delta S + \Delta X \circ S &= X \circ S - \mu I =: R_c\,. \end{aligned} \tag{20}$$

For $X \in \mathcal{S}^n_{++}$ we define the operator $\mathcal{L}_X \colon \mathcal{S}^n \longrightarrow \mathcal{S}^n$ by

$$\mathcal{L}_X(Z) := X \circ Z \quad (Z \in \mathcal{S}^n)\,.$$

With that the linearized system (20) can be written as:

$$\begin{aligned} \mathcal{A}^*\Delta y + \Delta S &= R_d \\ \mathcal{A}\Delta X &= r_p \\ \mathcal{L}_S(\Delta X) + \mathcal{L}_X(\Delta S) &= R_c \end{aligned} \tag{21}$$

For implementations we reformulate (21) identifying $\mathcal{S}^n$ with $\mathbb{R}^{\widetilde{n}}$ via

$$\operatorname{svec}\colon \mathcal{S}^n \longrightarrow \mathbb{R}^{\widetilde{n}},$$

where $\widetilde{n} := n(n+1)/2$ (cf. page 301 f). With

$$\begin{aligned} x &:= \operatorname{svec}(X), & s &:= \operatorname{svec}(S), & r_d &:= \operatorname{svec}(R_d), \\ \Delta x &:= \operatorname{svec}(\Delta X), & \Delta s &:= \operatorname{svec}(\Delta S), & r_c &:= \operatorname{svec}(R_c), \end{aligned}$$

the 'matrix representations' L_X, L_S of the operators $\mathcal{L}_X$, $\mathcal{L}_S$ — that is,

$$L_X v := \operatorname{svec}(\mathcal{L}_X(V)) \text{ and } L_S v := \operatorname{svec}(\mathcal{L}_S(V))$$

for $v \in \mathbb{R}^{\widetilde{n}}$ and $V \in \mathcal{S}^n$ with $v = \operatorname{svec}(V)$ — and the matrix

$$A := \begin{pmatrix} \operatorname{svec}(A_1)^T \\ \vdots \\ \operatorname{svec}(A_m)^T \end{pmatrix}$$

we get the equivalent linear system

$$\begin{pmatrix} 0 & A^T & I \\ A & 0 & 0 \\ L_S & 0 & L_X \end{pmatrix} \begin{pmatrix} \Delta x \\ \Delta y \\ \Delta s \end{pmatrix} = \begin{pmatrix} r_d \\ r_p \\ r_c \end{pmatrix}. \tag{22}$$

Remark *The matrices L_S and L_X are invertible.*

Proof: It suffices to show that $L_S v = 0$ for $v \in \mathbb{R}^{\widetilde{n}}$ only has the trivial solution. The symmetric positive definite matrix S can be written as QDQ^T, where Q is orthogonal and D diagonal with positive diagonal components $d_1, \ldots, d_n$. $L_S v = 0$ implies $0 = 2\mathcal{L}_S(V) = SV + VS$ and further $0 = Q^T(SV + VS)Q = D(Q^T V Q) + (Q^T V Q)D = DW + WD$, where $W := Q^T V Q$. For $i, j \in \{1, \ldots, \widetilde{n}\}$ we have

Chapter 7

$$0 = (DW + WD)_{ij} = d_i w_{ij} + w_{ij} d_j = w_{ij} \underbrace{(d_i + d_j)}_{>0},$$

thus $w_{ij} = 0$. From $W = 0$ we conclude that $V = 0$. □

With the additional assumption that $L_S^{-1} L_X$ is positive definite the unique solvability of system (22) can be obtained easily. We will therefore not repeat the details here (compare theorem 6.3.1).

If we start from $X \in \mathcal{F}_P^0$ and $y \in \mathcal{F}_D^0$, it is generally difficult to maintain the — nonlinear — third equation of (18) *exactly*. *Path-following methods* therefore require the iterates to satisfy the equation only in a suitable approximate sense. This can be stated in terms of *neighborhoods* of the central path.

There is a gap between the practical behavior of algorithms and the theoretical performance results, in favor of the practical behavior.

The considerations necessary for that closely follow those of chapter 6. They are, however, far more laborious with often tedious calculations. We therefore abstain from going into any more details here and refer the reader to the special literature on this topic, for example, the *Handbook of Semidefinite Programming* [WSV] or DE KLERK's monograph [Klerk].

7.8 Icing on the Cake: Pattern Separation via Ellipsoids

A *pattern* is a reliable sample of observable characteristics of an object. A specific pattern of n characteristics can be represented by a point in $\mathbb{R}^n$. *Pattern recognition*[9] is the art of classifying patterns.[10] We assume that two sets of points $\{p_1, \ldots, p_{m_1}\}$ and $\{q_1, \ldots, q_{m_2}\}$ are given, which we want to separate by ellipsoids written in the form

$$\mathcal{E} := \left\{ x \in \mathbb{R}^n \mid \left\langle x - x_c, X(x - x_c) \right\rangle \leq 1 \right\}$$

with center $x_c \in \mathbb{R}^n$ and a matrix $X \in \mathcal{S}_{++}^n$.

We want the best possible separation, that is, we want to optimize the *separation ratio* ϱ in the following way:

$$\left\{ \begin{array}{ll} \varrho \longrightarrow \max & \\ \left\langle p_i - x_c, X(p_i - x_c) \right\rangle \leq 1 & (i = 1, \ldots, m_1) \\ \left\langle q_j - x_c, X(q_i - x_c) \right\rangle \geq \varrho & (j = 1, \ldots, m_2) \end{array} \right. \tag{23}$$

This is a nonlinear optimization problem depending on the variables $X \in \mathcal{S}_{++}^n$, $x_c \in \mathbb{R}^n$ and $\varrho \in \mathbb{R}$.

[9] Confer [Gli] and [Bo/Va], p. 429ff.

[10] See exercise 20, chapter 2.

Chapter 7

We use the following maps

$$p_i \longmapsto \widetilde{p_i} := \begin{pmatrix} p_i \\ 1 \end{pmatrix}, \quad q_j \longmapsto \widetilde{q_j} := \begin{pmatrix} q_j \\ 1 \end{pmatrix}$$

to 'lift' each point to the hyperplane

$$H := \{x \in \mathbb{R}^{n+1} \mid x_{n+1} = 1\}.$$

The optimal ellipsoid of problem (23) can then be recovered as the intersection of H with the optimal *elliptic cylinder* in $\mathbb{R}^{n+1}$ (with center 0) containing the lifted points $\widetilde{p_i}$ and $\widetilde{q_j}$.

Visualization of 'Lifting'

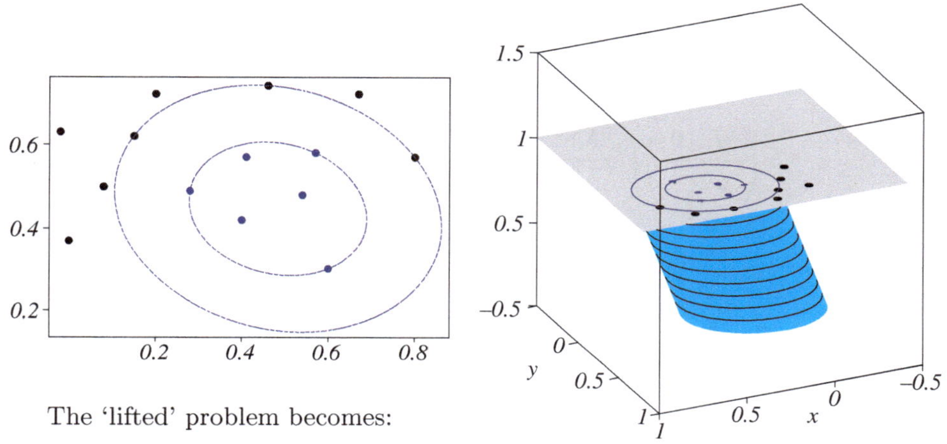

The 'lifted' problem becomes:

$$\begin{cases} \varrho \longrightarrow \max \\ \langle \widetilde{p_i}, \widetilde{X}\widetilde{p_i} \rangle \le 1 & (i = 1, \ldots, m_1) \\ \langle \widetilde{q_j}, \widetilde{X}\widetilde{q_j} \rangle \ge \varrho & (j = 1, \ldots, m_2) \\ \widetilde{X} \in S_+^{n+1}, \ \varrho \in \mathbb{R} \text{ with } \varrho > 1. \end{cases} \tag{24}$$

Due to

$$\langle \widetilde{p_i}, \widetilde{X}\widetilde{p_i} \rangle \overset{\checkmark}{=} \left\langle \widetilde{p_i}\widetilde{p_i}^T, \widetilde{X} \right\rangle, \quad \langle \widetilde{q_j}, \widetilde{X}\widetilde{q_j} \rangle = \left\langle \widetilde{q_j}\widetilde{q_j}^T, \widetilde{X} \right\rangle$$

and by introducing scalar slack variables, problem (24) can be transformed to the standard form of a primal SDO.

From the optimizers

$$\widetilde{X} = \left(\begin{array}{c|c} X & -b \\ \hline -b^T & \gamma \end{array} \right) \in S_+^{n+1} \text{ and } \varrho \in \mathbb{R} \text{ with } \varrho > 1,$$

we recover $X \in S_+^n$ and $x_c \in \mathbb{R}^n$ of the original problem: We have

Chapter 7

$$\left\langle \widetilde{x}, \widetilde{X}\widetilde{x} \right\rangle = \langle x, Xx\rangle - 2\langle b, x\rangle + \gamma \tag{25}$$

for $x \in \mathbb{R}^n$ and $\widetilde{x} := \begin{pmatrix} x \\ 1 \end{pmatrix}$. There exists a vector $x_c \in \mathbb{R}^n$ such that $Xx_c = b$, since for $y \in \text{kernel}(X)$ it follows:

$$\begin{pmatrix} y \\ 0 \end{pmatrix}^T \widetilde{X} \begin{pmatrix} y \\ 0 \end{pmatrix} = \begin{pmatrix} y \\ 0 \end{pmatrix}^T \begin{pmatrix} 0 \\ -\langle b, y\rangle \end{pmatrix} = 0$$

Therefore, $\widetilde{X}\begin{pmatrix} y \\ 0 \end{pmatrix} = 0$ and thus $\langle b, y\rangle = 0$. This shows the solvability of the above system of equations (cf. exercise 10). Now (25) yields

$$\left\langle \widetilde{x}, \widetilde{X}\widetilde{x} \right\rangle = \langle x, Xx\rangle - 2\langle x_c, Xx\rangle + \gamma = \langle x - x_c, X(x - x_c)\rangle + \delta$$

where $\delta := \gamma - \langle x_c, Xx_c\rangle$. With

$$\left(\begin{array}{c|c} I & 0 \\ \hline x_c^T & 1 \end{array}\right)\left(\begin{array}{c|c} X & -b \\ \hline -b^T & \gamma \end{array}\right)\left(\begin{array}{c|c} I & x_c \\ \hline 0 & 1 \end{array}\right) = \left(\begin{array}{c|c} X & 0 \\ \hline 0 & \delta \end{array}\right)$$

we get

$$\delta \geq 0 \text{ and } \det(\widetilde{X}) = \delta \cdot \det(X).$$

Since

$$\begin{aligned} \langle p_i - x_c, X(p_i - x_c)\rangle + \delta \leq 1 \qquad & (i = 1, \ldots, m_1), \\ \langle q_j - x_c, X(q_i - x_c)\rangle + \delta \geq \varrho \qquad & (j = 1, \ldots, m_2), \end{aligned} \tag{26}$$

we also have

$$\delta \leq 1 \text{ and } X \neq 0.$$

$\delta = 1$ implies $X(p_i - x_c) = 0$ $(i = 1, \ldots, m_1)$, that is, $p_1, \ldots, p_{m_1}$ lie in an affine subspace of at most dimension $n-1$. If we exclude this, then $\delta < 1$. The above inequalities can then be rewritten as

$$\begin{aligned} \left\langle p_i - x_c, \frac{1}{1-\delta} X(p_i - x_c)\right\rangle \leq 1 \qquad & (i = 1, \ldots, m_1) \\ \left\langle q_j - x_c, \frac{1}{1-\delta} X(q_i - x_c)\right\rangle \geq \tfrac{\varrho-\delta}{1-\delta} \quad & (j = 1, \ldots, m_2). \end{aligned}$$

If $\delta > 0$ this gives a contradiction to the maximality of ϱ since $\frac{\varrho-\delta}{1-\delta} > \varrho$. (26) now implies that (X, x_c, ϱ) is a solution to (23).

Remark

If $X \in S^n_{++}$ we thus obtain the desired separation of the two point sets by two concentric ellipsoids whose center x_c is uniquely determined.

If $\det(X) = 0$, then we get a 'parallel strip' instead of an 'ellipsoid'.

In appendix C we will find more information on how to solve this kind of problem using SeDuMi .

Exercises

1. *a)* Find a matrix in $S^2 \setminus S^2_+$ whose leading principal subdeterminants are nonnegative.

   b) Show that the nonsymmetric matrix $M := \begin{pmatrix} 1 & -1 \\ 1 & 1 \end{pmatrix}$ is positive definite, that is, $x^T M x > 0$ holds for all $x \in \mathbb{R}^2 \setminus \{0\}$.

   c) Find two matrices $A,\, B \in S^2$ whose product AB is not symmetric.

2. Let $A \in S^n_+$ and $k \in \mathbb{N}$. Show that there exists a uniquely determined matrix $A^{1/k} := R \in S^n_+$ such that $R^k = A$. This matrix is called the *k-th root of A*. Show furthermore that $\operatorname{rank}(A) = \operatorname{rank}(R)$ and $AR = RA$.

   Hint: Cf. the spectral theorem and applications [Hal], p. 156f and p. 166.

3. With $\alpha > 0$ let $\mathcal{L}^n_\alpha := \left\{(u,t) \in \times\mathbb{R}^{n-1} \times \mathbb{R} \mid t \geq \alpha \|u\|_2\right\}$.

   Verify: *a)* $\mathcal{L}^n_\alpha$ is a closed convex cone.

   b) $(\mathcal{L}^n_\alpha)^* = \mathcal{L}^n_{1/\alpha}$

4. *Logarithmic Barrier Function for S^n_+*

   Show: *a)* $\left(S^n_+\right)^\circ = S^n_{++}$

   b) The function $B\colon S^n_{++} \longrightarrow \mathbb{R}$ given by $B(X) := -\log\det(X)$ for $X \in S^n_{++}$ has the *barrier property:*

   For every sequence (X_k) in $\left(S^n_+\right)^\circ$, converging to an $X \in \partial S^n_+$ (boundary of S^n_+), it holds that $B(X_k) \longrightarrow \infty$ for $k \longrightarrow \infty$.

5. *a)* Consider the optimization problem

$$\begin{cases} \sum_{j=0}^{k} \langle c_j\,, x_j \rangle \longrightarrow \min \\ \sum_{j=0}^{k} A_j x_j \; = b \\ x_0 \in \mathbb{R}^{n_0}_+,\; x_j \in \mathcal{L}^{n_j} \;\; (\text{Lorentz cone}) \;\; (j = 1, \dots, k)\,, \end{cases}$$

   where $A_j \in \mathbb{R}^{m \times n_j}$, $c_j \in \mathbb{R}^{n_j}$ for $j = 0, \dots, k$ and $b \in \mathbb{R}^m$.

   Determine the corresponding dual problem.

   Hint: For $y \in \mathbb{R}^m$ minimize

$$L(x_0, \dots, x_k, y) := \sum_{j=0}^{k} \langle c_j\,, x_j \rangle + \Big\langle b - \sum_{j=0}^{k} A_j x_j,\, y \Big\rangle$$

   with respect to $x_0 \in \mathbb{R}^n_+$, $x_j \in \mathcal{L}^{n_j}$ $(j = 1, \dots, k)$.

   b) How will the dual problem change if we demand $x_0 \in \mathbb{R}^{n_0}$ instead of $x_0 \in \mathbb{R}^{n_0}_+$?

6. *Minimum Circumscribed Ball*

We have points $x^{(1)},\ldots,x^{(m)} \in \mathbb{R}^n$ and wish to find the smallest (euclidean) ball with center x and radius r containing them, that is, we have to solve the problem

$$r \longrightarrow \min$$
$$\|x^{(j)} - x\| \le r \quad \text{for} \quad j = 1,\ldots,m.$$

a) Formulate this problem as a *second-order cone problem.*

b) With SeDuMi [11], solve this problem for the points

$$x^{(1)} := \begin{pmatrix} 0 \\ 0 \end{pmatrix},\; x^{(2)} := \begin{pmatrix} 5 \\ -1 \end{pmatrix},\; x^{(3)} := \begin{pmatrix} 4 \\ 6 \end{pmatrix} \text{ and } x^{(4)} := \begin{pmatrix} 1 \\ 3 \end{pmatrix}$$

(cf. chapter 1, example 4). Visualize your solution and compare it with:

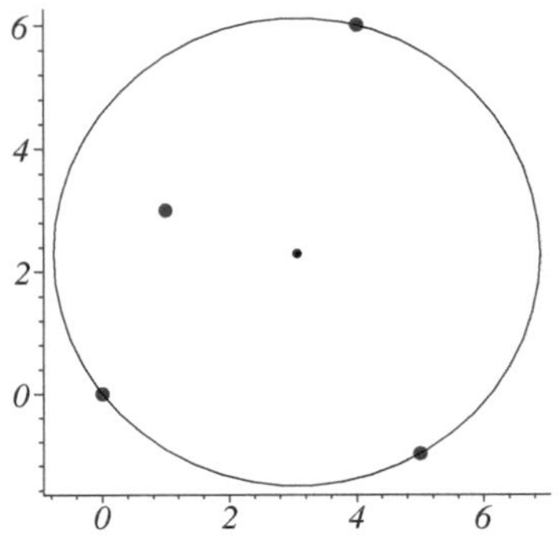

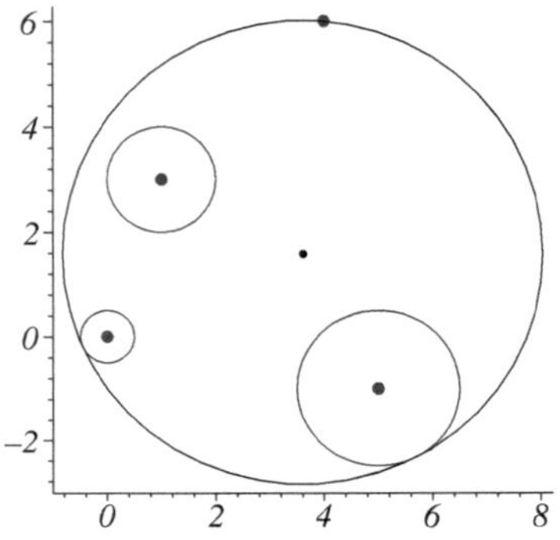

c) Test your program with 100 random points (generated with randn), then with 1000 random points.

d) Look at the more general problem of determining the minimum circumscribed ball to given *balls* with centers $x^{(j)}$ and radii ϱ_j for $j = 1,\ldots,m$. Solve this problem for the points from *b)* and the respective radii 0.5, 1.5, 0.0, 1.0.

7. *Shortest Distance between Two Ellipses* (cf. [An/Lu], example 14.5)

Solve the shortest distance problem

$$\begin{cases} \delta \longrightarrow \min \\ \text{with } \|u - v\| \le \delta, \\ u^T \begin{pmatrix} 1 & 0 \\ 0 & 4 \end{pmatrix} u - (2,0)\,u - 3 \le 0 \text{ and} \\ \frac{1}{2} v^T \begin{pmatrix} 5 & 3 \\ 3 & 5 \end{pmatrix} v - (22,26)\,v + 70 \le 0 \quad \text{for } u, v \in \mathbb{R}^2. \end{cases}$$

[11] Confer appendix C.

a) Let $z := \left(u_1, u_2, v_1, v_2, \delta\right)^T \in \mathbb{R}^5$ and formulate the above problem as a *second-order cone problem:*

$$\begin{aligned} \langle b, z \rangle &\longrightarrow \min \\ \|A_j^T z + c_j\|_2 &\leq \langle b_j, z \rangle + d_j \quad (j = 1, 2, 3) \end{aligned}$$

b) Solve this problem by means of SeDuMi (cf. example 4 in appendix C).

c) Determine the largest strip which separates the two ellipses. Compare your solution with our result:

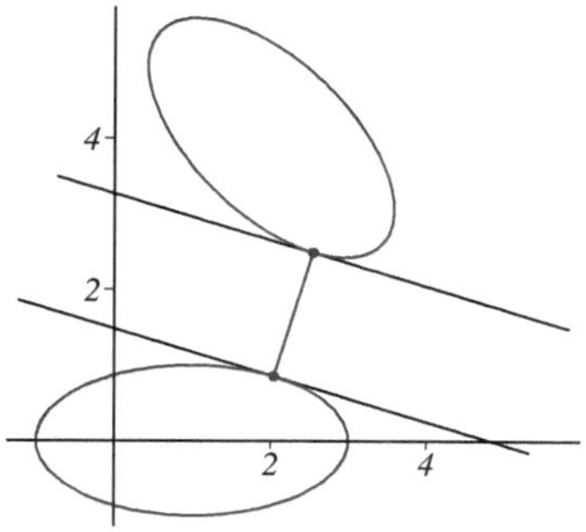

8. Let the following quadratic optimization problem be given:

$$\begin{aligned} \frac{1}{2}x^T Q x + p^T x &\longrightarrow \min \\ A^T x &\leq b \end{aligned}$$

with $Q \in \mathcal{S}^n_{++}$, $p \in \mathbb{R}^n$, $A \in \mathbb{R}^{n \times m}$ and $b \in \mathbb{R}^m$.

a) Show that this problem is equivalent to a second-order cone problem of the form

$$\begin{aligned} \tau &\longrightarrow \min \\ \|\widetilde{A}^T + \widetilde{c}\| &\leq \tau \\ A^T x &\leq b. \end{aligned}$$

Hint: The Cholesky decomposition $Q = LL^T$ gives $\widetilde{A} = L$ and $\widetilde{c} = L^{-1}p$.

b) Have another close look at exercise 8 in chapter 4 and solve this problem with SeDuMi .

9. *a)* Let $\mu > 0$ and $X, S \in \mathcal{S}^n_+$ with $XS + SX = 2\mu I$.
Verify: $X, S \in \mathcal{S}^n_{++}$ and $XS = \mu I$.

b) For $X := \begin{pmatrix} 1 & 0 \\ 0 & -1 \end{pmatrix}$ and $S := \begin{pmatrix} \mu & \sigma \\ \sigma & -\mu \end{pmatrix}$ with $\sigma \neq 0$ it holds that $XS + SX = 2\mu I$, but $XS \neq \mu I$.

Hint to a): Firstly show that X can WLOG be chosen as a diagonal matrix.

Chapter 7

10. *a)* Show for $A \in \mathcal{S}^n_+$ and $x \in \mathbb{R}^n$: If $\langle x, Ax \rangle = 0$, then $Ax = 0$.

b) Show for $A \in \mathbb{R}^{m \times n}$ and $b \in \mathbb{R}^m$: The system of equations $Ax = b$ has a solution if and only if $A^T y = 0$ always implies $b^T y = 0$.

11. *a)* Solve the following semidefinite optimization problem:

$$b^T y \longrightarrow \max$$
$$C - \mathcal{A}^* y \in \mathcal{S}^n_+ \,,$$

where $b = \begin{pmatrix} 11 \\ 9 \end{pmatrix}$, $y \in \mathbb{R}^2$ and

$$A_1 = \begin{pmatrix} 1 & 0 & 1 \\ 0 & 3 & 7 \\ 1 & 7 & 5 \end{pmatrix}, \quad A_2 = \begin{pmatrix} 0 & 2 & 8 \\ 2 & 6 & 0 \\ 8 & 0 & 4 \end{pmatrix}, \quad C = \begin{pmatrix} 1 & 2 & 3 \\ 2 & 9 & 0 \\ 3 & 0 & 7 \end{pmatrix}.$$

b) Find scalars y_1, y_2 such that the maximum eigenvalue of

$$C + y_1 A_1 + y_2 A_2$$

is minimized. Formulate this problem as an SDO problem. Visualize this problem and compare it with:

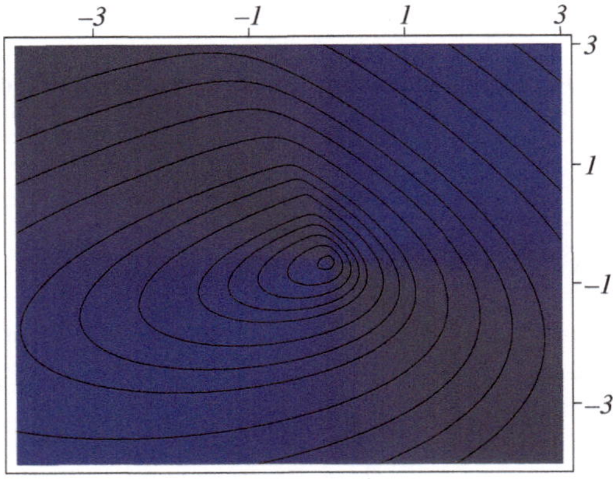

8

Global Optimization

Global optimization is concerned with the computation and characterization of *global* optimizers of — in general — nonlinear functions. It is an important task since many real-world questions lead to global rather than local problems. Global optimization has a variety of applications including, for example, chemical process design, chip layout, planning of just-in-time manufacturing, and pooling and blending problems.

In this chapter we will have a look at the special questions and approaches in global optimization as well as some of the particular difficulties that may arise. Our aim is by no means a comprehensive treatment of the topic which would go beyond the scope of our book, but rather a first introduction to this interesting branch of optimization suitable for our intended target group. Within the framework of our book this chapter is a good complement to the range of topics covered so far.

Readers interested in pursuing this topic further will find a variety of state-of-the-art approaches and suggestions in the *Handbook of Global Optimization* [Ho/Pa]. There also exists an extensive amount of further reading in which more advanced topics are covered.

8.1 Introduction

So far we have — apart from the convex case — primarily dealt with the computation of *local* minima. The computation of a *global* minimum is generally a much more complex problem. We take the simple special case of concave minimization to illustrate that there can exist a great number of local minima from which the global minima have to be — often laboriously — extracted.

W. Forst and D. Hoffmann, *Optimization—Theory and Practice*,
Springer Undergraduate Texts in Mathematics and Technology,
DOI 10.1007/978-0-387-78977-4_8,

Example 1 (PARDALOS, ROSEN (1987))

a) Let $0 < c < 1$. At first we consider the simple problem:

$$-(cx + \tfrac{1}{2}x^2) \longrightarrow \min$$
$$-1 \le x \le 1$$

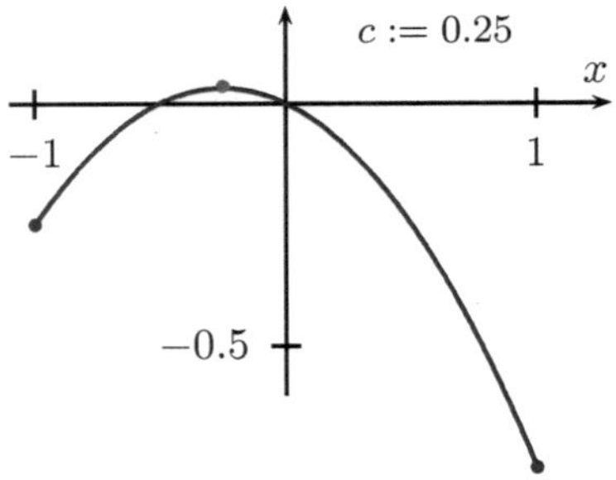

It is obvious (parabola opening downwards!) that: For $x = -c$ we get the maximum $1/2c^2$. The boundary point $x = -1$ gives a local minimum $c - 1/2$, the boundary point $x = 1$ the global minimum $-(c + 1/2)$.

With $g_1(x) := -1 - x$ and $\overline{g}_1(x) := x - 1$ we write the constraints $-1 \le x \le 1$ in the usual form $g_1(x) \le 0$, $\overline{g}_1(x) \le 0$ and illustrate the situation again with the help of the KARUSH–KUHN–TUCKER conditions (for a feasible point x). From

$$-(c + x) - \lambda_1 + \overline{\lambda}_1 = 0$$
$$\lambda_1(1 + x) = 0, \ \overline{\lambda}_1(x - 1) = 0$$
$$\lambda_1, \ \overline{\lambda}_1 \ge 0 \quad \text{we get:}$$

$x = 1$	$\lambda_1 = 0, \ \overline{\lambda}_1 = 1 + c$	global minimizer
$x = -1$	$\overline{\lambda}_1 = 0, \ \lambda_1 = 1 - c$	local minimizer
$-1 < x < 1$	$\lambda_1 = \overline{\lambda}_1 = 0, \ x = -c$	global maximizer

b) We now use the functions from *a)* as building blocks for the following problem in n variables. To given numbers $0 < c_i < 1$ $(i = 1, \ldots, n)$ and for $x = (x_1, \ldots, x_n)^T \in \mathbb{R}^n$ we consider

$$f(x) := -\sum_{i=1}^{n} \left(c_i x_i + \tfrac{1}{2}x_i^2\right) \longrightarrow \min$$
$$-1 \le x_i \le 1 \qquad (i = 1, \ldots, n)\,.$$

With $c := (c_1, \ldots, c_n)^T$ we get $f(x) = -\langle c, x\rangle - 1/2\,\langle x, x\rangle$ and further with $g_i(x) := -1 - x_i = -1 - \langle e_i, x\rangle$ and $\overline{g}_i(x) := x_i - 1 = \langle e_i, x\rangle - 1$ at first: $\nabla f(x) = -(c + x)$, $\nabla g_i(x) = -e_i$, $\nabla \overline{g}_i(x) = e_i$

Hence, in this case the KARUSH–KUHN–TUCKER conditions (for feasible points x) are

$$-(c + x) + \sum_{i=1}^{n}(-\lambda_i + \overline{\lambda}_i)e_i = 0$$

$$\left.\begin{array}{l}\lambda_i, \ \overline{\lambda}_i \ge 0 \\ \lambda_i(x_i + 1) = 0 = \overline{\lambda}_i(x_i - 1)\end{array}\right\} \quad (i = 1, \ldots, n).$$

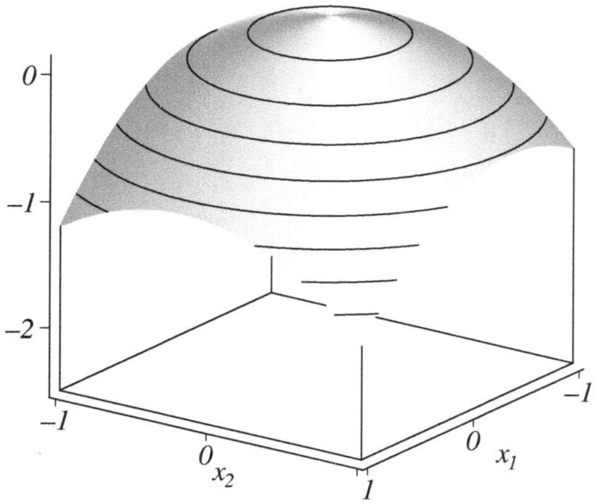

In this case we thus have 3^n KARUSH–KUHN–TUCKER points; they yield

2^n	local minimizers (at the vertices); among them
	one global minimizer at $x = (1, \dots, 1)^T$
1	global maximizer at $x = -c$
$3^n - 2^n - 1$	saddlepoints.

◁

We see: Global optimization only via KKT conditions may be very inefficient. For the following considerations we will need a weak form of the theorem of KREIN–MILMAN which gives an extremely important characterization of compact convex sets, and for that the term *extreme points.*

A point z of a convex set C is called an *extreme point to* C iff there are *no* distinct points $x, y \in C$ such that $z = \alpha x + (1 - \alpha) y$ for a suitable $\alpha \in (0, 1)$. In other words: A point z in C is an extreme point iff it is not an interior point of any nontrivial line segment lying in C.

$$\mathrm{ext}(C) := \{z \in C \mid z \text{ extreme point}\}$$

For example, in the plane a triangle has three extreme points, and a sphere has all its boundary points as extreme points.

Theorem (Krein–Milman)

Let K be a compact convex set in $\mathbb{R}^n$. Then K is the convex hull of its extreme points.

A *proof* can be found, for example, in [Lan].

Theorem 8.1.1

Let $K \subset \mathbb{R}^n$ be convex, compact and $f\colon K \longrightarrow \mathbb{R}$ concave. If f attains a global minimum on K, then this minimum will also be attained at an extreme point of K.

Proof: Assume that the point $\overline{x} \in K$ gives a global minimum of f. Since $K = \operatorname{conv}(\operatorname{ext}(K))$, there exists a representation $\overline{x} = \sum_{j=1}^{m} \lambda_j x_j$ with $m \in \mathbb{N}$, $\lambda_j \geq 0$, $\sum_{j=1}^{m} \lambda_j = 1$ and $x_j \in \operatorname{ext}(K)$. WLOG let

$$f(x_1) = \min\{f(x_j) \mid 1 \leq j \leq m\}\,.$$

From $f(\overline{x}) \geq \sum_{j=1}^{m} \lambda_j f(x_j) \geq f(x_1)$ then follows $f(\overline{x}) = f(x_1)$. □

Remark

Let $M \subset \mathbb{R}^n$ be convex and $f\colon M \longrightarrow \mathbb{R}$ concave. Assume furthermore that $\overline{x}$ yields a local minimum of f. Then it holds that:

1) If the function f is even strictly concave, then $\overline{x} \in \operatorname{ext}(M)$.

2) If $\overline{x}$ yields a strict local minimum, then $\overline{x} \in \operatorname{ext}(M)$.

Proof: 1): By assumption we can find an $\varepsilon > 0$ such that $f(\overline{x}) \leq f(x)$ for all $x \in M$ with $\|x - \overline{x}\|_2 < \varepsilon$.

From $\overline{x} \notin \operatorname{ext}(M)$ we deduce that there exist $x_1, x_2 \in M$ with $x_1 \neq x_2$ and $\alpha \in (0,1)$ with $\overline{x} = \alpha x_1 + (1-\alpha)x_2$. Then $\overline{x}$ can also be written as a convex combination of two distinct points v, w on the connecting line $\overline{x_1 x_2}$, for which $\|v - \overline{x}\|_2 < \varepsilon$ and $\|w - \overline{x}\|_2 < \varepsilon$ hold: $\overline{x} = \beta v + (1-\beta)w$ for some $\beta \in (0,1)$ thus leads to a *contradiction*:

$$f(\overline{x}) > \beta f(v) + (1-\beta) f(w) \geq \beta f(\overline{x}) + (1-\beta) f(\overline{x}) = f(\overline{x})$$

2): Follows 'analogously'; above we first use only the concavity of f for our estimate, and only afterwards do we utilize the fact that $\overline{x}$ yields a strict local minimum. □

Chapter 8

We begin with the definition of the *convex envelope* of a given function which is one of the basic tools used in the theory and algorithms of general global optimization.

Definition

Let $M \subset \mathbb{R}^n$ be a nonempty, convex set, and $f\colon M \longrightarrow \mathbb{R}$ a function bounded from below by a convex function. A function $f_M := F\colon M \longrightarrow \mathbb{R}$ is called the *convex envelope*[1] *of f on M* if and only if

[1] Also known as the *convex hull* in the literature.

$$\begin{cases} F \text{ is convex.} \\ F \le f\,, \quad \text{that is,} \quad F(x) \le f(x) \text{ for all } x \in M\,. \\ \text{For every convex function G with } G \le f \text{ it holds that } G \le F\,. \end{cases}$$

The convex envelope of a given function is the best convex 'underestimation' of this function over its domain.

In most cases it is not even mentioned that — strictly speaking — the *existence* of the convex envelope has to be proven. It of course follows from the obvious fact that the supremum of any family of convex functions (taken pointwise) is again convex. The *uniqueness* on the other hand follows immediately from the defining characteristics.

Geometrically, f_M is the function whose epigraph is the convex hull of the epigraph of the function f. It is the pointwise supremum over all convex underestimators of f over M.

Convex Envelope

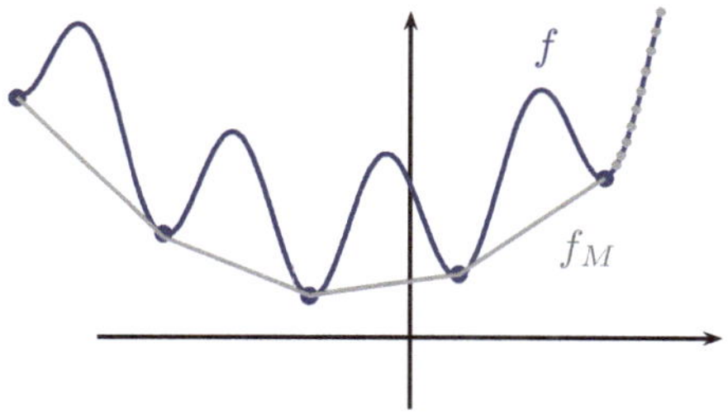

Remark

For nonempty convex sets A, B with $A \subset B \subset M$ it holds that $f_A(x) \ge f_B(x)$ for all $x \in A$.

Proof: ✓

Every optimization problem whose set of feasible points is convex is related to a convex problem with the same optimal value:

Theorem 8.1.2 (Kleibohm (1967))

Let $M \subset \mathbb{R}^n$ be nonempty and convex, and $f \colon M \longrightarrow \mathbb{R}$. Further assume that there exists an $\overline{x} \in M$ with $f(\overline{x}) \le f(x)$ for all $x \in M$. With the convex envelope f_M of f it holds that

$$f(\overline{x}) = \min_{x \in M} f(x) = \min_{x \in M} f_M(x) = f_M(\overline{x})\,.$$

Proof: By definition we have $f_M \le f$, in particular $f_M(\overline{x}) \le f(\overline{x})$. Therefore $\inf_{x \in M} f_M(x) \le \inf_{x \in M} f(x) = f(\overline{x})$. For the constant convex function G

defined by $G(x) := f(\overline{x})$ it holds that $G \le f$. It follows that $f(\overline{x}) = G(x) \le f_M(x)$ for all $x \in M$ and thus $f(\overline{x}) \le \inf_{x\in M} f_M(x) \le f_M(\overline{x})$. Hence, altogether, the desired result. □

This theorem might suggest that we should attempt to solve a general optimization problem by solving the corresponding convex problem where the new objective function is the convex envelope. The difficulty, however, is that — *in general* — finding the convex envelope of a function is at least as difficult as computing its global minimum. In addition, even though the theorem states that every global minimum of f is also a global minimum of f_M, simple one-dimensional examples (like $f(x) := \sin(x)$ for $x \in \mathbb{R}$) show that the inversion of the argument is seriously wrong.

We are now going to discuss two important special cases where convex envelopes can be explicitly described. If we have the special case that M is a *polytope*[2] and the function f *concave,* the convex envelope can be evaluated by solving a *linear* program:

Theorem 8.1.3 (Falk/Hoffman (1976), cf. [Fa/Ho])

Let the points $v_0, \ldots, v_k$ be the vertices of the polytope P, that is, $P = \operatorname{conv}\{v_0, \ldots, v_k\}$. Then the convex envelope f_P of a concave function $f \colon P \longrightarrow \mathbb{R}$ is given by

$$f_P(x) = \min\Big\{ \sum_{\kappa=0}^{k} \lambda_\kappa f(v_\kappa) \;\Big|\; \lambda \in \Lambda(x) \Big\},$$

where

$$\Lambda(x) := \Big\{ \lambda = (\lambda_0 \ldots, \lambda_k) \in [0,1]^{k+1} \;\Big|\; \sum_{\kappa=0}^{k} \lambda_\kappa = 1, \sum_{\kappa=0}^{k} \lambda_\kappa v_\kappa = x \Big\}.$$

Proof: We denote the right-hand side by G, that is,

$$G(x) := \min\Big\{ \sum_{\kappa=0}^{k} \lambda_\kappa f(v_\kappa) \;\Big|\; \lambda \in \Lambda(x) \Big\} \qquad (x \in P).$$

For $x, y \in P$ we choose $\lambda \in \Lambda(x)$ and $\mu \in \Lambda(y)$ with $G(x) = \sum_{\kappa=0}^{k} \lambda_\kappa f(v_\kappa)$ and $G(y) = \sum_{\kappa=0}^{k} \mu_\kappa f(v_\kappa)$. First, we prove that G *is convex:* For $\alpha \in [0,1]$ we have $\alpha\lambda + (1-\alpha)\mu \in \Lambda\big(\alpha x + (1-\alpha)y\big)$ and therefore $G\big(\alpha x + (1-\alpha)y\big) \le \sum_{\kappa=0}^{k} [\alpha\lambda_\kappa + (1-\alpha)\mu_\kappa] f(v_\kappa) = \alpha G(x) + (1-\alpha) G(y)$.

The concavity of f implies

$$G(x) = \sum_{\kappa=0}^{k} \lambda_\kappa f(v_\kappa) \le f\Big(\sum_{\kappa=0}^{k} \lambda_\kappa v_\kappa\Big) = f(x),$$

hence — with the convexity of G — we have $G \le f_P$. Now

[2] That is, the convex hull of a finite number of points in $\mathbb{R}^n$. Thereby polytopes are *compact* convex sets.

$$f_P(x) = f_P\Big(\sum_{\kappa=0}^{k} \lambda_\kappa v_\kappa\Big) \le \sum_{\kappa=0}^{k} \lambda_\kappa f_P(v_\kappa) \le \sum_{\kappa=0}^{k} \lambda_\kappa f(v_\kappa) = G(x)$$

implies $f_P = G$. □

Theorem 8.1.4 (FALK/HOFFMAN (1976), cf. [Fa/Ho])

Let S be a simplex[3] with vertices $v_0, \ldots, v_n$ and let $f\colon S \longrightarrow \mathbb{R}$ be a concave function. Then the convex envelope f_S of f over S is an affinely linear function, that is, $f_S(x) = \langle a, x\rangle + \alpha \quad (x \in S)$ with a suitable pair $(a, \alpha) \in \mathbb{R}^n \times \mathbb{R}$ which is uniquely determined by the system of linear equations

$$f(v_\nu) = \langle a, v_\nu \rangle + \alpha \qquad (\nu = 0, \ldots n).$$

Proof: The matrix $A := \begin{pmatrix} v_0^T & 1 \\ \vdots & \vdots \\ v_n^T & 1 \end{pmatrix}$ has the (maximal) rank $n+1$. Therefore the above system of linear equations has a uniquely determined solution $(a, \alpha) \in \mathbb{R}^n \times \mathbb{R}$. Via $\ell(x) := \langle a, x\rangle + \alpha$ for $x \in \mathbb{R}^n$ we define an affinely linear function ℓ — which is in particular convex. For $x \in S$ with a $\lambda \in \Lambda(x)$ it holds that

$$\ell(x) \stackrel{\checkmark}{=} \sum_{\nu=0}^{n} \lambda_\nu \ell(v_\nu) = \sum_{\nu=0}^{n} \lambda_\nu f(v_\nu) \le f\Big(\sum_{\nu=0}^{n} \lambda_\nu v_\nu\Big) = f(x)$$

and hence $\ell \le f_S$. It furthermore holds that

$$f_S(x) \le \sum_{\nu=0}^{n} \lambda_\nu f_S(v_\nu) \le \sum_{\nu=0}^{n} \lambda_\nu f(v_\nu) \underset{(\text{v. s.})}{=} \ell(x),$$

hence, altogether, $f_S = \ell$. □

8.2 Branch and Bound Methods

Branch and bound is a general search method for finding solutions to general global optimization problems. A branch and bound procedure or *successive partitioning* requires two tools: *Branching* refers to a successive partitioning of the set of feasible points, that is, the feasible region is divided into 'disjoint' subregions of the original,

Chapter 8

[3] If $n+1$ points $v_0, \ldots, v_n$ in $\mathbb{R}^n$ are *affinely independent*, which means that the vectors $v_1 - v_0, \ldots, v_n - v_0$ are linearly independent, then the generated set $S := \operatorname{conv}\{v_0, \ldots, v_n\}$ is called a *simplex* with the vertices $v_0, \ldots, v_n$.

which together cover the whole feasible region. This is called branching, since the procedure may be repeated recursively. *Bounding* refers to the determination of lower and upper bounds for the optimal value within a feasible subregion. The core of the approach is the simple observation (for a minimization problem) that a subregion R may be removed from consideration if the lower bound for it is greater than the upper bound for any other subregion. This step is called *pruning*. Branch and bound techniques differ in the way they define rules for partitioning and the methods used for deriving bounds.

The following considerations are based on the works of FALK–SOLAND (cf. [Fa/So]) and their generalization by KALANTARI–ROSEN (cf. [Ka/Ro]). As an introduction we consider a two-dimensional *concave* optimization problem with linear constraints:

$$(KP) \quad \begin{cases} f(x) \longrightarrow \min \\ A^T x \leq b \end{cases}$$

In this case let $m \in \mathbb{N}, A \in \mathbb{R}^{2\times m}$, $b \in \mathbb{R}^m$ and the objective function f be *concave* and *separable*, that is, assume that $f(x_1, x_2) = f_1(x_1) + f_2(x_2)$ holds, with suitable functions f_1 and f_2. At the beginning let $M = [a_1, b_1] \times [a_2, b_2]$ be an axis-parallel rectangle that contains the feasible region $\mathcal{F} := \{x \in \mathbb{R}^2 \mid A^T x \leq b\}$. The construction of the convex envelope is trivial for the concave functions f_j: The graph of the convex envelope is the line segment passing through the points $(a_j, f(a_j))$ and $(b_j, f(b_j))$. Hence we obtain the convex envelope f_M via *linear interpolation* of f at the vertices of M in the following way:

$$f_M(x) = \frac{f_1(b_1) - f_1(a_1)}{b_1 - a_1}(x_1 - a_1) + f_1(a_1) + \frac{f_2(b_2) - f_2(a_2)}{b_2 - a_2}(x_2 - a_2) + f_2(a_2)$$

Instead of (KP) we now solve — for a rectangle or, more generally, a polyhedral set M — the following *linear* program:

$$(LP) \quad \begin{cases} f_M(x) \longrightarrow \min \\ x \in \mathcal{F} \cap M \end{cases}$$

Assume that it has the *minimal point* $\omega(M) \in \mathcal{F} \cap M$ with *value* $\beta(M) := f_M(\omega(M))$. $\alpha(M) := f(\omega(M))$ then yields an upper bound for the global minimum of (KP).

a) If $\omega(M)$ is a *vertex* of M, we have

$$f(x) \geq f_M(x) \geq \beta(M) \underset{(8.1.1)}{=} \alpha(M) = f(\omega(M)) \text{ for all } x \in \mathcal{F} \cap M .$$

$\omega(M)$ either yields a new candidate for the global minimum or there is no point in $\mathcal{F} \cap M$ which could be a global minimizer.

b) Assume now that $\omega(M)$ *is not a vertex* of M. If $\beta(M)$ is strictly greater than the current minimal value, then $\mathcal{F} \cap M$ does not contain any candidate for a global minimum, as

$$\beta(M) = f_M(\omega(M)) \le f_M(x) \le f(x) \quad \text{for} \quad x \in \mathcal{F} \cap M .$$

Otherwise nothing can be said about the point $\omega(M)$. We therefore divide the rectangle M into smaller rectangles and solve the corresponding (LP). We continue this procedure until all rectangles constructed in this way allow a decision: either that there is no global minimizer in the respective rectangle or that a vertex yields an optimal value.

Example 2 (Kalantari, Rosen (1987))

$$(KP) \quad \begin{cases} f(x) = -x_1^2 - 4x_2^2 \longrightarrow \min & \\ x_1 + x_2 \le 10 & (1) \\ x_1 + 5x_2 \le 22 & (2) \\ -3x_1 + 2x_2 \le 2 & (3) \\ -x_1 - 4x_2 \le -4 & (4) \\ x_1 - 2x_2 \le 4 & (5) \end{cases}$$

Feasible Region with Contour Lines

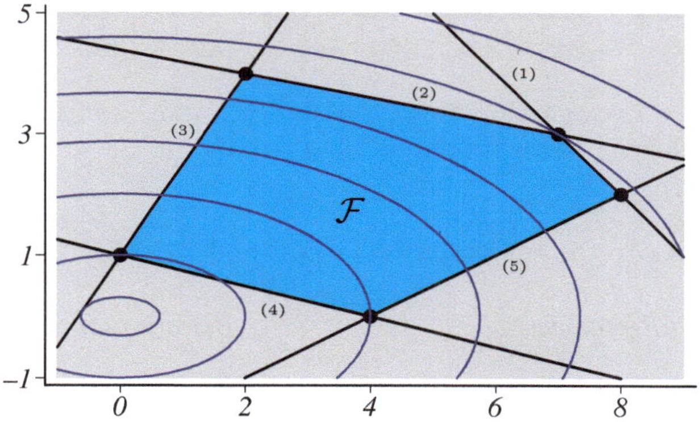

We start with the rectangle $M_0 := [0, 8] \times [0, 4]$; it is obviously the smallest axis-parallel rectangle that contains the feasible region $\mathcal{F}$ of (KP). We obtain the convex envelope f_{M_0} of f on M_0 via $f_{M_0}(x) = -8x_1 - 16x_2$.

We solve the following linear program:

$$(LP) \quad \begin{cases} f_{M_0}(x) \longrightarrow \min \\ \text{subject to the constraints (1) to (5).} \end{cases}$$

Objective Function and Convex Envelope

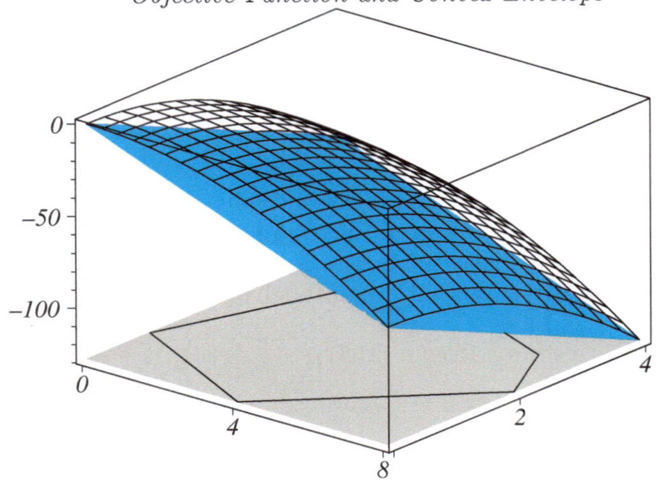

It has the minimizer $\omega(M_0) = (7,3)^T$ with the value $\beta(M_0) := f_{M_0}(\omega(M_0)) = -104$. The value $\alpha(M_0) := f(\omega(M_0)) = -85$ gives an upper bound for the global minimum of (KP). However, one cannot decide yet whether $\omega(M_0)$ gives a solution to (KP). Therefore, we *bisect* M_0 and get two rectangles $M_{1,1} := [0,8] \times [0,2]$ and $M_{1,2} = [0,8] \times [2,4]$. Via

$$f_{M_{1,1}}(x) \;=\; -8x_1 - 8x_2, \quad f_{M_{1,2}}(x) \;=\; -8x_1 + (32 - 24x_2)$$

we get the new convex envelopes $f_{M_{1,1}}$ and $f_{M_{1,2}}$ and — with their help — we solve the following linear programs for $j = 1,2$:

$$(LP) \quad \begin{cases} f_{M_{1,j}}(x) \longrightarrow \min \\ \text{subject to the constraints (1) to (5) and } x \in M_{1,j}\,. \end{cases}$$

Starting Rectangle M_0

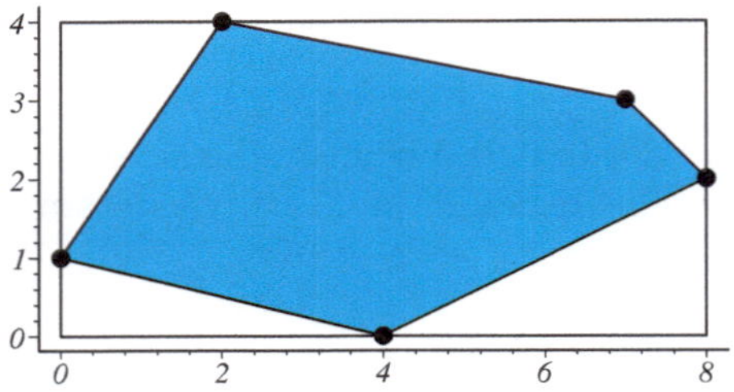

Bisection of M_0

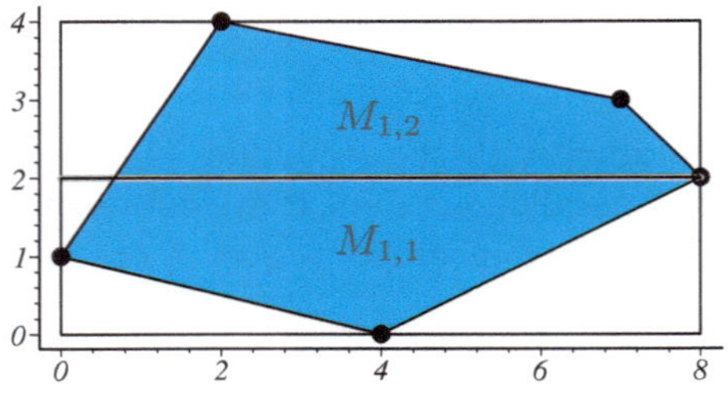

Hence, we get the minimal points: $\omega(M_{1,1}) = (8,2)^T$, $\omega(M_{1,2}) = (7,3)^T$ and to them $\alpha(M_{1,1}) = -80 = \beta(M_{1,1})$, $\alpha(M_{1,2}) = -85$, $\beta(M_{1,2}) = -96$.

Thus, $M_{1,1}$ cannot contain a global minimal point, and we cannot answer this question for $M_{1,2}$. Therefore, we divide the rectangle $M_{1,2}$ further into $M_{2,1} = [0,4] \times [2,4]$ and $M_{2,2} = [4,8] \times [2,4]$. The corresponding envelopes are

$$f_{M_{2,1}}(x) = -4x_1 + (32 - 24x_2) \quad \text{and} \quad f_{M_{2,2}}(x) = (32 - 12x_1) + (32 - 24x_2).$$

The linear programs

$$(LP) \quad \begin{cases} f_{M_{2,j}}(x) \longrightarrow \min \\ \text{subject to the constraints (1) to (5) and } x \in M_{2,j} \end{cases}$$

$(j = 1,2)$ have the minimizers $\omega(M_{2,1}) = (2,4)^T$ and $\omega(M_{2,2}) = (7,3)^T$ with $\alpha(M_{2,1}) = -68$, $\beta(M_{2,1}) = -72$ and $\alpha(M_{2,2}) = -85$, $\beta(M_{2,2}) = -92$. The rectangle $M_{2,1}$ hence does not contain any minimal points. We divide $M_{2,2}$ further into the four rectangles

$$M_{3,1} = [4,7] \times [2,3], \; M_{3,2} = [7,8] \times [2,3],$$

$$M_{3,3} = [4,7] \times [3,4], \; M_{3,4} = [7,8] \times [3,4]\,.$$

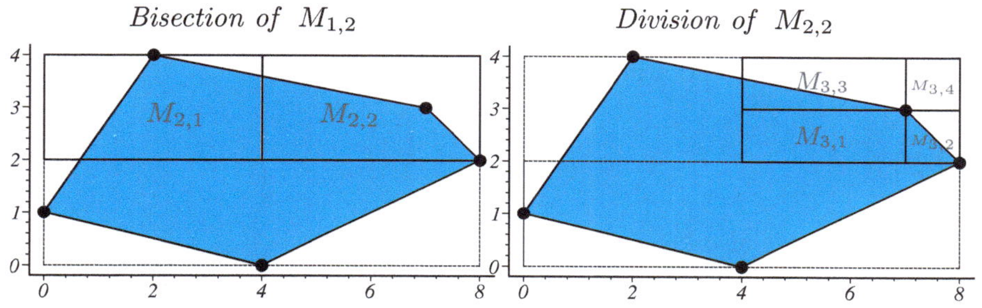

We get the following convex envelopes

$$f_{M_{3,1}}(x) = (28 - 11x_1) + (24 - 20x_2), \; f_{M_{3,2}}(x) = (56 - 15x_1) + (24 - 20x_2),$$

$$f_{M_{3,3}}(x) = (28 - 11x_1) + (48 - 28x_2), \; f_{M_{3,4}}(x) = (56 - 15x_1) + (48 - 28x_2).$$

The linear programs

$$(LP) \quad \begin{cases} f_{M_{3,j}}(x) \longrightarrow \min \\ \text{subject to the constraints (1) to (5) and } x \in M_{3,j} \end{cases}$$

$(j = 1,2,3,4)$ all have the minimizer $\omega(M_{3,j}) = (7,3)^T$ with $\beta(M_{3,j}) = \alpha(M_{3,j}) = -85$. Therefore $(7,3)^T$ gives the global minimum of (KP). ◁

More generally, let now a *concave quadratic minimization problem* be given:

$$(QP) \quad \begin{cases} f(x) := \langle p, x\rangle - \frac{1}{2}\langle x, Cx\rangle \longrightarrow \min \\ x \in \mathcal{F} := \{v \in \mathbb{R}^n \mid A^T v \le b\} \end{cases}$$

We hereby assume that $p \in \mathbb{R}^n, b \in \mathbb{R}^m, A \in \mathbb{R}^{n\times m}$, $C \in S^n_{++}$ and $\mathcal{F}$ *compact.* Assume furthermore that $u^{(1)}, \dots, u^{(n)}$ are *C-conjugate directions*, that is, it holds that

$$\left\langle u^{(\nu)}, C u^{(\mu)}\right\rangle = \begin{cases} 0 & , \quad \nu \neq \mu \\ \lambda_\nu > 0\,, & \nu = \mu \end{cases}$$

for $\nu, \mu = 1, \dots, n$.

The following considerations could be written even more clearly and concisely with the new scalar product $\langle\ ,\ \rangle_C$ defined by $\langle v, w\rangle_C := \langle v, Cw\rangle$ for $v, w \in \mathbb{R}^n$ and the normalization of the vectors $u^{(\nu)}$ (to $\lambda_\nu = 1$). We will, however, not go into the details here.

For $\nu = 1, \dots, n$ a maximizer of the linear program

$$(\overline{LP})_\nu \quad \begin{cases} \left\langle u^{(\nu)}, Cx\right\rangle \longrightarrow \max \\ x \in \mathcal{F} \end{cases}$$

is denoted by $\overline{x}^{(\nu)}$ and a minimizer of the linear program

$$(\underline{LP})_\nu \quad \begin{cases} \left\langle u^{(\nu)}, Cx\right\rangle \longrightarrow \min \\ x \in \mathcal{F} \end{cases}$$

by $\underline{x}^{(\nu)}$. Now we also define

$$\overline{\beta}_\nu := \frac{\left\langle u^{(\nu)}, C\overline{x}^{(\nu)}\right\rangle}{\lambda_\nu} \quad \text{and} \quad \underline{\beta}_\nu := \frac{\left\langle u^{(\nu)}, C\underline{x}^{(\nu)}\right\rangle}{\lambda_\nu}\,,$$

hence

$$P_\nu := \left\{x \in \mathbb{R}^n \mid \lambda_\nu \underline{\beta}_\nu \le \left\langle u^{(\nu)}, Cx\right\rangle \le \lambda_\nu \overline{\beta}_\nu\right\}$$

gives a *parallel strip* with the corresponding *parallelepiped*

$$P := \bigcap_{\nu=1}^{n} P_\nu\,.$$

Theorem 8.2.1 *It holds that:*

a) $\mathcal{F} \subset P$

b) $f\left(\sum_{\nu=1}^{n} \varrho_\nu u^{(\nu)}\right) = \sum_{\nu=1}^{n} f\big(\varrho_\nu u^{(\nu)}\big)$ *for* $\varrho_1, \dots, \varrho_n \in \mathbb{R}$.

c) With $U := \big(u^{(1)}, \dots, u^{(n)}\big)$,

$$h := \frac{1}{2}\left(\underline{\beta}_1 + \overline{\beta}_1, \dots, \underline{\beta}_n + \overline{\beta}_n\right)^T,$$

$$\alpha := \frac{1}{2}\sum_{\nu=1}^{n} \lambda_\nu \underline{\beta}_\nu \overline{\beta}_\nu \quad \textit{and} \quad a := p - CUh$$

the convex envelope f_P *of* f *on* P *is given by*

$$f_P(x) := \langle a, x \rangle + \alpha.$$

Proof:

a) For $x \in \mathcal{F}$ and $\nu = 1, \dots, n$ we have by the definition of $\underline{x}^{(\nu)}$ and $\overline{x}^{(\nu)}$

$$\lambda_\nu \underline{\beta}_\nu = \left\langle u^{(\nu)}, C\underline{x}^{(\nu)} \right\rangle \le \left\langle u^{(\nu)}, Cx \right\rangle \le \left\langle u^{(\nu)}, C\overline{x}^{(\nu)} \right\rangle = \lambda_\nu \overline{\beta}_\nu .$$

This shows $x \in P_\nu$; hence, $\mathcal{F} \subset \bigcap_{\nu=1}^{n} P_\nu = P$.

b)
$$\begin{aligned} f\left(\sum_{\nu=1}^{n} \varrho_\nu u^{(\nu)}\right) &= \left\langle p, \sum_{\nu=1}^{n} \varrho_\nu u^{(\nu)} \right\rangle - \frac{1}{2}\left\langle \sum_{\nu=1}^{n} \varrho_\nu u^{(\nu)}, C \sum_{\mu=1}^{n} \varrho_\mu u^{(\mu)} \right\rangle \\ &= \sum_{\nu=1}^{n} \langle p, \varrho_\nu u^{(\nu)} \rangle - \frac{1}{2}\sum_{\nu=1}^{n} \langle \varrho_\nu u^{(\nu)}, C \varrho_\nu u^{(\nu)} \rangle \\ &= \sum_{\nu=1}^{n} f\big(\varrho_\nu u^{(\nu)}\big) \end{aligned}$$

c) We consider the function F defined by $F(x) := \langle a, x \rangle + \alpha$ for $x \in \mathbb{R}^n$ with the above noted a and α:

(i) F is affinely linear, hence convex.

(ii) We show: $F(w) = f(w)$ for all *vertices* w of P:

A vector $w \in \mathbb{R}^n$ is a vertex of P iff there exists a representation $w = \sum_{\nu=1}^{n} \omega_\nu u^{(\nu)}$ with $\omega_\nu \in \{\underline{\beta}_\nu, \overline{\beta}_\nu\}$.

$$\begin{aligned} \langle CUh, u^{(\nu)} \rangle = \langle Uh, Cu^{(\nu)} \rangle &= \left\langle \sum_{\mu=1}^{n} \frac{1}{2}(\underline{\beta}_\mu + \overline{\beta}_\mu)\, u^{(\mu)}, Cu^{(\nu)} \right\rangle \\ &= \sum_{\mu=1}^{n} \frac{1}{2}(\underline{\beta}_\mu + \overline{\beta}_\mu)\, \langle u^{(\mu)}, Cu^{(\nu)} \rangle = \frac{1}{2}(\underline{\beta}_\nu + \overline{\beta}_\nu)\, \lambda_\nu \end{aligned}$$

$$\begin{aligned} F(w) = \langle a, w \rangle + \alpha &= \langle p, w \rangle - \langle CUh, w \rangle + \frac{1}{2}\sum_{\nu=1}^{n} \lambda_\nu \underline{\beta}_\nu \overline{\beta}_\nu \\ &= \langle p, w \rangle - \left\langle CUh, \sum_{\nu=1}^{n} \omega_\nu u^{(\nu)} \right\rangle + \frac{1}{2}\sum_{\nu=1}^{n} \lambda_\nu \underline{\beta}_\nu \overline{\beta}_\nu \\ &= \langle p, w \rangle - \frac{1}{2}\sum_{\nu=1}^{n} \lambda_\nu \underbrace{\big(\omega_\nu(\underline{\beta}_\nu + \overline{\beta}_\nu) - \underline{\beta}_\nu \overline{\beta}_\nu\big)}_{= \omega_\nu^2} \\ &\overset{\checkmark}{=} \sum_{\nu=1}^{n} f\big(\omega_\nu u^{(\nu)}\big) \underset{b)}{=} f(w) \end{aligned}$$

Chapter 8

With $N := 2^n$ let $w^{(1)}, \dots, w^{(N)}$ be the extreme points of P. For $x \in P$ there exist — by the theorem of KREIN–MILMAN — $\alpha_\nu \in \mathbb{R}_+$ with $\sum_{\nu=1}^{N} \alpha_\nu = 1$ and $x = \sum_{\nu=1}^{N} \alpha_\nu w^{(\nu)}$. Then it holds that

$$F(x) \overset{\checkmark}{\leq} \sum_{\nu=1}^{N} \alpha_\nu F(w^{(\nu)}) \underset{(ii)}{=} \sum_{\nu=1}^{N} \alpha_\nu f(w^{(\nu)}) \leq f\left(\sum_{\nu=1}^{N} \alpha_\nu w^{(\nu)}\right) = f(x).$$

(*iii*) Assume that $G\colon P \longrightarrow \mathbb{R}$ is a convex function with $G \leq f$. With the above-mentioned representation for an $x \in P$ we get

$$G(x) \leq \sum_{\nu=1}^{N} \alpha_\nu G(w^{(\nu)}) \leq \sum_{\nu=1}^{N} \alpha_\nu f(w^{(\nu)}) \underset{\text{(v.s.)}}{=} F(x).$$

□

8.3 Cutting Plane Methods

Cutting Plane Algorithm by Kelley

The *cutting plane algorithm by* KELLEY was developed to solve *convex* optimization problems of the form

$$\begin{aligned} &f(x) \longrightarrow \min \\ &g_i(x) \leq 0 \quad (i = 1, \dots, m) \end{aligned}$$

where $x \in \mathbb{R}^n$ and f as well as $g_1, \dots, g_m$ are *differentiable convex* functions. This problem is evidently equivalent to

$$\begin{aligned} &x_{n+1} \longrightarrow \min \\ &f(x) - x_{n+1} \leq 0, \quad g_i(x) \leq 0 \quad (i = 1, \dots, m) \end{aligned} \quad .$$

It therefore suffices to consider problems of the form

$$(P) \quad \begin{cases} \langle c, x \rangle \longrightarrow \min \\ g_i(x) \leq 0 \quad (i = 1, \dots, m) \end{cases}$$

where $c, x \in \mathbb{R}^n$ and $g_1, \dots, g_m$ *convex and differentiable.* Let the feasible region

$$\mathcal{F} := \{x \in \mathbb{R}^n \mid g_i(x) \leq 0 \quad (i = 1, \dots, m)\}$$

be nonempty and suppose that a *polyhedral set* P_0 with $\mathcal{F} \subset P_0$ is known.

A *polyhedral set* or *polyhedron* is the intersection of a finite family of closed half-spaces — $\{x \in \mathbb{R}^n \mid \langle a, x \rangle \leq \beta\}$ for some $a \in \mathbb{R}^n$ and $\beta \in \mathbb{R}$ — in $\mathbb{R}^n$. Using matrix notation, we can define a *polyhedron* to be the set

$$\{x \in \mathbb{R}^n \mid A^T x \leq b\},$$

where A is an (n, m)-matrix and $b \in \mathbb{R}^m$. Clearly a *polyhedron* is convex and closed. Without proof we note: A set in $\mathbb{R}^n$ is a *polytope* iff it is a bounded polyhedron.[4]

KELLEY's *cutting plane algorithm* now yields an antitone sequence of outer approximations $P_0 \supset P_1 \supset \cdots \supset \mathcal{F}$ in the following way:

1) *Solve* $\quad (LP) \quad \begin{cases} \langle c, x \rangle \longrightarrow \min \\ x \in P_0 \end{cases}$.

If $x^{(0)} \in P_0$ is a minimizer of (LP) with $x^{(0)} \in \mathcal{F}$, then we have found a solution to the convex optimization problem (P); otherwise go to 2).

2) *Hence, there exists an index i_0 such that $g_{i_0}(x^{(0)}) > 0$, for example, let $g_{i_0}(x^{(0)}) = \max_{1 \leq i \leq m} g_i(x^{(0)})$. Set*

$$P_1 := P_0 \cap \left\{x \in \mathbb{R}^n \mid g_{i_0}(x^{(0)}) + g_{i_0}'(x^{(0)})(x - x^{(0)}) \leq 0\right\},$$

go to 1) and continue with P_1 instead of P_0.

It holds that: $x^{(0)} \in P_0$, $x^{(0)} \notin P_1$, $P_0 \supset P_1$ as well as $P_1 \supset \mathcal{F}$; since for $x \in \mathcal{F}$ we have

$$g_{i_0}(x^{(0)}) + g_{i_0}'(x^{(0)})(x - x^{(0)}) \leq g_{i_0}(x) \leq 0.$$

The linear constraint

$$\ell(x) := g_{i_0}(x^{(0)}) + g_{i_0}'(x^{(0)})(x - x^{(0)}) \leq 0$$

reduces the given set P_0 in such a way that no feasible points are excluded. The *hyperplane* $\{x \in \mathbb{R}^n \mid \ell(x) = 0\}$ is called a *cutting plane* or simply a *cut*. Thus, a cut 'cuts off' the current optimizer to the LP 'relaxation', but no feasible point to (P).

Other cutting plane approaches differ in the generation of cuts and the way the set of feasible points is updated when a cut is chosen.

Concavity Cuts

Cutting plane methods of the kind described above are less suitable for *concave* objective functions.

Therefore, we will give some preliminary remarks which will enable us to describe another class of cutting plane methods. They are based on the iterative reduction of $P_0 = P$.

Chapter 8

[4] Some authors use just the opposite notation.

Given a *polyhedron* $P \subset R^n$ and
a *concave* function $f\colon \mathbb{R}^n \longrightarrow \mathbb{R}$,

we want to solve the problem:

$$\begin{aligned} & f(x) \longrightarrow \min \\ & x \in P \end{aligned}$$

For that let $x^{(0)}$ be a vertex of P with $f\bigl(x^{(0)}\bigr) =: \gamma$. Suppose that this vertex is nondegenerate in the sense that there are n 'edges' leading off from it in linearly independent directions $u^{(1)}, \ldots, u^{(n)}$. Furthermore assume that $f(z^{(\nu)}) \geq \gamma$ for $z^{(\nu)} := x^{(0)} + \vartheta_\nu u^{(\nu)}$ with $\vartheta_\nu > 0$ maximal for $\nu = 1, \ldots, n$. By the concavity of f, $x^{(0)}$ gives a local minimum of f in P.

There exists exactly one affinely linear mapping $\ell\colon \mathbb{R}^n \longrightarrow \mathbb{R}$ with $\ell(z^{(\nu)}) = 1$ for $\nu = 1, \ldots, n$ and $\ell(x^{(0)}) = 0$ (compare the proof of theorem 8.1.4). It is defined by $\ell(x) := \bigl\langle a, x - x^{(0)} \bigr\rangle$, where $a := Q^{-T} e$ with $e := (1, \ldots, 1)^T$ and $Q := (z^{(1)} - x^{(0)}, \ldots, z^{(n)} - x^{(0)})$; since

$$\ell(z^{(\nu)}) = \Bigl\langle Q^{-T} e, z^{(\nu)} - x^{(0)} \Bigr\rangle = \Bigl\langle e, Q^{-1}(z^{(\nu)} - x^{(0)}) \Bigr\rangle = \langle e, e_\nu \rangle = 1 .$$

Suppose that $w \in P$ is a maximizer to

$$(LP) \quad \begin{cases} \ell(x) \longrightarrow \max \\ x \in P \end{cases}$$

with $\mu := \ell(w)$. Then it holds that:

1) $\ell(x) > 1$ *for* $x \in P$ *with* $f(x) < \gamma$.

2) *If* $\mu \leq 1$, *then* $x^{(0)}$ *yields the global minimum of* f *on* P.

Proof: 1): It holds that (cf. exercise 5)
$P \subset K := x^{(0)} + \text{ cone } \bigl(u^{(1)}, \ldots, u^{(n)}\bigr)$. In addition

$$S := \text{conv } \bigl(x^{(0)}, z^{(1)}, \ldots, z^{(n)}\bigr) = K \cap \{x \in \mathbb{R}^n \mid \ell(x) \leq 1\}$$

can be easily verified as follows: To $w \in S$ there exist $\lambda_\nu \geq 0$ with $\sum_{\nu=0}^n \lambda_\nu = 1$ and

$$\begin{aligned} w = \lambda_0 x^{(0)} + \sum_{\nu=1}^n \lambda_\nu z^{(\nu)} &= \lambda_0 x^{(0)} + \sum_{\nu=1}^n \lambda_\nu \bigl(x^{(0)} + \vartheta_\nu u^{(\nu)}\bigr) \\ &= x^{(0)} + \sum_{\nu=1}^n \lambda_\nu \vartheta_\nu u^{(\nu)}. \end{aligned}$$

This shows $w \in K$ and $\ell(w) \overset{\checkmark}{=} \lambda_0 \underbrace{\ell\bigl(x^{(0)}\bigr)}_{=0} + \sum_{\nu=1}^n \lambda_\nu \underbrace{\ell\bigl(z^{(\nu)}\bigr)}_{=1} \leq 1$. Conversely, if w is an element of the right-hand side, we get $w = x^{(0)} +$

Chapter 8

$\sum_{\nu=1}^{n} \lambda_\nu u^{(\nu)}$ with suitable $\lambda_\nu \geq 0$ and $\ell(w) \leq 1$, consequently,

$$1 \geq \ell(w) = \Big\langle a, \sum_{\nu=1}^{n} \lambda_\nu u^{(\nu)} \Big\rangle = \sum_{\nu=1}^{n} \lambda_\nu \langle a, u^{(\nu)} \rangle = \sum_{\nu=1}^{n} \lambda_\nu \langle Q^{-T} e, u^{(\nu)} \rangle$$

$$= \sum_{\nu=1}^{n} \lambda_\nu \langle e, Q^{-1} u^{(\nu)} \rangle = \sum_{\nu=1}^{n} \frac{\lambda_\nu}{\vartheta_\nu} \langle e, Q^{-1}\big(z^{(\nu)} - x^{(0)}\big) \rangle \underset{\text{(v. s.)}}{=} \sum_{\nu=1}^{n} \frac{\lambda_\nu}{\vartheta_\nu}.$$

This yields

$$w = x^{(0)} + \sum_{\nu=1}^{n} \frac{\lambda_\nu}{\vartheta_\nu}\big(z^{(\nu)} - x^{(0)}\big) = \Big(1 - \sum_{\nu=1}^{n} \frac{\lambda_\nu}{\vartheta_\nu}\Big) x^{(0)} + \sum_{\nu=1}^{n} \frac{\lambda_\nu}{\vartheta_\nu} z^{(\nu)} \in S.$$

Then we get $f(x) \geq \min\big\{f(x^{(0)}), f(z^{(1)}), \ldots, f(z^{(n)})\big\} = \gamma$ for $x \in S$, and for $x \in P$ with $f(x) < \gamma$ we have $x \notin S$, hence, $\ell(x) > 1$.

2): If $\mu \leq 1$, then $\ell(x) \leq \ell(w) = \mu \leq 1$ for $x \in P$ shows $P \subset S$. For $x \in P$ we thus get $f(x) \geq \gamma = f\big(x^{(0)}\big)$. □

The above considerations give that

$$P^*(\gamma) := \big\{x \in P \mid f(x) < \gamma\big\} \subset P \cap \big\{x \in \mathbb{R}^n \mid \ell(x) \geq 1\big\}.$$

The affinely linear inequality $\ell(x) \geq 1$ is called a *γ-cut of* (f, P).

We may restrict our further search to the subset $P^*(\gamma)$ of P. To preserve the polyhedral structure, we consider the larger set $P \cap \big\{x \in \mathbb{R}^n \mid \ell(x) \geq 1\big\}$ instead.

Example 3 (Horst/Tuy, p. 200f)

$$\begin{aligned} f(x) = -(x_1 - 1.2)^2 - (x_2 - 0.6)^2 &\longrightarrow \min \\ -2x_1 + x_2 &\leq 1 \\ x_1 + x_2 &\leq 4 \\ x_1 - 2x_2 &\leq 2 \\ 0 \leq x_1 \leq 3,\ 0 \leq x_2 &\leq 2 \end{aligned}$$

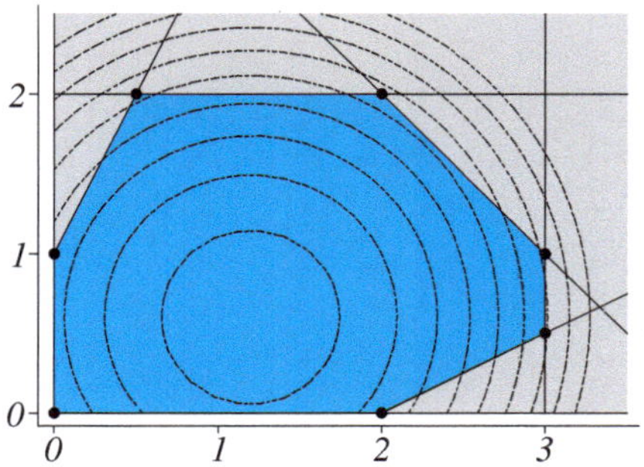

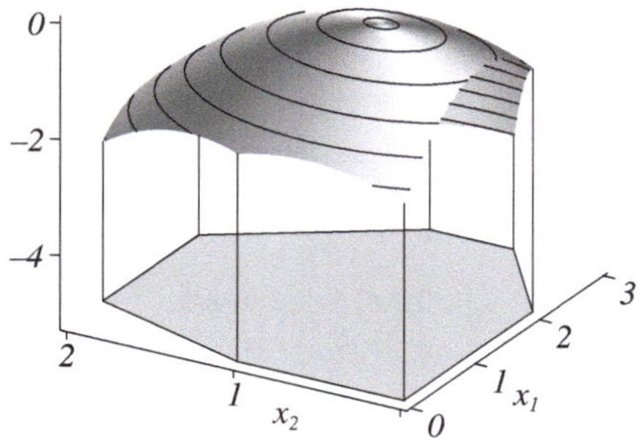

Initialization: $x^{(0)} := \binom{0}{0}$, $\gamma = f\big(x^{(0)}\big) = -1.8$, $P_0 := P$

First Iteration Cycle:

$k = 0:$ $u^{(1)} := \binom{1}{0}$, $u^{(2)} := \binom{0}{1}$

$$z^{(1)} = \binom{2.4}{0}, \; z^{(2)} = \binom{0}{1.2}\; ; \; f(z^{(1)}) = f(z^{(2)}) = -1.8$$

We then get $\ell_0(x) = \frac{5}{12}x_1 + \frac{5}{6}x_2$ with $\ell_0(z^{(1)}) = \ell_0(z^{(2)}) = 1$.
The linear program

$$(LP) \quad \begin{cases} \ell_0(x) \longrightarrow \max \\ x \in P_0 \end{cases}$$

is solved by $w_0 = \binom{2}{2}$ with $\ell(w_0) = \frac{5}{2} > 1$ and $f(w_0) = -2.6 < \gamma$.

In this simple example it is still possible to easily verify this by hand. In more complex cases, however, we prefer to work with the help of *Maple*® or *Matlab*®.

Replace P_0 by $P_1 := P_0 \cap \{x \in \mathbb{R}^n \mid \ell_0(x) \geq 1\}$.

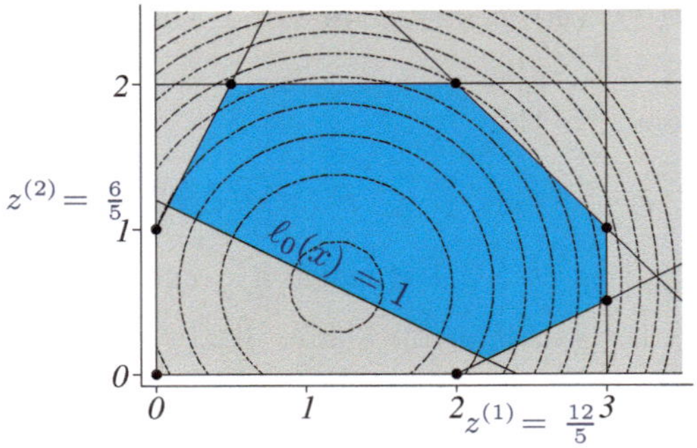

Start with $w_0 = \binom{2}{2}$ and calculate a vertex $x^{(1)}$ of P_1 which yields a local minimum of f in P_1 with $f\big(x^{(1)}\big) \leq f(w_0)$. Since $f(w_0) < \gamma$, we have $\ell_0(x^{(1)}) > 1$; hence, $x^{(1)}$ is also a vertex of P. In this case we obtain $x^{(1)} = \binom{3}{1}$ with $f(x^{(1)}) = -3.4$.

Set: $x^{(0)} := \binom{3}{1}$, $\gamma := -3.4$, $P_0 := P_1$.

Second Iteration Cycle:

$k = 0: \quad u^{(1)} = \binom{-1}{1}, \; u^{(2)} = \binom{0}{-1}$

$$z^{(1)} = \binom{1.6}{2.4}, \; z^{(2)} = \binom{3}{0.2}; \; f(z^{(1)}) = f(z^{(2)}) = -3.4$$

$$\ell_0(x) = \frac{50}{7} - \frac{55}{28}x_1 - \frac{5}{4}x_2; \; \ell_0(z^{(1)}) = \ell_0(z^{(2)}) = 1$$

The linear program

$$(LP) \quad \begin{cases} \ell_0(x) \longrightarrow \max \\ x \in P_0 \end{cases}$$

is solved by $w_0 = (0.08, 1.16)^T$ with $\ell_0(w_0) = 5.536$ and $f(w_0) = -1.568 > \gamma$.
Replace P_0 by the set $P_1 := P_0 \cap \big\{x \in \mathbb{R}^n \mid \ell_0(x) \geq 1\big\}$.

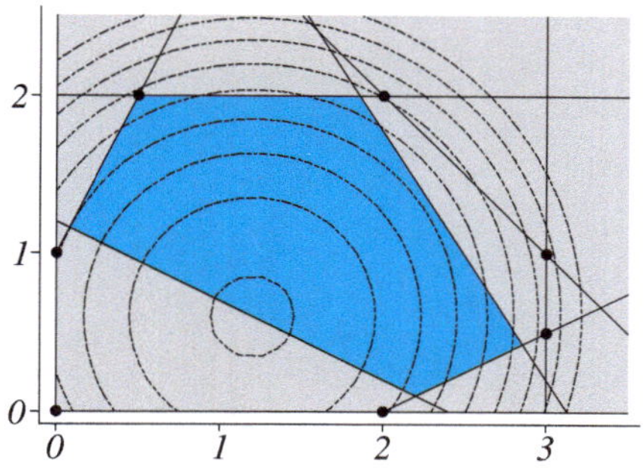

Start with $w_0 = (0.08, 1.16)^T$ and calculate a vertex $x^{(1)}$ of P_1 which yields a local minimum of f in P_1. This gives: $x^{(1)} = (1/2, 2)^T$, $f(x^{(1)}) = -2.45 > \gamma$.

$k = 1:\quad u^{(1)} = \binom{1}{0},\ u^{(2)} = \binom{-1}{-2}$

$z^{(1)} = \binom{2.4}{2},\ z^{(2)} = \binom{-0.5253}{-0.0506};$

$f(z^{(1)}) = f(z^{(2)}) = \gamma$

$\ell_1(x) = 1.238 + 0.5263x_1 - 0.7508x_2;\ \ \ell_1(z^{(1)}) = \ell_1(z^{(2)}) = 1$

The linear program

$$(LP)\quad \begin{cases} \ell_1(x) \longrightarrow \max \\ x \in P_1 \end{cases}$$

is solved by $w_1 = (2.8552, 0.4276)^T$ with $\ell_1(w_1) = 2.4202$ and $f(w_1) = -2.7693 > \gamma$.
Replace P_1 by $P_2 := P_1 \cap \{x \in \mathbb{R}^n \mid \ell_1(x) \geq 1\}$.

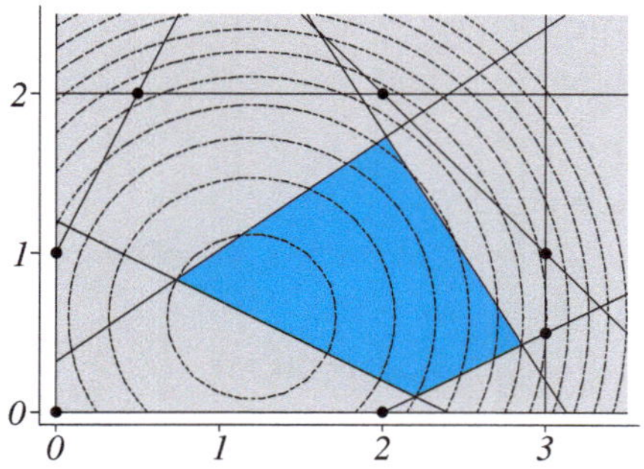

$x^{(2)} := w_1$

In the next two iteration steps we get:

$k = 2:\quad z^{(1)} = (1.6, 2.4)^T,\ z^{(2)} = (0.1578, -0.9211)^T;\ f(z^{(1)}) = f(z^{(2)}) = \gamma$

$\ell_2(x) = 1.2641 - 0.4735x_1 + 0.2056x_2;\ \ \ell_2(z^{(1)}) = \ell_2(z^{(2)}) = 1$

$w_2 = (0.7347, 0.8326)^T,\ \ell_2(w_2) = 1.087,\ f(w_2) = -0.2706 > \gamma$

$x^{(3)} := w_2$

Replace P_2 by $P_3 := P_2 \cap \{x \in \mathbb{R}^n \mid \ell_2(x) \geq 1\}$.

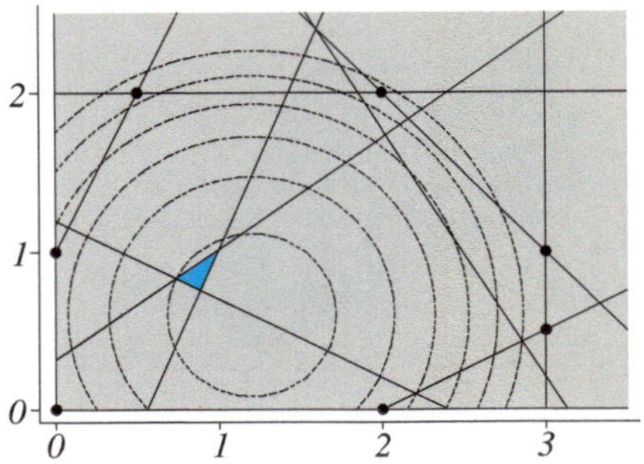

$k = 3: \quad z^{(1)} = (2.8492, -0.2246)^T, \; z^{(2)} = (2.4, 2)^T; \; f(z^{(1)}) = f(z^{(2)}) = \gamma$
$\ell_3(x) = -0.4749 + 0.5260 x_1 + 0.1062 x_2; \; \ell_3(z^{(1)}) = \ell_3(z^{(2)}) = 1$
$w_3 = (1.0002, 1.0187)^T, \; \mu := \ell_3(w_3) = 0.1594 < 1$

Thus, $x^{(0)} = \binom{3}{1}$ yields the global minimum. ◁

Summarizing the results we conclude with the formulation of a corresponding *cutting plane algorithm:*

Algorithm

0) Calculate a vertex $x^{(0)}$ which yields a local minimum of f in P such that the function values in the neighboring vertices are greater or equal to $f(x^{(0)})$. Set $\gamma := f(x^{(0)})$ and $P_0 := P$.

Iteration cycle: *For* $k = 0, 1, 2, \ldots$

1) Construct a γ-cut ℓ_k of (f, P_k) in $x^{(k)}$.

2) Solve $(LP) \quad \begin{cases} \ell_k(x) \longrightarrow \max \\ x \in P_k \end{cases}$

and get a maximizer w_k.
If $\ell_k(w_k) \le 1$: STOP; $x^{(0)}$ yields the global minimum. Otherwise:

3) $P_{k+1} := P_k \cap \{x \in \mathbb{R}^n \mid \ell_k(x) \ge 1\}$
Start with w_k and find a vertex $x^{(k+1)}$ of P_{k+1} which gives a local minimum of f in P_{k+1} in the above stated sense.
If $f(x^{(k+1)}) \ge \gamma$: Go to iteration $k+1$
If $f(x^{(k+1)}) < \gamma$: Set $\gamma := f(x^{(k+1)}$, $x^{(0)} := x^{(k+1)}$, $P_0 := P_{k+1}$ and start a new iteration cycle.

In [Ho/Tu], for example, so-called 'Φ-cuts' are considered which yield 'deeper' cuts and thus reach their goal faster. We, however, do not want to pursue this approach any further.

Exercises

1. Do some numerical experiments to test the efficiency of global optimization software. Take for example *Maple*® and compare the local solver Optimization[NLPSolve] with GlobalOptimization[GlobalSolve], choosing the options *method=multistart* (default), *method=singlestart*, *method= branchandbound* and *method=reducedgradient*:

   a) Consider the minimization problem of example 1:

   $$f(x) := -\sum_{i=1}^{n}\left(c_i x_i + \tfrac{1}{2}x_i^2\right) \longrightarrow \min$$
   $$-1 \le x_i \le 1 \qquad (i = 1, \ldots, n)$$

   Choose $c_i := \frac{1}{i+1}$ for $i = 1, \ldots, n$ and $n \in \{5, 10, 15, 20, 25\}$.

   b) *The* HAVERLY *Pooling Problem* (cf. [Hav])

   The following problem results from the so-called pooling problem which is one of the fundamental optimization problems encountered in the petroleum industry:

   $$\begin{aligned}
   &f(x) := 6x_1 + 16x_2 - 9x_5 + 10x_6 - 15x_9 \longrightarrow \min\\
   &x_1 + x_2 - x_3 - x_4 = 0\\
   &x_3 - x_5 + x_7 = 0\\
   &x_4 + x_8 - x_9 = 0\\
   &-x_6 + x_7 + x_8 = 0\\
   &-2.5x_5 + 2x_7 + x_3x_{10} \le 0\\
   &2x_8 - 1.5x_9 + x_4x_{10} \le 0\\
   &3x_1 + x_2 - (x_3 + x_4)x_{10} = 0\\
   &lb \le x \le ub,
   \end{aligned}$$

   where $lb = (0, 0, 0, 0, 0, 0, 0, 0, 0, 1)^T$ and
   $ub = (300, 300, 100, 200, 100, 300, 100, 200, 200, 3)^T$.

   Hint: The best value we know is $f(x^*) = -400$ attained for
   $x^* = (0, 100, 0, 100, 0, 100, 0, 100, 200, 1)^T$.

2. *a)* Let $M \subset \mathbb{R}^n$ be a nonempty convex set and $f\colon M \longrightarrow \mathbb{R}$ a function with the convex envelope $F = f_M$. Then for any *affinely linear* function $\ell\colon M \longrightarrow \mathbb{R}$ the function $f + \ell$ has the convex envelope $F + \ell$.

   Hint: $f_M + g_M \le (f + g)_M$ for arbitrary functions $f, g\colon M \longrightarrow \mathbb{R}$.

   b) Let the functions $f_j\colon M := [-1, 1] \longrightarrow \mathbb{R}$ be defined by $f_1(x) := 1 - x^2$, $f_2(x) := x^2$ and $f_3(x) := f_1(x) + f_2(x)$. Determine the convex envelopes to f_j on M $(j = 1, 2, 3)$.

3. Let $f\colon R \longrightarrow \mathbb{R}$ be defined by $f(x) := x_1 x_2$ on the rectangle $R := [a_1, b_1] \times [a_2, b_2]$. Show that the convex envelope of f over R is given by $F(x) := \max\{F_1(x), F_2(x)\}$, where
$F_1(x) := a_1 a_2 + a_2 (x_1 - a_1) + a_1 (x_2 - a_2)$
$F_2(x) := b_1 b_2 + b_2 (x_1 - b_1) + b_1 (x_2 - b_2)$.

Hint: Use the TAYLOR expansion and the interpolation properties of F_1 and F_2.

4. ZWART's *counterexample* (cf. [Fa/Ho])

$2 - (x_1 - 1)^2 - x_2^2 - (x_3 - 1)^2 \longrightarrow \min$
$A^T x \le b$,

$$\text{where } A = \begin{pmatrix} 1 & -1 & 12 & 12 & -6 & -1 & 0 & 0 \\ 1 & 1 & 5 & 12 & 1 & 0 & -1 & 0 \\ -1 & -1 & 12 & 7 & 1 & 0 & 0 & -1 \end{pmatrix}$$

and $b = \big(1, -1, 34.8, 29.1, -4.1, 0, 0, 0\big)^T$

Solve this problem by means of the *branch and bound method* (cf. theorem 8.2.1).

5. Let $P := \big\{x \in \mathbb{R}^n \mid A^T x \le b\big\}$ be a polytope with $A \in \mathbb{R}^{n \times m}$, $b \in \mathbb{R}^m$ and $m \ge n$.

a) $x^{(0)} \in P$ is a *vertex* of P iff the matrix A_J with $J := \mathcal{A}(x^{(0)})$ has rank n.

b) Let $x^{(0)}$ be a nondegenerate vertex of P, that is, $\big|\mathcal{A}(x^{(0)}\big| = n$. Then there exist exactly n neighbors $y^{(1)}, \ldots, y^{(n)}$ of $x^{(0)}$.

c) Let $x^{(0)}$ be a nondegenerate vertex of P and $y^{(1)}, \ldots, y^{(n)} \in P$ its neighbors. Show for $u^{(\nu)} := y^{(\nu)} - x^{(0)}$ $(\nu = 1, \ldots, n)$:
$P \subset \big\{x \in \mathbb{R}^n \mid A_J^T x \le b_J\big\} = x^{(0)} + \big\{u \in \mathbb{R}^n \mid A_J^T u \le 0\big\}$
$\big\{u \in \mathbb{R}^n \mid A_J^T u \le 0\big\} = \operatorname{cone}\big\{u^{(1)}, \ldots, u^{(n)}\big\}$

6. Solve the quadratic optimization problem $\begin{array}{l} f(x) \longrightarrow \min \\ A^T x \le b \end{array}$ from exercise 8 in chapter 4 by means of the KELLEY *cutting plane method.* Derive upper bounds U_k and lower bounds L_k for $p^* := \inf\{f(x) \mid A^T x \le b\}$ and terminate the calculation when $U_k - L_k < \varepsilon$.

Hint: The quadratic optimization problem is equivalent to

$t \longrightarrow \min$
$f(x) \le t,\ A^T x \le b$.

Start the iteration with $P_0 := \left\{\binom{x}{t} \in \mathbb{R}^3 \mid A^T x \le b,\, t \ge \alpha\right\}$ for a suitable $\alpha \in \mathbb{R}$. You can take $L_k := t^{(k)}$, $U_k := f(x^{(k)})$ where $\begin{pmatrix} x^{(k)} \\ t^{(k)} \end{pmatrix}$ is the solution to the linear program

$$\begin{aligned} &t \longrightarrow \min \\ &\tbinom{x}{t} \in P_k \end{aligned} \quad .$$

7. Generalize the implementation of SHOR's ellipsoid method (cf. chapter 3, exercise 2) to convex constrained problems

$$\begin{aligned} &f(x) \longrightarrow \min \\ &g_i(x) \le 0 \quad (i = 1, \ldots, m) \end{aligned} \quad .$$

Test the program with the quadratic problem from exercise 8 in chapter 4.

Hint: Construct the ellipsoids $\mathcal{E}^{(k)}$ in such a way that they contain the minimizer x^*.
Case 1: $x^{(k)} \in \mathcal{F}$; update $\mathcal{E}^{(k)}$ based on $\nabla f(x^{(k)})$ like in chapter 3.
Case 2: $x^{(k)} \notin \mathcal{F}$, say $g_j(x^{(k)}) > 0$; $g_j'(x^{(k)})(x - x^{(k)}) > 0$ implies $g_j(x) > 0$, that is, $x \notin \mathcal{F}$. Update $\mathcal{E}^{(k)}$ based on $\nabla g_j(x^{(k)})$ and eliminate the halfspace of infeasible points.

8. Solve the linearly constrained quadratic problem

$$\begin{aligned} &-(x_1 - 2.5)^2 - 4(x_2 - 1)^2 \longrightarrow \min \\ &x_1 + x_2 \le 10 \\ &x_1 + 5x_2 \le 22 \\ &-3x_1 + 2x_2 \le 2 \\ &x_1 - 2x_2 \le 4 \\ &x_1, \ge 0,\, x_2 \ge 0 \end{aligned}$$

by means of *concavity cuts*. Start the iteration at $x^{(0)} := \binom{0}{0}$.

Chapter 8

Appendices

A A Second Look at the Constraint Qualifications

The GUIGNARD constraint qualification $\mathcal{C}_\ell(x_0)^* = \mathcal{C}_t(x_0)^*$ in section 2.2 seems to somewhat come out of the blue. The correlation between this regularity condition and the corresponding 'linearized' problem which we will discuss now makes the matter more transparent.

The Linearized Problem

We again consider the general problem:

$$(P) \begin{cases} f(x) \longrightarrow \min \\ g_i(x) \le 0 \qquad \text{for } i \in \mathcal{I} := \{1, \ldots, m\} \\ h_j(x) = 0 \qquad \text{for } j \in \mathcal{E} := \{1, \ldots, p\}. \end{cases}$$

Here let $n \in \mathbb{N}, m, p \in \mathbb{N}_0$ (hence, $\mathcal{E} = \emptyset$ or $\mathcal{I} = \emptyset$ permitted), the real-valued functions $f, g_1, \ldots, g_m, h_1, \ldots, h_p$ defined on an open subset D of $\mathbb{R}^n$ and $p \le n$. The set

$$\mathcal{F} := \big\{ x \in D \mid g_i(x) \le 0 \text{ for } i \in \mathcal{I},\ h_j(x) = 0 \text{ for } j \in \mathcal{E} \big\}$$

W. Forst and D. Hoffmann, *Optimization—Theory and Practice*,
Springer Undergraduate Texts in Mathematics and Technology,
DOI 10.1007/978-0-387-78977-4,

was called the *feasible region* or *set of feasible points of* (P).

If we combine, as usual, the m functions g_i to a vector-valued function g and the p functions h_j to a vector-valued function h, respectively, then we are able to state problem (P) in the shortened form

$$(P) \begin{cases} f(x) \longrightarrow \min \\ g(x) \le 0 \\ h(x) = 0 \end{cases}$$

with the feasible region

$$\mathcal{F} = \{x \in D \mid g(x) \le 0,\, h(x) = 0\}.$$

If we assume the functions f, g and h to be differentiable at a point x_0 in D, we can 'linearize' them and with

$$\begin{aligned} \widetilde{f}(x) &:= f(x_0) + f'(x_0)(x - x_0) \\ \widetilde{g}(x) &:= g(x_0) + g'(x_0)(x - x_0) \\ \widetilde{h}(x) &:= h(x_0) + h'(x_0)(x - x_0) \end{aligned}$$

draw on the — generally simpler — 'linearized problem'

$$(P_\ell(x_0)) \begin{cases} \widetilde{f}(x) \longrightarrow \min \\ \widetilde{g}(x) \le 0 \\ \widetilde{h}(x) = 0 \end{cases}$$

with the feasible region

$$\mathcal{F}_\ell(x_0) := \{x \in \mathbb{R}^n \mid \widetilde{g}(x) \le 0,\, \widetilde{h}(x) = 0\}$$

for comparison and apply the known optimality conditions to it.

If a local minimizer x_0 of (P) also solves $(P_\ell(x_0))$, then the KKT conditions for $(P_\ell(x_0))$ are met and hence also for (P) since the gradients that occur are the same in both problems. In this case we get a *necessary* condition.

A simple example, however, shows that a minimizer x_0 of (P) does not necessarily yield a (local) solution to $(P_\ell(x_0))$:

Example 1

To illustrate this fact, let us look at the following problem (with $n = 2$):

$$(P) \begin{cases} f(x) := x_1 \longrightarrow \min & (x = (x_1, x_2)^T \in \mathbb{R}^2) \\ g_1(x) := -x_1^3 + x_2 \le 0 \\ g_2(x) := -x_2 \le 0 \end{cases}$$

Here we have $p = 0$, $m = 2$, and D can be chosen as $\mathbb{R}^2$. We thus obtain the following set of feasible points:

$$\mathcal{F} = \{x \in \mathbb{R}^2 \mid x_2 \geq 0,\ x_2 \leq x_1^3\}$$

Since $x_1 \geq 0$ holds for $x \in \mathcal{F}$, $x_0 := (0,0)^T$ yields the minimum of (P). As the linearized problem we get

$$(P_\ell(x_0)) \begin{cases} \widetilde{f}(x) = f(x) = x_1 \longrightarrow \min \\ \widetilde{g_1}(x) = x_2 \leq 0 \\ \widetilde{g_2}(x) = g_2(x) = -x_2 \leq 0 \end{cases}$$

with the set of feasible points

$$\mathcal{F}_\ell(x_0) = \{x \in \mathbb{R}^2 \mid x_2 = 0\}.$$

The function $\widetilde{f}$, however, is not even bounded from below on $\mathcal{F}_\ell(x_0)$. ◁

The question of in which cases a local minimizer x_0 to (P) also solves locally $\big(P_\ell(x_0)\big)$ leads to the *regularity conditions:*

We can rewrite problem $\big(P_\ell(x_0)\big)$ with the set of indices of the *active inequalities*

$$\mathcal{A}(x_0) := \{i \in I : g_i(x_0) = 0\}$$

as

$$(P_1(x_0)) \begin{cases} f'(x_0)(x - x_0) \longrightarrow \min \\ g_i'(x_0)(x - x_0) \leq 0 \quad \text{for } i \in \mathcal{A}(x_0) \\ h'(x_0)(x - x_0) = 0 \end{cases}$$

since $h(x_0) = 0$ and $g_i(x_0) = 0$ for $i \in \mathcal{A}(x_0)$.

In this case we were obviously able to drop the additive constant $f(x_0)$ in the objective function.

Lemma A.1

Let $x_0 \in \mathcal{F}$ and f, g, h differentiable in x_0. Then x_0 is a local minimizer of $\big(P_\ell(x_0)\big)$ if and only if x_0 is a local minimizer of problem $\big(P_1(x_0)\big)$.

Let us furthermore denote the set of feasible points of $\big(P_1(x_0)\big)$ by

$$\mathcal{F}_1(x_0) := \{x \in \mathbb{R}^n : h'(x_0)(x - x_0) = 0,\ g_i'(x_0)(x - x_0) \leq 0 \text{ for } i \in \mathcal{A}(x_0)\}.$$

Proof: If x_0 is a local minimizer of $\big(P_1(x_0)\big)$, then $f'(x_0)(x - x_0) \geq 0$ holds for $x \in \mathcal{F}_1(x_0)$ in a suitable neighborhood of x_0. Since $x_0 \in \mathcal{F}_\ell(x_0) \subset \mathcal{F}_1(x_0)$, x_0 gives locally a solution to $\big(P_\ell(x_0)\big)$.

If conversely x_0 is a local minimizer of $\bigl(P_\ell(x_0)\bigr)$, then $\widetilde{f}(x) \ge \widetilde{f}(x_0) = f(x_0)$ holds locally, hence $f'(x_0)(x - x_0) \ge 0$ for $x \in \mathcal{F}_\ell(x_0)$. To $x \in \mathcal{F}_1(x_0)$ we consider the vector $u(t) := x_0 + t\,(x - x_0)$ with $t > 0$. It holds that

$$h'(x_0)(u(t) - x_0) \;=\; t\,h'(x_0)(x - x_0) \;=\; 0$$

and for $i \in \mathcal{A}(x_0)$

$$g_i'(x_0)(u(t) - x_0) \;=\; t\,g_i'(x_0)(x - x_0) \;\le\; 0\,.$$

For $i \in \mathcal{I} \setminus \mathcal{A}(x_0)$ we have $g_i(x_0) < 0$ and thus for t sufficiently small

$$g_i(x_0) + g_i'(x_0)(u(t) - x_0) \;=\; g_i(x_0) + t\,g_i'(x_0)(x - x_0) \;<\; 0\,.$$

Hence, for such t the vector $u(t)$ is in $\mathcal{F}_\ell(x_0)$, consequently

$$f(x_0) + f'(x_0)(u(t) - x_0) \;=\; \widetilde{f}(u(t)) \;\ge\; \widetilde{f}(x_0) \;=\; f(x_0)$$

and thus

$$t\,f'(x_0)(x - x_0) \;=\; f'(x_0)(u(t) - x_0) \;\ge\; 0\,,$$

hence $f'(x_0)(x - x_0) \ge 0$. Since we had chosen $x \in \mathcal{F}_1(x_0)$ at random, x_0 yields a local solution to $\bigl(P_1(x_0)\bigr)$. □

With the transformation $d := x - x_0$, we obtain the following problem which is equivalent to $\bigl(P_1(x_0)\bigr)$:

$$\bigl(P_2(x_0)\bigr) \begin{cases} f'(x_0)\,d \longrightarrow \min \\ g_i'(x_0)\,d \le 0 \quad \text{for } i \in \mathcal{A}(x_0) \\ h'(x_0)\,d = 0\,. \end{cases}$$

From the lemma we have just proven we can now easily deduce

Lemma A.2

Let $x_0 \in \mathcal{F}$ and f, g, h differentiable in x_0. Then x_0 is a local minimizer of $\bigl(P_\ell(x_0)\bigr)$ if and only if $0\ (\in \mathbb{R}^n)$ minimizes locally $\bigl(P_2(x_0)\bigr)$.

The set of feasible points of $\bigl(P_2(x_0)\bigr)$

$$\mathcal{F}_2(x_0) := \bigl\{ d \in \mathbb{R}^n : h'(x_0)d = 0 \text{ and } g_i'(x_0)d \le 0 \text{ for } i \in \mathcal{A}(x_0) \bigr\}$$

is nothing else but the *linearizing cone* $C_\ell(x_0)$ to $\mathcal{F}$ in x_0. We see that this cone occurs here in a very natural way.

Let us summarize the results we have obtained so far:

Proposition

Let $x_0 \in \mathcal{F}$ and f, g, h differentiable in x_0. Then the following assertions are equivalent:

(a) x_0 gives a local solution to $\big(P_\ell(x_0)\big)$.
(b) $0\,(\in \mathbb{R}^n)$ gives a local solution to $\big(P_2(x_0)\big)$.
(c) $\nabla f(x_0) \in \mathcal{C}_\ell(x_0)^*$
(d) $\mathcal{C}_\ell(x_0) \cap \mathcal{C}_{dd}(x_0) = \emptyset$
(e) x_0 is a KKT point.

Here

$$\mathcal{C}_{dd}(x_0) = \{d \in \mathbb{R}^n \mid f'(x_0)d < 0\}$$

denotes the *cone of descent directions of f at x_0*.

Proof: Lemma A.2 gives the equivalence of (a) and (b). (b) means $\nabla f(x_0)d \geq 0$ for all $d \in \mathcal{F}_2(x_0) = \mathcal{C}_\ell(x_0)$, hence $\nabla f(x_0) \in \mathcal{C}_\ell(x_0)^*$, which gives (c). The equivalence of (c) and (d) can be deduced directly from the definition of $\mathcal{C}_{dd}(x_0)$. We had formulated this finding as lemma 2.2.2. Proposition 2.2.1 yields precisely the equivalence of (d) and (e). □

Recall that for *convex* optimization problems (with continuously differentiable functions f, g, h) every KKT point is a global minimizer (proposition 2.2.8).

Correlation to the Constraint Qualifications

For $x_0 \in \mathcal{F}$ we had defined the *cone of tangents* or *tangent cone* of $\mathcal{F}$ in x_0 by

$$\mathcal{C}_t(x_0) := \left\{d \in \mathbb{R}^n \mid \exists\, (x_k) \in \mathcal{F}^{\mathbb{N}} \;\; x_k \xrightarrow{d} x_0\right\}.$$

Here $x_k \xrightarrow{d} x_0$ meant: There exists a sequence (α_k) of positive numbers with $\alpha_k \downarrow 0$ and

$$\frac{1}{\alpha_k}(x_k - x_0) \longrightarrow d \text{ for } k \longrightarrow \infty.$$

Following lemma 2.2.3, it holds that $\mathcal{C}_t(x_0) \subset \mathcal{C}_\ell(x_0)$, hence

$$\mathcal{C}_\ell(x_0)^* \subset \mathcal{C}_t(x_0)^*.$$

Lemma 2.2.4 gives that for a minimizer x_0 of (P) it always holds that $\nabla f(x_0) \in \mathcal{C}_t(x_0)^*$.

If a local minimizer x_0 of (P) also yields a local solution to $\big(P_\ell(x_0)\big)$, the preceding proposition gives

$$\nabla f(x_0) \in \mathcal{C}_\ell(x_0)^*.$$

This leads to the demand $\mathcal{C}_t(x_0)^* \subset \mathcal{C}_\ell(x_0)^*$, hence $\mathcal{C}_t(x_0)^* = \mathcal{C}_\ell(x_0)^*$, which is named after GUIGNARD.

We now want to take up example 1 — or example 4 from section 2.2 — once more:

Example 2

With $\mathcal{A}(x_0) = \{1, 2\}$ we get

$$\mathcal{C}_\ell(P, x_0) = \{d \in \mathbb{R}^2 \mid \forall\, i \in \mathcal{A}(x_0)\;\; g_i'(x_0)d \le 0\} = \{d \in \mathbb{R}^2 \mid d_2 = 0\}.$$

In section 2.2 we had obtained for the cone of tangents (cf. page 50)

$$\mathcal{C}_t(x_0) = \{d \in \mathbb{R}^2 \mid d_1 \ge 0,\; d_2 = 0\},$$

thus it holds that $\mathcal{C}_t(x_0) \ne \mathcal{C}_\ell(P, x_0)$.

If we write the feasible region in a different way and consider

$$(P_0)\begin{cases} f(x) := x_1 \longrightarrow \min & (x = (x_1, x_2)^T \in \mathbb{R}^2) \\ g_1(x) := -x_1^3 + x_2 \le 0 \\ g_2(x) := -x_2 \le 0 \\ g_3(x) := -x_1 \le 0, \end{cases}$$

we obtain a linearization at the point $x_0 := (0, 0)^T$

$$\begin{aligned} \widetilde{f}(x) &= f(x) = x_1 \longrightarrow \min \\ \widetilde{g_1}(x) &= x_2 \le 0 \\ \widetilde{g_2}(x) &= g_2(x) = -x_2 \le 0 \\ \widetilde{g_3}(x) &= g_3(x) = -x_1 \le 0 \end{aligned}$$

with the set of feasible points

$$\mathcal{F}_\ell(x_0) = \{x \in \mathbb{R}^2 \mid x_1 \ge 0,\; x_2 = 0\}.$$

The additional condition $g_3'(x_0)d \le 0$ with $g_3'(x_0) = (-1, 0)$ gives $d_1 \ge 0$. It therefore holds that

$$\mathcal{C}_\ell(P_0, x_0) = \{d \in \mathbb{R}^2 \mid d_1 \ge 0,\; d_2 = 0\} = \mathcal{C}_t(x_0),$$

in particular $\mathcal{C}_\ell(P_0, x_0)^* = \mathcal{C}_t(x_0)^*$. ◁

B The Fritz John Condition

In this appendix we want to have a look at FRITZ JOHN's important paper from 1948 ([John]) in which he derives a necessary condition for a minimizer of a general optimization problem subject to an *arbitrary number of constraints.* JOHN then applies this condition in particular to the problem of finding *enclosing balls or ellipsoids of minimal volume.*

Theorem B.1 (JOHN, 1948)

Let $D \subset \mathbb{R}^n$ be open and Y a compact HAUSDORFF *space*[1]*. Consider the optimization problem:*

$$(J) \quad \begin{array}{l} f(x) \longrightarrow \min \ \text{subject to the constraint} \\ g(x,y) \le 0 \ \text{for all } y \in Y\,. \end{array}$$

Let the objective function $f\colon D \longrightarrow \mathbb{R}$ be continuously differentiable and the function $g\colon D \times Y \longrightarrow \mathbb{R}$, which gives the constraints, continuous as well as continuously differentiable in x, hence in the components $x_1, \ldots, x_n$.

Let $x_0 \in \mathcal{F} := \{x \in D \mid g(x,y) \le 0 \text{ for all } y \in Y\}$ yield a local solution to (J), that is, there exists a neighborhood U of x_0 such that $f(x_0) \le f(x)$ for all $x \in U \cap \mathcal{F}$.

Then there exist an $s \in \mathbb{N}_0$ with $0 \le s \le n$, scalars $\lambda_0, \ldots, \lambda_s$, which are not all zero and $y_1, \ldots, y_s \in Y$ such that

(i) $g(x_0, y_\sigma) = 0$ *for* $1 \le \sigma \le s$

(ii) $\lambda_0 \ge 0,\ \lambda_1 > 0, \ldots, \lambda_s > 0$ *and*

(iii) $\lambda_0 \nabla f(x_0) + \sum\limits_{\sigma=1}^{s} \lambda_\sigma \nabla_x g(x_0, y_\sigma) = 0\,.$

Each $y \in Y$ gives a constraint, which means that an arbitrary number of constraints is permitted. If we compare assertions *(i)–(iii)* to the KARUSH–KUHN–TUCKER conditions (cf. 2.2.5), we firstly see that λ_0 occurs here as an additional Lagrangian multiplier. If $\lambda_0 = 0$, assertion *(iii)* is not very helpful since the objective function f is irrelevant in this case. Otherwise, if $\lambda_0 > 0$, *(iii)* yields precisely the KKT condition for the Lagrangian multipliers $\lambda'_\sigma := \lambda_\sigma/\lambda_0$ of $g(\,\cdot\,, y_\sigma)$ for $\sigma = 1, \ldots, s$. Positive — compared to theorem 2.2.5, for example — is the fact that no constraint qualifications are necessary. However, we have to strengthen the smoothness conditions. In special types of problems — for example the problem of finding the minimum volume enclosing ellipsoid — it is possible to deduce that λ_0 is positive. In these cases JOHN's theorem is more powerful than theorem 2.2.5.

In preparation for the proof of the theorem we firstly prove the following two lemmata.

Lemma B.2

Let $A \in \mathbb{R}^{s \times n}$. Then exactly one of the following two assertions holds:

1) There exists a $z \in \mathbb{R}^n$ such that $Az < 0$.

2) There exists a $\lambda \in \mathbb{R}^s_+ \setminus \{0\}$ such that $\lambda^T A = 0$.

[1] If you are not familiar with this term, think of a compact subset of an $\mathbb{R}^k$, for example.

Proof: *1)* and *2)* cannot hold at the same time since we get a *contradiction* from $Az < 0$ and $\lambda^T A = 0$ for a $\lambda \in \mathbb{R}^s_+ \setminus \{0\}$ by $0 = (\lambda^T A) z = \lambda^T (Az) < 0$. If *1)* does *not* hold, there does not exist any $\binom{v}{z} \in (-\infty, 0) \times \mathbb{R}^n$ with $ve - Az \geq 0$ for $e := (1, \ldots, 1)^T \in \mathbb{R}^s$ either. Hence, the inequality $(1, 0, \ldots, 0)\binom{v}{z} \geq 0$ follows from $(e, -A)\binom{v}{z} \geq 0$ for all $\binom{v}{z} \in \mathbb{R} \times \mathbb{R}^n$. Following the Theorem of the Alternative (FARKAS, cf. p. 43) there thus exists a $\lambda \in \mathbb{R}^s_+$ with $\lambda^T(e, -A) = (1, 0, \ldots, 0)$. Consequently $\lambda \neq 0$ and $\lambda^T A = 0$, hence *2)*. □

Lemma B.3

Let K be a nonempty, convex and compact subset of $\mathbb{R}^n$. If for all $z \in K$ there exists a $u \in K$ such that $\langle u, z \rangle = 0$, then it holds that $0 \in K$.

Proof: Let u_0 be an element of K with minimal norm. We have to show that $u_0 = 0$. By assumption there exists a $u_1 \in K$ to u_0 with $\langle u_0, u_1 \rangle = 0$. For that the relation

$$\|u_0\|^2 \leq \|(1-\lambda)u_0 + \lambda u_1\|^2 = (1-\lambda)^2 \|u_0\|^2 + \lambda^2 \|u_1\|^2$$

follows for all $\lambda \in [0, 1]$. For $\lambda \neq 0$ we obtain from the above: $2\|u_0\|^2 \leq \lambda\left(\|u_0\|^2 + \|u_1\|^2\right)$. Passage to the limit as $\lambda \longrightarrow 0$ yields $\|u_0\| = 0$, hence, $u_0 = 0$. □

Proof of theorem B.1:

The set $Y(x_0) := \{y \in Y \mid g(x_0, y) = 0\}$ — as the pre-image of the closed set $\{0\}$ under the continuous function $g(x_0, \cdot)$ — is closed and therefore compact since it is a subset of the compact set Y.

We prove the following *auxiliary assertion:*

There exists no $z \in \mathbb{R}^n$ with

$$\langle \nabla f(x_0), z \rangle < 0 \text{ and } \langle \nabla_x g(x_0, y), z \rangle < 0 \text{ for all } y \in Y(x_0). \tag{1}$$

For the *proof of the auxiliary assertion* we assume (1) to have a solution z. The continuity of ∇f and $\nabla_x g$, and the compactness of $Y(x_0)$ give: There exist $\delta, \eta > 0$ such that the neighborhood $U := \{x \in \mathbb{R}^n \mid \|x - x_0\| \leq \eta\}$ of x_0 lies completely in D, as well as an open set $Y_0 \supset Y(x_0)$ with

$$\langle \nabla f(x), z \rangle \leq -\delta \text{ and } \langle \nabla_x g(x, y), z \rangle \leq -\delta \text{ for all } y \in Y_0 \text{ and all } x \in U.$$

Since $Y \setminus Y_0$ is compact and $g(x_0, y) < 0$ for all $y \in Y \setminus Y_0$, there exists furthermore an $\varepsilon > 0$ with $g(x_0, y) \leq -\varepsilon$ for such y. Now choose a $t \in (0, 1)$ with $x_0 + tz \in U$ and $t|\langle \nabla_x g(x, y), z \rangle| \leq \varepsilon/2$ for all $(x, y) \in U \times (Y \setminus Y_0)$. With suitable $\vartheta_0 \in (0, 1)$ and $\vartheta_y \in (0, 1)$ for $y \in Y$ it holds that

$$\begin{aligned} f(x_0 + tz) &= f(x_0) + t\langle \nabla f(x_0 + \vartheta_0 tz), z \rangle \text{ and} \\ g(x_0 + tz, y) &= g(x_0, y) + t\langle \nabla_x g(x_0 + \vartheta_y tz, y), z \rangle . \end{aligned}$$

With $x_0 + tz$ the points $x_0 + \vartheta_0 tz$ and $x_0 + \vartheta_y tz$ for $y \in Y$ also lie in U and we therefore get with the above estimates:

$$\begin{aligned} f(x_0 + tz) &\le f(x_0) - t\delta, \\ g(x_0 + tz, y) &\le g(x_0, y) - t\delta \text{ for } y \in Y_0 \text{ and} \\ g(x_0 + tz, y) &\le g(x_0, y) + t\,|\langle \nabla_x g(x_0 + \vartheta_y tz, y), z\rangle| \le -\frac{\varepsilon}{2} \text{ for } y \in Y \setminus Y_0 . \end{aligned}$$

Consequently, $x_0 + tz \in \mathcal{F} \cap U$ with $f(x_0 + tz) < f(x_0)$ is a *contradiction* to the local minimality of f in x_0 which proves the auxiliary assertion.

For the rest of the proof we distinguish two cases:

Case 1: $s := |Y(x_0)| \le n$, that is, there exist $y_1, \ldots, y_s \in Y$ with $Y(x_0) = \{y_1, \ldots, y_s\}$. If we set $A^T := (\nabla f(x_0), \nabla_x g(x_0, y_1), \ldots, \nabla_x g(x_0, y_s)) \in \mathbb{R}^{n \times (s+1)}$ and apply lemma B.2, the desired result follows immediately from the unsolvability of (1).

Case 2: $|Y(x_0)| \ge n+1$: Set $T := \{\nabla f(x_0)\} \cup \{\nabla_x g(x_0, y) \mid y \in Y(x_0)\}$. The continuity of $\nabla_x g(x_0, \cdot)$ on the compact set $Y(x_0)$ gives the compactness of T. Since (1) is unsolvable, it holds that: For all $z \in \mathbb{R}^n$ there exist $u_1, u_2 \in T$ with $\langle u_1, z\rangle \le 0 \le \langle u_2, z\rangle$ and thus a $u \in \operatorname{conv}(T)$ with $\langle u, z\rangle = 0$. By CARATHÉODORY's lemma (cf. exercise 7, chapter 2) the convex hull $\operatorname{conv}(T)$ of T is given by

$$\left\{ \sum_{\sigma=1}^{s} \varrho_\sigma u_\sigma \,\middle|\, u_\sigma \in T, \varrho_\sigma > 0 \ (\sigma = 1, \ldots, s), \sum_{\sigma=1}^{s} \varrho_\sigma = 1;\ 1 \le s \le n+1 \right\}.$$

Since it is compact (cf. exercise 7, chapter 2), we can apply lemma B.3 and obtain $0 \in \operatorname{conv}(T)$. With the above representation we get that there exist $\varrho_\sigma \ge 0$ and $u_\sigma \in T$ for $\sigma = 1, \ldots, n+1$ with

$$\sum_{\sigma=1}^{n+1} \varrho_\sigma = 1 \text{ and } \sum_{\sigma=1}^{n+1} \varrho_\sigma u_\sigma = 0, \tag{2}$$

where WLOG $\varrho_1, \ldots, \varrho_s > 0$ for an s with $1 \le s \le n+1$.

If $\nabla f(x_0)$ is one of these u_σ, we are done. If otherwise $s \le n$ holds, the assertion follows with $\lambda_0 := 0$. Hence, only the case $s = n+1$ remains: In this case, however, the $n+1$ vectors $u_\sigma - \nabla f(x_0)$ are linearly dependent and therefore there exists an $\alpha := (\alpha_1, \ldots, \alpha_{n+1})^T \in \mathbb{R}^{n+1} \setminus \{0\}$ such that $\sum_{\sigma=1}^{n+1} \alpha_\sigma (u_\sigma - \nabla f(x_0)) = 0$ and WLOG $\sum_{\sigma=1}^{n+1} \alpha_\sigma \le 0$. Setting $\alpha_0 := -\sum_{\sigma=1}^{n+1} \alpha_\sigma$, gives $\sum_{\sigma=1}^{n+1} \alpha_\sigma u_\sigma + \alpha_0 \nabla f(x_0) = 0$. For the ϱ_σ given by (2) and an arbitrary $\tau \in \mathbb{R}$ it then holds that $\tau\alpha_0 \nabla f(x_0) + \sum_{\sigma=1}^{n+1} (\varrho_\sigma + \tau\alpha_\sigma) u_\sigma = 0$. If we now choose

$$\tau := \min\{-\varrho_\sigma/\alpha_\sigma \mid 1 \le \sigma \le n+1,\ \alpha_\sigma < 0\}$$

and set $\lambda_0 := \tau\alpha_0$ and $\lambda_\sigma := \varrho_\sigma + \tau\alpha_\sigma$ for all $\sigma = 1, \ldots, n+1$, we have

$\lambda_0 \geq 0$, $\lambda_1 \geq 0, \ldots, \lambda_{n+1} \geq 0$, where — by definition of τ — for at least one $\sigma \in \{1, \ldots, n+1\}$ the relation $\lambda_\sigma = 0$ holds. Therefore, we also obtain the assertion in this case. □

At this point it would be possible to add interesting discussions and powerful algorithms for the generation of minimum volume enclosing ellipsoids, and as the title page of the book shows, the authors do have a certain weakness for this topic. We, however, abstain from a further discussion of these — rather more complex — issues.

C Optimization Software for Teaching and Learning

Matlab® Optimization Toolbox

The Optimization Toolbox is a collection of functions that extend the numerical capabilities of *Matlab*®. The toolbox includes various routines for many types of optimization problems. In the following tables we give a short overview for the main areas of application treated in our book.

One-Dimensional Minimization

fminbnd	Minimization with Bounds	Golden-Section Search or Parabolic Interpolation

Unconstrained Minimization

fminsearch	Unconstrained Minimization	Nelder–Mead Method
fminunc	Unconstrained Minimization	Steepest Descent, BFGS, DFP or Trust Region Method
\	Unconstrained Linear Least Squares	$Ax = b: \quad x = A \backslash b$
fminimax	Minimax Optimization	SQP Method
lsqnonlin lsqcurvefit fsolve	Nonlinear Least Squares Nonlinear Curve Fitting Nonlinear Systems of Equations	Trust Region, Levenberg–Marquardt or Gauss–Newton Method

Constrained Minimization

linprog	Linear Programming	Primal-Dual Interior Point Method, Active Set Method or Simplex Algorithm
quadprog lsqnonneg lsqlin	Quadratic Programming Nonnegative Linear Least Squares Constrained Linear Least Squares	Active Set Method
fmincon	Constrained Nonlinear Minimization	SQP Method
fminimax lsqnonlin lsqcurvefit	Constrained Minimax Optimization Constrained Nonlinear Least Squares Constrained Nonlinear Curve Fitting	

The User Guide [MOT] gives in-depth information about the Optimization Toolbox. *Matlab*® includes a powerful Help facility which provides an easy online access to the documentation. We therefore content ourselves here with an overview of the methods accessible with the important option LargeScale (= on/off) for the functions of the Toolbox and the restrictions that hold for them.

Function	Option	Method/Remarks
fminbnd		Golden-Section Search or Parabolic Interpolation
fmincon	LargeScale = on LargeScale = off	Subspace Trust Region Method expects GradObj = on **either:** bound constraints $\ell \le x \le u,\ \ell < u$ **or:** linear equality constraints SQP Method allows inequality and/or equality constraints

fminimax		$\max_{1\le i\le m} f_i(x) \longrightarrow \min$ $\max_{1\le i\le m} \lvert f_i(x)\rvert \longrightarrow \min$ with or without constraints uses an SQP method
fminsearch		NELDER–MEAD Polytope Method
fminunc	LargeScale = on (default) LargeScale = off	Subspace Trust Region Method expects GradObj = on HessUpdate = bfgs (default) HessUpdate = dfp HessUpdate = steepdesc
fsolve	LargeScale = on LargeScale = off (default)	Subspace Trust Region Method NonlEqnAlgorithm = dogleg: (default) NonlEqnAlgorithm = lm: LEV.–MARQUARDT NonlEqnAlgorithm = gn: GAUSS–NEWTON
linprog	LargeScale = on (default) LargeScale = off	Linear Interior-Point Solver (LIPSOL) based on MEHROTRA's Predictor-Corrector Method Simplex = off: Active Set Method Simplex = on: Simplex Algorithm
lsqcurvefit	LargeScale = on (default) LargeScale = off	Subspace Trust Region Method bound constraints $\ell \le x \le u$ where $\ell < u$ LevenbergMarquardt = on: (default) LevenbergMarquardt = off: GAUSS–NEWTON no bound constraints
lsqlin	LargeScale = on (default) LargeScale = off	Subspace Trust Region Method bound constraints $\ell \le x \le u$ where $\ell < u$ allows linear $\{\le, =\}$-constraints $\rightsquigarrow$ based on quadprog

lsqnonlin	LargeScale = on (default)	Subspace Trust Region Method bound constraints $\ell \le x \le u$ where $\ell < u$
	LargeScale = off	LevenbergMarquardt = on: (default) LevenbergMarquardt = off: GAUSS–NEWTON no bound constraints ⇝ otherwise use fmincon
lsqnonneg		Active Set Method using svd ⇝ lsqlin
quadprog	LargeScale = on (default)	Subspace Trust Region Method **either:** bound constraints $\ell \le x \le u$, $\ell < u$ **or:** linear equality constraints
	LargeScale = off	Active Set Method allows linear $\{\le, =\}$-constraints

SeDuMi: An Introduction by Examples

'SeDuMi' — an acronym for 'Self-Dual Minimization' (cf. [YTM]) — is a *Matlab*®-based free software package for linear optimization over self-dual cones like $\mathbb{R}^n_+$, $\mathcal{L}^n$ and $\mathcal{S}^n_+$.

How to Install SeDuMi

Download *SeDuMi* from http://sedumi.mcmaster.ca/ . After uncompressing the distribution in the folder of your choice, follow the SeDuMi 1.1 installation instructions in the file Install.txt. Start *Matlab*® and change the current folder to where you put all the files. Type install_sedumi. This will build all the binary files and add the directory to the *Matlab*® path. You are now ready to use SeDuMi and can start by typing help sedumi or by reading the User Guides in the /doc directory. You can find there in particular the user guide [Stu] by J. STURM, the developer of SeDuMi.

LP Problems

It is possible to formulate an LP problem in either the primal standard form (P) or the dual standard form (D) (cf. p. 242 f). The solution of the primal problem can be obtained by x = sedumi(A,b,c). The solutions of both the primal and the dual problem can be obtained by means of the command [x, y, info] = sedumi(A,b,c) .

Example 3 (cf. chapter 4, example 1)

```
A = [2 1 1; 1 2 3; 2 2 1]; m = size(A,1);
A = [A, eye(m)]; b = [2; 5; 6];
c = [ -3, -1, -3, zeros(1,m)]';
x = sedumi(A,b,c)
[x,y,info] = sedumi(A,b,c)
```

Instead of requiring $x \in \mathbb{R}^n_+$, in SeDuMi, it is possible to restrict the variables to a LORENTZ *cone*[1] $\mathcal{L}^n$ or to the cone S^n_+ of positive semidefinite matrices. More generally, we can require $x \in \mathcal{K}$, where $\mathcal{K}$ is a Cartesian product of $\mathbb{R}^n_+$, Lorentz cones and cones of positive semidefinite matrices.

Second-Order Cone Programming

Let the problem

$$(D) \quad \begin{cases} \langle b, y\rangle \longrightarrow \max \\ A^T y \le c \\ \|A_j^T y + c_j\|_2 \le \langle b_j, y\rangle + d_j \quad (j = 1, \ldots, k) \end{cases}$$

be given with $y,\, b,\, b_j \in \mathbb{R}^m;\ A \in \mathbb{R}^{m\times n_0},\ c \in \mathbb{R}^{n_0};$
$A_j \in \mathbb{R}^{m\times(n_j-1)},\ c_j \in \mathbb{R}^{n_j-1},\ d_j \in \mathbb{R} \quad (j = 1, \ldots, k).$

According to exercise 5 in chapter 7 problem (D) can be interpreted as the dual problem to

$$(P) \quad \begin{cases} c^T x_0 + \sum\limits_{j=1}^{k} \begin{pmatrix} d_j \\ c_j \end{pmatrix}^T x_j \longrightarrow \min \\ A x_0 + \sum\limits_{j=1}^{k} \left[-\,(b_j \mid A_j)\right] x_j = b \\ x_0 \in \mathbb{R}^{n_0}_+,\ x_j \in \mathcal{L}^{n_j} \quad (j = 1, \ldots, k). \end{cases}$$

Example 4 *Shortest Distance of Two Ellipses* (cf. exercise 7 in chapter 7)

```
F1 = [1 0; 0 4]; P1 = sqrtm(F1);
F2 = [5/2 3/2; 3/2 5/2]; P2 = sqrtm(F2);
g1 = [-1; 0]; g2 = [-11; -13];
c0 = [0; 0]; c1 = P1 \ g1; c2 = P2 \ g2;
gamma1 = -3; gamma2 = 70; d0 = 0;
```

[1] Recall the definition of $\mathcal{L}^n$ in SeDuMi (cf. p. 313).

```
d1 = sqrt(c1'*c1-gamma1); d2 = sqrt(c2'*c2-gamma2);
A0 = [zeros(2,1) eye(2) -eye(2)]';
A1 = [zeros(2,1) P1 zeros(2,2)]';
A2 = [zeros(2,1) zeros(2,2) P2]';
b0 = [1 0 0 0 0]'; b1 = zeros(5,1); b2 = b1;
At0 = -[b0 A0]; At1 = -[b1 A1]; At2 = -[b2 A2]
At = [At0 At1 At2];
bt = [-1 0 0 0 0]';
ct0 = [d0; c0]; ct1 = [d1; c1]; ct2 = [d2; c2]
ct = [ct0; ct1; ct2];
K = []; K.q = [size(At0,2) size(At1,2) size(At2,2)];
[xs,ys,info] = sedumi(At,bt,ct,K);
x = ys;
d = x(1) % Distance: d = norm(u-v,2)
u = x(2:3)
v = x(4:5)
```

In the above call to SeDuMi, we see a new input argument K. This argument makes SeDuMi solve the primal problem (P) and the dual problem (D), where the cone $\mathcal{K} = \mathcal{L}^{n_1} \times \cdots \times \mathcal{L}^{n_k}$ is described by the structure K. Without the fourth input argument K, SeDuMi would solve an LP problem.

If in addition there are variables $x_0 \in \mathbb{R}_+^{n_0}$, the field $K.\ell$ equals the number of nonnegative variables.

Semidefinite Programming

In this case we deal with the *primal problem*

$$\begin{cases} \langle c, x \rangle + \langle C, X \rangle \longrightarrow \min \\ Ax + \mathcal{A}(X) = b \\ x \in \mathbb{R}_+^{n_0}, \ X \in \mathcal{S}_+^m \end{cases} \qquad A \in \mathbb{R}^{m \times n_0}, \ c \in \mathbb{R}^{n_0}, \quad \mathcal{A}(X) := \begin{pmatrix} \langle A_1, X \rangle \\ \vdots \\ \langle A_m, X \rangle \end{pmatrix}$$

and its *dual problem*:

$$\begin{cases} \langle b, y \rangle \longrightarrow \max \\ c - A^T y \geq 0 \\ C - \mathcal{A}^*(y) \in \mathcal{S}_+^{n_1} \end{cases} \qquad \mathcal{A}^*(y) = y_1 A_1 + \cdots + y_m A_m$$

With

```
ct = [c; vec(C)];
At = [A [vec(A_1); ... ; vec(A_m)]];
K.l = size(A,2); K.s = size(C,2);
```

the command [xs, ys, info] = sedumi(At,b,ct,K); yields the solutions (x, X) of the primal problem and y of the dual problem in the following way:

```
x = xs(1:K.l);
X = mat(xs(K.l+1:end), K.s);
y = ys;
```

Example 5 *Pattern Separation via Ellipses* (cf. section 7.8)

```
clc; clear all; close all;

% Data Set 1: cf. the picture on p. 335
%P1 = [6.1 2.8; 5.9 2.92; 5.78 2.99; 6.04 2.98; 5.91 3.07; 6.07 3.08]'
%P2 = [5.5 2.87; 5.58 3; 5.48 3.13; 5.65 3.12; 5.7 3.22; 5.96 3.24; ...
%       6.17 3.22; 6.3 3.07]'
% Data Set 2: parallel strip
%P1 = [-2 1; -2 3; 0 3; 0 5; 2 5; 2 7]'
%P2 = [-2 0; -2 4; 0 2; 0 6; 2 4; 2 8]'
% Data Set 3: random points (cf. [Bo/Va], p. 429 ff)
n = 2; N = 100;
P1 = randn(n,N);  P1 = P1*diag((0.95*rand(1,N))./sqrt(sum(P1.^2)));
P2 = randn(n,N);  P2 = P2*diag((1.05+rand(1,N))./sqrt(sum(P2.^2)));
T = [1 -1; 2 1];  P1 = T*P1;  P2 = T*P2;

% Solution via SeDuMi
[n,m1] = size(P1); m2 = size(P2,2); m = m1+m2;
P  = [P1 P2]; At = [];
for p = P; At = [At; vec([p; 1]*[p; 1]')']; end
A = [eye(m1)         zeros(m1,m2)  zeros(m1,1);
     zeros(m2,m1) -eye(m2)        -ones(m2,1)  ];
At = [A At];
b  = [ones(m1,1); zeros(m2,1)];
C  = zeros(n+1,n+1); c = [zeros(m,1); -1]; ct = [c; vec(C)];
K  = []; K.l = size(A,2); K.s = size(C,2);
[xs,ys,info] = sedumi(At,b,ct,K);
rho = xs(K.l)
Z = mat(xs(K.l+1:end),K.s)
X = Z(1:n,1:n)
xc =  - X \ Z(1:n,n+1) % center of 'ellipsoid'

% Displaying results
if n > 2; break; end;
F = @(x,y) X(1,1)*(x-xc(1)).^2 + 2*X(1,2)*(x-xc(1)).*(y-xc(2)) + ...
           X(2,2)*(y-xc(2)).^2;
XX = linspace(min(P2(1,:)),max(P2(1,:)),50);
YY = linspace(min(P2(2,:)),max(P2(2,:)),50);
[X1,Y1] = ndgrid(XX,YY); F1 = F(X1,Y1);
FF1 = contour(X1,Y1,F1,[1 1],'b-'); hold on;
FF2 = contour(X1,Y1,F1,[rho rho],'k-'); hold on;
plot(P1(1,:),P1(2,:),'b+',P2(1,:),P2(2,:),'k*');
title('Pattern Separation via Ellipses');
```

Pattern Separation via Ellipses

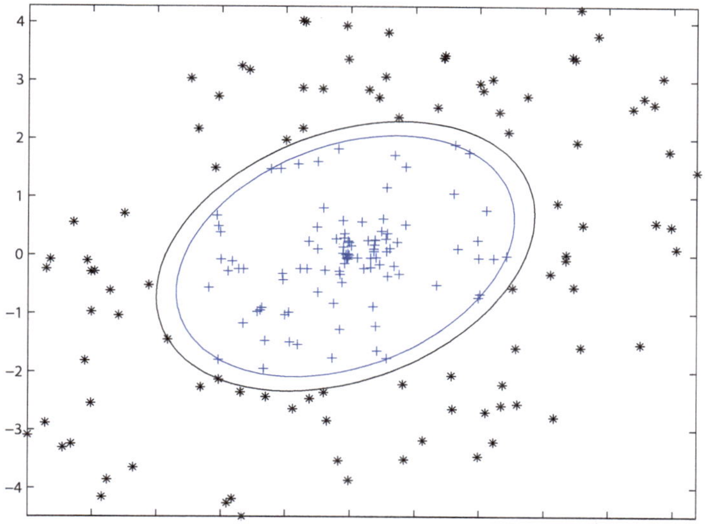

Maple® Optimization Tools

Maple® — besides other computer algebra systems such as *Mathematica*® — is one of the most mature products and in many respects a powerful tool for scientific computing in mathematics, natural sciences, engineering and economics. Its benefits do not only come from the combination of symbolic and numerical computing but also from its visualization abilities of functions and geometric 2D/3D objects. These components blend together to a powerful working environment for research and teaching. The power of *Maple*® is enhanced by numerous add-on packages. Since version 9 *Maple*® offers the *Optimization* package, which consists of highly efficient algorithms for local optimization coming from the NAG library and at the same time gives full access to the power of *Maple*®.

In 2004 Maplesoft introduced the so-called *Professional Toolboxes* that can be purchased separately as add-on products to extend the scope and functionality of *Maple*® in key application areas. The first of these add-on products has been the *Global Optimization Toolbox* (GOT) the core of which is based on the LGO (Lipschitz Global Optimization) solver developed by Pintér Consulting Services. LGO runs on several other platforms, for example in its *Mathematica*® or GAMS implementation. For a review of the GOT for *Maple*® we refer the reader to [Hen]. More details are presented in the e-book [Pin 2], which can be viewed as a hands-on introduction to the GOT.

The LGO solver offers the following four optional solution strategies:

- branch-and-bound based global search
- global adaptive random search (single-start)
- random multi-start based global search
- generalized reduced gradient algorithm based local search

For technical details we refer the reader to the elaborate monograph [Pin 1].

Invoking the *GlobalOptimization* package, we get two options to use the GOT. *GlobalSolve* is its command-line usage, *Interactive* uses the interactive Maplet graphical interface. The interplay between *Maple*® and the underlying optimization tools, for example the visualization of the objective functions by means of the *plot3d* command with the default option "style=patch", does not seem to be adequate in this context. We suggest using the options "style=patchcontour, shading=zhue" which yield a multicolored graph and contour lines of the surface.

The *Maple*® *help system* gives more details about the GOT. Furthermore we can get additional information during the solution process by using higher information levels, and there are numerous options whose default settings can be adapted. In contrast to the *Optimization* package, the GOT does not rely on the *objectivegradient*, *objectivejacobian* and *constraintjacobian* options, since the GOT requires only computable model function values of continuous functions. The global search options do not rely on derivatives, and the local search option applies central finite difference based gradient estimates.

Bibliography

[Ab/Wa] A. Abdulle, G. Wanner (2002): *200 Years of Least Squares Method.* Elemente der Mathematik 57, pp. 45–60

[Alt] W. Alt (2002): *Nichtlineare Optimierung.* Vieweg, Braunschweig, Wiesbaden

[An/Lu] A. Antoniou, W. Lu (2007): *Practical Optimization: Algorithms and Engineering Applications.* Springer, Berlin, Heidelberg, New York

[Avr] M. Avriel (1976): *Nonlinear Programming — Analysis and Methods.* Prentice-Hall, Englewood Cliffs

[Ba/Do] E.W. Barankin, R. Dorfman (1958): *On Quadratic Programming.* Univ. of California Publ. in Statistics 2, pp. 285–318

[Bar 1] M. C. Bartholomew-Biggs (2005): *Nonlinear Optimization with Financial Applications.* Kluwer, Boston, Dordrecht, London

[Bar 2] M. C. Bartholomew-Biggs (2008): *Nonlinear Optimization with Engineering Applications.* Springer, Berlin, Heidelberg, New York

[Be/Ne] A. Ben-Tal, A. Nemirovski (2001): *Lectures on Modern Convex Optimization: Analysis, Algorithms, and Engineering Applications.* SIAM, Philadelphia

[Bha] M. Bhatti (2000): *Practical Optimization Methods with* Mathematica *Applications.* Springer, Berlin, Heidelberg, New York

[Bl/Oe 1] E. Blum, W. Oettli (1972): *Direct Proof of the Existence Theorem for Quadratic Programming.* Oper. Research 20, pp. 165–167

[Bl/Oe 2] E. Blum, W. Oettli (1975): *Mathematische Optimierung.* Springer, Berlin, Heidelberg, New York

[Bo/To] P. T. Boggs, J.W. Tolle (1995): *Sequential Quadratic Programming.* Acta Numerica 4, pp. 1–51

[Bonn] J. F. Bonnans et al. (2003): *Numerical Optimization.* Springer, Berlin, Heidelberg, New York

[Bo/Le] J. Borwein, A. Lewis (2000): *Convex Analysis and Nonlinear Optimization.* Springer, Berlin, Heidelberg, New York

[Bo/Va] S. Boyd, L. Vandenberghe (2004): *Convex Optimization.* Cambridge University Press, Cambridge

W. Forst and D. Hoffmann, *Optimization—Theory and Practice*,
Springer Undergraduate Texts in Mathematics and Technology,
DOI 10.1007/978-0-387-78977-4,

[Br/Ti] J. BRINKHUIS, V. TIKHOMIROV (2005): *Optimization: Insights and Applications.* Princeton University Press, Princeton

[BSS] M. S. BAZARAA, H. D. SHERALI, C. M. SHETTY (2006): *Nonlinear Programming — Theory and Algorithms.* Wiley, Hoboken

[Ca/Ma] V. CANDELA, A. MARQUINA (1990): *Recurrence Relations for Rational Cubic Methods II: The Chebyshev Method.* Computing 45, pp. 355–367

[Cau] A. CAUCHY (1847): *Méthode générale pour la résolution des systèmes d'équations simultanées.* Comptes Rendus Acad. Sci. Paris 25, pp. 536–538

[Co/Jo] R. COURANT, F. JOHN (1989): *Introduction to Calculus and Analysis II/1.* Springer, Berlin, Heidelberg, New York

[Co/We] L. COLLATZ, W. WETTERLING (1971): *Optimierungsaufgaben.* Springer, Berlin, Heidelberg, New York

[Col] A.R. COLVILLE (1968): *A Comparative Study on Nonlinear Programming Codes.* IBM New York Scientific Center Report 320-2949

[Cr/Sh] N. CRISTIANINI, J. SHAWE-TAYLOR (2000): *Introduction to Support Vector Machines.* Cambridge University Press, Cambridge

[Dav] W. C. DAVIDON (1991): *Variable Metric Method for Minimization.* SIAM J. Optimization 1, pp. 1–17

[Deb] G. DEBREU (1952): *Definite and Semidefinite Quadratic Forms.* Econometrica 20, pp. 295–300

[De/Sc] J.E. DENNIS, R.B. SCHNABEL (1983): *Numerical Methods for Unconstrained Optimization and Nonlinear Equations.* Prentice-Hall, Englewood Cliffs

[DGW] J.E. DENNIS, D.M. GAY, R.E. WELSCH (1981): *An Adaptive Nonlinear Least-Squares Algorithm.* ACM Trans. Math. Software 7, pp. 348–368

[Elst] K.-H. ELSTER ET AL. (1977): *Einführung in die nichtlineare Optimierung.* Teubner, Leipzig

[Erik] J. ERIKSSON (1980): *A Note on Solution of Large Sparse Maximum Entropy Problems with Linear Equality Constraints.* Math. Progr. 18, pp. 146–154

[Fa/Ho] J.E. FALK, K.R. HOFFMAN (1976): *A Successive Underestimation Method for Concave Minimization Problems.* Math. Oper. Research 1, pp. 251–259

[Fa/So] J. E. FALK, R. M. SOLAND (1969): *An Algorithm for Separable Nonconvex Programming Problems.* Management Science 15, pp. 550–569

[Fi/Co] A. FIACCO, G. MCCORMICK (1990): *Nonlinear Programming — Sequential Unconstrained Minimization Techniques.* SIAM, Philadelphia

[Fle] R. FLETCHER (2006): *Practical Methods of Optimization.* Wiley, Chichester

[Fl/Pa] C. A. FLOUDAS, P. M. PARDALOS (1990): *A Collection of Test Problems for Constrained Global Optimization Algorithms.* Springer, Berlin, Heidelberg, New York

[Fl/Po] R. FLETCHER, M. J. D. POWELL (1963): *A Rapidly Convergent Descent Method for Minimization.* Computer J. 6, pp. 163–168

[Fo/Ho 1] W. FORST, D. HOFFMANN (2002): *Funktionentheorie erkunden mit* Maple®. Springer, Berlin, Heidelberg, New York

[Fo/Ho 2] W. FORST, D. HOFFMANN (2005): *Gewöhnliche Differentialgleichungen — Theorie und Praxis.* Springer, Berlin, Heidelberg, New York

[Fra] J. FRANKLIN (1980): *Methods of Mathematical Economics.* Springer, Berlin, Heidelberg, New York

[Ga/Hr] W. GANDER, J. HREBICEK (EDS.) (2004): *Solving Problems in Scientific Computing using* MAPLE *and* MATLAB. Springer, Berlin, Heidelberg, New York. Chapter 6

[Gau] C. F. GAUSS (1831): *Brief an* SCHUMACHER. Werke Bd. 8, p. 138

[Ge/Ka 1] C. GEIGER, C. KANZOW (1999): *Numerische Verfahren zur Lösung unrestringierter Optimierungsaufgaben.* Springer, Berlin, Heidelberg, New York

[Ge/Ka 2] C. GEIGER, C. KANZOW (2002): *Theorie und Numerik restringierter Optimierungsaufgaben.* Springer, Berlin, Heidelberg, New York

[Gli] F. GLINEUR (1998): *Pattern Separation via Ellipsoids and Conic Programming.* Mémoire de D.E.A. (Master's Thesis), Faculté Polytechnique de Mons, Belgium

[GLS] M. GRÖTSCHEL, L. LOVASZ, A. SCHRIJVER (1988): *Geometric Algorithms and Combinatorial Optimization.* Springer, Berlin, Heidelberg, New York

[GMW] P. E. GILL, W. MURRAY, M. H. WRIGHT (1981): *Practical Optimization.* Academic Press, London, New York

[GNS] I. Griva, S. G. Nash, A. Sofer (2009): *Linear and Nonlinear Programming.* SIAM, Philadelphia

[Go/Id] D. Goldfarb, A. Idnani (1983): *A Numerically Stable Dual Method for Solving Strictly Convex Quadratic Programs.* Math. Progr. 27, pp. 1–33

[GQT] S. M. Goldfeld, R. E. Quandt, H. F. Trotter (1966): *Maximization by Quadratic Hill-Climbing.* Econometrica 34, pp. 541–551

[Gr/Te] C. Grossmann, J. Terno (1993): *Numerik der Optimierung.* Teubner, Stuttgart

[Hai] E. Hairer (2001): *Introduction à l'Analyse Numérique.* Lecture Notes, Genève, p. 158

[Hal] P. R. Halmos (1993): *Finite-Dimensional Vector Spaces.* Springer, Berlin, Heidelberg, New York

[Hav] C. A. Haverly (1978): *Studies of the Behavior of Recursion for the Pooling Problem.* ACM SIGMAP Bulletin 25, pp. 19–28

[Hel] C. Helmberg (2002): *Semidefinite Programming.* European J. Operational Research 137, pp. 461–482

[Hen] D. Henrion (2006): *Global Optimization Toolbox for* Maple. IEEE Control Systems Magazine, October, pp. 106–110

[HKR] M. Halická, E. de Klerk, C. Roos (2002): *On the Convergence of the Central Path in Semidefinite Optimization.* SIAM J. Optimization 12, pp. 1090–1099

[HLP] G. H. Hardy, J. E. Littlewood, G. Pólya (1967): *Inequalities.* Cambridge University Press, Cambridge

[Ho/Pa] R. Horst, P. M. Pardalos (eds.) (1995): *Handbook of Global Optimization Vol. 1.* Kluwer, Boston, Dordrecht, London

[Ho/Sc] W. Hock, K. Schittkowski (1981): *Test Examples for Nonlinear Programming Codes.* Springer, Berlin, Heidelberg, New York

[Ho/Tu] R. Horst, H. Tuy (1996): *Global Optimization: Deterministic Approaches.* Springer, Berlin, Heidelberg, New York

[HPT] R. Horst, P. M. Pardalos, N. V. Thoai (2000): *Introduction to Global Optimization.* Kluwer, Boston, Dordrecht, London

[HRVW] C. Helmberg et al. (1996): *An Interior-Point Method for Semidefinite Programming.* SIAM J. Optimization 6, pp. 342–361

[Ja/St] F. Jarre, J. Stoer (2004): *Optimierung.* Springer, Berlin, Heidelberg, New York

[John] F. JOHN (1948): *Extremum Problems with Inequalities as Subsidiary Conditions. In: Studies and Essays. Courant Anniversary Volume.* Interscience, New York, pp. 187–204

[Ka/Ro] B. KALANTARI, J. F. ROSEN (1987): *An Algorithm for Global Minimization of Linearly Constrained Concave Quadratic Functions.* Mathematics of Operations Research 12, pp. 544–561

[Kar] W. KARUSH (1939): *Minima of Functions of Several Variables with Inequalities as Side Conditions.* Univ. of Chicago Master's Thesis

[Kel] C. T. KELLEY (1999): *Detection and Remediation of Stagnation in the* NELDER-MEAD *Algorithm Using a Sufficient Decrease Condition.* SIAM J. Optimization 10, pp. 43–55

[Klei] K. KLEIBOHM (1967): *Bemerkungen zum Problem der nichtkonvexen Programmierung.* Zeitschrift für Operations Research (formerly: Unternehmensforschung), pp. 49–60

[Klerk] E. DE KLERK (2002): *Aspects of Semidefinite Programming: Interior Point Algorithms and Selected Applications.* Kluwer, Boston, Dordrecht, London

[Kopp] J. KOPP (2007): *Beiträge zum Hüllkugel- und Hüllellipsoidproblem.* University of Konstanz Diploma Thesis

[Kos] P. KOSMOL (1991): *Optimierung und Approximation.* de Gruyter, Berlin, New York

[Kru] S. O. KRUMKE (2004): *Interior Point Methods.* Lecture Notes, Kaiserslautern

[KSH] M. KOJIMA, S. SHINDOH, S. HARA (1997): *A note on the* NESTEROV-TODD *and the* KOJIMA-SHINDOH-HARA *Search Directions in Semidefinite Programming.* SIAM J. Optimization 7, pp. 86–125

[Lan] S. LANG (1972): *Linear Algebra.* Addison-Wesley, Reading

[Lev] K. LEVENBERG (1944): *A Method for the Solution of Certain Nonlinear Problems in Least Squares.* Quart. Appl. Math. 2, 164–168

[Lin] C. LINDAUER (2008): *Aspects of Semidefinite Programming: Applications and Primal-Dual Path-Following Methods.* University of Konstanz Diploma Thesis

[Lu/Ye] D. LUENBERGER, Y. YE (2008): *Linear and Nonlinear Programming.* Springer, Berlin, Heidelberg, New York

[LVBL] M. S. LOBO ET AL. (1998): *Applications of Second Order Cone Programming.* Linear Algebra and its Applications 284, pp. 193–228

[LWK] C.-J. LIN, R. C. WENG, S. S. KEERTHI (2008): *Trust Region Newton Methods for Large-Scale Logistic Regression.* J. Machine Learning Research 9, 627–650

[Man] O. L. MANGASARIAN (1969): *Nonlinear Programming.* McGraw-Hill, New York

[Mar] D. W. MARQUARDT (1963): *An Algorithm for Least-Squares Estimation of Nonlinear Parameters.* SIAM J. Appl. Math. 11, pp. 431–441

[Meg] N. MEGIDDO (1989): *Pathways to the Optimal Set in Linear Programming. In:* N. MEGIDDO (ED.), *Progress in Mathematical Programming: Interior-Point Algorithms and Related Methods.* Springer, Berlin, Heidelberg, New York , pp. 131–158

[Meh] S. MEHROTRA (1992): *On the Implementation of a Primal-Dual Interior-Point Method.* SIAM J. Optimization 2, pp. 575–601

[Me/Vo] J. A. MEIJERINK, H. A. VAN DER VORST (1977): *An Iterative Solution Method for Linear Systems of which the Coefficient Matrix is a Symmetric M-Matrix.* Math. Computation 31, pp. 148–162

[Mon] R. D. C. MONTEIRO (1997): *Primal-Dual Path-Following Algorithms for Semidefinite Programming.* SIAM J. Optimization 7, pp. 663–678

[MOT] MATLAB (2003): *Optimization Toolbox, User's Guide.* The Mathworks Inc.

[MTY] S. MIZUNO, M. J. TODD, Y. YE (1993): *On Adaptive-Step Primal-Dual Interior-Point Algorithms for Linear Programming.* Mathematics of Operations Research 18, pp. 964–981

[Nah] P. NAHIN (2004): *When Least is Best.* Princeton University Press, Princeton

[Ne/Ne] Y. NESTEROV, A. NEMIROVSKI (1994): *Interior-Point Polynomial Algorithms in Convex Programming.* SIAM, Philadelphia

[Ne/To] Y. NESTEROV, M. J. TODD (1997): *Self-Scaled Barriers and Interior Point Methods for Convex Programming.* Mathematics of Operations Research 22, pp. 1–42

[No/Wr] J. NOCEDAL, S. J. WRIGHT (2006): *Numerical Optimization.* Springer, Berlin, Heidelberg, New York

[Pa/Ro] P. M. PARDALOS, J. B. ROSEN (1987): *Constrained Global Optimization.* Springer, Berlin, Heidelberg, New York

[Ped] P. PEDREGAL (2004): *Introduction to Optimization.* Springer, Berlin, Heidelberg, New York

[Pin 1] J. D. PINTÉR (1996): *Global Optimization in Action.* Springer, Berlin, Heidelberg, New York

[Pin 2] J. D. PINTÉR (2006): *Global Optimization with* MAPLE. Maplesoft, Waterloo, and Pintér Consulting Services, Halifax

[Pol] E. POLAK (1997): *Optimization — Algorithms and Consistent Approximations.* Springer, Berlin, Heidelberg, New York

[Pow 1] M. J. D. POWELL (1977): *Variable Metric Methods for Constrained Optimization.* Report DAMTP 77/NA6, University of Cambridge

[Pow 2] M. J. D. POWELL (1977): *A Fast Algorithm for Nonlinearly Constrained Optimization Calculations. In: Numerical Analysis, Dundee 1977.* Lecture Notes in Mathematics 630. Springer, Berlin, Heidelberg, New York, pp. 144–157

[Pow 3] M. J. D. POWELL (1986): *Convergence Properties of Algorithms for Nonlinear Optimization.* SIAM Review 28, pp. 487–500

[Rock] R. T. ROCKAFELLAR (1970): *Convex Analysis.* Princeton University Press, Princeton

[Ro/Wo] C. ROOS, J. VAN DER WOUDE (2006): *Convex Optimization and System Theory.* http://www.isa.ewi.tudelft.nl/~roos/courses/WI4218/

[RTV] C. ROOS, T. TERLAKY, J. P. VIAL (2005): *Interior Point Methods for Linear Optimization.* Springer, Berlin, Heidelberg, New York

[Shor] N. Z. SHOR (1985): *Minimization Methods for Non-Differentiable Functions.* Springer, Berlin, Heidelberg, New York

[Son] G. SONNEVEND (1986): *An 'Analytical Center' for Polyhedrons and New Classes of Global Algorithms for Linear (Smooth, Convex) Programming.* Lecture Notes in Control and Information Sciences 84. Springer, Berlin, Heidelberg, New York , pp. 866–875

[Sou] R. V. SOUTHWELL (1940): *Relaxation Methods in Engineering Science.* Clarendon Press, Oxford

[Spe] P. SPELLUCCI (1993): *Numerische Verfahren der nichtlinearen Optimierung.* Birkhäuser, Basel, Boston, Berlin

[St/Bu] J. STOER, R. BULIRSCH (1996): *Introduction to Numerical Analysis.* Springer, Berlin, Heidelberg, New York

[Stu] J. F. STURM (1999): *Using* SEDUMI 1.02, *A* MATLAB *Toolbox for Optimization over Symmetric Cones.* Optimization Methods and Software 11-12 , pp. 625–653

[Todd 1] M. J. TODD (1999): *A Study of Search Directions in Primal-Dual Interior-Point Methods for Semidefinite Programming.* Optimization Methods and Software 11 , pp. 1–46

[Todd 2] M. J. Todd (2001): *Semidefinite Optimization.* Acta Numerica 10, pp. 515–560

[Toh] K.C. Toh (1999): *Primal-Dual Path-Following Algorithms for Determinant Maximization Problems with Linear Matrix Inequalities.* Computational Optimization and Applications 14, pp. 309–330

[VBW] L. Vandenberghe, S. Boyd, S.P. Wu (1998): *Determinant Maximization with Linear Matrix Inequality Constraints.* SIAM J. on Matrix Analysis and Applications 19, pp. 499–533

[Vogt] J. Vogt (2008): *Primal-Dual Path-Following Methods for Linear Programming.* University of Konstanz Diploma Thesis

[Wer 1] J. Werner (1992): *Numerische Mathematik — Band 2: Eigenwertaufgaben, lineare Optimierungsaufgaben, unrestringierte Optimierungsaufgaben.* Vieweg, Braunschweig, Wiesbaden

[Wer 2] J. Werner (1998): *Nichtlineare Optimierungsaufgaben.* Lecture Notes, Göttingen

[Wer 3] J. Werner (2001): *Operations Research.* Lecture Notes, Göttingen

[Wil] R.B. Wilson (1963): *A Simplicial Algorithm for Concave Programming.* PhD Thesis, Harvard University

[Wri] S. J. Wright (1997): *Primal-Dual Interior-Point Methods.* SIAM, Philadelphia

[WSV] H. Wolkowicz, R. Saigal, L. Vandenberghe (eds.) (2000): *Handbook of Semidefinite Programming: Theory, Algorithms and Applications.* Kluwer, Boston, Dordrecht, London

[YTM] Y. Ye, M.J. Todd, S. Mizuno (1994): *An $O(\sqrt{n}\,L)$-Iteration Homogeneous and Self-Dual Linear Programming Algorithm.* Mathematics of Operations Research 19, pp. 53–67

[YYH] H. Yamashita, H. Yabe, K. Harada (2007): *A Primal-Dual Interior Point Method for Nonlinear Semidefinite Programming.* Technical Report, Department of Math. Information Science, Tokyo

Index of Symbols

W. Forst and D. Hoffmann, *Optimization—Theory and Practice*,
Springer Undergraduate Texts in Mathematics and Technology,
DOI 10.1007/978-0-387-78977-4,

Mathfrak

Greek

Cones

Norms and inner products

Miscellaneous

Subject Index

W. Forst and D. Hoffmann, *Optimization—Theory and Practice*,
Springer Undergraduate Texts in Mathematics and Technology,
DOI 10.1007/978-0-387-78977-4,

D

E

F

G

H

M

N

O

P

Q

R

T

U

V

W

Z